Methods in Enzymology

Volume 264
MITOCHONDRIAL BIOGENESIS AND GENETICS
Part B

METHODS IN ENZYMOLOGY

EDITORS-IN-CHIEF

John N. Abelson   Melvin I. Simon

DIVISION OF BIOLOGY
CALIFORNIA INSTITUTE OF TECHNOLOGY
PASADENA, CALIFORNIA

FOUNDING EDITORS

Sidney P. Colowick and Nathan O. Kaplan

Methods in Enzymology

Volume 264

Mitochondrial Biogenesis and Genetics

Part B

EDITED BY

Giuseppe M. Attardi

DIVISION OF BIOLOGY
CALIFORNIA INSTITUTE OF TECHNOLOGY
PASADENA, CALIFORNIA

Anne Chomyn

DIVISION OF BIOLOGY
CALIFORNIA INSTITUTE OF TECHNOLOGY
PASADENA, CALIFORNIA

ACADEMIC PRESS

San Diego   New York   Boston   London   Sydney   Tokyo   Toronto

Academic Press, Inc.
A Division of Harcourt Brace & Company
525 B Street, Suite 1900, San Diego, California 92101-4495

United Kingdom Edition published by
Academic Press Limited
24-28 Oval Road, London NW1 7DX

International Standard Serial Number: 0076-6879

International Standard Book Number: 0-12-182165-X

PRINTED IN THE UNITED STATES OF AMERICA
96 97 98 99 00 01 MM 9 8 7 6 5 4 3 2 1

Table of Contents

Section I. *In Vivo* and *in Vitro* Approaches to Study of Mitochondrial Gene Expression

Section II. Mitochondrial Genetics and Gene Manipulation

Section III. Mitochondrial Diseases and Aging

Contributors to Volume 264

Article numbers are in parentheses following the names of contributors.
Affiliations listed are current.

BRUCE N. AMES (38), *Division of Biochemistry and Molecular Biology, University of California, Berkeley, Berkeley, California 94720*

CHANDRAMOHAN V. AMMINI (2, 3), *Departments of Molecular Genetics and Medical Microbiology, University of Florida, Gainesville, Florida 32610*

NORMAN ARNHEIM (36), *Molecular Biology Program, University of Southern California, Los Angeles, California 90089*

GIUSEPPE ATTARDI (1, 12, 15, 17, 19, 27, 28, 37, 41), *Division of Biology, California Institute of Technology, Pasadena, California 91125*

CARLO AUSENDA (11), *Institute of Clinical Neurology, University of Milan, 20122 Milan, Italy*

KENNETH B. BECKMAN (38), *Division of Biochemistry and Molecular Biology, University of California, Berkeley, Berkeley, California 94720*

SCOTT M. BELCHER (24), *Department of Biochemistry, University of Texas Southwestern Medical Center, Dallas, Texas 75235*

GARY BELLUS (4), *Center for Medical Genetics, Johns Hopkins University School of Medicine, Baltimore, Maryland 21205*

HERMAN A. C. M. BENTLAGE (20), *Department of Pediatrics, Academic Hospital, 6500 HB Nijmegen, The Netherlands*

EDUARDO BONILLA (43), *Department of Neurology, College of Physicians and Surgeons, Columbia University, New York, New York 10032*

LOUISE BOULET (40), *Molecular Neurogenetics, Montréal Neurological Institute and Hospital, Montréal, Québec, Canada H3A 2B4*

JOHN E. BOYNTON (25), *Developmental, Cellular and Molecular Biology Group, Departments of Botany and Zoology, Duke University, Durham, North Carolina 27708*

BARBARA BRUNNER NEUENSCHWANDER (47), *Department of Botany/Developmental, Cellular, and Molecular Biology Group, Duke University, Durham, North Carolina 27708*

RONALD A. BUTOW (24), *Department of Biochemistry, University of Texas Southwestern Medical Center, Dallas, Texas 75235*

ANNE CHOMYN (11, 18, 20, 29), *Division of Biology, California Institute of Technology, Pasadena, California 91125*

MANUEL G. CLAROS (34), *Ecole Normale Superieure Laboratoire de Génétique Moléculaire, CNRS URA 1302, 75230 Paris, France*

DAVID A. CLAYTON (13, 14), *Department of Developmental Biology, Stanford University School of Medicine, Stanford, California 94305*

YAN LI DANG (9), *Division of Neuroscience, Baylor College of Medicine, Houston, Texas 77030*

NANCY D. DENSLOW (22), *Department of Biochemistry and Molecular Biology, Health Science Center, University of Florida, Gainesville, Florida 32610*

RODNEY J. DEVENISH (33), *Department of Biochemistry and Molecular Biology, and Centre for Molecular Biology and Medicine, Monash University, Clayton, Victoria 3168, Australia*

JOSE ANTONIO ENRÍQUEZ (6, 17), *Division of Biology, California Institute of Technology, Pasadena, California 91125*

MARY FARWELL (23), *Department of Biology, East Carolina University, Greenville, North Carolina 27858*

PATRICIO FERNÁNDEZ-SILVA (1, 12, 15), *Division of Biology, California Institute of Technology, Pasadena, California 91125*

THOMAS D. FOX (21), *Section of Genetics and Development, Cornell University, Ithaca, New York 14853*

GEORGES C. FRECH (10), *Howard Hughes Medical Institute, Research Laboratories, University of California, Los Angeles, Los Angeles, California 90024*

GEORGE L. GAINES III (5), *Isogenetics, Inc., Chicago Technology Park, Chicago, Illinois 60612*

GUO-JIAN GAO (9), *Department of Biochemistry, University of Louisville School of Medicine, Louisville, Kentucky 40292*

STEVEN C. GHIVIZZANI (2, 3), *Departments of Molecular Genetics and Medical Microbiology, University of Florida, Gainesville, Florida 32610*

NICHOLAS W. GILLHAM (25), *Developmental, Cellular and Molecular Biology Group, Departments of Botany and Zoology, Duke University, Durham, North Carolina 27708*

BRADLEY GOEHRING (4), *Department of Molecular, Cellular, and Developmental Biology, University of Colorado, Boulder, Colorado 80309*

ROBYN E. GRAY (33), *Department of Biochemistry and Molecular Biology, and Centre for Molecular Biology and Medicine, Monash University, Clayton, Victoria 3168 Australia*

H. CAROL GRIFFIN (47), *Department of Genetics, North Carolina State University, Raleigh, North Carolina 27695*

KATHLEEN R. GROOM (9), *Department of Biochemistry, University of Louisville School of Medicine, Louisville, Kentucky 40292*

WILLIAM W. HAUSWIRTH (2, 3), *Departments of Molecular Genetics and Medical Microbiology, University of Florida, Gainesville, Florida 32610*

MIKA HAYAKAWA (35), *Department of Biomedical Chemistry, Faculty of Medicine, University of Nagoya, Nagoya, 466, Japan*

R. MICHAEL HENKE (24), *Department of Biochemistry, University of Texas Southwestern Medical Center, Dallas, Texas 75235*

GÖTZ HOFHAUS (41), *Institut für Biochemie der Heinrich-Heine Universität, 40225 Düsseldorf, Germany*

CLAUDE JACQ (34), *Ecole Normale Superieure, Laboratoire de Génétique Moléculaire, CNRS URA 1302, 75230 Paris, France*

JUDITH A. JAEHNING (7), *Department of Biochemistry, Biophysics, and Genetics, University of Colorado Health Sciences Center, Denver, Colorado 80262*

TIM JOHNS (40), *Molecular Neurogenetics, Montréal Neurological Institute and Hospital, Montréal, Québec, Canada H3A 2B4*

ALBERT S. JUN (42), *Department of Genetics and Molecular Medicine, Emory University School of Medicine, Atlanta, Georgia 30322*

YOON L. KIM (42), *Department of Genetics and Molecular Medicine, Emory University School of Medicine, Atlanta, Georgia 30322*

MICHAEL P. KING (16, 27, 28, 30), *Department of Neurology, College of Physicians and Surgeons, Columbia University, New York, New York 10032*

PHILIP J. LAIPIS (31), *Department of Biochemistry and Molecular Biology, University of Florida College of Medicine, Gainesville, Florida 32610*

RUBY H. P. LAW (33), *Department of Biochemistry and Molecular Biology, and Centre for Molecular Biology and Medicine, Monash University, Clayton, Victoria 3168, Australia*

CHARLES S. LEVINGS III (47), *Department of Genetics, North Carolina State University, Raleigh, North Carolina 27695*

HUA-XIN LIAO (23), *Arthritis Center, Duke University Medical Center, Durham, North Carolina 27710*

PAOLA LOGUERCIO POLOSA (19), *Dipartimento di Biochimica de Biologia Molecolare, Universita degli Studi, 70126 Bari, Italy*

MANUEL J. LÓPEZ-PÉREZ (6), *Departamento de Bioquímica y Biología Molecular y Celular, Facultad de Veterinaria, Universidad de Zaragoza, E-50013 Zaragoza, Spain*

YAN CHUN LOU (9), *Department of Biochemistry, University of Louisville School of Medicine, Louisville, Kentucky 40292*

CORT S. MADSEN (2, 3), *Departments of Molecular Genetics and Medical Microbiology,*

University of Florida, Gainesville, Florida 32610

DAVID A. MANGUS (7), *Department of Molecular Genetics and Microbiology, University of Massachusetts Medical School, Worcester, Massachusetts 01655*

NANCY C. MARTIN (9), *Department of Biochemistry, University of Louisville School of Medicine, Louisville, Kentucky 40292*

DMITRI A. MASLOV (10), *Howard Hughes Medical Institute, Research Laboratories, University of California, Los Angeles, Los Angeles, California 90024*

ETSUKO T. MATSUURA (32), *Department of Biology, Ochanomizu University, Bunkyo-ku, Tokyo 112, Japan*

EDWARD E. McKEE (4), *SBCME, Indiana University School of Medicine, and University of Notre Dame, Notre Dame, Indiana 46556*

VICENTE MICOL (1, 12, 15), *Department of Biochemistry and Molecular Biology, Faculty of Veterinary Medicine, University of Murcia, 30071 Murcia, Spain*

JULIO MONTOYA (6), *Departamento de Bioquímica y Biología Molecular y Celular, Facultad de Veterinaria, Universidad de Zaragoza, E-50013 Zaragoza, Spain*

CARLOS T. MORAES (44), *Departments of Neurology and Cell Biology and Anatomy, University of Miami, Miami, Florida 33136*

RÉJEAN MORAIS (26), *Département de Biochimie, Université de Montréal, Montréal, Québec, Canada H3C 3J7*

MICHAEL J. MORALES (9), *Department of Neurobiology and Anatomy, Washington University, St. Louis, Missouri 63110*

JOHN V. MORAN (24), *Department of Biochemistry, University of Texas Southwestern Medical Center, Dallas, Texas 75235*

J. MÜLLER-HÖCKER (45), *Institute of Pathology, Ludwig-Maximilians-Universität, München, 80337 München, Germany*

PHILLIP NAGLEY (33), *Department of Biochemistry and Molecular Biology and Centre for Molecular Biology and Medicine, Monash University, Clayton, Victoria 3168, Australia*

MARCEL R. NELEN (2), *Department of Molecular Genetics, and Medical Microbiology, University of Florida, Gainesville, Florida 32610*

YUZO NIKI (32), *Department of Biology, Ibaraki University, Bunkyo, Mito 310, Japan*

THOMAS W. O'BRIEN (22), *Department of Biochemistry and Molecular Biology, Health Science Center, University of Florida, Gainesville, Florida 32610*

TAKAYUKI OZAWA (35), *Department of Biomedical Chemistry, Faculty of Medicine, University of Nagoya, Nagoya, 466, Japan*

JAVIER PEREA (34), *Ecole Normale Superieure, Laboratoire de Génétique Moléculaire, CNRS URA 1302, 75230 Paris, France*

ACISCLO PÉREZ-MARTOS (6), *Departamento de Bioquímica y Biología Molecular y Celular, Facultad de Veterinaria, Universidad de Zaragoza, E-50013 Zaragoza, Spain*

PHILIP S. PERLMAN (8, 24), *Department of Biochemistry, University of Texas Southwestern Medical Center, Dallas, Texas 75235*

MIRCEA PODAR (8), *Department of Biochemistry, University of Texas Southwestern Medical Center, Dallas, Texas 75235*

ROBERT O. POYTON (4), *Department of Molecular, Cellular, and Developmental Biology, University of Colorado, Boulder, Colorado 80309*

DAVID M. RHOADS (47), *Department of Botany/Developmental, Cellular, and Molecular Biology Group, Duke University, Durham, North Carolina 27708*

BRIAN H. ROBINSON (39), *Departments of Pediatrics and Biochemistry, University of Toronto, Toronto, Ontario, Canada M5G 1X8, and The Research Institute, The Hospital for Sick Children, Toronto, Ontario, Canada M5G 1X8*

S. SCHÄFER (45), *Institute of Pathology, Ludwig-Maximilians-Universität, München, 80337 München, Germany*

HERMANN SCHÄGGER (46), *Zentrum der Biologischen Chemie, Universitätsklinikum Frankfurt, Theodor-Stern-Kai 7, 60590 Frankfurt am Main, Germany*

ERIC A. SCHON (44), *Departments of Neurology and Genetics and Development, Columbia University, New York, New York 10027*

CARYL J. SCHWARTZBACH (23), *Burroughs Wellcome Company, Research Triangle Park, North Carolina 27709*

MONICA SCIACCO (43), *Department of Neurology, College of Physicians and Surgeons, Columbia University, New York, New York 10032*

KEVIN A. SEVARINO (4), *Division of Molecular Psychiatry, Department of Psychiatry, Yale University School of Medicine, New Haven, Connecticut 06508*

GERALD S. SHADEL (13, 14), *Department of Developmental Biology, Stanford University School of Medicine, Stanford, California 94305*

REBECCA M. SHAKELEY (41), *Division of Biology, California Institute of Technology, Pasadena, California 91125*

ERIC A. SHOUBRIDGE (40), *Molecular Neurogenetics, Montréal Neurological Institute and Hospital, Montréal, Québec, Canada H3A 2B4*

JAMES N. SIEDOW (47), *Department of Botany/Developmental, Cellular and Molecular Biology Group, Duke University, Durham, North Carolina 27708*

LARRY SIMPSON (10), *Howard Hughes Medical Institute, Research Laboratories, University of California, Los Angeles, Los Angeles, California 90024*

NAY-WEI SOONG (36), *Department of Gene Therapy, University of Southern California, Norris Cancer Hospital 610, Los Angeles, California 90033*

LINDA L. SPREMULLI (23), *Department of Chemistry, University of North Carolina, Chapel Hill, North Carolina 27599*

MASASHI TANAKA (35), *Department of Biomedical Chemistry, Faculty of Medicine, University of Nagoya, Nagoya, 466, Japan*

YOSHINORI TANNO (37), *Department of Neurology, Brain Research Institute, Niigata University, Niigata, 951, Japan*

IAN A. TROUNCE (42), *Department of Genetics and Molecular Medicine, Emory University School of Medicine, Atlanta, Georgia 30322*

SHOJI TSUJI (37), *Department of Neurology, Brain Research Institute, Niigata University, Niigata, 951, Japan*

DOUGLAS C. WALLACE (42), *Department of Genetics and Molecular Medicine, Emory University School of Medicine, Atlanta, Georgia 30322*

CAROL A. WISE (9), *Department of Biochemistry, University of Texas Southwestern Medical Center, Dallas, Texas 75235*

MAKOTO YONEDA (37), *Department of Biomedical Chemistry, Faculty of Medicine, University of Nagoya, Nagoya, 466, Japan*

Preface

Since publication in 1983 of Volume 97 of *Methods in Enzymology,* which covered mitochondrial biogenesis and genetics, striking progress has been made in these areas and there have been exciting new developments. These are exemplified by the discovery of RNA editing, RNA import into mitochondria, and mitochondrial DNA-encoded histocompatibility antigens; by the detailed molecular dissection of protein import into mitochondria and of the active role of introns in RNA splicing and intron transposition; by the high-resolution structural analysis of oxidative phosphorylation enzyme complexes; by the introduction of powerful methods for mitochondrial gene manipulation; and by the discovery of the role of mitochondrial DNA mutations in human disease.

As in other fields of science, progress in mitochondrial research has been facilitated by the development of new methods and by critical improvement of existing methods. This progress has stimulated new technological advances. A wealth of *in vivo* and *in vitro* approaches, involving genetic, biochemical, molecular, biophysical, immunological, and immunohistochemical techniques, are being utilized on a variety of organisms to understand the processes of mitochondrial biogenesis and genetics. It was, therefore, timely and appropriate to assemble and present in an integrated fashion in Volumes 260 and 264 of *Methods in Enzymology* the latest "know-how" pertaining to the study of these processes. This volume, *Mitochondrial Biogenesis and Genetics,* Part B, covers methodology used in mitochondrial gene expression, mitochondrial genetics and gene manipulation, and mitochondrial diseases and aging. Volume 260 (Part A) covers methodology used to study the structure and function of the oxidative phosphorylation complexes, import of proteins and RNA into mitochondria, ion and metabolite transport, and mitochondrial inheritance and turnover.

We are confident that these volumes will be very useful to investigators working in mitochondrial research and will help promote rapid developments in this area. With the widening recognition of the important role of mitochondrial function in many physiological, developmental, and pathological processes, one can anticipate that the methodology presented will also be valuable to investigators outside the circle of mitochondria specialists.

The assembly of chapters presented in these two volumes was greatly facilitated by the cooperation and enthusiasm of the many colleagues who were asked to write articles in the areas of their expertise, and to them we

are most grateful. We are indeed gratified by the excellent response we
received and by the quality of the chapters. We are also greatly indebted
to several contributors for their valuable advice concerning the content of
these volumes. Finally, we want to acknowledge the efficient and expert
help of Stephanie Canada in managing the various aspects of the editorial
work and to express our thanks to the staff of Academic Press.

GIUSEPPE M. ATTARDI
ANNE CHOMYN

METHODS IN ENZYMOLOGY

VOLUME XVII. Metabolism of Amino Acids and Amines (Parts A and B)
Edited by HERBERT TABOR AND CELIA WHITE TABOR

VOLUME XVIII. Vitamins and Coenzymes (Parts A, B, and C)
Edited by DONALD B. MCCORMICK AND LEMUEL D. WRIGHT

VOLUME XIX. Proteolytic Enzymes
Edited by GERTRUDE E. PERLMANN AND LASZLO LORAND

VOLUME XX. Nucleic Acids and Protein Synthesis (Part C)
Edited by KIVIE MOLDAVE AND LAWRENCE GROSSMAN

VOLUME XXI. Nucleic Acids (Part D)
Edited by LAWRENCE GROSSMAN AND KIVIE MOLDAVE

VOLUME XXII. Enzyme Purification and Related Techniques
Edited by WILLIAM B. JAKOBY

VOLUME XXIII. Photosynthesis (Part A)
Edited by ANTHONY SAN PIETRO

VOLUME XXIV. Photosynthesis and Nitrogen Fixation (Part B)
Edited by ANTHONY SAN PIETRO

VOLUME XXV. Enzyme Structure (Part B)
Edited by C. H. W. HIRS AND SERGE N. TIMASHEFF

VOLUME XXVI. Enzyme Structure (Part C)
Edited by C. H. W. HIRS AND SERGE N. TIMASHEFF

VOLUME XXVII. Enzyme Structure (Part D)
Edited by C. H. W. HIRS AND SERGE N. TIMASHEFF

VOLUME XXVIII. Complex Carbohydrates (Part B)
Edited by VICTOR GINSBURG

VOLUME XXIX. Nucleic Acids and Protein Synthesis (Part E)
Edited by LAWRENCE GROSSMAN AND KIVIE MOLDAVE

VOLUME XXX. Nucleic Acids and Protein Synthesis (Part F)
Edited by KIVIE MOLDAVE AND LAWRENCE GROSSMAN

VOLUME XXXI. Biomembranes (Part A)
Edited by SIDNEY FLEISCHER AND LESTER PACKER

VOLUME XXXII. Biomembranes (Part B)
Edited by SIDNEY FLEISCHER AND LESTER PACKER

VOLUME XXXIII. Cumulative Subject Index Volumes I–XXX
Edited by MARTHA G. DENNIS AND EDWARD A. DENNIS

VOLUME XXXIV. Affinity Techniques (Enzyme Purification: Part B)
Edited by WILLIAM B. JAKOBY AND MEIR WILCHEK

VOLUME XXXV. Lipids (Part B)
Edited by JOHN M. LOWENSTEIN

VOLUME 233. Oxygen Radicals in Biological Systems (Part C)
Edited by LESTER PACKER

VOLUME 234. Oxygen Radicals in Biological Systems (Part D)
Edited by LESTER PACKER

VOLUME 235. Bacterial Pathogenesis (Part A: Identification and Regulation of Virulence Factors)
Edited by VIRGINIA L. CLARK AND PATRIK M. BAVOIL

VOLUME 236. Bacterial Pathogenesis (Part B: Integration of Pathogenic Bacteria with Host Cells)
Edited by VIRGINIA L. CLARK AND PATRIK M. BAVOIL

VOLUME 237. Heterotrimeric G Proteins
Edited by RAVI IYENGAR

VOLUME 238. Heterotrimeric G-Protein Effectors
Edited by RAVI IYENGAR

VOLUME 239. Nuclear Magnetic Resonance (Part C)
Edited by THOMAS L. JAMES AND NORMAN J. OPPENHEIMER

VOLUME 240. Numerical Computer Methods (Part B)
Edited by MICHAEL L. JOHNSON AND LUDWIG BRAND

VOLUME 241. Retroviral Proteases
Edited by LAWRENCE C. KUO AND JULES A. SHAFER

VOLUME 242. Neoglycoconjugates (Part A)
Edited by Y. C. LEE AND REIKO T. LEE

VOLUME 243. Inorganic Microbial Sulfur Metabolism
Edited by HARRY D. PECK, JR., AND JEAN LEGALL

VOLUME 244. Proteolytic Enzymes: Serine and Cysteine Peptidases
Edited by ALAN J. BARRETT

VOLUME 245. Extracellular Matrix Components
Edited by E. RUOSLAHTI AND E. ENGVALL

VOLUME 246. Biochemical Spectroscopy
Edited by KENNETH SAUER

VOLUME 247. Neoglycoconjugates (Part B: Biomedical Applications)
Edited by Y. C. LEE AND REIKO T. LEE

VOLUME 248. Proteolytic Enzymes: Aspartic and Metallo Peptidases
Edited by ALAN J. BARRETT

VOLUME 249. Enzyme Kinetics and Mechanism (Part D: Developments in Enzyme Dynamics)
Edited by DANIEL L. PURICH

VOLUME 250. Lipid Modifications of Proteins
Edited by PATRICK J. CASEY AND JANICE E. BUSS

VOLUME 251. Biothiols (Part A: Monothiols and Dithiols, Protein Thiols, and Thiyl Radicals)
Edited by LESTER PACKER

VOLUME 252. Biothiols (Part B: Glutathione and Thioredoxin; Thiols in Signal Transduction and Gene Regulation)
Edited by LESTER PACKER

VOLUME 253. Adhesion of Microbial Pathogens
Edited by RON J. DOYLE AND ITZHAK OFEK

VOLUME 254. Oncogene Techniques
Edited by PETER K. VOGT AND INDER M. VERMA

VOLUME 255. Small GTPases and Their Regulators (Part A: Ras Family)
Edited by W. E. BALCH, CHANNING J. DER, AND ALAN HALL

VOLUME 256. Small GTPases and Their Regulators (Part B: Rho Family)
Edited by W. E. BALCH, CHANNING J. DER, AND ALAN HALL

VOLUME 257. Small GTPases and Their Regulators (Part C: Proteins Involved in Transport)
Edited by W. E. BALCH, CHANNING J. DER, AND ALAN HALL

VOLUME 258. Redox-Active Amino Acids in Biology
Edited by JUDITH P. KLINMAN

VOLUME 259. Energetics of Biological Macromolecules
Edited by MICHAEL L. JOHNSON AND GARY K. ACKERS

VOLUME 260. Mitochondrial Biogenesis and Genetics (Part A)
Edited by GIUSEPPE M. ATTARDI AND ANNE CHOMYN

VOLUME 261. Nuclear Magnetic Resonance and Nucleic Acids
Edited by THOMAS L. JAMES

VOLUME 262. DNA Replication
Edited by JUDITH L. CAMPBELL

VOLUME 263. Plasma Lipoproteins (Part C: Quantitation)
Edited by WILLIAM A. BRADLEY, SANDRA H. GIANTURCO, AND JERE P. SEGREST

VOLUME 264. Mitochondrial Biogenesis and Genetics (Part B)
Edited by GIUSEPPE M. ATTARDI AND ANNE CHOMYN

VOLUME 265. Cumulative Subject Index Volumes 228, 230–262 (in preparation)

VOLUME 266. Computer Methods for Macromolecular Sequence Analysis (in preparation)
Edited by RUSSELL F. DOOLITTLE

VOLUME 267. Combinatorial Chemistry (in preparation)
Edited by JOHN N. ABELSON

VOLUME 268. Nitric Oxide (Part A: Sources and Detection of NO; NO Synthase) (in preparation)
Edited by LESTER PACKER

Section I

In Vivo and *in Vitro* Approaches to Study of Mitochondrial
Gene Expression

[1] *In Vivo* Footprinting of Human Mitochondrial DNA in Cultured Cell Systems

By VICENTE MICOL, PATRICIO FERNÁNDEZ-SILVA, and GIUSEPPE ATTARDI

Introduction

Several footprinting techniques have been developed for the molecular analysis of specific protein–DNA interactions. DNase I footprinting has allowed the detection of interactions between purified or partially purified proteins and their specific target DNA sequences *in vitro*.[1] The more recently developed footprinting technique known as genomic footprinting[2] has permitted the identification of protein–DNA interactions in intact cells, yielding important information concerning developmentally regulated processes.

Most footprinting studies concerned the genomic DNA of eukaryotes, and as a consequence the original techniques exhibited intrinsic difficulties. The high complexity of eukaryotic genomic DNA and its association with structural proteins, such as histones, have raised doubts as to whether the results of *in vitro* footprinting experiments with naked DNA can be extrapolated to the *in vivo* situation. Furthermore, *in vivo* footprinting of a single-copy gene in a large genome has been difficult to achieve because of the large cell number required and because of the very high signal-to-noise ratio. The latter problem has been eliminated by using the ligation-mediated polymerase chain reaction technique (LMPCR), based on an exponential amplification of a relatively short DNA segment, which has been successfully used by many investigators.[3–5]

Mammalian mitochondrial DNA (mtDNA) has been studied by footprinting techniques, in particular by applying the DNase I protection technique to isolated DNA[6,7] and the methylation protection technique to

[1] D. J. Galas and A. Schmitz, *Nucleic Acids Res.* **5,** 3157 (1978).
[2] A. Ephrussi, G. M. Church, S. Tonegawa, and W. Gilbert, *Science* **227,** 134 (1985).
[3] P. R. Mueller and B. Wold, *Science* **246,** 780 (1989).
[4] P. G. Pfeiffer, S. Steigerwald, P. R. Mueller, B. Wold, and A. D. Riggs, *Science* **246,** 810 (1989).
[5] I. K. Hornstra and T. P. Yang, *Anal. Biochem.* **213,** 179 (1993).
[6] R. P. Fischer and D. A. Clayton, *J. Biol. Chem.* **260,** 11330 (1985).
[7] B. Kruse, N. Narasimhan, and G. Attardi, *Cell (Cambridge, Mass.)* **58,** 391 (1989).

isolated organelles.[8,9] Knowledge about the proteins that interact with the mitochondrial genome in mammalian cells is still incomplete, and, so far, just a few proteins have shown specific binding to mtDNA sequences. The mitochondrial transcription factor A (mtTFA),[6,10,11] which binds to sequences upstream of the L-strand promoter and the H-strand rRNA-specific promoter of mtDNA, has shown a certain flexibility in its sequence specificity. The mitochondrial transcription termination factor (mTERF),[7,12] which protects a 28-base pair region within the tRNA^{Leu} (UUR) gene at a position immediately adjacent and downstream of the 3' end of the 16 S rRNA gene, has been shown to be involved in termination of transcription of the mitochondrial rRNA genes in human cells. A 48-kDa protein has been found to protect specifically a TAS (termination-associated sequence) element located close to the D-loop 3' end, and it has been suggested to be involved in the control of mtDNA copy number.[8] Finally, a single-strand binding protein has been described in *Xenopus laevis*,[13] rats,[14] and humans[15] that stabilizes the displaced single strand of normal and expanded D-loops during mtDNA replication.

The simplicity of the structure of mammalian mtDNA, in particular the absence of histones and its high copy number (from several hundreds or thousands in cultured mouse and human cells[16–18] to one to a few hundreds of thousands in mammalian oocytes[19,20]), makes this system particularly suitable for applications involving the methylation protection technique, using dimethyl sulfate (DMS), for *in vivo* footprinting assays. The application of *in vivo* footprinting to the study of DNA–protein interactions in mtDNA can answer questions concerning the physiological significance of these interactions in situations such as different environmental conditions, different stages of cell growth or differentiation, or various disease-causing mtDNA mutations. This technique is very useful, not only when the functional importance of particular DNA–protein interactions (previously known to exist from *in vitro* footprinting assays) needs to be studied, but

[8] C. S. Madsen, S. C. Ghivizzani, and W. W. Hauswirth, *Mol. Cell. Biol.* **13,** 2162 (1993).
[9] S. C. Ghivizzani, C. S. Madsen, and W. W. Hauswirth, *J. Biol. Chem.* **268,** 8675 (1993).
[10] R. P. Fischer, J. N. Topper, and D. A. Clayton, *Cell (Cambridge, Mass.)* **50,** 247 (1987).
[11] M. A. Parisi and D. A. Clayton, *Science* **252,** 965 (1991).
[12] A. Daga, V. Micol, D. Hess, R. Aebersold, and G. Attardi, *J. Biol. Chem.* **268,** 8123 (1993).
[13] B. Mignotte, M. Barat, and J. C. Mounolou, *Nucleic Acids Res.* **13,** 1703 (1985).
[14] P. A. Pavco and G. C. Van Tuyle, *J. Cell Biol.* **100,** 258 (1985).
[15] V. Tiranti, M. Rocchi, S. DiDonato, and M. Zeviani, *Gene* **126,** 219 (1993).
[16] E. D. Robin and R. Wong, *J. Cell. Physiol.* **136,** 507 (1988).
[17] D. Bogenhagen and D. A. Clayton, *J. Biol. Chem.* **249,** 7991 (1974).
[18] M. P. King and G. Attardi, *Science* **246,** 500 (1989).
[19] G. S. Michaels, W. W. Hauswirth, and J. P. Laipis, *Dev. Biol.* **94,** 246 (1982).
[20] X. Chen, R. Prosser, S. Simonetti, and E. Schon, *Neurology* **44,** A336 (1994).

also when a complete screening of a given regulatory DNA region must be carried out to search for the potential presence of unknown DNA-binding proteins.

Principle of Method

The methodology of *in vivo* footprinting applied to the study of mtDNA-binding proteins is based on previously described techniques, with some modifications (for detailed description of the background of the procedure, refer to previous papers[2,5,21]). The first step in the *in vivo* DMS footprinting technique is treatment of intact cells with DMS to methylate unprotected guanines and adenines by a standard Maxam and Gilbert[22] chemical modification (Fig. 1). Once the *in vivo* methylation has been produced, total nucleic acids are extracted, and the sample is then incubated with piperidine in order to produce cleavage preferentially at the methylated guanine residues in DNA and to obtain a sequence ladder.

There are two methods to visualize the footprint generated by DMS. One involves the digestion of the purified DNA with an appropriate restriction endonuclease to create defined DNA ends, followed by size fractionation of the DNA fragments in a sequencing gel; thereafter, electrotransfer onto a membrane and hybridization with an appropriate probe are performed.[2,5,21] The alternate method is amplified primer extension (APEX),[23–25] in which the piperidine-cleaved DNA fragments are subjected to multiple rounds of primer extension reaction by PCR with a ^{32}P 5'-end-labeled specific oligodeoxynucleotide primer, using *Taq* polymerase. This amplification step increases the sensitivity of detection, allowing the use of smaller quantities of DNA template. Furthermore, the primer extension method gives sharper bands in the autoradiograms, as compared to those obtained by the blot hybridization method, and also avoids the time-consuming hybridization step.

Using a variant of the latter method outlined above, applied to the study of the mitochondrial genome, we have been able to map regions located between 50 and 60 to 300 nucleotides from the 5' end of the designed oligodeoxynucleotide, obtaining high-quality mtDNA ladders from as little as 5 μg total DNA after a few hours of exposure time.

[21] P. B. Becker, F. Weih, and G. Schutz, this series, Vol. 218, p. 568.
[22] A. M. Maxam and W. Gilbert, this series, Vol. 65, p. 499.
[23] D. J. Axelrod and J. Majors, *Nucleic Acids Res.* **17,** 171 (1989).
[24] H. P. Saluz and J. P. Jost, *Proc. Natl. Acad. Sci. U.S.A.* **86,** 2602 (1989).
[25] J. M. Huibregtse, D. R. Engelke, and D. J. Thiele, *Proc. Natl. Acad. Sci. U.S.A.* **86,** 65 (1989).

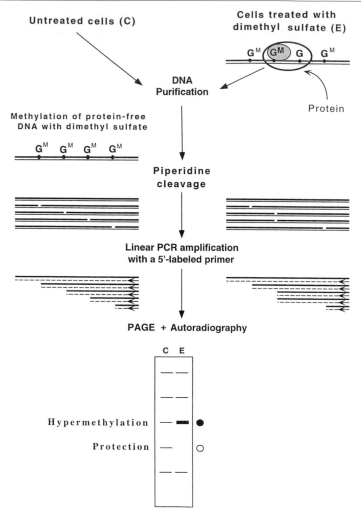

FIG. 1. Schematic representation of the *in vivo* footprinting method. G, Guanine residues; G^M, methylated guanine residues. Arrows represent the 5'-labeled oligodeoxynucleotide used for the asymmetric PCR reactions. For details see text.

Procedures

Reagents and Buffers

Dimethyl sulfate (DMS; Sigma, St. Louis, MO): Store in the dark at 4° under nitrogen. DMS is a potent carcinogenic agent, and thus all solutions, as well as the plasticware containing the reagent, should be treated with 5 M NaOH to inactivate it.

Piperidine (grade I, Sigma: 10 M stock): Store at 4° in the dark; immediately before use, make a 1 M dilution with distilled water.

dNTPs (stock solution of 25 mM each, neutralized with NaOH)

Phosphate-buffered saline (PBS): 140 mM NaCl, 2.5 mM KCl, 8.1 mM NaH$_2$PO$_4$, 1.5 mM KH$_2$PO$_4$, pH 7.5

DNA harvest buffer: 1 mM Tris-HCl (pH 7.5 at 25°), 400 mM NaCl, 2 mM EDTA, 0.2% w/v sodium dodecyl sulfate (SDS), 0.2 mg/ml proteinase K (Boehringer Mannheim, Indianapolis, IN; 20 units/mg)

TE buffer: 10 mM Tris-HCl (pH 8.0), 1 mM EDTA

DMS stop buffer: 1.5 M sodium acetate, pH 7.0, 1 M 2-mercaptoethanol, 100 μg/ml yeast tRNA

10× T4 polynucleotide kinase buffer: 500 mM Tris-HCl, pH 7.6, 100 mM MgCl$_2$, 50 mM dithiothreitol (DTT)

10× PCR reaction buffer (Promega, Madison, WI): 100 mM Tris-HCl, pH 9.0, 500 mM KCl, 1% Triton X-100

Phenol/chloroform/isoamyl alcohol (25 : 24 : 1, v/v): This mixture, containing 0.1% 8-hydroxyquinoline w/v, is equilibrated with 10 mM Tris-HCl, pH 7.4, 100 mM NaCl, 0.5% SDS, 1 mM EDTA. Store the mixture at 4° in a brown glass bottle with screw cap.

Mineral oil (Sigma)

1-Butanol

25 mM MgCl$_2$ solution

Formamide loading buffer: 95% formamide v/v, 20 mM EDTA, 0.05% bromphenol blue w/v, and 0.05% xylene cyanol w/v FF

In Vivo Methylation of DNA

Cells grown on a solid substrate to be used for an *in vivo* footprinting experiment should not be more than 60–70% confluent, and the medium should be changed 2–3 hr before DMS treatment. The protocol described below is for one 10-cm plate, which yields enough DNA to perform several footprinting experiments. The cells are treated with DMS at 37° directly on the plate by replacing the medium of the disk [e.g., Dulbecco's modified Eagle's medium (DMEM) with 5 or 10% v/v calf or fetal bovine serum] with 10 ml of fresh medium equilibrated at 37° and containing 1 μl/ml DMS (added immediately before use). The DMS is allowed to react in the dish for 2 min, the DMS-containing medium is then removed, and the dish is rinsed three times with cold PBS buffer for 1 min each, under gentle shaking. The optimal *in vivo* DMS treatment (DMS concentration and length of treatment) must be determined empirically for each case in order to avoid over- and undermethylation. For this purpose, the extracted DNA, after treatment with piperidine (see below), must be sized on alkaline agarose gels, where it should produce a distribution of single-stranded DNA fragments averaging approximately 600 bases. We have obtained an optimal degree of methylation of mtDNA using a final concentration of DMS of 0.1% v/v for lengths of treatment varying from 2 to 4 min.

Cell lysis is performed on the plate by adding 1.5 ml DNA harvest buffer containing proteinase K and incubating the plate with shaking for 6 hr at room temperature. Proteinase K treatment must be performed at room temperature in order to avoid depurination of nucleotides, which occurs in methylated DNA at higher temperatures under neutral buffer conditions. After incubation with proteinase K, the clear and viscous nucleic acid solution is extracted twice with 1 volume of phenol/chloroform/isoamyl alcohol (25:24:1, v/v) and once with chloroform/isoamyl alcohol (24:1, v/v). The *in vivo* methylated DNA is then mixed with 1/4 volume DMS stop buffer and precipitated twice with 3 volumes cold ethanol at −20°. At this stage, the DNA can be stored at −20°, if so desired. Alternatively, the pellet is washed with 70% (v/v) ethanol, dissolved in TE buffer, and, after the addition of NaCl to 0.2 M and ethanol, precipitated to give a pellet of approximately 200 μg total nucleic acids for the subsequent piperidine cleavage. If scaling up of the procedure is needed, the proteinase K-digested lysate derived from different plates can be combined for the subsequent phenol extraction.

A similar protocol can be followed to methylate mtDNA derived from cells growing in suspension, such as HeLa cells. About 10^7 to 10^8 cells are collected by centrifugation at 300 g_{av} for 5 min in a 50-ml conical tube. The cells are suspended in 25 ml fresh medium (such as high-phosphate-containing DMEM supplemented with 5% calf serum) containing DMS at a final concentration of 0.1% and warmed to 37°. The DMS is allowed to react for 2 min, and the cells are then promptly centrifuged at 1000 g for 30 sec. The DMS-treated cells are suspended in 50 ml ice-cold PBS and centrifuged again. The washing step is repeated twice more. After removing the third PBS wash, the cell pellet is suspended in 5 ml DNA harvest buffer, and the sample is digested with proteinase K at room temperature for 6 hr. Phenol extraction and ethanol precipitation are performed as described above.

In Vitro Dimethyl Sulfate Treatment of Naked DNA

DNA for *in vitro* DMS treatment is prepared in the same way as described above, except that the *in vivo* DMS treatment is omitted. After the cells on the plate are washed with PBS buffer, lysis, proteinase K digestion, phenol extraction, and ethanol precipitation are carried out as detailed above.

Samples containing 200 μg non-DMS-treated total nucleic acids are suspended in 200 μl TE, and 23 μl freshly prepared 1% DMS is added; the methylation reaction is allowed to proceed 2.5 min at 37°. The reaction is stopped by the addition of 50 μl ice-cold DMS stop buffer and 3 volumes chilled ethanol. The samples are then treated in parallel with the *in vivo* samples. As mentioned for the *in vivo* treatment, the conditions for optimal *in vitro* DMS treatment of the naked DNA must be determined empirically for each case in terms of time and concentration of DMS, to match the level of methylation of the *in vivo* samples.

Piperidine Cleavage of DNA

Pellets containing approximately 200 μg total nucleic acids each are suspended in 200 μl freshly prepared 1 M piperidine in microcentrifuge tubes and heated to 90° for 30 min in a heating block. After heating, the samples are frozen in an ethanol/dry ice mixture for 5 min and dried in a vacuum concentrator (Speed-Vac; Savant, Hicksville, NY) for at least 2 hr. The nucleic acid pellets are then dissolved in 100 μl distilled water and vacuum dried for an additional 1 hr. The samples are dissolved in 50 μl distilled water, transferred to fresh tubes, and vacuum-dried a third time for 1 hr. The final pellets, containing single-stranded DNA, are suspended in 50 μl TE buffer, and the concentration of each sample is determined by absorbance measurements.

5'-End-Labeling of Primer with [γ-^{32}P]rATP

The oligodeoxynucleotide primer should be checked for purity in a 20% polyacrylamide gel. If the primer is obtained as a pellet directly from chemical synthesis, it must be dissolved in distilled water and extracted three times with 1 volume of 1-butanol (collect the lower phase). After ethanol precipitation, the primer concentration must be determined by absorbance measurements.

For the kinase reaction, the following mixture must be prepared in a microcentrifuge tube: 16 μl distilled water, 2.5 μl of 10× T4 polynucleotide kinase buffer, 2.5 μl of the 10 μM primer stock solution (25 pmol), 2 μl (~60 pmol) [γ-^{32}P]rATP (>5000 Ci/mmol, >150 mCi/ml; Amersham, Arlington Heights, IL), and 2 μl T4 polynucleotide kinase (New England BioLabs, Beverly, MA; 10 units/μl). The reaction is carried out at 37° for 45 min. The enzyme is then inactivated at 65° for 20 min; the mixture is phenol-extracted and ethanol-precipitated in the presence of 10 μg yeast tRNA. The pellet is dissolved in 100 μl distilled water. The labeled primer should have a specific activity exceeding 10^6 counts/min (cpm)/pmol.

Primer Extension of Dimethyl Sulfate-Treated DNA

Five micrograms of either *in vivo* or *in vitro* DMS-treated total nucleic acids (containing ~0.015 pmol mtDNA) is used for primer extension analysis. From our experience, this is the minimal amount required to obtain a satisfactory footprinting signal of mtDNA, using total DNA from human cells as a template. In the PCR amplification, 0.5 pmol ^{32}P 5'-end-labeled primer is added to the DNA sample in a 50-μl reaction mixture containing 2.5 units *Taq* DNA polymerase (Promega; 5 units/μl), 5 μl of 10$\times$ PCR reaction buffer, 3 μl of 25 mM MgCl$_2$, and dNTPs to a final concentration of 100 μM. The samples are covered with 50 μl mineral oil, heated to 94° for 30 sec, incubated at an annealing temperature compatible with the T_m of the utilized oligodeoxynucleotide, and extended at 72° for 2 min, using 20 cycles in a DNA Thermal Cycler 480 (Perkin-Elmer, Norwalk, CT).

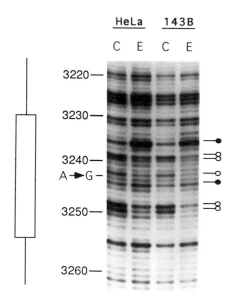

FIG. 2. *In vivo* footprinting pattern of the L-strand in the termination region of mtDNA from HeLa and 143B cell lines. Black circles represent hypermethylated residues, and open circles represent protected residues. The box on the left-hand side shows the position of the 28-bp region protected by mTERF, as determined by DNase I protection assays.[11] The position of the mtDNA mutation associated with mitochondrial myopathy, encephalopathy, lactic acidosis, and stroke-like episodes (MELAS) is shown (A → G).

After the amplification, the samples are extracted with phenol/chloroform/ isoamyl alcohol and ethanol-precipitated. The DNA pellets are washed with ice-cold 70% (v/v) ethanol, suspended in 10 μl formamide loading buffer, denatured for 3 min at 90°, and run in a standard 0.4-mm-thick, 6% (19:1) acrylamide–bisacrylamide/7.7 M urea sequencing gel in TBE (89 mM Tris base, 89 mM boric acid, 2 mM EDTA).

Concluding Remarks

In vivo footprinting is a very useful technique to obtain information on the functional behavior of mtDNA-binding proteins, not only when studying normal cells subjected to different environmental conditions or grown under different physiological situations, but also in the analysis of some disease-causing mutations affecting regulatory sequences of mtDNA. In particular, the protocol described has been successfully applied in our laboratory to the *in vivo* study of the interactions between mTERF and mtDNA (Fig. 2), as well as to the investigation of protein–DNA interactions in the regions of the H-strand and L-strand transcription promoters. In particular, we have established that the degree of occupancy of the mTERF binding sites *in vivo* is very high and that the *in vivo* footprinting pattern is very similar to that observed after *in organello* mtDNA footprinting.

The simplicity and rapidity of the method and the limited number of cells required for an experiment make the protocol described here the method of choice to study protein–DNA interactions involving regulatory sequences of mtDNA for the purpose of understanding their role in transcription and replication.

Acknowledgments

This work was supported by National Institutes of Health Grant GM-11726 (to G.A.) and scholarships from the Fulbright/Spanish Ministry of Education and Science Visiting Scholar Program (to V.M.) and from the California Institute of Technology (Gosney Fellowship, to P.F.-S.).

[2] Genomic Footprinting of Mitochondrial DNA: I. *In Organello* Analysis of Protein–Mitochondrial DNA Interactions in Bovine Mitochondria

By Cort S. Madsen, Steven C. Ghivizzani,[1] Chandramohan V. Ammini, Marcel R. Nelen, and William W. Hauswirth

Introduction

The study of molecular mechanisms that regulate mammalian mitochondrial (mt) replicative and transcriptional processes has been limited to *in vitro* analyses. Although much information resulted from these studies, *in vitro* systems are often difficult to develop and may not accurately reflect biological processes occurring *in vivo*. We have developed the technique of *in organello* footprinting[2,3] to analyze protein–mtDNA interactions in an *in vivo*-like environment. This procedure involves the use of dimethyl sulfate (DMS) in a methylation protection assay, first used to analyze protein–nuclear DNA interactions *in vivo*.[4,5] Dimethyl sulfate is a small molecule that methylates guanine residues at the N-7 position and, to a lesser extent, adenines at the N-3 position, making them sensitive to subsequent cleavage at alkaline pH and elevated temperatures. When bound to specific DNA residues, proteins can decrease or intensify purine reactivity to DMS relative to naked DNA.[6] The ability of DMS to permeate mitochondrial membranes readily allows detection of protein–mtDNA interaction within the organelle. During development of the *in organello* technique, we used bovine brain cortex mitochondria to monitor protein–mtDNA interactions at several domains within the mitochondrial genome considered important in regulating transcriptional and/or replicative processes. An outline of the *in organello* footprinting technique is presented in Table I.

---

[1] The first two authors made an equal contribution to the work reported in this chapter.
[2] S. C. Ghivizzani, C. S. Madsen, and W. W. Hauswirth, *J. Biol. Chem.* **268,** 8675 (1993).
[3] C. S. Madsen, S. C. Ghivizzani, and W. W. Hauswirth, *Mol. Cell. Biol.* **13,** 2162 (1993).
[4] G. M. Church and W. Gilbert, *Proc. Natl. Acad. Sci. U.S.A.* **81,** 1991 (1984).
[5] G. M. Church, A. Ephrussi, W. Gilbert, and S. Tonegawa, *Nature (London)* **313,** 798 (1985).
[6] W. Gilbert, A. Maxam, and A. Mirzabekov, *in* "Control of Ribosome Synthesis" (N. O. Kjeldgaard and O. Maaloe, eds.), p. 139. Munksgaard, Copenhagen.

TABLE I

In Organello FOOTPRINTING IN BOVINE MITOCHONDRIA[a]

1. To 1.7 ml pelleted freshly prepared mitochondria, add equal volume PBS (room temperature), suspend by gentle pipetting, and divide (200-μl aliquots) into 16 Eppendorf tubes (labeled as 8 controls and 8 tests).

2. Add 1 μl DMS (final concentration 0.5%) to test samples, mix by vortexing briefly, and incubate 3 min at room temperature. Add 1 ml ice-cold PBS and pellet mitochondria by centrifuging 1 min at 10,000 g; repeat PBS wash twice more.

3. Add 400 μl lysis buffer to mitochondrial pellet, vortex, and lyse mitochondria by adding 25 μl of 10% SDS. After vortexing, extract mtDNA twice with phenol, twice with phenol–chloroform–isoamyl alcohol (25:24:1, v/v), and once with chloroform. Precipitate nucleic acids with ethanol.

4. Digest mtDNA with restriction enzyme of choice in 100-μl volume. After 3-hr incubation, electrophorese 5 μl sample on 1% agarose gel to test for complete digestion and to estimate DNA concentration.

5. Add 5 μl water and 100 μl of 100 mM sodium cacodylate (pH 7.0) to each sample. To control samples add 1 μl DMS (final concentration 0.5%) and incubate 3 min (room temperature). Terminate reaction by adding 50 μl ice-cold DMS stop buffer, and precipitate mtDNA by adding 2.5 volumes ethanol.

6. Resuspend pellet in 180 μl H$_2$O, add 20 μl piperidine, and incubate 30 min at 90°. Place sample on ice and ethanol-precipitate using 3 M sodium acetate. Remove residual amounts of piperidine by reprecipitating 3 times or by adding 200 μl water and lyophilizing. Suspend sample in 20 μl formamide gel-loading dye.

7. Resolve samples (1.5 μl) on 6% sequencing gel, transfer mtDNA to nylon membrane by vacuum blotting for 45 min on standard gel dryer, and UV-cross-link mtDNA to membrane. Wet membrane in tray of distilled, deionized water and remove all pieces of polyacrylamide.

8. Prehybridize membrane 30 min (65°) in 30 ml hybridization solution. Discard hybridization solution and add 10 ml same solution plus probe and continue hybridization overnight. Pour off hybridization/probe solution and wash blot in 500 ml wash buffer (60°). Wash membrane 1 hr at 60° with buffer changes every 20 min, rinse membrane final time, and expose to X-ray film.

[a] This step-by-step protocol is designed for the simultaneous preparation of 8 control (naked mtDNA) and 8 test (protein–mtDNA) samples to be analyzed by Southern hybridization. Except for the timing of DMS addition (steps 2 and 5), all procedures should be applied to both control and test samples alike. All ethanol precipitations should include a 70% (v/v) ethanol wash step. PBS, Phosphate-buffered saline; SDS, sodium dodecyl sulfate.

Procedures

Construction of Mitochondrial DNA Clones and Probe Synthesis

We have employed two techniques, Southern hybridization and primer extension, to visualize patterns of altered methylation *in organello*. Because the entire sequence of the bovine mitochondrial genome is known,[7] either procedure can be implemented to analyze virtually any mtDNA domain for protein interaction. For Southern hybridization, a sequence map encompassing the targeted domain is required, first for selection of a restriction enzyme site to be used as a reference point for footprinting, and then for construction of a probe for hybridization.

In choosing the best restriction site for footprint analysis, several factors should be considered. First, the distance between the restriction site and the domain of interest should be between 70 and 300 bp. DNA fragments smaller than 70 bp are often difficult to detect using standard Southern hybridization conditions, and G ladders greater than 300 bp in length may be difficult to resolve. Also, longer G ladders may require optimization of the methylation reaction conditions to generate a uniform distribution of DNA fragments. Second, the restriction enzyme must recognize only one site near the domain of interest. If a second recognition site is present within a few hundred base pairs on the other side of the domain, then, depending on the size of the radiolabeled probe, two G ladders may be superimposed on each other, making an informative analysis impossible. Third, if possible, avoid selecting a reference site separated from the domain of interest by unusually G-rich regions. Homopolymer runs of G residues or G-rich sequences of DNA often require extra manipulation of methylation reaction conditions (see below) and are also potential hot spots for heteroplasmic sequences in mitochondria[8] which can result in unreadable G ladders.

Following selection of a reference restriction site, sequence information is also necessary for construction of DNAs used in the synthesis of radiolabeled probes. Oligodeoxynucleotide primer pairs, bracketing about 150–200 bp of sequence extending from the restriction site, are synthesized and combined with unmethylated bovine mtDNA in a standard polymerase chain reaction (PCR) to generate double-stranded mtDNA fragments (see Fig. 1). The amplified mtDNA fragment (~150 to 200 bp) is then cloned into the pBS(+) phagemid vector (Stratagene, La Jolla, CA) using standard methods. Radiolabeled RNA probes complementary to the region and strand of interest are generated using either T7 or T3 RNA polymerase (United States Biochemical, Cleveland, OH) and the corresponding promoter site of the appropriate linearized template clone. We usually verify riboprobe synthesis by electrophoresis in 4% nondenaturing polyacrylamide gels and autoradiography. A small gel slice containing the labeled

[7] S. Anderson, M. H. L. de Bruijn, A. R. Coulson, I. C. Eperon, F. Sanger, and I. G. Young, *J. Mol. Biol.* **156,** 683 (1982).
[8] W. W. Hauswirth, M. J. Van de Walle, P. J. Laipis, and P. D. Olivo, *Cell (Cambridge, Mass.)* **37,** 1001 (1984).

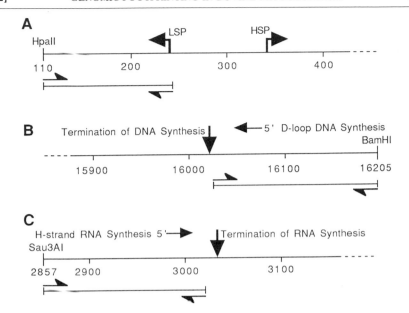

FIG. 1. Strategies for analyzing three bovine mtDNA regulatory domains by Southern analysis. For each region analyzed, oligonucleotide primers (noted by half-arrows) were synthesized and used to generate PCR-amplified fragments that were subsequently cloned into the pBS(+) vector. Radiolabeled riboprobes complementary to either DNA strand were generated using the appropriate linearized cloned template and either T7 or T3 RNA polymerase. For each region, the restriction enzyme site used as a reference point for footprinting is indicated. (A) For *in organello* footprint analysis of the upstream portion of the D-loop region containing the transcription initiation sites from the L-strand promoter (LSP) and H-strand promoter (HSP), probes extended from genomic positions 111 to 258. (B) For analyzing the termination-associated sequence (TAS) located at the D-loop 3' end, probes extended from 16,205 to 16,022. (C) For analyzing sequences of the bovine analog of the human mitochondrial transcription termination site (mTERF), probes extended from 2861 to 3017.

RNA is excised, crushed, and added to the hybridization solution (see below).

For primer extension analysis, restriction digestion of the mtDNA is unnecessary. Knowledge of the DNA sequence is essential, though, for synthesis of oligodeoxynucleotide primers complementary to sequences about 60 to 70 nucleotides upstream of the region of interest on both DNA strands.

In Organello Methylation of Mitochondrial DNA

Mitochondria are purified from bovine brain tissue by differential centrifugation and sucrose step gradient banding using standard methods.[9]

[9] G. L. Hehman and W. W. Hauswirth, *Proc. Natl. Acad. Sci. U.S.A.* **89,** 8562 (1992).

After the final centrifugation step, the pelleted mitochondria (~1.7 ml) are resuspended in an equal volume of room temperature phosphate-buffered saline (PBS) [137 mM NaCl, 2.7 mM KCl, 4.3 mM sodium phosphate, 1.4 mM potassium phosphate (pH 7.4)], and 200-μl aliquots are placed into 1.5-ml Eppendorf tubes. This amount of mitochondria is sufficient for 8 test and 8 control samples, about the maximum number that can be easily manipulated at one time. As one of the most difficult steps in a methylation-based footprinting technique is the production of equally intense G ladders in control (naked DNA) and test (DNA–protein) samples, we find that dividing the purified mitochondria into multiple control and test samples allows evaluation of a range of DMS concentrations and incubation times, as well as testing of the reproducibility of results. To each of the 8 test samples add DMS to a final concentration between 0.50 and 0.062%, mix by vortexing 2–3 sec, and incubate for 3 min at room temperature. The reaction is stopped by diluting the DMS with 1 ml ice-cold PBS. The mitochondria are pelleted by centrifugation for 1 min at 10,000 g and washed twice more with cold PBS to further dilute the DMS. If desired, mitochondrial pellets can be stored at this point at $-80°$ for further processing at a more convenient time.

Purification of Nucleic Acids and Restriction Digestion

To lyse the mitochondria, individual test and control mitochondrial pellets are suspended in 400 μl lysis buffer [50 mM Tris-HCl (pH 8.0), 25 mM EDTA, 250 mM NaCl], and 25 μl of 10% SDS is added. The solution is vortexed vigorously until it clears. If there is difficulty achieving complete lysis, SDS may be added up to a final concentration of 2%. The samples are deproteinized by extracting twice with phenol, twice with phenol–chloroform–isoamyl alcohol (25 : 24 : 1, v/v), and once with chloroform and then precipitated with 2.5 volumes ethanol. The addition of sodium acetate is unnecessary owing to the presence of NaCl in the lysis buffer.

For Southern hybridization analysis, the samples are resuspended in a 100-μl reaction volume and digested with the restriction enzyme of choice (this step is unnecessary for primer extension analyses). For convenience, double restriction digests can be performed on the same mtDNA sample for the simultaneous analysis of different domains, assuming that the two domains are sufficiently distant from one another. Following digestion, 5 μl of each sample is resolved on a 1% agarose gel. Following ethidium bromide staining of the gel, the samples are visualized using ultraviolet light. This procedure allows for verification of digestion and provides a means for estimating relative DNA concentrations among samples (Fig. 2). Generally, if distinct mtDNA bands are visible on the gel, then there

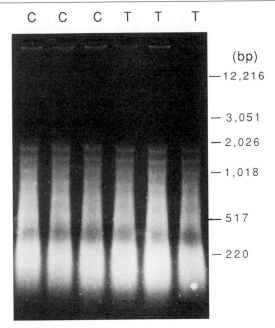

FIG. 2. Visualization of control and test mtDNA samples. Three control (C lanes) and three test samples (T lanes) of bovine mtDNA were digested with *Sau*3AI and *Kpn*I. Five percent of each control and test sample was subjected to electrophoresis alongside a molecular weight size marker on a 1% agarose gel. Nucleic acids were visualized with ethidium bromide staining and UV illumination. DNA bands that correspond to known mitochondrial restriction sites are evident, and the amounts of sample loaded per lane appear to be equal. Positions of DNA size standards are indicated at right.

is sufficient sample for at least 15 different primer extension reactions or gel loadings for Southern analysis.

Methylation of Control Samples

Following digestion, all samples are brought to a volume of 200 μl with the addition of 5 μl water and 100 μl of 100 mM sodium cacodylate (pH 7.0). For primer extension analysis, test and control mtDNA pellets are suspended in 200 μl of 50 mM sodium cacodylate (pH 7.0). Dimethyl sulfate is then added to the control samples only, to a final concentration between 0.50 and 0.062% (any alterations in the methylation conditions that may have been applied to the test samples should likewise be applied to the control samples), and the samples are incubated 3 min at room temperature. Fifty microliters of 4° DMS stop buffer [1.5 M sodium acetate (pH 7.4), 1 M 2-mercaptoethanol] is then added to all the samples, and the mtDNAs

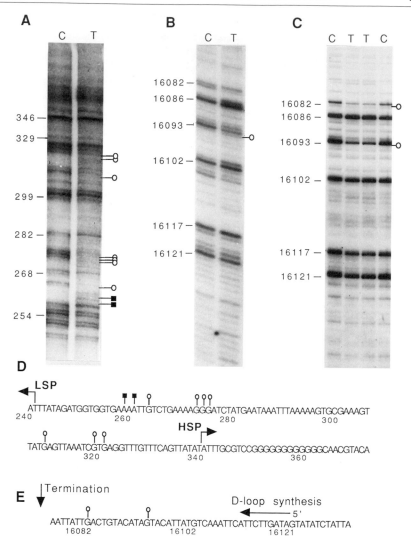

FIG. 3. Examples of *in organello* footprinting using electroblotting (A), primer extension (B), and vacuum blotting (C). (A) Isolated mitochondria from bovine brain tissue were treated with 0.5% DMS and the mtDNA extracted (test DNA, lane T). Naked mtDNA from the same animal was also treated with DMS to generate a control G ladder (lane C). Test and control DNAs were digested with *Hpa*II, cleaved with piperidine, resolved on a 6% polyacrylamide sequencing gel, and electroblotted onto a nylon membrane. The DNA blot was hybridized to a radiolabeled RNA probe complementary to the mtDNA H strand (see Fig. 1A). Sites of *in organello* methylation protection (open circles) and methylation hypersensitivity (filled boxes) are evident at both the LSP and HSP. Mitochondrial genomic positions are indicated. (B) Control and test samples were prepared as described for (A), except for

are precipitated with ethanol. To cleave the mtDNAs at methylated residues,[10] the DNA pellets are resuspended in 200 μl of 10% piperidine and incubated 30 min at 90°. Afterward, the samples are placed on ice, an equal volume of 7.5 M ammonium acetate is added, and the samples are ethanol-precipitated. To ensure sharp resolution of adjacent DNA bands in subsequent sequencing gels, it is critical to remove all traces of piperidine from the DNA samples. Generally, three successive ethanol precipitations with ammonium acetate are sufficient. Alternatively, following an initial ethanol precipitation, the samples may be suspended in 200 μl water and lyophilized. Samples to be analyzed by Southern hybridization are then suspended in 20 μl formamide loading dye. Those intended for primer extension are suspended in 50 μl TE [10 mM Tris-HCl (pH 8.0), 1 mM EDTA].

Analysis of Mitochondrial DNA Methylation Products
  by Primer Extension

For primer extension analysis, an oligodeoxynucleotide complementary to sequences near the region of interest is first end-labeled using T4 polynucleotide kinase (United States Biochemical). The labeled primer is then mixed in 100-μl reaction volumes containing control and test mtDNAs (~1/20 of each sample), reaction buffer [10 mM Tris-HCl (pH 9.0)], 50 mM KCl, 0.1% Triton X-100, 2.5 mM MgCl$_2$], and a thermostable DNA polymerase and incubated for 10 cycles (1 min at 94°, 2 min at 55°, and 5 min at 72°/cycle) in a thermocycler. The samples are then extracted with phenol–chloroform, ethanol-precipitated, and resolved on a 6% sequencing gel.

In using primer extension analyses we have occasionally noted the appearance of bands that do not correspond to G or A residues. Such a band can be seen above the G residue at nucleotide 16,086 in Fig. 3B (lane T). If such bands were to arise at positions where G residues are normally found, their presence could hinder the detection of altered methylation

---

[10] A. M. Maxam and W. Gilbert, *Proc. Natl. Acad. Sci. U.S.A.* **74,** 560 (1977).

---

the omission of the restriction digestion step. Protein–mtDNA interaction near the D-loop 3′ terminus (16,022) was visualized by primer extension of mtDNA fragments as described in the text. (C) Control and test samples were prepared and analyzed as described for (A), except that *Bam*HI was used to digest the mtDNA, and vacuum blotting was used to transfer the DNA to the nylon membrane. The probe used to visualize the same protein–mtDNA interaction (B) is shown in Fig. 1. (D, E) Sequence summaries of the sites of *in organello* protein–DNA interaction at transcription initiation and D-loop DNA termination domains, respectively.

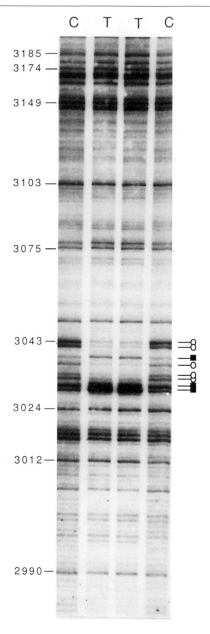

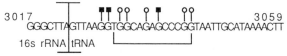

reactivities. This may account for the lack of methylation protection observed by primer extension at the 16,082 G residue (compare Fig. 3B,C). As discussed in [3] of this volume, many of these types of problems can be alleviated by optimizing reaction conditions. In general, primer extension analysis has proved to be extremely valuable when mtDNA samples are present in limited quantities. However, for *in organello* footprinting of large animal mitochondria, where obtaining large amounts of sample is usually not a problem, Southern analysis is the method of choice.

Southern Analysis of Mitochondrial DNA Methylation Products

Equal amounts of test and control mtDNAs (~1/20 of each sample) are loaded onto adjacent lanes of a 6% polyacrylamide, 7 *M* urea sequencing gel and subjected to electrophoresis using standard conditions. Following electrophoresis, one of the most difficult steps in Southern analysis is the efficient, equal transfer of all DNA fragments to a nylon membrane. When numerous control and test samples are electrophoresed simultaneously, the size and fragility of the gel make handling particularly awkward. This is especially true for electroblotting, where several manipulations of the gel and membrane are required. As handling difficulties increase so do problems with uneven transfer and smearing of the DNA bands. An example is shown in Fig. 3A.

The frequency of poor transfer can be greatly reduced by employing a vacuum blotting technique.[11] Briefly, following electrophoresis a precut piece of dry Hybond-N$^+$ nylon transfer membrane (Amersham, Chicago, IL) is placed directly on the polyacrylamide gel and allowed to sit for 5 min. Because the gel sticks to the membrane, it can then be easily peeled back and positioned, membrane-side down, on a single sheet of Whatman 3MM paper (Whatman Paper Ltd., Clifton, NJ). After placing the gel, Whatman paper-side down, onto a standard gel-drying apparatus, it is covered with plastic wrap, and the DNA is transferred to the membrane by applying a

[11] M. C. Lopez, J. B. Smerage, and H. V. Baker, *BioTechniques* **15**, 362 (1993).

FIG. 4. *In organello* footprinting at the bovine analog of the mitochondrial transcription termination site. Control and test samples were digested with *Sau*3AI, cleaved with piperidine, resolved on a polyacrylamide sequencing gel, and vacuum-blotted onto nylon membrane. The L strand of this region was visualized using the radiolabeled probe shown in Fig. 1C. Symbols used to indicate sites of altered methylation reactivity have the same meanings as in Fig. 3. Mitochondrial genomic positions are given, and a sequence summary is provided below the autoradiograph. The bovine analog of the tridecamer transcription termination site is bracketed. The junction of the 16 S rRNA and tRNA^{Leu} genes is indicated. H-strand transcription proceeds from left to right.

vacuum (without heat) for approximately 45 min. This technique is fairly easy and yields a sharp, precise transfer of DNA bands (Figs. 3C and 4). As a note of caution, we find that the use of newer, more efficient vacuum pumps during DNA transfer may result in poor transfer of higher molecular weight fragments (above 200 bp). However, placing an additional sheet of Whatman paper under the membrane often corrects this problem. The extra sheet most likely functions to slow the drying of the polyacrylamide gel, thus allowing more time for DNA transfer.

Once the mtDNA has been transferred, it is cross-linked to the membrane by UV irradiation. With the gel still attached, the membrane is placed gel-side down on a transilluminator at full power for 5 min. Afterward, the polyacrylamide is removed by soaking the membrane in a tray of water. While still in the tray, the membrane is rolled up and placed into a rolling hybridization bottle. To prehybridize the membrane, 30 ml hybridization solution, containing 1% bovine serum albumin, 1 mM EDTA, 7% SDS, and 0.5 M Na$_2$HPO$_4$ buffer [7.1% (w/v) anhydrous Na$_2$HPO$_4$, titrated with phosphoric acid to pH 7.2; this solution is 1 M with respect to sodium ion], is added, and the bottle is placed in a rolling bottle hybridization oven (Biometra, Tampa, FL) for 30 min at 65°. Discard hybridization solution and replace it with 10 ml fresh hybridization solution including the crushed polyacrylamide gel slice containing the labeled probe. The DNA blots are then hybridized overnight at 65°. The hybridization solution is decanted (may be saved for reuse if desired), and the membrane placed in a container filled with 500 ml of 60° wash buffer (1 mM EDTA, 40 mM Na$_2$HPO$_4$ buffer, 1% SDS). Perform three 20-min washes at 60° with gentle agitation. Briefly rinse the membrane once more in wash buffer and expose to X-ray film. When each step is performed successfully, *in organello* footprinting allows analysis of protein–mtDNA interactions over a large span of nucleotides. As shown in Fig. 4, analysis of a ~200 bp region the containing 16 S rRNA and tRNA^{Leu} genes, demonstrates significant protein—mtDNA interaction within the bovine analog of the human mitochondrial transcription termination factor (mTERF) binding sequence.[12]

---

[12] B. Kruse, N. Narasimhan, and G. Attardi, *Cell* (*Cambridge, Mass.*) **58,** 391 (1989).

[3] Genomic Footprinting of Mitochondrial DNA: II. *In Vivo* Analysis of Protein–Mitochondrial DNA Interactions in *Xenopus laevis* Eggs and Embryos

By CHANDRAMOHAN V. AMMINI, STEVEN C. GHIVIZZANI, CORT S. MADSEN, and WILLIAM W. HAUSWIRTH

Introduction

Genomic footprinting is an established and invaluable tool in analyzing gene regulation. Since its initial application to study protein–DNA interactions of a B-cell-specific immunoglobin enhancer *in vivo*,[1] genomic footprinting has been used in a wide variety of experimental situations to answer crucial questions in gene expression. This technique has been adapted in our laboratory to analyze protein–mitochondrial (mt) DNA interactions in purified bovine brain mitochondria (*in organello* footprinting)[2] and is described in detail in [2] of this volume. Here, we describe an extension of mitochondrial genomic footprinting to single-cell and multicell *Xenopus laevis* eggs and embryos.

Southern blotting is the method of choice for visualizing genomic footprints in the example above, or in other experimental situations where a large amount of tissue is available for mtDNA isolation. Such *in organello* footprint analyses of both bovine[2] and human[3] mtDNA have furnished important new insights into mitochondrial genome organization and regulation. However, further advances into our understanding of mitochondrial gene regulation await investigations of how protein–mtDNA interactions correlate with important metabolic activities of the mitochondria. Of the various model systems for studying mitochondrial metabolism in different experimental situations, the *Xenopus laevis* embryogenesis system offers a unique natural model of regulated mitochondrial gene expression.[4,5] The primary advantage of these systems is the feasibility of performing *in vivo* methylation of mtDNA in intact cells, making true *in vivo* footprinting of mtDNA possible. The major bottleneck, however, is one of sensitivity where Southern blotting generally fails due to the problem of low tissue

[1] A. Ephrussi, G. M. Church, S. Tonegawa, and W. Gilbert, *Science* **227,** 134 (1985).

[2] S. C. Ghivizzani, C. S. Madsen, and W.W. Hauswirth, *J. Biol. Chem.* **268,** 8675 (1993).

[3] S. C. Ghivizzani, C. S. Madsen, M. R. Nelen, C. V. Ammini, and W. W. Hauswirth, *Mol. Cell Biol.* **14,** 7717 (1994).

[4] J. W. Chase and I. B. Dawid, *Dev. Biol.* **27,** 504 (1972).

[5] A. E. Meziane, J. C. Callen, and J. C. Mounolou, *EMBO J.* **8,** 1649 (1989).

availability, exacerbated by the inevitable DNA losses during DNA transfer to hybridization membranes. Two PCR (polymerase chain reaction)-based innovations that overcome this signal-to-noise problem are (1) linear amplification of footprint ladders with radiolabeled primers[6,7] (primer extension footprinting or PEF) and (2) exponential amplification of footprint ladders after an intermediate ligation step with an asymmetric linker DNA (ligation-mediated PCR or LMPCR footprinting).[8,9] Described here is the application of primer extension footprinting to analyze DNA–protein interactions in the mitochondria of *Xenopus laevis* oocytes.

The mitochondrial genome of *Xenopus laevis* has an organization similar to those of higher vertebrates, exhibiting all known conserved regulatory features.[10,11] Molecular studies have also highlighted the potential of the system for studying mitochondrial gene regulation.[4,5] Distinct embryonic stages with induction or shutdown of mtDNA replication and transcription have been identified.[5] The PEF protocol was adapted to mtDNA to exploit this natural model of regulated gene expression in mitochondria. It is hoped that these studies will offer fresh insights into the genomic organization and regulation of higher vertebrates.

The feasibility of LMPCR footprinting in this system was also investigated. This method, however, is significantly more demanding technically, and its reliability depends on extensive empirical optimization of multiple steps. Another reason for preferring PEF is that the amount of mtDNA available for footprinting is far in excess of that required for an ultrasensitive technique like LMPCR. Additionally, the PEF protocol is a simple, one-step procedure that is highly reproducible and provides high-resolution sequence information for multiple regions from a limited DNA sample.

Principle

Sequence-specific DNA-binding proteins frequently contact DNA in the major groove into which the N-7 position of guanines is directed. Thus, the principle of *in vivo* footprinting using dimethyl sulfate (DMS) is based on the interference to guanine methylation caused by the occupancy of DNA-binding proteins at their cognate binding sites. Protein binding can also induce localized conformational changes in DNA, leading to additional changes in the reactivity to DMS. These altered methylation interactions

[6] H. P. Saluz and J. P. Jost, *Nature (London)* **338,** 277 (1989).
[7] H. P. Saluz and J. P. Jost, *Proc. Natl. Acad. Sci. U.S.A.* **86,** 2602 (1989).
[8] P. R. Mueller and B. Wold, *Science* **246,** 780 (1989).
[9] I. K. Hornstra and T. P. Yang, *Anal. Biochem.* **213,** 179 (1993).
[10] B. A. Roe, D. P. Ma, R. K. Wilson, and J. F. H. Wong, *J. Biol. Chem.* **260,** 9759 (1985).
[11] S. C. Cairns and D. F. Bogenhagen, *J. Biol. Chem.* **261,** 8481 (1986).

appear on a sequencing gel as either protected or hypersensitive bands when compared to controls.

After achieving partial methylation of genomic DNA, sequence ladders are generated using standard Maxam and Gilbert G + A sequencing chemistry.[12] This chapter describes the use of primer extension to visualize footprint ladders. In this technique, the design of genomic primers is based on the published sequence of the *X. laevis* mitochondrial genome.[10,11,13,14] The primers are radiolabeled to a high specific activity and used to perform a linear amplification of the footprint ladder with *Taq* DNA polymerase in a thermal cycler (see Fig. 1B). The main advantage of this method, in addition to sensitivity over Southern blotting, is that it is not necessary to perform restriction digestion of the DNA to generate reference points, saving time and making possible the analysis of multiple domains of DNA using the same DNA template. Following primer extension, the heterogeneous DNA population is size-resolved on a sequencing gel and autoradiographed to obtain *in vivo* footprinting patterns. Comparison of the intensity of the bands in the control and experimental lanes allows the identification of sites of protein occupancy. Naked genomic DNA, methylated *in vitro*, is the most commonly used control. The most desirable control, however, would be *in vivo* methylation of cells of the same lineage caught in a different biochemical state (either induced experimentally or available naturally). The *X. laevis* embryogenesis model provides such internal controls for comparing footprinting action. Some tissue culture systems may also fulfill this need.

Procedures

Materials

Frogs: Sexually mature male and female *X. laevis* frogs are purchased from Xenopus I (Ann Arbor, MI) and maintained in separate tanks using standard procedures[15]

Human chorionic gonadotropin (hCG; Sigma CG-5, St. Louis, MO):

---

[12] A. M. Maxam and W. Gilbert, this series, Vol. 65, p. 499.

[13] The nucleotide numbering used in the text is based on the published sequence of the whole genome of *X. laevis* mitochondria,[10] with the following modifications. The sequence of the primer and those cited in the text and figures are based on the sequence of Dunon-Bluteau *et al.*[14] and corrections made by Cairns and Bogenhagen.[11] The numbering is corrected for a 31-nucleotide insertion in the termination-associated sequence (TAS) region (between nucleotides 78 and 79).[10]

[14] D. Dunon-Bluteau, M. Voltovitch, and G. Brun, *Gene* **36,** 65 (1985).

[15] M. Wu and J. Gerhart, *Methods Cell Biol.* **36,** 3 (1991).

A

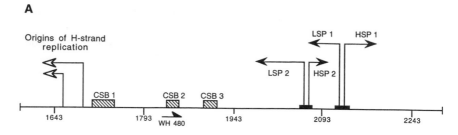

B

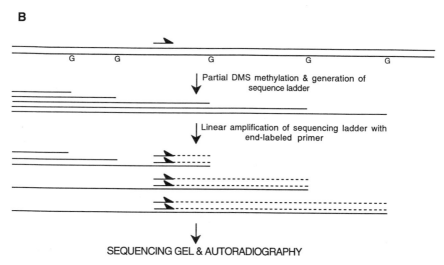

SEQUENCING GEL & AUTORADIOGRAPHY

FIG. 1. (A) Schematic representation of the important conserved cis-elements in the D-loop region of *X. laevis*. Transcription from the L-strand promoter (LSP) is likely to provide the primers for H-strand DNA replication, for which two major origins have been mapped. The conserved sequence blocks (CSB) are thought to be important for mediating the generation of RNA primers for H-strand DNA replication. (B) Primer extension footprinting. A hypothetical example with five guanosine residues is shown to illustrate the principle.

Reconstitute in water at 1000 U/ml and store at 4° (good for 3–4 weeks)

10× OR-2: 825 mM NaCl, 25 mM KCl, 10 mM CaCl$_2$, 10 mM MgCl$_2$, 10 mM Na$_2$HPO$_4$, 50 mM HEPES, adjusted to pH 7.8 with NaOH (the divalent and monovalent ions are made up as separate stock solutions and mixed in water to the desired concentration just before use)

10× F-1: 412.5 mM NaCl, 12.5 mM KCl, 2.5 mM CaCl$_2$, 0.625 mM

$MgCl_2$, 5 mM Na_2HPO_4, 25 mM HEPES, adjusted to pH 7.8 with NaOH (the divalent and monovalent ions are made up as separate stock solutions and mixed in water to the desired concentration just before use)

Phosphate-buffered saline (PBS): 140 mM NaCl, 2.5 mM KCl, 10 mM Na_2HPO_4, 1.5 mM KH_2PO_4, pH 7.5

2% (w/v) Cysteine solution: (cysteine, free base Sigma), make up within 1 hr of use, and titrate to pH 7.8 with NaOH (4.5 ml of 1 M NaOH per 100 ml yields the desired pH)

Dimethyl sulfate (DMS; Aldrich, Milwaukee, WI): All manipulations involving DMS should be done in a chemical hood, and all solutions containing DMS as well as used plasticware should be inactivated with 5 M NaOH

Primers: (DNA synthesis core, University of Florida, Gainesville). Primers are gel-purified using standard procedures.[16] Primer WH 480 with the sequence 5'-CGCTAAACCCCCCTACCCCCC-3', corresponding to positions 1830–1851,[13] was used to analyze the H strand of the promoter region near the D-loop 5' end (see Fig. 1A).

10× *Taq* DNA polymerase reaction buffer (BRL, Gaithersburg, MD): 200 mM Tris-HCl (pH 8.4), 500 mM KCl

10× *Pfu* DNA polymerase reaction buffer (Stratagene, La Jolla, CA): 200 mM Tris-HCl (pH 8.75), 100 mM KCl, 100 mM $(NH_4)_2SO_4$, 20 mM $MgSO_4$, 1% v/v Triton X-100, 1 mg/ml bovine serum albumin (BSA).

10× Vent DNA polymerase reaction buffer (New England Biolabs, Beverly, MA): 100 mM KCl, 100 mM $(NH_4)_2SO_4$, 200 mM Tris-HCl (pH 8.8), 20 mM $MgSO_4$, 1% Triton X-100

2 mM dNTP mix: Make from 100 mM dNTP stocks from Pharmacia (Piscataway, NJ)

In Vitro Fertilization of Eggs

Fertilized embryos are obtained using the *in vitro* fertilization procedure of Hollinger and Corton.[17] The female frogs are injected with 700 U hCG into the peritoneal cavity 10 hr before eggs are required. Male frogs are also primed with 200 U hCG 1 day earlier. After spontaneous egg laying begins, eggs are gently squeezed from the female onto 100 × 15 mm petri dishes containing approximately 20 ml of 1× OR-2, in batches of 400–500

[16] J. Sambrook, E. F. Fritsch, and T. Maniatis, "Molecular Cloning." 2nd ed. Cold Spring Harbor Laboratory, Cold Spring Harbor, New York, 1989.

[17] T. G. Hollinger and G. L. Corton, *Gamete Res.* **3,** 45 (1980).

eggs. Sperm is obtained by dissecting testes from a male frog (anesthetized on ice) and gently macerating the tissue in 1.5× OR-2 to release the sperm. Sperm motility is tested under a light microscope after diluting with water to 0.5× OR-2. The eggs are then drained of excess 1× OR-2, and 10–12 drops of the sperm suspension in 1.5× OR-2 are layered on top of the eggs. Then 15–20 ml of 1× F-1 is quickly added to dilute the sperm suspension. After being covered with lids, the petri dishes are gently swirled to distribute the sperm evenly and to spread out the eggs into a monolayer; then the plates are left undisturbed. Fertilization can be monitored after 30 min and unfertilized or damaged eggs discarded. A separate batch of eggs left unfertilized in 1× OR-2 serves as the unfertilized egg control, and another batch can be used to prepare mtDNA (naked DNA control) for *in vitro* methylation.

In Vivo Methylation of Eggs and Embryos

Eggs and embryos are dejellied by gently swirling in 2% (w/v) cysteine solution for 2–3 min, followed by repeated washings in 1× OR-2. After draining the OR-2, float the eggs or embryos in PBS containing 0.125% DMS and gently swirl for 3 min. The DMS solution is rapidly emptied after 3 min, and the eggs or embryos are repeatedly washed with 1× OR-2 to remove traces of DMS. Mitochondria are then isolated by differential centrifugation using standard methods,[18] except that banding on sucrose gradients is not performed. The mitochondrial pellets are quickly frozen and stored at −80°. Typically, one mitochondrial pellet from a batch of 400 eggs or embryos yields sufficient mtDNA to perform 10–12 PEF reactions.

Preparation of Mitochondrial DNA for Footprinting Reactions

Isolation of mtDNA from control and test samples, methylation of control DNA samples, and piperidine cleavage of methylated samples are done exactly as described in [2] in this volume with the following modifications. (1) Restriction digestion of the DNA samples is unnecessary in the primer extension protocol as the sequencing reference is provided by the site of hybridization of the primer. (2) Agarose gel electrophoresis to estimate the DNA concentration is not desirable in this system, as a substantial fraction of the DNA would be lost in attempting to visualize bands by ethidium bromide staining. On average, about 0.8 to 1 μg mtDNA is recovered from 300–400 eggs, assuming approximately 4 ng mtDNA/egg.[19] (3)

---

[18] G. L. Hehman and W. W. Hauswirth, *Proc. Natl. Acad. Sci. U.S.A.* **89,** 8562 (1992).
[19] B. Gurdon and L. Wakefield, *in* "Microinjection and Organelle Transplantation Techniques," p. 269. Academic Press, London, 1986.

Similar patterns of methylation in control DNA can be achieved by 0.125% DMS treatment for 30–45 sec, compared to the *in vivo* reaction time of 3 min at 0.125%. (4) Traces of piperidine are removed from the cleaved DNA samples by lyophilization of DNA samples resuspended in 500 μl sterile water after a single ethanol precipitation and 80% (v/v) ethanol wash. (5) The cleaved control and test DNA samples are resuspended in 20 μl TE [10 mM Tris-HCl (pH 7.5), 1 mM EDTA].

Primer Extension Footprinting Reactions

The most important factor for obtaining accurate and reproducible sequences is optimal selection of genomic primers. In theory, a primer can be as close as 10–30 nucleotides to the domain of interest. However, in practice, we have found that primers located about 100 nucleotides away from the region to be analyzed work best in providing reproducible genomic ladders which are essentially free of low molecular weight background. In general, primers which fall within 70–150 bp of the target region can be used successfully. A primer length of 21 nucleotides is usually sufficient, except for exceptionally AT-rich domains, where a longer primer is preferable.

5′-End-Labeling. Gel-purified primers are 5′-end-labeled to a high specific activity with [γ-^{32}P]ATP (ICN Biomedicals, Costa Mesa, CA; 7000 Ci/ mmol) and a T4 polynucleotide kinase kit (USB Corporation–Amersham Life Science, Cleveland, OH) in a standard 50-μl reaction. Free label is removed by two sequential ethanol precipitations in the presence of 2.5 M ammonium acetate.

Optimization of Primer Extension Footprinting Reaction Conditions

Primer and Template Concentrations. A number of preliminary experiments indicated that about one-tenth of the piperidine-cleaved DNA sample (about 50–100 ng) is sufficient for obtaining a usable PEF signal. Increasing the primer concentration beyond 1 pmol per reaction does not further enhance signal intensity and leads to higher background.

Cycles of Linear Amplification. The results of an experiment designed to determine the number of linear amplification cycles necessary in this system to obtain reproducible results without sacrificing accuracy are shown in Fig. 2. Amplifying for 10–15 cycles provides adequate signal intensity, whereas nonspecific bands begin to appear with a higher number of cycles. It is important to note that some of the nonspecific bands appearing with higher numbers of cycles are 1–2 nucleotides longer than the authentic bands. Therefore, it seemed that the well-established ability of *Taq* DNA

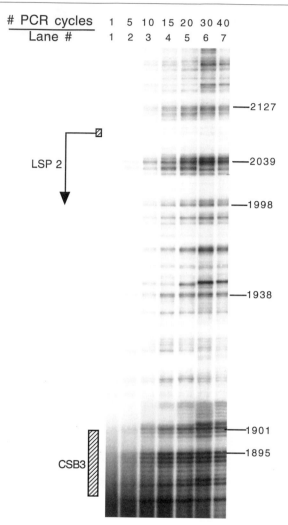

FIG. 2. Experiment to determine the number of linear amplification cycles necessary to obtain reproducible results. *In organello* methylated mtDNA isolated from *X. laevis* ovaries was used as a template, with 1 pmol end-labeled primer WH 480, 1× *Taq* buffer, 1.75 m*M* MgCl$_2$, 100 μ*M* dNTPs, and 2.5 U *Taq* DNA polymerase in a 50-μl reaction. The samples were denatured at 94° for 5 min, followed by independent cycling of each reaction for 1, 5, 10, 15, 20, 30, and 40 cycles with the following profile: 94° for 1 min, 60° for 1 min, and 72° for 5 min. Note the appearance of longer bands, especially in the samples cycled at 30 and 40 cycles.

polymerase to add nontemplated nucleotides to the 3' end of duplex DNA[20] was elongating accurate DNA ends made in earlier cycles. Thermostable polymerases having a 3'–5' proofreading ability, such as Vent DNA polymerase or *Pfu* DNA polymerase, have been used to alleviate this problem in LMPCR protocols.[21] The results of an experiment with *Pfu* DNA polymerase and Vent DNA polymerase in a PEF reaction using the manufacturer's suggested buffer for primer extension are shown in Fig. 3. Neither polymerase primer extended accurately in this system and should be compared to the reproducible pattern produced by *Taq* DNA polymerase (compare lanes 4–8 with lanes 1–3 and lanes 9 and 10 in Fig. 3). The manufacturers suggest lowering the annealing temperature of primers for these polymerases because of the lower ionic strength of the buffers. However, these and a variety of other conditions were tried with these polymerases without success (data not shown). Possibly these polymerases could be adapted to this system with extensive optimization of buffer and PCR conditions, as has been noted for Vent DNA polymerase in the LMPCR footprinting reaction.[9] However, the need for a reliable thermostable polymerase that requires minimal buffer optimizations is ably fulfilled by *Taq* DNA polymerase, and we have used this polymerase in all the other experiments described.

Nucleotide Concentration. A final dNTP concentration of 20 μM was found to be optimum in enhancing specificity and enabling reproducibility of results (Fig. 3, lanes 4–8, and data not shown). Lowering the dNTP concentration below this level sacrificed both fidelity and signal intensity.

Primer Extension Time. Most current protocols for primer extension up to 300 or 400 bases recommend extension times of 10 min to allow completion of all initiated primers.[9,21] We have investigated the feasibility of lowering the extension time, which would enhance fidelity and drastically reduce total reaction time, enabling faster processing of samples. An extension time of 3 min has been found to be sufficient for a number of primer–template combinations without compromising fidelity or sensitivity.

Typical Primer Extension Footprinting Reaction

On the basis of the above considerations, a typical 100-μl PEF reaction should contain the following:

50–100 ng piperidine-cleaved DNA
1 pmol γ-^{32}P 5'-end-labeled primer

[20] J. M. Clark, *Nucleic Acids Res.* **16**, 9677 (1988).
[21] P. A. Garrity and B. J. Wold, *Proc. Natl. Acad. Sci. U.S.A.* **89**, 1021 (1992).

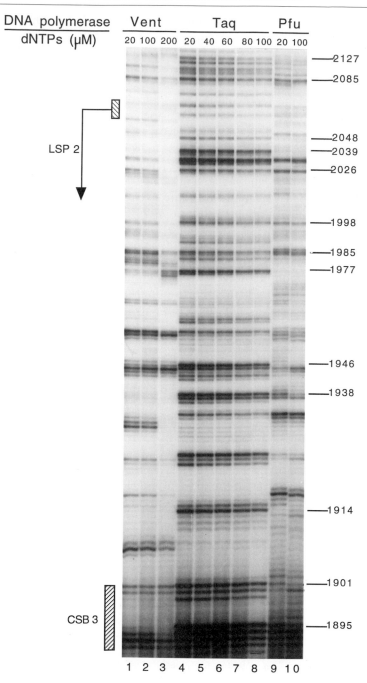

$1\times$ *Taq* DNA polymerase buffer [20 mM Tris-HCl (pH 8.4), 50 mM KCl]

2 mM MgCl$_2$

20 μM each of dATP, dTTP, dGTP, and dCTP

2.5 U *Taq* DNA polymerase

The reaction is assembled on ice and cycled in a thermal cycler as follows: denaturing at 94° for 5 min, followed by 10–15 cycles with the profile denaturing at 94° for 1 min, annealing at 45–65° (based on primer T_m) for 1 min, and primer extension at 72° for 3 min. The tubes are then immediately chilled on ice, 10 μg carrier tRNA is added, and DNA is ethanol-precipitated in the presence of 2.5 M ammonium acetate. An organic extraction prior to DNA precipitation is usually not necessary. The precipitate is suspended in 5 μl sequencing dye, and 1.5 μl of each reaction is resolved in a 6% polyacrylamide–7 M urea gel as a preliminary test to optimize signal intensities. Using visual inspection or phosphor imager quantitation, the signals in each sample are equalized in a final gel and autoradiographed to obtain the footprint pattern. Autoradiography at room temperature for 12–48 hr is usually sufficient to generate adequate signal intensity. Figure 4 shows the result of a footprinting reaction to analyze the H strand of the promoter region of *X. laevis* mitochondrial DNA. Footprinting is seen at multiple sites between the promoters and the conserved sequence blocks (CSB), as in both bovine[2] and human[3] mtDNA, and there is no apparent difference in footprinting pattern between the unfertilized egg and the 8-hr embryo. Comparison of footprinting among different biochemically distinct stages of embryogenesis[5] is currently under investigation.

Concluding Remarks

The procedure described above is a simple footprinting protocol which yields reproducible, high-resolution genomic footprints from small amounts of mtDNA templates. This should be generally applicable to study mitochondrial gene regulation in a wide variety of experimental samples where

---

Fig. 3. Experiment to compare the performance of *Taq* DNA polymerase with Vent DNA polymerase and *Pfu* DNA polymerase, two thermostable polymerases that possess 3′–5′ proofreading ability. The template in each reaction was mtDNA isolated from the mitochondria of *X. laevis* ovaries, methylated *in organello*. Each 100-μl reaction was assembled in the respective 1× buffers, with 2 mM Mg^{2+}, 1 pmol end-labeled primer, the indicated dNTP concentrations, and 2–2.5 units of the enzyme. The samples were denatured at 94° for 5 min, followed by 5 cycles with the profile 94° for 2 min, 65° for 1 min, and 72° for 5 min, and 10 cycles with the profile 94° for 2 min, 68° for 30 sec, and 72° for 3 min.

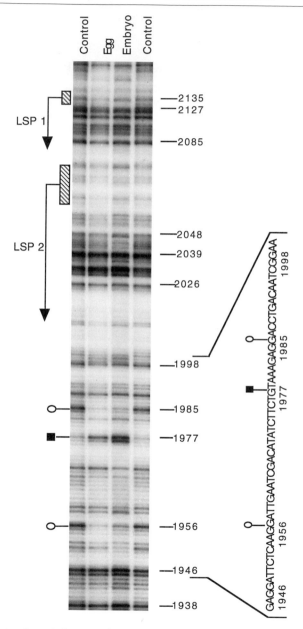

FIG. 4. *In vivo* footprinting experiment with *X. laevis* unfertilized egg and 8-hr embryo mitochondria. Eggs from a single female frog were used to obtain the stages shown. Eggs were fertilized *in vitro* (see text) and the embryos treated at the 8-hr stage with 0.125% DMS for 3 min, along with a batch of unfertilized eggs. Mitochondrial DNA purified from eggs

mitochondrial metabolism can be modified at will. As the list of cis-acting elements and trans-acting factors involved in mitochondrial DNA metabolism and nucleomitochondrial interactions continues to grow,[22–25] this approach should gain in importance. Hence genomic footprinting of mitochondrial DNA is expected to be an important tool in understanding mitochondrial gene expression. Further refinements in footprinting strategies will surely be made with the advent of new technologies, making the protocol easier and more reliable. The advantage of using newer versions of recombinant *Taq* polymerase with higher processivity is currently being investigated.

One important factor in obtaining useful footprint information is the uniformity of the cell or tissue sample. Heterogeneity in cell lineage or cell stage may well lead to masking or overrepresentation of footprint data, leading to inaccurate conclusions. Hence, in the *X. laevis* example above, it is absolutely essential to obtain all the stages of embryo, unfertilized egg, and control DNA from the same frog to make meaningful comparisons possible. Otherwise, artifactual bands cannot be differentiated from bands arising due to intraspecific DNA polymorphisms because of the highly heterozygous nature of commercially available *X. laevis*. Another important consideration is to remove as much nuclear DNA as possible before starting the footprinting protocol, to obtain clean sequencing ladders. With mitochondria, this is easily achieved by the recommended organelle purification scheme using differential centrifugation.

Finally, we note that this technique is readily adaptable to other systems where the amount of DNA available does not warrant developing a LMPCR protocol for genomic footprinting. This PEF protocol has been used to perform *in vitro* DMS footprinting using recombinant plasmids and mitochondrial protein fractions (data not shown). Footprint information in this system can be derived from 50 ng plasmid DNA. Hence, this method could

---

[22] H. Suzuki, Y. Hosokawa, M. Nishikimi, and T. Ozawa, *J. Biol. Chem.* **266,** 2333 (1991).
[23] C. A. Virbasius, J. V. Virbasius, and R. C. Scarpulla, *Genes Dev.* **7,** 2431 (1993).
[24] G. S. Shadel and D. A. Clayton, *J. Biol. Chem.* **268,** 16083 (1993).
[25] A. Chomyn and G. Attardi, *in* "Molecular Mechanisms in Bioenergetics" (L. Ernster, ed.), p. 483. Elsevier, Amsterdam, 1992.

---

was methylated *in vitro* with 0.125% DMS for 45 sec to obtain the naked DNA controls. The methylated and piperidine-cleaved DNA samples from the three preparations were used as template in PEF reactions, set up as described for a typical PEF reaction. The samples were initially denatured at 94° for 5 min, followed by amplification for 15 cycles with the following profile: 94° for 2 min, 60° for 1 min, and 72° for 3 min. Protected residues are indicated by open circles and hypersensitive sites by filled squares.

also be used to study protein–DNA interactions in transfected/transformed plasmids of interacting systems.

Acknowledgments

The authors would like to especially thank Dr. Tom Hollinger for help and expertise in conducting the *in vitro* fertilization. We also thank Drs. Alfred S. Lewin, Maurice S. Swanson, Carl M. Feldherr, and Thomas C. Rowe for helpful suggestions.

[4] *In Organello* Mitochondrial Protein and RNA Synthesis Systems from *Saccharomyces cerevisiae*

By ROBERT O. POYTON, GARY BELLUS, EDWARD E. MCKEE, KEVIN A. SEVARINO, and BRADLEY GOEHRING

Introduction

The mitochondrial genome of *Saccharomyces cerevisiae* encodes eight polypeptide subunits of respiratory proteins (cytochrome-*c* oxidase subunits I, II, and III, cytochrome *b*, and ATP synthase subunits 6, 8, and 9), var1 (a ribosomal protein), several proteins involved in RNA splicing, large and small rRNAs, 24 tRNAs, and an RNA component of an RNase P enzyme involved in tRNA processing.[1-4] Although this genome is genetically tractable, the yeast mitochondrial translation system has not been readily amenable to biochemical analysis *in vitro*.[5,6] This results in part from the following: some codon assignments in mitochondrial genetic systems differ from the universal genetic code[7]; isolated yeast mitochondrial ribosomes per se do not properly recognize the initiation codon[6]; and translation of yeast mitochondrial mRNAs require a number of mRNA-specific translational activators that are encoded by nuclear genes.[3,5,8] In contrast, isolated yeast mitochondrial (*in organello*) translation and transcription systems have been

[1] G. Attardi and G. Schatz, *Annu. Rev. Cell Biol.* **4**, 289 (1988).
[2] L. A. Grivell, *Eur. J. Biochem.* **182**, 477 (1989).
[3] L. Pon and G. Schatz, in "The Molecular Biology of the Yeast *Saccharomyces cerevisiae*: Genome Dynamics, Protein Synthesis, and Energetics" (J. R. Broach, J. R. Pringle, and E. W. Jones, eds.), p. 333. Cold Spring Harbor Laboratory, Cold Spring Harbor, New York.
[4] H. J. Pel and L. A. Grivell, *Mol. Biol. Rep.* **18**, 1 (1993).
[5] H. J. Pel and L. A. Grivell, *Mol. Biol. Rep.* **19**, 183 (1994).
[6] P. J. T. Dekker, B. Papadopoulou, and L. A. Grivell, *Curr. Genet.* **23**, 22 (1993).
[7] T. D. Fox, *Annu. Rev. Genet.* **21**, 67 (1987).
[8] C. L. Dieckmann and R. R. Staples, *Int. Rev. Cytol.* **152**, 145 (1994).

developed[9-12] for studying various aspects of mitochondrial gene expression *in vitro*. These systems have been used to determine the stoichiometry of transcription and translation of the major mitochondrial gene products[10-12]; to determine that mitochondrial transcription and translation are not obligately coupled[12]; to identify a preprotein precursor to cytochrome-*c* oxidase subunit II[13,14]; to establish that mitochondrial translation products are inserted into the inner mitochondrial membrane cotranslationally[15,16]; to examine the kinetic relationships between mitochondrial protein import and mitochondrial translation[17,18]; to examine the assembly of respiratory proteins[17,19,20]; and to study proteolytic turnover of mitochondrial translation products.[21,22] In this chapter we describe yeast *in organello* systems that have been optimized for mitochondrial protein and RNA synthesis.

Procedures

Isolation of Translation- and Transcription-Competent Yeast Mitochondria

Mitochondria from wild-type or respiratory-deficient strains of *Saccharomyces cerevisiae* can be used.[9,10] For routine studies we use *S. cerevisiae* strain D273-10B [*MAT* α(rho$^+$), ATCC, Rockville, MD]. Cells are grown in a semisynthetic medium containing, per liter, the following: 3 g yeast extract (Difco, Detroit, MI), 10 g galactose, 0.8 g $(NH_4)_2SO_4$, 0.7 g $MgSO_4 \cdot 7H_2O$, 0.5 g NaCl, 1.0 g KH_2PO_4, 0.4 g anhydrous $CaCl_2$, and 5 mg $FeCl_3 \cdot 6H_2O$. Cells are harvested in mid-exponential growth at densities between 5×10^7 and 1×10^8 by centrifugation (5 min at 2500 g_{max}) and

[9] P. Boerner, T. L. Mason, and T. D. Fox, *Nucleic Acids Res.* **9,** 6379 (1981).

[10] E. E. McKee and R. O. Poyton, *J. Biol. Chem.* **259,** 9320 (1984).

[11] E. E. McKee, J. E. McEwen, and R. O. Poyton, *J. Biol. Chem.* **259,** 9332 (1984).

[12] G. A. Bellus, Ph.D. Dissertation, University of Connecticut, Farmington (1985).

[13] K. A. Sevarino and R. O. Poyton, *Proc. Natl. Acad. Sci. U.S.A.* **77,** 142 (1980).

[14] E. E. McKee, K. A. Sevarino, G. Bellus, and R. O. Poyton, *in* "Current Developments in Yeast Research" (G. Stewart and I. Russell, eds.), p. 357. Pergamon, Ontario.

[15] G. H. Clarkson and R. O. Poyton, *J. Biol. Chem.* **264,** 10114 (1989).

[16] R. O. Poyton, K. A. Sevarino, E. E. McKee, D. J. M. Duhl, V. Cameron, and B. Goehring, *Adv. Cell Biol.* **18,** in press (1996).

[17] D. J. M. Duhl Ph.D. Dissertation, University of Colorado, Boulder (1991).

[18] D. J. M. Duhl, G. H. Clarkson, A. Cassidy-Stone, and R. O. Poyton, in preparation (1995).

[19] D. J. M. Duhl, A. Cassidy-Stone, and R. O. Poyton, in preparation (1996).

[20] R. H. P. Law, R. J. Devenish, and P. Nagley, *Eur. J. Biochem.* **188,** 421 (1990).

[21] C. C. Black-Schaefer, J. D. McCourt, R. O. Poyton, and E. E. McKee, *Biochem. J.* **274,** 199 (1991).

[22] A. Pajic, R. Tauer, H. Feldmann, W. Neupert, and T. Langer, *FEBS Lett.* **353,** 201 (1994).

washed once with distilled water. The washed cells are suspended at 0.2 g wet weight/ml in 0.1 M Tris–2.5 mM dithiothreitol (DTT, pH 9.0) and incubated at 30° for 20 min. The cells are harvested, washed twice with sterile distilled water, and suspended at 0.5 g wet weight/ml in spheroplasting buffer (1.35 M sorbitol–0.1 M Na$_2$EDTA, pH 7.4). Cells are spheroplasted using Zymolyase 20T (ICN ImmunoBiologicals, Costa Mesa, CA) added to a final concentration of 3 mg/g wet weight of cells. Zymolyase 20T is made as a fresh stock suspension (5 mg per milliliter spheroplasting buffer). The yeast suspension is incubated with Zymolyase 20T at 30° with gentle shaking.

Spheroplasting is monitored both by phase-contrast microscopy and by diluting 10-μl aliquots of the suspension into 1 ml distilled water, allowing 2 min for lysis, and measuring the optical density (OD) at 650 nm. Initial values are typically between 1.2 and 1.5 OD units. When more than 90% of the cells have been converted to spheroplasts (as deduced from a drop in optical density at 650 nm to between 0.05 and 0.1 OD units), the suspension is centrifuged at 25° (5 min at 2500 g_{max}), gently suspended in 1.35 M mannitol, and washed carefully by centrifugation as above. Spheroplasting generally takes 20 to 25 min. The washed spheroplasts are allowed to recover from cell wall removal by incubation at 30° for 60 min in growth medium (described above) supplemented with 1 M sorbitol as follows. Spheroplasts are suspended at a concentration of 10 to 15 g (original wet weight of cells) per liter and shaken gently (100 rpm) in a water bath shaker. (Note: This recovery period is essential for optimal translation and transcription. Shorter recovery times give lower levels of transcription and translation; longer recovery times do not increase the levels of transcription or translation.)

After the recovery period, spheroplasts are harvested by centrifugation (5 min at 2500 g_{max}) and washed twice with 1.0 M sorbitol at 4°. All subsequent steps are performed on ice. Spheroplasts are suspended at 0.4 g wet weight/ml in a buffer containing 0.6 M mannitol–1 mM Na$_2$EDTA (pH 6.7), lysed in a Sorvall Omnimixer at full speed in two 25-sec bursts separated by 30 sec, and centrifuged 5 min at 1900 g_{max} to pellet unbroken cells, nuclei, and debris. The supernatant is decanted and centrifuged at 13,000 g_{max} for 10 min to pellet the mitochondria. The mitochondrial pellet is suspended in 0.6 M mannitol (0.5 to 1.0 g original wet weight/ml), homogenized with a Teflon–glass homogenizer, and purified by differential centrifugation: first for 5 min at 1900 g_{max} and then by centrifugation at 20,000 g_{max}. The final mitochondrial pellet is suspended by homogenization to a concentration of 5 mg mitochondrial protein per milliliter 0.6 M mannitol adjusted to pH 7.2 with KOH.

Mitochondria may be used immediately or frozen for future use as follows. Dilute the suspension of mitochondria (5 mg mitochondrial protein/ ml in 0.6 M mannitol at pH 7.2) 2-fold with ice-cold 0.6 M mannitol—40% (v/v) glycerol (ultrapure), pH 7.2. The diluted suspension is divided into 2-mg mitochondrial protein aliquots, frozen in liquid N_2, and stored at $-80°$. When ready for use, frozen mitochondria are thawed on ice, diluted 10-fold with ice-cold 0.6 M mannitol adjusted to pH 7.2 with KOH, and pelleted by centrifugation for 10 min at 13,000 g_{max} and 4°. The resulting pellet is resuspended gently in a small volume of ice-cold 0.6 M mannitol, pH 7.2, and adjusted to 5 mg mitochondrial protein/ml.

Translation in Isolated Mitochondria

A typical reaction is carried out in a final volume of 0.5 ml. Protein synthesizing medium (Table I)[23] is made up as a 1.25× stock solution containing all components except pyruvate kinase, mitochondria, and the radiolabeled amino acid. An aliquot (0.4 ml) of the 1.25× stock is placed in a disposable plastic centrifuge tube (95 × 16 mm) on ice. Immediately before incubation, 50 μl pyruvate kinase [0.2 mg (500 to 700 units per milligram protein)/ml 10 mM KCl, prepared fresh daily] and 50 μl mitochondria (5 mg/ml in 0.6 M mannitol, pH 7.2) are added. Protein synthesis is initiated by transferring the tubes to a vigorously shaking water bath (200 rpm) at 25°. After a 10-min preincubation period, radiolabeling is initiated by the addition of 0.02 to 0.025 mCi/ml of L-[^{35}S]methionine (>1000 Ci/mmol).

Measurement of Mitochondrial Protein Synthesis

Aliquots (20 μl) of the translation mix are spotted onto 2.3-cm-diameter Whatman (Clifton, NJ) 3MM disks and allowed to air dry (several seconds). The disks are then dropped into ice-cold 5% (v/v) trichloroacetic acid–5 mM methionine. After a minimum of 5 min the solution is decanted and a fresh 5% (v/v) trichloroacetic acid–5 mM methionine solution, heated to 90°, is added. The disks are incubated in the solution at 90° for 5 min, then washed twice (2 to 5 min per wash) in ice-cold 5% (v/v) trichloroacetic acid–5 mM methionine, and twice in ice-cold ethanol–ether (3:1, v/v). They are then air-dried and subjected to scintillation counting.[9]

Analysis of Mitochondrial Translation Products

At the end of the incubation period nascent radiolabeled polypeptides are runoff by diluting the translation mix with an equal volume of 0.6 M

[23] G. A. Bellus and R. O. Poyton, in preparation (1996).

TABLE I

COMPOSITION OF MEDIA OPTIMIZED FOR MITOCHONDRIAL
PROTEIN SYNTHESIS AND RNA SYNTHESIS *in Organello*

Component	PSM[a]	RSM[b]
Sorbitol	600 mM	600 mM
KCl	150 mM	10 mM
KH$_2$PO$_4$	15 mM	15 mM
MgSO$_4$	12.5 mM	5 mM
ATP	4 mM	0.25 mM
GTP	0.5 mM	0.2 mM
UTP	—	0.1 mM
CTP	—	0.3 mM
α-Ketoglutarate	5 mM	5 mM
Phosphoenolpyruvate	5 mM	—
Pyruvate kinase	10 units/ml	—
Amino acids mix[c]	0.1 mM	—
Tris base	20 mM	20 mM
Bovine serum albumin	3 mg/ml	3 mg/ml
pH	7.2	7.2

[a] Formulation for optimal protein synthesis medium (PSM)
as described by McKee and Poyton.[10]

[b] RNA synthesis medium (RSM) has been optimized for
maximal incorporation of [α^{32}P]UTP as well as fidelity of
transcription and RNA processing.[12,23]

[c] The amino acid mixture contains equimolar amounts of
alanine, arginine hydrochloride, aspartic acid, cysteine hydrochloride, glutamic acid, glycine, histidine, isoleucine,
leucine, lysine hydrochloride, phenylalanine, proline, serine, threonine, tryptophan, tyrosine, and valine. Methionine is deleted from the medium when [^{35}S]methionine is
used as the labeling amino acid.

mannitol–0.2 M methionine, pH 7.2, and samples are incubated for 10 min
at 25° with shaking. Mitochondria are then pelleted from the translation
mix by centrifugation for 5 min at 12,000 g max and 4° in an Eppendorf
microcentrifuge. The supernatant is discarded, and the mitochondrial pellet
is washed by suspension in a buffer containing 0.6 M mannitol, 1 mM
Na$_2$EDTA, and 5 mM methionine. The mitochondria are pelleted as above
and the pellet dissolved in a protein dissociation buffer containing 15%
(v/v) glycerol, 2% sodium dodecyl sulfate (SDS), 1% 2-mercaptoethanol,
10 mM sodium phosphate (pH 7.0), 1 mM Na$_2$EDTA, and a few drops of
0.25% (v/v) bromphenol blue. This solution is incubated at 37° for 30 min
and then boiled for 2 min. The samples may be analyzed immediately by
SDS–polyacrylamide gel electrophoresis (SDS–PAGE) or kept frozen at

$-20°$ until ready for use. The SDS–PAGE step is performed in a discontinuous buffer system as described.[9,15,24]

Transcription in Isolated Mitochondria

Mitochondria are isolated as described above except that the prespheroplasting incubation in 0.1 M Tris–2.5 mM DTT is eliminated and the harvested, washed cells are suspended at 0.5 g/ml in a spheroplasting buffer consisting of 1.35 M sorbitol–0.1 M EDTA–3 mM DTT.

RNA synthesis reactions are carried out in 0.5 to 1.0 ml RNA-synthesizing medium (RSM; Table I) and contain isolated mitochondria at a concentration of 1 mg (total protein)/ml. All incubations are carried out at room temperature ($20°$ to $22°$) in a gyratory shaker (200 rpm). Reactions are initiated in one of two ways: either [α-^{32}P]UTP (0.2–1.45 mCi/ml, 4–10 mCi/mmol) is added to the mitochondrial suspension in RSM and the tube placed at room temperature in a gyratory shaker, or the tube containing all components, including radioisotope, is transferred from an ice bath to a shaker at room temperature. Typically, incorporation of isotope into RNA is linear for 40 min.

Measurement of Mitochondrial RNA Synthesis

The kinetics of incorporation of [α-^{32}P]UTP into RNA are measured by removing 10-μl aliquots from the reaction mixture and precipitating the labeled RNA with 500 μl ice-cold 5% trichloroacetic acid. The precipitates are collected on glass fiber filters. (Whatman GFC), then washed three times with 5 ml ice-cold 5% trichloroacetic acid and once with 5 ml ice-cold 95% (v/v) ethanol. The filters are dried under a heat lamp for 30 min. They are transferred to scintillation fluid and counted.

Analysis of Mitochondrial Transcripts

After transcription reactions are finished, the reaction mixture is transferred to a chilled ($4°$) 1.5-ml centrifuge tube containing 25 μl of 100 mM UTP and 25 μl of 100 mM aurintricarboxylic acid. The mitochondria are pelleted by centrifugation at 12,000 g max for 2.5 min at $4°$ in an Eppendorf microcentrifuge, washed with 500 μl of 0.6 M mannitol–5 mM UTP, and pelleted once more. The RNA is extracted by dissolving the pelleted mitochondria in 300 μl of 2% SDS–10 mM Tris-Cl–10 mM EDTA–5 mM adenosine vanadate, pH 8.0, followed by extraction with 300 μl phenol/

---

[24] R. O. Poyton, B. Goehring, M. Droste, K. A. Sevarino, L. A. Allen, and X.-J. Zhao, this series, Vol. 260, [8].

chloroform/isoamyl alcohol (50:50:1, v/v). The aqueous phase is removed, sodium acetate is added to 0.35 M, and the RNA is precipitated with 2.5 volumes of ice-cold ethanol. The RNA is stored at $-20°$ as an ethanol precipitate. It is stable for several weeks. Mitochondrial RNA is prepared for electrophoresis and run on 1.5% agarose–2.2 M formaldehyde gels as described by Maniatis et al.[25] or on 1.5% agarose–6 M urea gels as described by Locker.[26] The gels are run at 5 V/cm for 8 hr, or until the bromphenol blue dye front has migrated 13–14 cm. The gel is cooled to $4°$ during the run, and the buffer of the formaldehyde gels is recirculated. After electrophoresis, the gels are fixed with methanol for at least 1 hr, dried onto filter paper, and exposed to Kodak (Rochester, NY) XAR or SB5 X-ray film with or without a DuPont (Boston, MA) Cronex Lightning Plus Intensifying screen.

Identification of Mitochondrial Transcripts by Hybrid Selection

Mitochondrial transcripts are identified by hybrid selection with cloned fragments of mitochondrial genes. Plasmids containing mitochondrial gene fragments are linearized and bound to small nitrocellulose filters ($\sim$9 mm^2). Reactions are carried out in 1.5-ml centrifuge tubes containing 250,000 to 1,000,000 counts/min (cpm) ^{32}P-labeled mitochondrial RNA in a volume of 30–40 μl hybridization buffer. The hybridization buffer consists of 65% deionized formamide–0.2% SDS–0.4 M NaCl–100 μg/ml yeast tRNA–20 mM 1,4-piperazinediethanesulfonic acid (PIPES), pH 6.4. Three nitrocellulose filters are used per reaction, and hybridization is allowed to proceed overnight at $50°$. The filters are washed 10 times at $65°$ with 1 ml of 0.15 M NaCl–1 mM EDTA–0.5% SDS–10 mM Tris, pH 7.6, and then twice more with 1 ml of the same buffer without SDS. The RNA is eluted in 300 μl distilled water containing 15 μg carrier tRNA by boiling for 1 min, followed by snap freezing in an ethanol dry ice bath. The hybrid selected RNA is ethanol-precipitated and analyzed by gel electrophoresis as described above.

Acknowledgment

This work was supported by National Institutes of Health Grants GM 39324 and 30228.

[25] T. Maniatis, E. F. Fritsch, and J. Sambrook, "Molecular Cloning: A Laboratory Manual." Cold Spring Harbor Laboratory, Cold Spring Harbor, New York.
[26] J. Locker, Anal. Biochem. **98,** 358 (1979).

[5] *In Organello* RNA Synthesis System from HeLa Cells

By GEORGE L. GAINES III

Introduction

Interest in the areas of enzyme regulation and metabolic pathways has been rekindled by research on the molecular and genetic relationships among development, homeostasis, and oncogenesis. Coordination of the metabolic processes needed for homeostasis and appropriate development is achieved through regulation at multiple levels, including control of interactions between the two genetic systems in animal cells, the nucleus and the mitochondrion.

The similarity of mitochondrial genomes within a species and near homogeneity within an organism does not exclude changes in the morphology and function of the organelles.[1-5] Differences in the numbers and shapes of mitochondria occur during the development and differentiation of a specific tissue. These changes are accompanied by variations in transcription rates and respiratory capacities of the mitochondria.[6-8] Abnormal differentiation resulting in oncogenesis can also cause functional and structural alterations in these organelles. Mitochondria from tumors differ from those of normal tissues in their membranous lipid and protein content, the type of substrate they oxidize during respiration, the magnitude of their acceptor control ratio, their capacity to accumulate Ca^{2+}, and their rate of protein synthesis and turnover.[5,9-15] As the changes in mitochondrial

[1] G. Attardi, *Int. Rev. Cytol.* **93**, 93 (1985).
[2] W. M. Brown, *Proc. Natl. Acad. Sci. U.S.A.* **77**, 3605 (1980).
[3] C. F. Aquadro and B. O. Greenberg, *Genetics* **103**, 287 (1983).
[4] W. W. Hauswirth, M. J. Van De Walle, P. J. Laipis, and P. D. Olive, *Cell (Cambridge, Mass.)* **37**, 1001 (1984).
[5] A. Tzagoloff, "Mitochondria." Plenum, New York and London, 1982.
[6] J. M. England and G. Attardi, *J. Neurochem.* **16**, 617 (1976).
[7] J. Himms-Hagen, *FASEB J.* **4**, 2890 (1990).
[8] D. A. Clayton, *Int. Rev. Cytol.* **141**, 217 (1992).
[9] P. L. Pedersen, *Prog. Exp. Tumor Res.* **22**, 190 (1978).
[10] L. H. Augenlicht and B. G. Heerdt, *J. Cell Biochem.* **16G**, 151 (1992).
[11] D. Burk, K. M. Woods, and J. Junter, *J. Natl. Cancer Inst.* **38**, 839 (1967).
[12] J. H. Hochman, M. Schindler, J. G. Lee, and S. Ferguson-Miller, *Proc. Natl. Acad. Sci. U.S.A.* **79**, 6866 (1982).
[13] L. V. Johnson, I. C. Summerhayes, and L. B. Chen, *Cell (Cambridge, Mass.)* **28**, 7 (1982).
[14] T. Kadowaki and Y. Kitagawa, *Exp. Cell Res.* **192**, 243 (1991).
[15] G. Woldegiorgis and E. Shrago, *J. Biol. Chem.* **260**, 7585 (1985).

morphology and function seen during normal and abnormal development are not the consequence of dramatic changes in the mitochondrial genomes, this heterogeneity must result in part from nuclear and cytoplasmic interactions with the mitochondria.[10,14,16–18]

Alterations in mitochondrial gene expression occurs in response to energetic conditions within the cell and possibly cytoplasmic and nuclear control factors. Hence, to simplify investigations of the potential control provided by the energetic states of the cell and cytoplasmic factors on mitochondrial DNA (mtDNA) transcription and processing of the mtRNA, a set of experimental protocols was developed that physically separates the mitochondria from the remaining cellular constituents.[19–21] This separation was achieved by mechanically lysing HeLa cells, isolating the intact mitochondria by centrifugation, and removing the nonmitochondrial RNA through enzyme treatment and washings. Simultaneously, the crude cytoplasmic fraction free of mitochondria and nuclei was retained on ice or frozen for reconstitution experiments. The isolated mitochondria could then be assayed for transcriptional capacity and processing fidelity by analyzing the incorporation of radioactive nucleotides into their RNA using gel electrophoretic techniques. As described below, it is possible to isolate mitochondria easily and quickly that are competent in transcription events and the processing of RNA, comparable to that found in the *in vivo* situation. It is therefore possible to study the effects of changes in the energetic or metabolic environments as well as the influences of purified cytoplasmic factors on the synthesis and maturation of the mtRNA. The molecular analysis of these effects is possible because of the wealth of preexisting knowledge about HeLa cell mitochondrial DNA and its *in vivo* transcription products.[1,8]

Cell Growth and Mitochondrial Isolation

HeLa cells are grown in suspension in modified Eagle's medium supplemented with 5% calf serum with a continuous flow of 5% (v/v) CO_2 in air to keep the pH constant. Exponentially growing cells are harvested by centrifugation, washed twice with 1 mM Tris-HCl (pH 7.0), 0.13 M NaCl, 5 mM KCl, and 7.5 mM MgCl$_2$. The cell pellet is resuspended in half the cell volume with 1/10× IB [4 mM Tris (pH 7.4), 2.5 mM NaCl, 0.5 mM

[16] R. Taylor, *J. NIH Res.* **4,** 62 (1992).
[17] C. Richter, *Mutat. Res.* **275,** 249 (1992).
[18] S. C. Ghivizzani, C. S. Madsen, and W. W. Hauswirth, *J. Biol. Chem.* **268,** 8675 (1993).
[19] G. Gaines and G. Attardi, *Mol. Cell. Biol.* **4,** 1605 (1984).
[20] G. Gaines and G. Attardi, *J. Mol. Biol.* **172,** 451 (1984).
[21] G. Gaines, C. Rossi, and G. Attardi, *J. Biol. Chem.* **262,** 1907 (1987).

MgCl$_2$], and the cells are broken using a Thomas homogenizer (Thomas Scientific, Swedesboro, NJ) with a motor-driven pestle. A rapid downward stroke is utilized, resulting in a distinct pop as the pestle exits the tube. The homogenate is immediately mixed with one-ninth of the packed cell volume of 10× IB [400 mM Tris (pH 7.4)], 250 mM NaCl, 50 mM MgCl$_2$], resulting in a lysate that is about two times the original packed cell volume in a buffer concentration of roughly 1× IB [40 mM Tris (pH 7.4), 25 mM NaCl, 5 mM MgCl$_2$]. The unbroken cells and nuclei are removed by two consecutive low-speed centrifugations (2000 rpm for 3 min). The remaining supernatant is divided into 1.5-ml Eppendorf tubes (0.5 ml supernatant/tube) and centrifuged at full speed in an Eppendorf microcentrifuge for 1 min. The supernatant from this spin is either saved on ice or frozen at −80° as the cytoplasmic fraction. The pellet containing the crude mitochondrial fraction is suspended in GIBB [10% glycerol, 40 mM Tris (pH 7.4), 25 mM NaCl, 5 mM MgCl$_2$, 2 mg/ml bovine serum albumin (BSA)] and centrifuged for 1 min. The resulting mitochondrial pellet is then suspended in an appropriate incubation medium for the transcription assay.

In Organello Labeling and Isolation of Mitochondrial Nucleic Acids

Samples of the mitochondrial fraction, derived from about 0.25 g HeLa cells (~0.4 mg mitochondrial protein) are resuspended in 0.5 ml of the appropriate incubation medium in Eppendorf tubes as described above. The standard base medium contains GIBB and varying amounts of the cytoplasmic fraction or other compounds, such as oxidizable substrates or inhibitors of mitochondrial functions. In general, ATP is added to 1 mM concentration in the mixtures, and 5–10 μCi [α-^{32}P]UTP (available from several vendors, 400–600 Ci/mmol) is added to start the incubation. The reactions are allowed to continue for 30–60 min at 37°. The incubation conditions are very flexible; however, the incubation conditions outlined here are optimized for maximum yield of labeled mtRNA without compromising the integrity of the mitochondria. As shown in Table I, there is considerable variation in the acid-precipitable radioactivity from experiment to experiment. The increase (-fold) represents the increase in acid-precipitable radioactivity produced by mitochondria in GIBB with the specific compound over that produced by mitochondria resuspended in only GIBB. In addition, the three transcription units behave differently depending on the added constituents.[21] These variations are probably due to many factors, including the growth state of the cells, the speed at which the mitochondria are isolated, and the intactness and purity of the organelles. Hence direct comparisons of effects of specific reagents or parameters

TABLE I
OPTIMIZED PARAMETERS FOR *in Organello* MITOCHONDRIAL DNA TRANSCRIPTION[a]

Compound	Optimized concentration	Increase in labeling[b] (-fold)
MgCl$_2$	5 mM	3–5 fold
NaCl	0–50 mM	None
ATP	1 mM	10–50
ADP	1 mM	5–20
ADP/pyruvate/phosphate	1 mM/1–2 mM/10 mM	25–50
Succinate	1–2 mM	5–10
Pyruvate	1–2 mM	5–10
Citrate	1–2 mM	5–10
Glycerol	10%	1–10
BSA	2 mg/ml	None

[a] As measured by trichloroacetic acid-precipitable radioactivity. Temperature: 30°–37°; pH: 7.0–7.7, Tris buffer; time: 30–60 min.
[b] Compared with GIBB without the compound in question.
[c] Highly dependent on state of the mitochondria and other compound present; for example, cytoplasm is able to compensate for the lack of an osmoregulator like glycerol.

should be done simultaneously with mitochondria isolated from the same batch of cells.

After the incubation, the mitochondria are pelleted at 12,700 g in the microcentrifuge for 1 min and washed twice with GIBB. The pellets are suspended in 1 ml of MNB [10% glycerol, 10 mM Tris (pH 8.0), 1 mM CaCl$_2$] and incubated with 200 units of micrococcal nuclease at room temperature for 20 min. The nuclease degrades the remaining cytoplasmic nucleic acids without disturbing the intramitochondrial RNA, allowing for much cleaner gel analysis. The incubation period may be reduced if rapid processing of the organelles is required, albeit with a corresponding increase in cytoplasmic RNA contamination. After the nuclease treatment, the mitochondria are pelleted and washed two more times with GTE [10% glycerol, 10 mM Tris (pH 7.5), 1 mM EDTA] to inactivate and remove the micrococcal nuclease. The mitochondrial samples are then lysed with 0.35 ml SDS buffer [0.5% sodium dodecyl sulfate (SDS), 10 mM Tris (pH 7.4), 0.15 M NaCl, 1 mM EDTA] and incubated with 100 μg pronase for 15 min at 30°. Phenol extraction and CH$_3$HgOH–agarose or polyacrylamide gel electrophoretic analysis are accomplished as described previously.[19–21] Quantitation of the autoradiograms of the gels is carried out using a Joyce-Loebl double-beam densitometer and analyzing the resulting peaks with a digitizer.

Energetic Effects on Mitochondrial *In Organello* Transcription

The association of the mtDNA (and hence its transcription and RNA processing) with the inner mitochondrial membrane[22] suggests that the control of mitochondrial gene expression might be sensitive to the energetic state of the organelles. Oxidizable substrates, such as pyruvate, succinate, and citrate, stimulate the labeling of mitochondrial RNA (Table I). It is interesting that the concentration of ATP found *in vivo*[23] is similar to the experimentally measured optima for both the *in organello* (Table I) and *in vitro* transcription systems (as described elsewhere in this volume). However, there is a strikingly different dependence of mRNA synthesis versus rRNA synthesis on the energetic state of the mitochondria in the *in organello* system.[21] Changing the concentrations of ATP and oxidizable substrates in the *in organello* transcription system results in different relative labeling of mitochondrial mRNA and rRNA encoded on the H strand of the mtDNA, as well as a change in the overall transcription levels. The addition of 1 mM ADP to the GIBB medium increases the overall RNA labeling by roughly 10-fold. The combination of 1 mM ADP with 1 mM pyruvate and 10 mM phosphate causes a further stimulation by about 2.5-fold, to a level similar to that induced by 1 mM ATP alone. Furthermore, addition of the pyruvate and phosphate with the ATP increases the level of labeling again by roughly 2.5-fold over that of ATP alone.[21] The stimulation of labeling by pyruvate and phosphate over that observed with ATP or ADP alone may be the result of subsequent phosphorylation of ADP, thereby raising the intramitochondrial ATP concentration. These stimulatory effects of ADP and an oxidizable substrate are substantially blocked by the introduction of respiration and oxidative phosphorylation inhibitors. In addition, differences have been noted between the two transcription units when other energetic parameters have been studied.[21]

Cytoplasmic Effects on Mitochondrial *in Organello* Transcription

Compared with buffer conditions alone, the presence of a crude cytoplasmic fraction has profound effects on the incorporation of [α-^{32}P]UTP into mtRNA by the isolated mitochondria. The total amount of labeling is dramatically decreased when cytoplasm is substituted for buffer during incubation, partly because of the high concentration of UTP in cytoplasm

---

[22] M. Albring, J. Griffith, and G. Attardi, *Proc. Natl. Acad. Sci. U.S.A.* **74**, 1348 (1977).
[23] R. K. Bestwick, G. L. Moffett, and C. E. Mathews, *J. Biol. Chem.* **257**, 9300 (1982).

$(200 \, \mu M)$.[23] As seen by comparing the first and last lanes in Fig. 1, cytoplasm also reduces the labeling of rRNAs to a level at or below that of the mRNAs. (Densitometry gives an artificially high estimation of the labeling of the rRNAs, 12S and 12S*, owing to the comigration of mRNAs 13 and 12 with the small rRNAs in the agarose gel systems.) Despite the differential

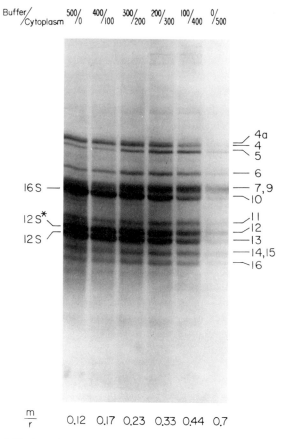

FIG. 1. Effect of dilution of cytoplasm on mtRNA labeling in isolated organelles. Varying amounts of incubation buffer and cytoplasm were mixed as noted at the top of the lanes and added to isolated mitochondria for the transcription reactions. In addition, 1 mM ATP and 10 μCi [α-³²P]UTP were added, and the transcription reactions were carried out for 90 min at 30°. Total RNA was extracted and electrophoretically separated as described in the text. The amounts electrophoresed were 2, 2, 4, 8, 15, and 20 μl, respectively, left to right. The autoradiogram was scanned with a Joyce-Loebl double-beam densitometer and the areas of the peaks integrated. At bottom, m is the sum of band integrations corresponding to mRNAs 14, 15, and 16, and r is the sum of band integrations corresponding to rRNAs 12S* (12) and 12S (13) for each lane.

effect cytoplasm has on transcription of the two H-strand transcription units[1] in this system, little change was seen for processing and polyadenylation, with all transcripts being efficiently processed and roughly 10% of the mRNA binding to oligo(dT)-cellulose.

The ability of the cytoplasmic fraction to affect *in organello* RNA labeling was extremely sensitive to dilution, as shown in Fig. 1. The labeled RNA from mitochondria, incubated in various concentrations of cytoplasm and GIBB, was extracted and electrophoresed through a 1.4% agarose–CH_3HgOH gel and the gel autoradiographed. The reduction in specific activity of the $[\alpha\text{-}^{32}P]UTP$ by the unlabeled cytoplasmic pool of UTP results in a decrease in the total labeling at all concentrations of cytoplasm commensurate with the dilution of the label. An additional decrease in labeling occurs at high concentrations of cytoplasm which cannot be explained by the reduction in the specific activity of $[\alpha\text{-}^{32}P]UTP$. This effect is very noticeable considering the 10-fold greater amount of mitochondrial RNA applied to the 100% cytoplasm lane (0/500) as compared with the 0% (500/0) and 20% (400/100) lanes. The ratio of mRNA to rRNA labeling (*m/r* in Fig. 1) also changes dramatically at higher concentrations. Because the extent of rRNA labeling is very sensitive to the available ATP, either added or mitochondrially synthesized, the reduction in rRNA labeling may be caused by a rapid removal or hydrolysis of ATP. Alternatively, the alterations in transcriptional activity in the presence of high concentrations of cytoplasm may be caused by factors that are directly responsible for the control of initiation or continuation of transcription in the isolated organelles. Preliminary studies (G. Gaines and C. Rossi, unpublished results, 1985) suggest that there are low molecular weight cytoplasmic species (<2000) responsible in part for these dramatic effects. The purification and identification of the factors are beyond the scope of this chapter.

This work provides the basis for further investigations into the nature of the mitochondrial, cytoplasmic, and nuclear interactions in the transcription of mitochondrial DNA. The ability to isolate the mitochondria from other cellular constituents and reconstitute cytoplasmic and nuclear interactions will continue to play an important role in the study of the physiology of the cell.

[6] *In Organello* RNA Synthesis System from Mammalian Liver and Brain

By José A. Enríquez, Acisclo Pérez-Martos, Manuel J. López-Pérez, and Julio Montoya

Introduction

Transcription of the mammalian mitochondrial genome has been extensively studied in cell culture systems, mainly in HeLa cells.[1-3] Isolation of the discrete transcription products and characterization of their structural and metabolic properties have provided important information on the mechanisms of mitochondrial DNA (mtDNA) transcription and RNA processing.[4-9]

In vivo study of the regulation of mtDNA expression is hampered by the inherent difficulty of manipulating mammalian mitochondrial genes *in vivo* and of discriminating between direct regulatory phenomena on mtDNA transcription and RNA metabolism, and secondary actions dependent on nuclear control. A way to overcome these difficulties could be the use of isolated mitochondria. Gaines and Attardi developed a successful system that faithfully reproduced *in vivo* mtDNA transcription in HeLa cell mitochondria.[10-12] Based on this experimental approach, several attempts have been made to apply this system to the study of mitochondrial RNA synthesis in mammalian organs.[13-16] However, under the incubation

[1] G. Attardi, *in* "Processing of RNA" (D. Apirion, ed.), p. 227. CRC Press, Boca Raton, Florida.

[2] G. Attardi, *Int. Rev. Cytol.* **93,** 683 (1985).

[3] G. Attardi and G. Schatz, *Annu. Rev. Cell Biol.* **4,** 289 (1988).

[4] F. Amalric, C. Merkel, R. Gelfand, and G. Attardi, *J. Mol. Biol.* **118,** 1 (1978).

[5] J. Montoya, D. Ojala, and G. Attardi, *Nature (London)* **290,** 465 (1981).

[6] D. Ojala, J. Montoya, and G. Attardi, *Nature (London)* **290,** 470 (1981).

[7] R. Gelfand and G. Attardi, *Mol. Cell. Biol.* **1,** 497 (1981).

[8] J. Montoya, T. Christianson, D. Levens, M. Rabinowitz, and G. Attardi, *Proc. Natl. Acad. Sci. U.S.A.* **79,** 7195 (1982).

[9] J. Montoya, G. L. Gaines, and G. Attardi, *Cell (Cambridge, Mass.)* **34,** 151 (1983).

[10] G. Gaines and G. Attardi, *J. Mol. Biol.* **172,** 451 (1984).

[11] G. Gaines and G. Attardi, *Mol. Cell. Biol.* **4,** 1605 (1984).

[12] G. Gaines, C. Rossi, and G. Attardi, *J. Biol. Chem.* **262,** 1907 (1987).

[13] P. Cantatore, P. Loguercio-Polosa, A. Musstich, V. Petruzzella, and M. N. Gadaleta, *Curr. Genet.* **14,** 477 (1988).

[14] A. Mutvei, S. Kuzela, and B. D. Nelson, *Eur. J. Biochem.* **180,** 235 (1989).

conditions described for the organelles isolated from HeLa cells, the mitochondria isolated from mammalian tissues do not reproduce closely the *in vivo* processes of mtDNA transcription and RNA processing, and the efficiency of transcription is very low.

In this chapter, we describe a system to study RNA synthesis by using isolated mitochondria from mammalian brain and liver. In this system, mtDNA transcription and RNA processing are supported in a way that closely reproduces the *in vivo* process.[17-19]

Experimental Procedures

Preparation of Intact Mitochondria from Mammalian Brain and Liver

Animals. Male Wistar rats fed *ad libitum* and weighing 200–300 g are used.

Isolation of Rat Liver Mitochondria. In this preparation,[18] the animal is stunned and sacrificed by decapitation. The liver is rapidly removed, weighed, and dropped into a beaker containing cold isolation medium (0.32 M sucrose, 1 mM potassium EDTA, 10 mM Tris-HCl, pH 7.4). The tissue is then cut into small pieces with a pair of scissors and rinsed several times with the same medium. All further operations are carried out at 2°–4° using sterile solutions and glassware. The chopped tissue is suspended in isolation medium (~4 ml/g tissue) and homogenized in a loose-fitting Potter–Elvejhem homogenizer (motor-driven Teflon pestle) by using three to four up-and-down strokes, after the pestle has reached the bottom of the vessel. The homogenate is transferred to 50-ml centrifuge tubes and centrifuged at 1000 g for 5 min to sediment tissue fragments, unbroken cells, nuclei, and large cytoplasmic debris. The pellet is discarded and the supernatant divided into microcentrifuge tubes and centrifuged at full speed (13,000 g) for 2 min in a microcentrifuge. The resultant supernatants are discarded, the pellets are resuspended in 500 μl isolation medium, the material from two tubes is combined into one tube, and these are filled with isolation medium. After centrifugation at 13,000 g for 2 min, this cycle is repeated

[15] P. Fernández-Sílva, V. Petruzzella, F. Fracasso, M. N. Gadaleta, and P. Cantatore, *Biochem. Biophys. Res. Commun.* **176,** 645 (1991).

[16] W. B. Coleman and C. C. Cunningham, *Biochim. Biophys. Acta* **1058,** 178 (1991).

[17] J. A. Enriquez, M. J. Lopez-Perez, and J. Montoya, *FEBS Lett.* **280,** 32 (1991).

[18] J. A. Enriquez, A. Pérez-Martos, P. Fernández-Sílva, M. J. López-Pérez, and J. Montoya, *FEBS Lett.* **304,** 285 (1992).

[19] J. A. Enriquez, A. Pérez-Martos, P. Fernández-Sílva, M. J. López-Pérez, and J. Montoya, *Int. J. Biochem.* **25,** 1951 (1993).

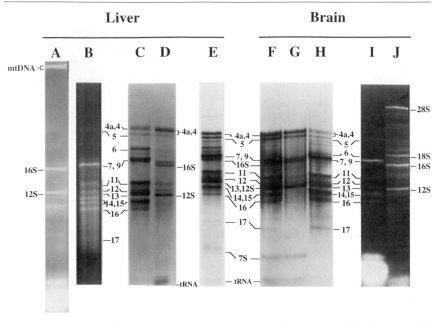

FIG. 1. Electrophoretic patterns in agarose–CH₃HgOH slab gels of rat liver[18] and brain[17] mitochondrial RNA synthesized *in organello* (autoradiogram; lanes C–H) and *in vivo* (ethidium bromide staining; lanes A, B, I, and J). Lanes A, E, and F: Total RNA; lanes B, C, H, and I: polyadenylated RNA; lanes D, G, and J: nonpolyadenylated RNA. Lane A shows ethidium bromide staining of the total RNA in the sample used for *in organello* RNA synthesis. The very intense bands at the bottom of lanes A and I correspond to tRNA carrier. Adapted from Enriquez *et al.*[17,18]

twice. Finally, the mitochondrial pellets are washed with the appropriate incubation buffer (see below), pelleted, and suspended in 1 ml of incubation buffer.

The number of washes employed in this preparation eliminates completely any trace of cytoplasmic RNA (estimated by the absence of 28 S and 18 S rRNA in the electrophoretic patterns of the ethidium bromide-stained gels; Fig. 1A), making unnecessary the micrococcal nuclease treatment of the mitochondrial fraction to eliminate contaminating extramitochondrial nucleic acids.[20] The concentration of mitochondrial protein is determined by the Waddel method.[21] Briefly, 10 µl mitochondrial suspension is added to 3 ml recently prepared 3.3 mM NaOH. After mixing, the absorbance is measured at 225 and 215 nm in an spectrophotometer, using NaOH as the blank. The absorbance at 225 nm is subtracted from that

[20] G. Attardi and J. Montoya, this series, Vol. 97, p. 435.
[21] W. J. Waddel, *J. Lab. Clin. Med.* **48,** 311 (1956).

at 215 mm. This difference, divided by 0.0374, gives directly the protein concentration (expressed in mg/ml) in the mitochondrial suspension. The yield is approximately 10 mg/ml suspension (4 mg/g rat liver).

Isolation of Rat Brain Mitochondria. For the isolation of mitochondria from rat brain,[22,23] the forebrains of at least six rats are rapidly removed, chilled in medium A (0.32 M sucrose, 1 mM potassium EDTA, 10 mM Tris-HCl, pH 7.4), and finely cut with a pair of scissors. All further operations are carried out at 2°–4° using sterile solutions and glassware. After rinsing to remove as much blood as possible, the tissue is resuspended in 6 volumes of medium A and homogenized manually with a Dounce-type glass homogenizer by using 15 up-and-down strokes. The homogenate is centrifuged at 1000 g for 5 min. The pellet thus obtained is discarded, and the supernatant, transferred to 50-ml tubes, is spun at 12,000 g for 10 min to pellet a crude mitochondrial fraction containing synaptosomes. The pellet is suspended in 20 ml medium B (0.32 M sorbitol, 5 mM potassium phosphate, pH 7.5) and centrifuged at 12,000 g for 10 min. The final pellet is suspended in a small volume of medium B to a final weight of 2 g.

Purification of free (nonsynaptic) mitochondria is carried out by partitioning in an aqueous two-phase system composed of dextran T-500 and polyethylene glycol (PEG) 4000.[22,23] For this, the 2 g of the crude mitochondrial fraction suspension in medium B is added to 14 g of a two-phase mixture to give final concentrations of 6.4% (w/w) dextran T-500, 6.4% (w/w) polyethylene glycol, 0.32 M sorbitol, 0.1 mM EDTA, 5 mM potassium phosphate, pH 7.8.[23] After 20 inversions of the tubes, the two-phase system is spun at 1500 g for 1 min to decrease the time of phase settling. The lower phase, containing free mitochondria, is then diluted with 20 ml medium A and centrifuged at 17,000 g for 10 min. The mitochondrial pellet is washed by suspending in the appropriate incubation buffer (see below) and, after portioning the suspension into Eppendorf tubes, centrifuged at full speed (13,000 g) for 1 min in a microcentrifuge. Finally, the mitochondrial pellet is suspended in 750 μl of incubation buffer. The yield is approximately 10 mg/ml suspension (0.8–1 mg/g rat brain), determined as described above.

To prepare mitochondria from different brain regions, animals larger than rats are used.[19]

In Organello Labeling and Isolation of Mitochondrial RNA

Samples of the mitochondrial fractions from rat liver and brain and from different sheep brain regions are incubated at a final concentration of 2 mg mitochondrial protein/ml in 0.5 ml incubation buffer, which contains

[22] V. Corbatón, P. Fernández-Sílva, M. J. López-Pérez, and J. Montoya, *Neurochem. Res.* **15**, 711 (1990).
[23] M. J. López-Pérez, this series, Vol. 228, p. 403.

25 mM sucrose, 75 mM sorbitol, 100 mM KCl, 10 mM K$_2$HPO$_4$, 50 μM EDTA, 5 mM MgCl$_2$, 1 mM ADP, 10 mM glutamate, 2.5 mM malate, 10 mM Tris-HCl, pH 7.4, 1 mg/ml bovine serum albumin (BSA), and 20 μCi of [α-^{32}P]UTP (400–600 Ci/mmol) in 1.5-ml Eppendorf tubes.[17,19] Incubation is at 37° for 60 min in a rotary shaker (12 rpm). The smooth movement of the rotating wheel with the consequent inversion of the tubes produces good oxygenation of the mitochondrial suspension, which is of great importance for the labeling of mitochondrial RNA.

After the incubation, the mitochondrial samples are pelleted at 13,000 g for 1 min and washed twice with 1 ml of 10% glycerol, 10 mM Tris-HCl, pH 6.8, 0.15 mM MgCl$_2$ to eliminate unincorporated nucleotides. The pellets are suspended in 400 μl pronase buffer (10 mM Tris-HCl, pH 7.4, 0.15 M NaCl, 1 mM EDTA) and incubated 5 min at 37° with 100 μg/ml pronase (Boehringer Mannheim, Indianapolis, IN; previously digested at 37° for 3 hr at a concentration of 2 mg/ml). The mitochondria are then lysed by addition of 2% w/v sodium dodecyl sulfate (SDS) and incubated 15 min at 37°. Twenty micrograms tRNA carrier is added before lysis of the organelle to protect RNA and control the amount of RNA recovered in the process. The RNA is extracted by adding an equal volume of a phenol/chloroform/ isoamyl alcohol mixture (25:25:1, by volume) the phenol should be equilibrated with 10 mM Tris-HCl, pH 7.4, 0.1 M NaCl, 1 mM EDTA, 0.5% SDS before using) and shaking for 1 min in a vortex at room temperature. After centrifugation at 13,000 g for 3–4 min, the aqueous phase is transfered to a clean tube, and phenol extraction is repeated. The RNA is then precipitated by the addition of 2 volumes of ethanol to the aqueous phase, and samples are centrifuged at full speed in a microcentrifuge for 30 min. The pelleted nucleic acids are dissolved in 10 mM Tris-HCl, pH 7.4, 1 mM EDTA (TE).

Cold mitochondrial RNA is prepared from 16 rat brains and 2 livers as previously described.[20,22]

RNA Fractionation and Analysis

Total mitochondrial RNA is fractionated by affinity chromatography through oligo(dT)-cellulose to separate the polyadenylated RNA from non-polyadenylated RNA.[20,22] The total mitochondrial RNA fraction (in TE) is subjected to a denaturation step, by heating at 65° for 5 min and fast cooling, and, after addition of NaCl to 0.15 M, passed through an oligo(dT)-cellulose column (Boehringer Mannheim) equilibrated with 10 mM Tris-HCl, pH 7.4, 0.15 M NaCl, 1 mM EDTA. Ten milliliters of the flow-through material containing the nonbound RNA is collected and the column washed with 10 ml of the same buffer. The bound fraction is then eluted from the column with 10 ml of 10 mM Tris-HCl, pH 7.4, 1 mM EDTA, heat-denatured again as described above, and rerun through the oligo(dT)-cellulose column. The final bound poly(A)-containing RNA, to which 20 μg tRNA carrier and NaCl to 0.2 M have been added, and the nonbound poly(A)$^-$ RNA fractions are then collected by ethanol precipitation, centrifuged, and the RNA pellets dissolved in TE.

The total, polyadenylated, and nonpolyadenylated mitochondrial RNA fractions are analyzed under denaturing conditions by vertical electrophoresis through 1.4% agarose–5 mM methylmercury hydroxide slab gels in 50 mM boric acid, 5 mM Na$_2$B$_4$O$_7 \cdot$ 10 H$_2$O, 10 mM Na$_2$SO$_4$, 0.1 mM EDTA, pH 8.2. Because methylmercury hydroxide obtained from commercial sources may not be of satisfactory quality, it is advisable to treat commercial methylmercury hydroxide solutions with anion- and cation-exchange resins, such as Amberlite (Sigma, St. Louis, MO) or AG501-X8 (Bio-Rad, Richmond, CA), to recover its denaturing capacity.[24] All work with methylmercury hydroxide is carried out in a fume hood. The RNA samples are dissolved in 20–40 μl TE, subjected to a denaturation step as above, mixed with 5–10 μl of 33% w/v Ficoll, 0.2% w/v bromphenol blue, and layered onto the gel. Electrophoresis is carried out at 120 V (6.0 V/cm) until the bromphenol blue band has migrated approximately 17 cm (5–6 hr) with circulation of the buffer and use of a fan to cool the gel. With this run all the RNA species, from the tRNAs to the high molecular weight bands, are visible in one gel. The gel is first stained with ethidium bromide in 0.2 M ammonium acetate and photographed under UV light, then dried and exposed for autoradiography, either at $-70°$ with a DuPont (Boston, MA) screen intensifier or at room temperature (Fig. 1). The definition of the bands allows densitometer quantification of the mtDNA or RNA from the ethidium bromide-stained gels and the autoradiograms, respectively.

Properties of *in Organello* RNA Synthesizing System

The incubation medium described in this work differs from that previously employed in HeLa cell[11] and rat liver[13] mitochondria in that it is isotonic and the energy requirements are provided by exogenous ADP in the presence of an oxidizable substrate, instead of being hypertonic and with an external source of ATP. The composition of the incubation medium and the incubation conditions were optimized to allow the maximum incorporation of radioactivity. It is noteworthy that *in organello* RNA synthesis is highly dependent on the ADP concentration present in the incubation medium and that it is not necessary to add nucleotides other than the

[24] P. Fernández-Sílva, J. A. Enriquez, and J. Montoya, *BioTechniques* **12**, 480 (1992).

labeled one. Ortho[^{32}P]phosphate can be used instead of [α-^{32}P]UTP as the labeled precursor of nucleic acids, but lower labeling of RNA and higher background are obtained after electrophoretic analysis.

The electrophoretic patterns of the *in organello* synthesized RNA, in particular those of the polyadenylated RNA fractions (Fig. 1C,H), are very similar to the electrophoretic pattern of a cold mitochondrial RNA preparation (Fig. 1B,I). In these patterns it is possible to recognize the characteristic set of transcripts previously described in the mitochondria of HeLa cells.[4,9] The physiological function of these RNA species is also identical to that already described in HeLa cells.[3,20] The labeled RNA species are characterized by a high degree of polyadenylation of the mRNAs, a lower level of synthesis of rRNAs, and the accumulation of rRNA precursor.

The most striking characteristic of the *in organello* RNA synthesizing system is that mtDNA transcription and RNA processing can be maintained for very long periods of time, the incorporation of [α-^{32}P]UTP into RNA at 37° remains linear for up to 3 hr and then is maintained for times greater than 6 hr. This allows pulse–chase experiments to be performed.

Comments

1. For RNA synthesis it is essential to keep mitochondria metabolically active and well coupled.

2. The methods followed for the preparation of the mitochondria have been chosen for their rapidity and high degree of purification of the organelles, features which are of great importance for the efficiency of the *in organello* labeling of RNA. However, other purification procedures may be used. The maintenance of cold temperatures during the isolation procedure is important, particularly for brain mitochondria.

3. In our hands the best results are obtained using a rotating wheel for the oxygenation of the mitochondria as indicated above. However, shaking the organelle suspension in a water bath is also satisfactory.

4. The method of RNA labeling is applicable for mitochondria isolated from any mammalian organ, but the isolation procedure may vary.

5. Formaldehyde–agarose gels can also be used for the analysis of the RNA. However, the level of definition of the RNA bands is higher when the strong denaturing agent methylmercury hydroxide is used. The anion- and cation-exchange resin AG501-X8 (Bio-Rad) is used to deionize commercial CH$_3$HgOH.

6. The purity of the mitochondrial preparation, free of cytoplasmic contaminants, allows the direct visualization of the mtDNA in the gel stained with ethidium bromide (Fig. 1A). The synthesized RNA can be

equated to the amount of DNA in the sample, and the steady-state level of 16 S and 12 S rRNA can be quantified.

7. One of the main advantages of an *in organello* RNA synthesizing system is the possibility of studying physiological effects predetermined *in vivo.*[18,19]

8. This *in organello* system can also be used for the synthesis of mtDNA by changing the labeled precursor to [α-^{32}P]TTP, adding cold deoxynucleotides, and increasing the time of incubation to 5 hr.[25]

Acknowledgment

This work was supported by a research grant from the Spanish Dirección General de Investigación Científica y Técnica (DGICYT No. PB90-0915).

[25] J. A. Enriquez, J. Ramos, A. Pérez-Martos, M. J. López-Pérez, and J. Montoya, *Nucleic Acids Res.* **22,** 1861 (1994).

[7] Transcription *in Vitro* with *Saccharomyces cerevisiae* Mitochondrial RNA Polymerase

By DAVID A. MANGUS and JUDITH A. JAEHNING

Introduction

Expression of the yeast mitochondrial genome is mediated by a nuclear-encoded RNA polymerase consisting of two subunits: a 145-kDa catalytic core (encoded by the *RPO41* gene), which resembles the enzymes from bacteriophage T7 and T3, and a 43-kDa specificity factor required for promoter recognition (encoded by the *MTF1* gene), which is similar to members of the bacterial sigma (σ) factor family.[1-3] Despite their low abundance, the subunits of the RNA polymerase have been purified to homogeneity from yeast cells.[4,5] The protocols for isolation of these proteins are arduous, requiring large quantities of cells and many chromatographic steps and yielding only small amounts of protein. We describe the purifica-

[1] J. A. Jaehning, *Mol. Microbiol.* **8,** 1 (1993).

[2] G. S. Shadel and D. A. Clayton, *J. Biol. Chem.* **268,** 16083 (1993).

[3] S.-H. Jang and J. A. Jaehning, *in* "Transcription: Mechanisms and Regulation" (R. C. Conaway and J. W. Conaway, eds.) p. 171, Raven, New York, 1994.

[4] J. L. Kelly and I. R. Lehman, *J. Biol. Chem.* **261,** 10340 (1986).

[5] S.-H. Jang and J. A. Jaehning, *J. Biol. Chem.* **266,** 22671 (1991).

tion of both the core polymerase, from an overproducing yeast strain, and a functional recombinant form of the specificity factor, providing substantial amounts of enzyme for biochemical analysis. We have also optimized conditions under which the core polymerase and specificity factor can be reconstituted for use in *in vitro* transcription reactions. The reconstituted enzyme has been used to demonstrate that Rpo41p and Mtf1p are necessary and sufficient for initiation, to analyze the fate of the subunits after initiation, and to test the properties of mutant forms of Mtf1p.[6,7]

RNA Polymerase Assay Methods

Core Polymerase: Nonselective Transcription Reactions

The core mitochondrial RNA polymerase binds DNA nonspecifically, although with a preference for A- and T-rich sequences,[8] and efficiently transcribes a poly[d(AT)] template.[9]

For a final reaction volume of 20 μl, prepare buffer cocktail that will give a final concentration of 20 mM Tris-HCl, pH 7.9, 23°, 10 mM MgCl$_2$, 500 μM ATP, CTP, and GTP, 10 μM [α-^{32}P]UTP [3000 to 8000 cpm (counts per minute)/pmol], and 60 μg/ml poly[d(AT)] (Sigma, St. Louis, MO). A total of 2 μl core polymerase is added to the buffer cocktail and incubated at 37° for 10 min. Concentrated fractions of core polymerase can be diluted prior to assaying as described below. The entire reaction is spotted onto DEAE paper disks, washed six times for 1 min with shaking in 0.5 M Na$_2$HPO$_4$, two times for 1 min in distilled water, rinsed in 95% (v/v) ethanol, and dried under a heat lamp. Incorporation is measured in a scintillation counter. A unit of core polymerase activity is defined as 1 nmol of UMP incorporated in 10 min under these reaction conditions.

This reaction is highly sensitive to salt, so it is preferable to add as small a volume of enzyme as possible. The assay can reach saturation with highly concentrated samples of the core polymerase. Therefore, it is important to titrate the reaction with dilutions of the core polymerase (see below) to determine the linear range of the assay.

[6] D. A. Mangus, S.-H. Jang, and J. A. Jaehning, *J. Biol. Chem.* **269,** 26568 (1994).
[7] S.-H. Jang, personal communication.
[8] A. H. Schinkel, M. J. A. Groot-Koerkamp, A. W. R. H. Teunissen, and H. F. Tabak, *Nucleic Acids Res.* **16,** 9147 (1988).
[9] C. S. Winkley, M. J. Keller, and J. A. Jaehning, *J. Biol. Chem.* **260,** 14214 (1985).

Holoenzyme: Selective Transcription Reactions

To assay for the presence of the specificity factor, or to reconstitute functional mitochondrial RNA polymerase, runoff transcription assays are performed on templates containing a mitochondrial promoter.[9–11]

Final reaction conditions are as follows: 50 mM Tris-HCl, pH 7.9, 23°, 20 mM $MgCl_2$, 1 mM dithiothreitol (DTT), 500 μM ATP, CTP, and GTP, 100 μM [α-^{32}P]UTP (1000–5000 cpm/pmol), and a promoter-containing template (see below) at 20 μg/ml in 20 μl. Prepare a 4× buffer cocktail containing the Tris-HCl, $MgCl_2$, DTT, and template. Also prepare a separate 4× nucleoside triphosphate (NTP) mix containing ATP, CTP, and GTP, and [α-^{32}P]UTP. Add the buffer cocktail to tubes on ice. Add salt-free buffer equivalent to the buffer that proteins are in [D(0) or T(0), see below], holoenzyme, or core polymerase and specificity factor in a total volume of 10 μl. Concentrated samples of the core polymerase and the specificity factor can be diluted prior to use as described below. For reconstitution reactions, purified core polymerase and specificity factor are incubated together for 5 min on ice in the buffer and buffer cocktail prior to initiation. Transcription is initiated with the addition of the NTP mix and gentle vortex mixing. The reactions are incubated 10 min at 30° and terminated with the addition of 90 μl of 1 mg/ml tRNA, 0.4% w/v sodium dodecyl sulfate (SDS), and 0.1 M sodium acetate, pH 5.5. The samples are extracted with phenol–chloroform, ethanol-precipitated, and subjected to electrophoresis on 7 M urea–polyacrylamide gels as described previously.[12]

Purification of Subunits of Mitochondrial RNA Polymerase

Purification of Core RNA Polymerase from Overproducing Yeast Cells

The yeast strain YJJ473 [*MATa, his3Δ200, leu2, ura3-52,* YEp(*RPO41*)][13] contains the *RPO41* gene on an episomal plasmid and produces high levels of Rpo41p when cells are grown on glucose. The overexpressed protein is localized to the mitochondria,[14] suggesting that it is properly folded and functional. The yield is approximately 4 μg/g cell wet weight.

[10] T. K. Biswas and G. S. Getz, *J. Biol. Chem.* **262,** 13690 (1987).
[11] G. T. Marczynski, P. W. Schultz, and J. A. Jaehning, *Mol. Cell. Biol.* **9,** 3193 (1989).
[12] J. Sambrook, E. F. Fritsch, and T. Maniatis, "Molecular Cloning: A Laboratory Manual," 2nd Ed. Cold Spring Harbor Laboratory, Cold Spring Harbor, New York, 1989.
[13] T. L. Ulery, S. H. Jang, and J. A. Jaehning, *Mol. Cell. Biol.* **14,** 1160 (1994).
[14] T. L. Ulery, Ph.D. Thesis, Indiana University, Bloomington (1993).

Cell Growth. Overexpression of Rpo41p causes an increased frequency of respiratory-incompetent cells and loss of the plasmid.[14] For that reason, cells are grown in selective medium without leucine as long as possible to minimize plasmid loss, thus maximizing expression of Rpo41p. Grow a 2-liter culture of YJJ473 in SD − Leu,[15] containing 2% w/v glucose to mid-log phase. Transfer the culture to a 100-liter fermentor and grow for four doublings to 150 Klett units (6×10^7 cells/ml) in 1% yeast extract containing 2% w/v glucose, 12.5 μg/ml tetracycline, and 50 μg/ml ampicillin. Concentrate the cells using a 0.2-μm cross-flow filter (Microgon, Laguna Hills, CA) and then harvest by centrifugation at 5000 g at 4° for 10 min. Suspend the cells in an equal volume of extraction buffer [20 mM Tris-HCl, pH 7.9, 23°, 10% glycerol, 1 mM EDTA, 10 mM MgCl$_2$, 0.3 M (NH$_4$)$_2$SO$_4$, 1 mM phenylmethylsulfonyl fluoride (PMSF), 0.5 mM DTT]. Freeze the samples by pipetting the cell slurry into liquid N$_2$ and store at −70°.

Yeast Whole-Cell Extract Preparation. All extraction and chromatography steps are carried out between 0° and 4°. Weigh out and thaw the required amount of YJJ473 cells in extraction buffer. Add Nonidet P-40 (NP-40) to a final concentration of 0.1% and lyse using a Dynomill grinding mill (Glen Mills, Clifton, NJ). Remove cell debris by centrifugation at 40,000 g for 60 min. Add hemoglobin-Sepharose to a final concentration of 0.01 ml/ml to the supernatant and stir for 5 min. Add polyethyleneimine (PEI) to a final concentration of 0.042% and stir for 10 min. Remove the precipitated nucleic acids by centrifuging at 18,000 g for 10 min. To the supernatant, add solid (NH$_4$)$_2$SO$_4$ to 38% saturation slowly over the course of 1 hr and continue stirring for an additional 30 min. Recover the precipitated protein by centrifugation at 30,000 g for 45 min and suspend the pellet in D(0) buffer [20 mM Tris-HCl, pH 7.9, 23°, 5% glycerol, 1 mM EDTA, 1 mM EGTA, 0.1 mM DTT, 1 mM PMSF, 0.35 μg/ml bestatin, 0.4 μg/ml pepstatin, and 0.5 μg/ml leupeptin, with values in parentheses indicating millimolar amounts of (NH$_4$)$_2$SO$_4$]. Dilute the sample to a conductance equivalent to 50–100 mM KCl.

DEAE and Phosphocellulose Chromatography. Load the whole-cell extract onto coupled DEAE (1 ml bed volume/g cells) and phosphocellulose (0.5 ml bed volume/g cells) columns previously equilibrated in D(75) buffer and wash with 5 column volumes of the same buffer. Disconnect the columns and elute the protein from the phosphocellulose column with a 5 column volume linear gradient from 0.15 to 0.4 M KCl in M buffer [M buffers are the same as D buffers but with KCl instead of (NH$_4$)$_2$SO$_4$], followed by 2 column volumes of M(400) buffer. Rpo41p elutes from the phosphocellulose column between 200 and 250 mM KCl.

[15] C. Guthrie and G. R. Fink, this series, Vol. 194.

Mono Q Chromatography. Pool the fractions containing core polymerase activity and dialyze against M(0) buffer until the conductance is equivalent to 15 mM KCl. Load the sample on a Mono Q HR 10/10 (Pharmacia, Piscataway, NJ) column equilibrated with M(0) buffer. Wash with 5 column volumes of M(0) buffer. Elute the protein with a 6 column volume linear gradient from 0 to 1 M KCl in M buffer at a flow rate of 1 ml/min. Rpo41p elutes between 40 and 100 mM KCl.

Phenyl-Superose Chromatography. Pool the peak activity fractions from the Mono Q column and dialyze the sample against D(1500) buffer containing 2% glycerol until the conductance is equivalent to 1 M (NH$_4$)$_2$SO$_4$. Load the sample onto a phenyl-Superose HR 5/5 (Pharmacia) column equilibrated with D(1000) buffer containing 2% glycerol. Wash with 5 column volumes of the same buffer. Elute the proteins with a 6 column volume linear gradient from 1 M (NH$_4$)$_2$SO$_4$ and 2% glycerol v/v to 20 mM (NH$_4$)$_2$SO$_4$ and 20% glycerol at a flow rate of 0.2 ml/min. Rpo41p elutes between 500 and 200 mM (NH$_4$)$_2$SO$_4$.

Superose 6 Chromatography. Pool the peak activity fractions from the phenyl-Superose column and dialyze against saturating (NH$_4$)$_2$SO$_4$. Collect the protein precipitate by centrifugation at 30,000 g for 30 min. Suspend the pellet in a small volume of D(0) buffer and load it onto a Superose 6 (Pharmacia) column, equilibrated in D(1000) buffer, at a flow rate of 0.33 ml/min. Core polymerase purified from overproducing cells by this protocol contains a small amount of Mtf1p.[6] Therefore, a concentrated preparation of the protein shows some promoter-specific holoenzyme activity. Dilution of the core polymerase sample (see below) reduces the holoenzyme activity below the limits of detection.

Purification of a Recombinant Form of the Specificity Factor

The entire *MTF1* gene has been subcloned into the expression vector pet5a and subsequently transformed into BL21(DE3) to generate strain pJJ525.[6] This strain utilizes an inducible T7 polymerase to express Mtf1p from a T7 promoter.[16] The yield is approximately 80 μg/g cell wet weight.

Induction of rMtf1p. Grow a culture of pJJ525 to 30 Klett units (red filter) in LB with 100 μg/ml ampicillin. Induce rMtf1p expression by adding isopropyl-β-D-thiogalactopyranoside (IPTG) to a final concentration of 0.2 mM. Allow the cells to grow for 6 to 8 hr and then harvest at 10,000 g for 10 min. Suspend the cell pellets in 4 volumes of T(50) buffer (30 mM Tris-HCl, pH 7.9, 23°, 2 mM EDTA, 5% glycerol, 10 mM MgCl$_2$, 0.1 mM DTT,

---

[16] F. W. Studier, A. H. Rosenberg, J. J. Dunn, and J. W. Dubendorff, this series, Vol. 185, p. 60.

1 mM PMSF, 0.35 μg/ml bestatin, 0.4 μg/ml pepstatin, 0.5 μg/ml leupeptin, and 20 μg/ml benzamidine, with values in parentheses indicating millimolar amounts of KCl).

Cell Extract Preparation. All extraction and chromatography steps are carried out between 0° and 4°. Lyse the cells in a French press (cell pressure 20,000 psi), collecting the sample dropwise. Remove the cell debris by centrifugation at 30,000 g for 15 min.

DEAE and Phosphocellulose Chromatography. Load the whole-cell extract onto coupled DEAE (2 ml bed volume/g cells) and phosphocellulose (1 ml bed volume/g cells) columns previously equilibrated in T(50) buffer and wash with 5 column volumes (phosphocellulose volume) of T(40) buffer. Separate the columns and elute the protein from the phosphocellulose column with a 5 column volume step with T(250) buffer. Analyze samples containing peak protein on Coomassie blue-stained SDS–polyacrylamide gels.[17] rMtf1p elutes from the phosphocellulose column between 100 and 200 mM KCl.

Mono S Chromatography. Pool the peak fractions containing rMtf1p (as determined from gel electrophoresis) and precipitate the protein with the addition of solid $(NH_4)_2SO_4$ to 80% saturation. Collect the protein precipitate by centrifugation at 140,000 g for 30 min. Suspend the pellet in H(50) buffer [50 mM HEPES–SO_4, pH 7.6, 23°, 1 mM EDTA, 5% glycerol, 0.1 mM DTT, 1 mM PMSF, 0.35 μg/ml bestatin, 0.4 μg/ml pepstatin, 0.5 μg/ml leupeptin, and 20 μg/ml benzamidine, with values in parentheses indicating millimolar amounts of $(NH_4)_2SO_4$] and dialyze the sample to a conductance equivalent to 50 mM $(NH_4)_2SO_4$. Load the sample onto a Mono S HR 5/5 (Pharmacia) column equilibrated in H(50) buffer and wash with 5 column volumes of the same buffer. Elute the protein with a 10 column volume linear gradient from 50 to 500 mM $(NH_4)_2SO_4$ at a flow rate of 0.33 ml/min. rMtf1p elutes between 100 and 300 mM $(NH_4)_2SO_4$.

Transcription Templates

A simple 9-bp sequence, 5'-ATATAAGTA(+1)-3', has been well established as the consensus for the yeast mitochondrial promoters.[10,11,18] This sequence functions as the promoter of the mitochondrial 14 S rRNA[19] gene but is also found in the nuclear Sc4816 DNA fragment containing the *GAL10* promoter.[9] In both cases, this sequence functions as a strong mitochondrial promoter in selective transcription reactions, although we have

[17] U. K. Laemmli, *Nature (London)* **227,** 680 (1970).
[18] T. Christianson and M. Rabinowitz, *J. Biol. Chem.* **258,** 14025 (1983).
[19] J. C. Edwards, D. Levens, and M. Rabinowitz, *Cell (Cambridge, Mass.)* **31,** 337 (1982).

consistently observed that the 14 S rRNA template is a more effective template, relative to Sc4816. We recommend that all templates be prepared by twice banding on cesium chloride gradients as described.[12]

Several investigators have taken advantage of the high AT richness of mitochondrial DNA to help assay mitochondrial RNA polymerase function. Neither the Sc4816 nor the 14 S rRNA promoter templates codes for a CTP until many nucleotides (47 nucleotides for Sc4816 and 109 nucleotides for the 14 S rRNA templates) have been incorporated into the nascent RNA chain.[9] Leaving CTP out of the transcription reaction stalls the core polymerase on the template, preventing multiple rounds transcription.[6] These reactions are also useful in analyzing transcripts from supercoiled templates since they create products of a defined length, thus eliminating the need to perform primer extension reactions on the products. Performing reactions without CTP changes the characteristics of the transcription reactions, however, so it is necessary to reoptimize the assay conditions. Reactions can also be performed on templates generated by PCR (polymerase chain reaction) amplification of the region surrounding a promoter element. By designing appropriate primers, short DNA templates can be generated that are useful for gel mobility shift analyses,[6] DNase footprinting, or transcription reactions (Fig. 1). An example of an experiment where the RNA polymerase was allowed to initiate and then proceed short, defined distances down a PCR-generated DNA template is shown in Fig. 1.[20] In the absence of GTP a 3-base transcript (AAU) is generated (Fig. 1, lane 1). In the absence of CTP a 13-base transcript is made (Fig. 1, lane 2), and with all four nucleoside triphosphates present a runoff transcript of 49 bases is synthesized from this short template. By analyzing the polypeptides associated with the DNA template in the stalled complexes corresponding to the reactions shown in Fig. 1, it has been established that Mtf1p is released from the elongating core polymerase in the window between 3 and 13 bases of RNA being synthesized.[6]

Dilution and Storage

Fractions of both the core polymerase and the specificity factor prepared as described above are very concentrated and can be carefully diluted in T(50) buffer containing 50 μg/ml bovine serum albumin (BSA) prior to use. The core polymerase in this buffer is very stable when stored at $-70°$ and can be frozen and thawed. The recombinant specificity factor is stable at high concentrations at $-70°$ but should be diluted just prior to use.

[20] D. A. Mangus, Ph.D. Thesis, Indiana University, Bloomington (1994).

A

B

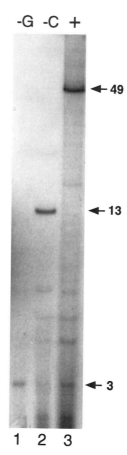

FIG. 1. *In vitro* transcripts synthesized by the reconstituted yeast mitochondrial RNA polymerase. (A) Template for the transcription reactions. The production of the PCR fragment from a recombinant plasmid was as described.[6] The nucleotide sequence of the relevant portion of the nontemplate strand is shown, the promoter is boxed, and the +1 nucleotide, as established by primer extension analysis,[20] is indicated. The position of transcriptional pauses in reactions lacking GTP (+3) or CTP (+13) are indicated. The full-length runoff product of 49 nucleotides is also indicated. (B) Paused and runoff transcripts. Selective transcription reactions using 0.34 pmol core RNA polymerase (Superose 6 fraction), combined with 0.16 pmol specificity factor (Mono S fraction) were performed on the template shown

In Vitro Transcription with Reconstituted Holoenzyme

The highly purified subunits of the mitochondrial RNA polymerase described above behave somewhat differently in in vitro transcription reactions when compared with the crude fractions previously described. For that reason it was important to optimize the selective transcription reaction conditions. The two most critical parameters in this assay are salt and template concentration. The KCl optimum for partially purified mitochondrial RNA polymerase holoenzyme was determined to be between 60 and 80 mM.[9] However, KCl above 40 mM is inhibitory to the reconstituted system, affecting both the Sc4816 and 14 S rRNA promoter templates equally.[20] Dilution of the subunits in low salt buffer prior to use, as described above, is an effective solution to this problem.

The specificity factor of the mitochondrial RNA polymerase functions in a fashion analogous to that of the bacterial sigma factors: it promotes specific initiation of the core polymerase, is released from the elongating polymerase shortly after initiation, and is then available to be used by another core polymerase molecule.[6] Because the specificity factor is recycled in the transcription cycle, it is important that it be the limiting component of the transcription reaction. Careful titrations of the specificity factor are critical for the interpretation of mechanistic experiments. For most applications, it is important that the transcription reactions be performed under conditions in which the promoter template is in excess. Careful titrations of the template of choice should be performed to determine a range in which the mitochondrial polymerase is not saturating.

The number of active molecules in the transcription reactions can be readily determined by calculating the number of transcripts made per molecule of RNA polymerase. The range of these values from several experiments has shown that between 5 and 10% of the core polymerases and 10–20% of the specificity factor are utilized under optimal conditions.[6] This suggests that reconstitution of transcription is very efficient.

We have used these methods to begin to characterize the role of the subunits in initiation of mitochondrial transcription; however, many important questions remain unanswered as to how this process is mediated. The specificity factor fails to bind DNA on its own but is required for selective transcription. Are both subunits required for promoter recogni-

in (A) under the reaction conditions described in the text except that the reactions contained 100 μM each of the NTPs described below with UTP at 10 μM. Reactions in lane 1 contained ATP and UTP only, lane 2 contained all NTPs except CTP, and lane 3 contained all NTPs. Transcripts were analyzed on a 7 M urea–25% polyacrylamide gel.[12] Arrows on the right-hand side of the gel indicate the size of the transcripts synthesized in nucleotides.

tion, or does the interaction of the specificity factor with the core polymerase allow the factor to recognize the promoter as demonstrated for bacterial σ^{70}? Which regions of the two subunits are involved in the protein–protein interactions that create the holoenzyme form of the RNA polymerase, and how does this interaction allow for the cyclic reuse of the specificity factor? Do other factors stimulate or repress the activity of the enzyme? We are presently using the reconstituted enzyme described in this chapter to characterize mutant forms of both subunits to better describe their role in the initiation and regulation of mitochondrial transcription.

Acknowledgments

We thank Sei-Heon Jang, Keith Otto, and Terrie Ulery for input into this work. This research was supported by grants from the National Institutes of Health (RO1 GM 36692, K04 AI 00874) to J. A. J.

[8] Reactions Catalyzed by Group II Introns *in Vitro*

By PHILIP S. PERLMAN and MIRCEA PODAR

Introduction

Group II introns are found in organelle DNAs of fungi and photosynthetic eukaryotes and in bacterial genomes.[1–3] The conserved secondary structures of group II introns consist of six substructures or domains, D1 through D6 (Fig. 1). There are two main subgroups, IIA and IIB, distinguished by differences in some of the domain substructures.[3] Some introns, mostly of subgroup IIA, have a reading frame inserted within the loop of domain 4. The proteins encoded by two such introns in yeast mitochondria have been shown to play roles in splicing and intron mobility (via maturase and reverse transcriptase functions).[4–6] Most group II introns share this overall secondary structure, but there are some exceptions.

[1] J.-L. Ferat and F. Michel, *Nature (London)* **264,** 358 (1993).
[2] J.-L. Ferat, M. Le Gouar, and F. Michel, *C. R. Acad. Sci. Paris* **317,** 141 (1994).
[3] F. Michel, K. Umesono, and H. Ozeki, *Gene* **82,** 5 (1989).
[4] J. V. Moran, K. L. Mecklenburg, P. Sass, S. M. Belcher, D. Mahnke, A. Lewin, and P. Perlman, *Nucleic Acids Res.* **22,** 2057 (1994).
[5] J. M. Moran, S. Zimmerly, R. Eskes, J. C. Kennell, A. M. Lambowitz, R. A. Butow, and P. S. Perlman, *Mol. Cell. Biol.* **15,** 2828 (1995).
[6] G. Carignani, O. Groudinsky, D. Frezza, E. Schiavon, E. Bergantino, and P. P. Slonimski, *Cell (Cambridge, Mass.)* **35,** 733 (1983).

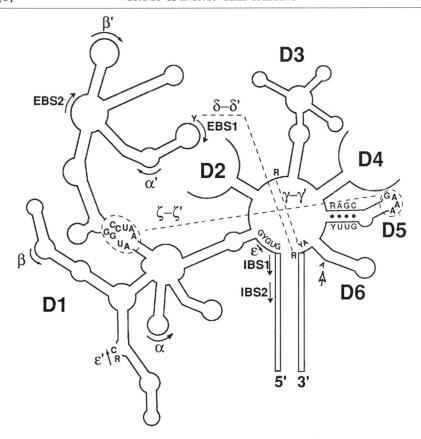

Fig. 1. Diagram of the secondary structure of a group II intron. The proposed secondary structure of the group IIB intron aI5γ is shown in diagram form. The six major substructures (domains 1–6) are labeled. Most of the sequences shown are highly conserved and are known to influence splicing activity. Three established tertiary interactions are indicated by dashed lines connecting the interacting sequences; sequences involved in the α–α', ε–ε', EBS1–IBS1, and EBS2–IBS2 interactions are labeled.

The finding that some group II introns self-splice *in vitro,* in the absence of any protein, showed that the catalytic functions required for this process reside within the RNA molecule itself.[7–9] This has allowed the design of *in vitro* systems to study aspects of the mechanism of group II intron

[7] C. L. Peebles, P. S. Perlman, K. L. Mecklenburg, M. L. Petrillo, J. H. Tabor, K. A. Jarrell, and H.-L. Chen, *Cell (Cambridge, Mass.)* **44,** 213 (1986).

[8] C. Schmelzer and R. J. Schweyen, *Cell (Cambridge, Mass.)* **46,** 557 (1986).

[9] R. van der Veen, A. C. Arnberg, G. van der Horst, L. Bonen, H. F. Tabak, and L. A. Grivell, *Cell (Cambridge, Mass.)* **44,** 225 (1986).

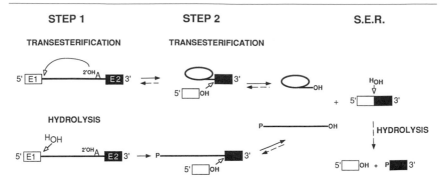

FIG. 2. Reactions catalyzed by self-splicing group II introns. Splicing by transesterification in both reaction steps is diagrammed in the first line. The course of splicing obtained in reactions containing at least $0.5\ M$ KCl (plus 100 mM Mg^{2+}) is shown in the second line; under that condition the first splicing reaction occurs by hydrolysis while the second reaction appears to occur by transesterification. In KCl-containing reactions, the excised intron (linear or lariat forms, as shown) catalyze the hydrolysis of the splice junction between the exons, resulting in the accumulation of free E1 and E2 RNAs.

RNA catalysis. The self-splicing mechanism involves two transesterification reactions leading to the formation of spliced exons and an excised intron RNA lariat that has a $2'$–$5'$ phosphodiester bond (Fig. 2) (reviewed in Refs. 10 and 11). The first reaction step involves the nucleophilic attack of the $5'$ splice junction by the $2'$-OH of the ribose moiety of an adenosine located in intron domain 6; this reaction releases the first exon and forms the lariat intron–second exon intermediate. The second reaction step involves the nucleophilic attack of the $3'$ splice junction by the $3'$-OH of the ribose of the last nucleotide of the first exon. In terms of reaction mechanism and products, group II intron splicing closely resembles that of spliceosomal introns of eukaryotes.[12]

Just a few features of the three-dimensional organization of the group II intron RNA that is responsible for the formation of the catalytic core are understood. A number of tertiary interactions are indicated in Fig. 1. Those interactions involve base pairing between different regions of the pre-mRNA and were deduced based on sequence covariations among group II introns. All of the ones shown are supported by some *in vitro* evidence,

[10] T. R. Cech, *in* "The RNA World" (R. F. Gesteland and J. F. Atkins, eds.), p. 239. Cold Spring Harbor Laboratory, Cold Spring Harbor, New York, 1993.
[11] F. Michel and J. L. Ferat, *Annu. Rev. Biochem.* **64,** 435 (1995).
[12] A. Jacquier, *Trends Biochem. Sci.* **15,** 351 (1990).

TABLE I
SELF-SPLICING GROUP II INTRONS

Intron	Subtype	Gene/genome	Organism	Refs.
COI I1	IIA	*COXI*/mtDNA	*Podospora anserina*	a
aI1	IIA	*COXI*/mtDNA	*Saccharomyces cerevisiae*	b
aI2	IIA	*COXI*/mtDNA	*Saccharomyces cerevisiae*	c
aI5γ	IIB	*COXI*/mtDNA	*Saccharomyces cerevisiae*	7, 9
bI1	IIB	*COB*/mtDNA	*Saccharomyces cerevisiae*	8
LSUrRNA IVS	IIB	rRNA/mtDNA	*Scenedesmus obliquus*	d
Int B	IIA	?/bacterial chromosome	*Escherichia coli*	2
Cal X1	IIA	?/cyanobacterial chromosome	*Calothrix* sp.	1

a U. Schmidt, B. Riederer, M. Morl, C. Schmelzer, and U. Stahl, *EMBO J.* **9,** 2289 (1990).
b S. K. Hebbar, S. M. Belcher, and P. S. Perlman, *Nucleic Acids Res.* **20,** 1747 (1992).
c R. Eskes, J. V. Moran, and P. S. Perlman, unpublished observations (1994).
d U. Kück, I. Godehardt, and U. Schmidt, *Nucleic Acids Res.* **18,** 2691 (1990).

chiefly with aI5γ[13–16]; we have obtained some *in vivo* support for the ε–ε′ interaction in aI1.[17]

Although over 100 group II introns have been reported[11] (excluding group II and group III introns from euglenoid organisms[18]), only 8 are known to self-splice (Table I). Although there are few specific reports of negative findings, it is clear that many of the other introns do not self-splice under any of the conditions that are presently used. In this chapter we summarize a number of experimental methods that are used in studies of self-splicing group II introns.

Preparing Transcripts

The RNAs for self-splicing reactions are generally made as *in vitro* transcripts of plasmid templates containing whole introns or fragments of them. A potential problem with those approaches is that many of the RNAs so made begin and/or end with sequences encoded by the plasmid (e.g.,

[13] M. Costa and F. Michel, *EMBO J.* **14,** 1276 (1995).
[14] A. Jacquier and F. Michel, *Cell (Cambridge, Mass)* **50,** 17 (1987).
[15] A. Jacquier and F. Michel, *J. Mol. Biol.* **213,** 437 (1990).
[16] A. Jacquier and N. Jacquesson-Breuleux, *J. Mol. Biol.* **219,** 415 (1991).
[17] S. C. Boulanger, Ph.D. Dissertation, University of Texas Southwestern Medical Center, Dallas (1995).
[18] R. B. Hallick, L. Hong, R. G. Drager, M. R. Favreau, A. Monfort, B. Orsat, A. Spielmann, and E. Stutz, *Nucleic Acids Res.* **21,** 3537 (1993).

polylinker). Although we are not aware that any such extensions have influenced the outcomes of specific experiments, it is preferable to minimize or exclude extraneous sequences from these RNAs whenever possible. This is usually achieved by a careful subcloning strategy and the choice of the restriction site used to linearize the plasmid template.

A powerful and fast alternative for obtaining the DNA templates for transcription is PCR (polymerase chain reaction) amplification of an intron fragment or an entire group II intron from a preexisting plasmid. One oligonucleotide primer contains the 17 nucleotides of the T7 RNA polymerase promoter (5'-TAATACGACTCACTATA) followed by the first 18–22 nucleotides of the sequence to be transcribed.[19,20] Because transcription has to start with a G nucleotide, either the 5' end of the RNA is chosen to be a G normally present in the template or a single extraneous nucleotide is inserted after the promoter. The second primer (with sequence complementary to that of the RNA) determines where the transcript ends.

The amplification is carried out according to standard PCR procedures and has to be optimized for each template. As a starting point we use 10–20 ng target plasmid DNA per 100 μl of a standard PCR mixture, with 1.5 mM Mg^{2+}, 2.5–5 units of Taq DNA polymerase; 0.2–0.5 units of a thermostable DNA polymerase with 3' $\rightarrow$ 5'-exonuclease activity is included to eliminate the 3' overhangs generated by Taq DNA polymerase. The following PCR cycle is often suitable: 30–45 sec at 94°, 30–45 sec at 58°, and 30 sec–2 min at 72°. Using these conditions repeated for 28–29 cycles, we have successfully amplified with good yields various fragments of the group II intron aI5γ. Gel purification of the PCR products is not necessary [we actually found the glass beads plus sodium iodide technique (kit from Bio101, La Jolla, CA) to be deleterious for the subsequent transcription]. This PCR strategy offers the advantages of eliminating the need for subcloning and yielding transcription templates with few undesired nucleotides. However, the transcription efficiency is sometimes lower than when using plasmid DNA. Including the +1 to +5 sequence of the T7 promoter region (GGGAG) is expected to improve the yield but introduces a short extraneous sequence at the 5' end of the RNA.[20] A cost-effective solution to this problem is to prepare large samples of highly concentrated RNA polymerase in the laboratory so that the yield of transcripts can be driven by a high level of enzyme.

For the synthesis of relatively short RNAs for use in self-splicing experiments, the protocol from the Uhlenbeck laboratory is used to make com-

[19] J. F. Milligan, D. R. Groebe, G. W. Witherel, and O. C. Uhlenbeck, *Nucleic Acids Res.* **15,** 8783 (1987).
[20] J. F. Milligan and O. C. Uhlenbeck, this series, Vol. 180, p. 51.

pletely synthetic, partially double-stranded DNA templates containing the T7 RNA polymerase promoter.[19,20] This approach has been used to make a family of 36 nucleotide long domain 5 RNAs.[21]

The labeling of the RNA depends on the details of the experiment. For internal labeling we use mostly $[\alpha\text{-}^{32}\text{P}]$UTP during transcription, as group II introns of yeast mitochondrial (mt)DNA are very uracil-rich. Alternatively, after synthesizing unlabeled RNA or with trace internal labeling, single high specific activity labels can be introduced at the 5' end using $[\gamma\text{-}^{32}\text{P}]$ATP and T4 polynucleotide kinase or at the 3' end with $[^{32}\text{P}]$Cp and T4 RNA ligase, according to published methods. End labels allow for the direct enzymatic sequencing of RNA products and also simplify some quantitative determinations; however, end-labeling restricts the number of radioactive product species detectable in a particular reaction.

The *in vitro* transcription reactions are done according to standard procedures and at scales depending on the needs of the experiment.[20,22] We have used several commercial sources of bacteriophage-encoded RNA polymerases but have begun using only preparations of T7 RNA polymerase made in our laboratory according to published procedures.[23,24] After transcription, the template DNA is removed with RNase-free DNase (especially when using PCR and oligonucleotide templates), and the RNA is purified by denaturing polyacrylamide gel electrophoresis (PAGE). The RNA band is identified by autoradiography or UV-shadowing, excised from the gel, crushed in a siliconized tube, and passively eluted overnight in 0.5–1 ml elution buffer [50 mM HEPES–KOH, pH 7.3, 200 mM NaCl, 10 mM EDTA, and 0.1% sodium dodecyl sulfate (SDS)] with rocking at 4°. To recover the RNA, the slurry is gently pushed through a 0.4- to 0.8-μm syringe filter (e.g., Millex, Millipore, Bedford, MA). The flow-through liquid is extracted with phenol–chloroform, then with chloroform, and finally the RNA is precipitated with 3 volumes ethanol. The labeled RNA is dissolved in 5 mM Tricine, pH 7.5, 1 mM EDTA and stored at −20° for several days in the case of RNAs with high specific activity internal labeling, or for weeks in the case of RNAs of low specific activity or end labeling. The concentration of the RNAs is determined either from the specific activity or in the case of nonlabeled RNAs by spectrophotometry, using an average nucleotide molar extinction coefficient of 7500 M^{-1} cm^{-1} at 260 nm. To reduce possible misfolding that can occur during handling and storage,

[21] C. L. Peebles, M. Zhang, P. S. Perlman, and J. F. Franzen, *Proc. Natl. Acad. Sci. U.S.A.* **92,** 4422 (1995).

[22] I. D. Povrovskaya and V. V. Gurevich, *Anal. Biochem.* **220,** 420 (1994).

[23] V. Zadowski and H. J. Gross, *Nucleic Acids Res.* **19,** 1948 (1991).

[24] J. Grodberg and J. J. Dunn, *J. Bacteriol.* **179,** 1245 (1988).

before each experiment we heat the RNA aliquot at 85°–90° for 2 min and then cool to room temperature.

For some purposes, it is not necessary to gel-purify the transcripts. However, some pre-mRNA transcripts (e.g., containing aI5γ or bI1) can self-splice to some extent in the transcription buffer. The presence of splicing products in the starting material is avoided by carrying out the transcription at a reduced temperature (30°) for a relatively short time (30 min).

The technique of DNA-mediated RNA ligation developed by Moore and Sharp is an effective means of introducing unique labels or modified nucleotides at defined positions within large RNA molecules.[25] Using this technique, two or three separate RNA molecules, obtained by transcription or chemical synthesis and containing defined modifications (at the nucleotide base, ribose, or phosphate groups) or radioactive labels, can be ligated. We have used this approach to introduce single internal radioactive labels and stereochemically defined phosphorothioate substitutions at the splice sites of aI5γ.[26,27] The efficiency of these ligations is greatly improved (up to 75% efficiency), especially when ligating short RNA oligonucleotides (7–12 nucleotides) to long, potentially folded RNAs, by performing the reaction in a thermal cycler ramped slowly over a 12-hr period from 20° to 10° followed by a return to 20° over another 12 hr.

Forward Splicing of Group II Introns

Low Salt Reaction Conditions

Using standard pre-mRNA substrates, the rate of self-splicing varies widely depending on the reaction conditions. The original reaction cocktail, often referred to as a "low salt" reaction condition, used to define the reactions of the first self-splicing group II intron (intron 5γ of the yeast mitochondrial *COXI* gene), contained 10 mM magnesium acetate, 2 mM spermidine chloride, and 40 mM Tris–acetate, pH 7.6.[7] The splicing reaction yielded readily detectable products after incubations of 30 min or more. Splicing had a sharp temperature optimum of 45° and was blocked by sodium acetate at concentrations higher than 100 mM.

Several other low salt cocktails are used more or less interchangeably; the variables include using $MgCl_2$ instead of magnesium acetate and using Tris-Cl instead of Tris–acetate.[9,28] In the initial study, it was shown that levels of magnesium acetate higher than 10 mM (up to 100 mM) support

[25] M. J. Moore and P. A. Sharp, *Science* **256,** 992 (1992).

[26] R. A. Padgett, M. Podar, S. C. Boulanger, and P. S. Perlman, *Science* **266,** 1685 (1994).

[27] M. Podar, P. S. Perlman, and R. A. Padgett, *Mol. Cell. Biol.* **15,** 4466 (1995).

[28] A. Jacquier and M. Rosbash, *Science* **234,** 1099 (1986).

comparable levels of self-splicing. In a later study, using $MgCl_2$, it was shown that spermidine is not required when 100 mM Mg^{2+} is present[29,30]; in some current studies, the solutions containing up to 100 mM Mg^{2+} plus a buffering agent are referred to as "low salt" reaction conditions[31,32] (see below for a definition of "high salt" conditions). The second self-splicing group II intron found, intron 1 of the cytochrome b gene of yeast mtDNA (bI1), splices weakly in low salt buffers, and most other self-splicing group II introns are unreactive under that condition (see Table I).

The precursor RNAs used in these *in vitro* reactions are approximations of the precursor RNAs that splice *in vivo*. In general, the *in vitro* substrate RNAs are truncated both 5' and 3' relative to the natural precursor; also, other introns are usually excluded. Several studies of the group I intron of the *Tetrahymena* ribosomal RNA gene indicate that the length of the exons in the model precursor RNA can influence the rate of splicing.[33,34] In an early study of the group II intron aI5γ, RNAs containing upstream exons as long as 548 nucleotides and as short as 13 nucleotides were compared.[28] It was reported that constructs with exon 1 at least 53 nucleotides long support about the same rate of splicing under low salt conditions, whereas ones with a first exon 35, 24, or 13 nucleotides long were somewhat inhibited. The precursor with the shortest 5' exon was the most inhibited and, because it accumulated detectable levels of free 5' exon and lariat intron–3' exon RNAs, was especially defective in the second splicing reaction. In a trans reaction (see Fig. 3B, and below) substrates containing the two shortest exons were completely inactive. It is known that base pairing between some nucleotides of the first exon, and the EBS1 and EBS2 sequences of domain 1 of these introns play a crucial role in 5' splice site definition.[14] Because the 13-nucleotide long exon 1 contains all of IBS1 and part of IBS2 and the others contain both sequences, it is not clear why they are partially defective. It was suggested that there is some other sequence between −35 and −53 that influences splicing. In the laboratory of Jacquier, the standard aI5γ precursor RNA has a 53 nucleotide long first exon, and the precursor with 13 nucleotides of E1 was used to test for effects of mutations on the

[29] C. L. Peebles, E. A. Benatan, K. A. Jarrell, and P. S. Perlman, *Cold Spring Harbor Symp. Quant. Biol.* **52,** 223 (1987).
[30] K. A. Jarrell, C. L. Peebles, R. C. Dietrich, S. L. Romiti, and P. S. Perlman, *J. Biol. Chem.* **263,** 3432 (1988).
[31] C. L. Peebles, S. M. Belcher, M. Zhang, R. C. Dietrich, and P. S. Perlman, *J. Biol. Chem.* **268,** 11929 (1993).
[32] S. C. Boulanger, S. M. Belcher, U. Schmidt, S. D. Dib-Hajj, T. Schmidt, and P. S. Perlman, *Mol. Cell. Biol.* **15,** 4479 (1995).
[33] S. A. Woodson and T. R. Cech, *Biochemistry* **30,** 2042 (1991).
[34] S. A. Woodson, *Nucleic Acids Res.* **20,** 4027 (1992).

second splicing reaction[15,16,35]; in our laboratory nearly all studies have employed a precursor with a first exon 296 nucleotides long. In both cases, the natural exon sequences are preceded by a short sequence encoded by the vector.

The study of self-splicing cis reactions has the advantage of being insensitive to RNA concentration. It is a convenient way to analyze the catalytic properties of a particular allele of a group II intron. However, cis reactions have limitations for the study of particular reaction steps that are not rate limiting. As a result, some mutations can appear to have little effect on the overall self-splicing pathway even though they may have a substantial effect on some step that is not rate limiting as assayed *in cis*. The phenotype resulting from a given mutation is often best determined by studying trans reactions that can distinguish between different steps involved in binding and catalytic events (see below).

High Salt Reaction Conditions

A remarkable feature of group II introns is that monovalent cation salts influence the rate of splicing and the nature of the reaction products. Peebles *et al.,*[29] with further details provided in Jarrell *et al.,*[30] showed that more rapid splicing is obtained when certain salts are added to a self-splicing solution containing 100 mM Mg^{2+}. For example, addition of 0.5–1.5 M $(NH_4)_2SO_4$ increases the apparent rate of splicing by at least 10-fold without changing the array of products. This rate enhancement is obtained with as little as 60 mM Mg^{2+}. The effect of temperature on self-splicing is much less dramatic in high salt reactions; the optimum remains approximately 45°, but substantial splicing still occurs at lower temperatures (where products are not observed in low salt reactions).

In solutions containing greater than 60 mM Mg^{2+}, KCl also stimulates the rate of splicing. Unlike $(NH_4)_2SO_4$, however, added KCl does not support efficient first step transesterification; instead the first step occurs mostly by hydrolysis.[30] Splicing without transesterification has not been observed with any other type of intron. The KCl-containing reactions yield mostly linear excised intron RNA rather than excised intron lariat (see Fig. 2). In this pathway, which probably is catalyzed by the same reaction center as the first step reaction by transesterification, a water molecule is activated and serves as a nucleophile instead of the normal 2'-OH from domain 6.

Another different outcome of KCl-stimulated reactions is that little of the spliced exons product accumulates. Reconstruction experiments, in which excised intron and spliced exon RNAs formed in reactions lacking

---

[35] G. Chanfreau and A. Jacquier, *EMBO J.* **12**, 5173 (1993).

KCl were isolated and incubated in KCl-containing buffers, showed that the conversion of spliced exons to free exons can occur as a postsplicing reaction.[30] That bimolecular reaction is a sequence-specific, intron-dependent hydrolysis reaction that has been called the spliced exons reopening (SER) reaction (Figs. 2 and 6C). There is little direct evidence that all or nearly all of the free second exon RNA formed in these cis-splicing reactions proceeds through a spliced exon RNA intermediate. Some kinetic evidence has been reported consistent with that interpretation: it was reported that there is more spliced exons product at short times in KCl reactions than at later times, indicating that the spliced exons product is consumed by a postsplicing reaction, presumably SER.[36] It is conceivable that some of the free 3' exon RNA formed in cis reactions in KCl-containing buffers occurs by direct hydrolysis of the 3' splice site; however, that reaction has not been demonstrated. It is clear, however, that the formation of linear intron RNA in KCl reactions is not a postsplicing reaction, as incubating bona fide lariat RNA in KCl buffer does not result in hydrolysis of the branch point (e.g., debranching).[30]

Splicing by transesterification and splicing without it are two extremes of a continuum of outcomes that can be obtained in various high salt buffers. Chloride ion is an important factor that promotes hydrolysis reactions. For example, when 0.5 M $(NH_4)_2SO_4$ reactions are carried out with no added chloride ion (brought in with the Mg^{2+} or the buffering agent), there is no hydrolysis; however, if 100 mM $MgCl_2$ is used or if Tris-Cl is the buffering agent, some SER occurs. Also, all other things kept equal, increasing the $(NH_4)_2SO_4$ concentration from 0.5 to 1.5 M increases the amount of the SER reaction that occurs.[37] At this point, the most reliable way to avoid the SER reaction is to use HEPES or MOPS as buffers, 60–100 mM $MgSO_4$, and a low concentration of ammonium sulfate (≤ 0.5 M). Chloride ion is not, however, sufficient for efficient hydrolysis reactions: the monovalent cation also influences hydrolysis reactions. For example, whereas 0.5 M KCl leads to about 80% of first reactions by hydrolysis and nearly all of the exon products to accumulate as free exons, neither 0.5 M NaCl nor 0.5 M LiCl promotes detectable hydrolysis reactions.[30] Also, in the presence of 0.5 M NH_4Cl, about half of the first reactions proceed by hydrolysis and about half of the spliced exon product is hydrolyzed to yield free exons.

In buffers containing at least 0.5 M KCl, a number of other hydrolysis reactions occur that in long incubations can consume the main products

---

[36] S. D. Dib-Hajj, S. C. Boulanger, S. K. Hebbar, C. L. Peebles, J. F. Franzen, and P. S. Perlman, *Nucleic Acids Res.* **21,** 1797 (1993).
[37] J. L. Koch, S. C. Boulanger, S. D. Dib-Hajj, S. K. Hebbar, and P. S. Perlman, *Mol. Cell. Biol.* **12,** 1950 (1992).

formed at short times of reaction. Several of these reaction sites have been mapped, one in domain 2 and another in the 3' exon; each one occurs just following a short sequence that closely resembles the 3' end of the first exon.[30] Thus, it appears that some sequences capable of pairing with the EBS1 site in domain 1, to form a pseudo EBS–IBS pairing, can be hydolyzed (see also Ref. 27).

It is not clear whether the change in the reaction pathway from primarily transesterification to primarily hydrolysis is determined by some specific effect of KCl on the RNA structure, somehow masking the branch site. Because both linear and lariat intron RNAs carry out the SER reaction apparently equivalently,[30] it is possible that the failure to branch reflects one effect of the salt and that the activation of the SER reaction reflects some other effect. Also, some intron mutations influence whether hydrolysis reactions occur. For example, deleting most or all of domain 6, which contains the branch point used for step 1 transesterification, from the standard precursor RNA does not abolish step 1 in ammonium sulfate-stimulated reactions (no KCl added)[16,37]; also, deleting domain 3 from the original construct (with long exons) abolishes the SER reaction in 0.5 M KCl reactions.[37] It is not known if any of the hydrolysis reactions summarized here occur *in vivo*.

The linear intron RNA resulting from first step hydrolysis was found to migrate identically to broken intron lariat (which can result from random nicking of the lariat, especially during prolonged incubations).[7,30] In several of our earliest studies we demonstrated several ways of distinguishing linear intron RNA from broken lariat. For example, for aI5γ those two RNAs are readily distinguished by RNase T1 digests.[30] Alternatively, treatment of broken lariat RNA with nuclear lariat debranching enzyme converts the single, sharp band into a smear of randomly broken linear RNAs while authentic linear intron RNA is unaffected by that enzyme treatment.[7] In some of the early papers in this field the gel position occupied by those species was labeled IVS-BL (for broken lariat) or IVS-LIN (for linear intron); it was shown that essentially all of the material at that gel position of KCl reactions is linear RNA, whereas from low salt or $(NH_4)_2SO_4$ reactions all of that material is broken lariat.[7,30]

We have reinvestigated the effects of temperature and pH on the self-splicing of aI5γ, using an RNA that contains shorter exons (70-nucleotide E1 and 60-nucleotide E2)[38] than the earlier pJD20 clone (296-nucleotide E1 and 335-nucleotide E2).[39] The rate of splicing of this RNA in buffer containing 100 mM MgSO$_4$ and 0.5 M $(NH_4)_2SO_4$ is about 30-fold higher

[38] M. Podar, P. S. Perlman, and R. A. Padgett, in preparation (1995).
[39] K. A. Jarrell, R. C. Dietrich, and P. S. Perlman, *Mol. Cell. Biol.* **8,** 2361 (1988).

than that of the original clone ($t_{1/2}$ ~30 sec). Under that reaction condition, substrate RNAs with shorter exons appear to avoid possible aberrant foldings of one or both exons. Using gel-purified samples of this RNA, we have found that a preincubation in buffered 0.5 M $(NH_4)_2SO_4$ for 5–10 min stimulates the initial rate by about 2- to 3-fold. The temperature of this preincubation is also important, having an optimum at 35°, whereas the reaction after adding the Mg^{2+} has an optimum at 42°. We believe that in the absence of Mg^{2+} the RNA is in a folding equilibrium and that the added Mg^{2+} locks the correct structure and initiates the reaction. Using buffers at different pH values, we have also found that there is an exponential increase in the reaction rate between pH 4.5 and pH 6.0 consistent with a reaction chemically limited by the formation of the nucleophile, followed by a plateau above pH 6.0.[38] Those findings suggest that at the pH values normally used (≥ 7.0), cis splicing is limited by conformational events. Considering that there probably is a conformational rearrangement between the two steps, the catalytic and self-folding capacities of some group II introns are quite remarkable.

Salt Effects on Self-Splicing by Other Group II Introns

The above paragraphs summarize effects of various salts on splicing reactions on aI5γ, the group II intron that has been most thoroughly characterized to date. The other group IIB intron from yeast mtDNA, bI1, responds to the ionic constituents of the reaction medium similarly.[8]

Self-splicing of group IIA introns is somewhat different from that of the IIB introns. First, the IIA introns analyzed to date (see Table I) do not self-splice in low salt buffers. They are only reactive in high salt buffers and are not as reactive as the IIB introns under any condition. The aI1 intron is only slightly reactive in the presence of as much as 1.5 M $(NH_4)_2SO_4$ plus 100 mM Mg^{2+} and splices optimally in 100 mM Mg^{2+} plus at least 1 M NH_4Cl.[40] Substantial splicing is also obtained when the NH_4Cl is replaced by at least 1 M KCl, but in those reactions almost no full-length intron lariat accumulates. Unlike the group IIB introns, aI1 does not carry out the SER reaction even in samples containing 1.5 M KCl. Interestingly, in KCl reactions the excised linear intron undergoes a postsplicing branching reaction that forms several species of shorter lariat RNAs plus short 5' fragments of the intron RNA.

Other Details

The self-splicing reactions are typically stopped by the addition of 1 volume (v/v) of 95% formamide or 8 M urea containing gel tracking dyes,

[40] S. K. Hebbar, S. M. Belcher, and P. S. Perlman, *Nucleic Acids Res.* **20,** 1747 (1992).

after which the samples are directly loaded on polyacrylamide TBE (90 mM Tris-borate, pH 8.3, 1 mM EDTA) gels containing 8 M urea. The gel concentration and cross-linking ratios depend on the sizes of the products. For relatively short introns such as aI5γ or bI1 (<1000 nucleotides), a 4–5% (w/v) gel with a 19:1 ratio of acrylamide to bisacrylamide allows the lariat RNA to enter the gel and all the main products to be separated. For larger introns (>2000 nucleotides) such as aI1 or aI2, different gels have to be used (3.5–4% with a 39:1 ratio). These gels are usually run at 10–15 V/cm for 12–24 hr depending on the resolution needed.

Usually we do not take any steps to reduce the amount of salt loaded on the above gels. Where all of the samples contain the same amount of salt, no mobility artifacts are evident. However, the mobility of these RNAs is affected by the salt loaded with the RNA; thus, in experiments where different amounts of salt are present in adjacent samples, some artifacts are evident. We avoid this problem by adding salt after the reaction has been stopped so that all samples have about the same amount of salt.

Effects of Intron Mutations on Splicing *in Vitro* and *in Vivo*

Many mutations of the intron aI5γ, and some of bI1 and aI1, have been analyzed for their effects on self-splicing. These include deletions of whole or parts of intron domains,[37,41,42] single and multiple base substitutions,[14–16,31,32,35,37,43] and, more recently, atomic mutations.[26,27] In general, mutations affecting highly conserved nucleotides, or pairs of nucleotides that covary, block or substantially inhibit self-splicing under low salt conditions. When assayed in cis-splicing reactions, the splicing defect of many mutations, including deletions of domains 2, 3, 4, and 6, and many point mutations of highly conserved nucleotides, is generally suppressed, sometimes completely, by high salt reaction conditions. Thus, in high salt conditions it is only the rare mutation that inhibits splicing by more than 100-fold; examples of mutations that inhibit splicing at least 100-fold under high salt conditions include the deletion of domain 5, deletion of the C1 substructure of domain 1, multinucleotide mutations of the EBS1 or IBS1 sequences, deletion of one or both nucleotides of the bulge of domain 5, shortening the length of the D5 helix, an *Rp*-phosphorothioate at the 5' or 3' splice site, or an *Rp*-phosphorothioate at either of two positions in D5. Mutations of nucleotides that have more modest effects on self-splicing in high salt buffers when assayed *in cis* include the first intron base (G),

[41] J. H. J. M. Kwakman, D. Konings, H. J. Pel, and L. A. Grivell, *Nucleic Acids Res.* **17,** 4205 (1989).
[42] J. Bachl and C. Schmelzer, *J. Mol. Biol.* **212,** 113 (1990).
[43] R. van der Veen, J. H. J. M. Kwakman, and L. A. Grivell, *EMBO J.* **6,** 1079 (1987).

the nucleotides involved in most of the tertiary interactions shown in Fig. 1 (except EBS1–IBS1), and the highly conserved paired nucleotides of D5.

The vast majority of group II intron mutations have been analyzed only under *in vitro* conditions. There are a few group II intron mutants of yeast mtDNA that were first obtained *in vivo* and were later studied *in vitro*.[4,8,42,44] For those mutations, there is no detectable spliced mRNA or excised intron RNA *in vivo* (at least a block of 100-fold), and the *in vitro* splicing block is less substantial (in high salt buffers). Only a few site-directed mutations have been analyzed *in vivo*[5,31,32]; doing so required transformation of plasmid DNA into yeast mtDNA using methods reviewed elsewhere in this volume (Butow et al.). So far there is a generally good correlation between *in vitro* and *in vivo* phenotypes. Both mutations that block self-splicing, even in high salt buffers, block splicing *in vivo*. Most mutations that can self-splice in low salt buffers splice well *in vivo*, and most mutations that splice poorly in high salt buffers do not splice detectably *in vivo*. There is no *a priori* reason that *in vivo* and *in vitro* phenotypes of all mutants must agree. Where a mutation affects some feature of the intron that directly influences the reaction center of the ribozyme, it makes sense that a splicing block *in vitro* should be matched with a similar defect *in vivo*. However, where a mutation blocks splicing *in vivo* because it influences an interaction with some protein that assists splicing *in vivo*, then it makes sense that there would be no strong splicing defect *in vitro*.

Reverse Splicing

Because both group I and group II introns self-splice by transesterification reactions, it was predicted that their splicing should be reversible. This was first demonstrated for group I introns[45] and was later observed for group IIB introns.[46,47] Figure 3 outlines reactions that have been used to study the reverse splicing of bI1. The basic idea is to react labeled spliced exon RNA with an excess of purified intron lariat and analyze the products on gels, looking for the transfer of counts from the fast migrating position of spliced exons to the much slower migrating position of intron lariat–second exon intermediate or linear pre-mRNA containing both exons. In the initial report, using 0.5 M $(NH_4)_2SO_4$ and 45°, those products were detected, but quite weakly. In a later study, it was found that carrying out reverse splicing reactions at 26°–30° yielded higher levels of the products

[44] C. Schmelzer, C. Schmidt, K. May, and R. J. Schweyen, *EMBO J.* **2,** 2047 (1983).
[45] S. A. Woodson and T. R. Cech, *Cell (Cambridge, Mass.)* **57,** 335 (1989).
[46] M. Morl and C. Schmelzer, *Cell (Cambridge, Mass.)* **60,** 629 (1990).
[47] S. Augustin, M. W. Mueller, and R. J. Schweyen, *Nature (London)* **343,** 383 (1990).

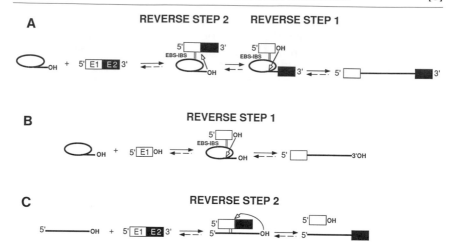

FIG. 3. Reverse splicing. Three assays for reverse splicing by group II introns are shown. (A) Reactants that can carry out the reversal of both splicing reactions; (B) assay for the reverse of step 1; (C) assay for the reverse of step 2.

of reverse splicing, presumably because the reverse splicing reaction has a lower temperature optimum than does forward splicing.[48] Reverse splicing of aI5γ was first detected by Harris-Kerr,[49] and relevant experiments have been published by us.[27] The reverse splicing reactions are detected in buffers containing ammonium sulfate and using 10–100 nM lariat intron RNA and 100 nM to 1 μM spliced exons RNA. The activation of the hydrolysis pathways by KCl or NH$_4$Cl will rapidly eliminate the reverse splicing products due to cleavage of the spliced exons by SER as well as internal cleavage of the intron.

Reversal of the first and second splicing reactions have been detected individually (see Fig. 3B, C). For example, linear intron RNA can catalyze the reversal of the second step of splicing when reacted with spliced exons RNA; however, because that intron RNA lacks the 2'–5' branch, reversal of the first splicing reaction is blocked. The reversal of the first splicing reaction has been assayed by mixing exon 1 and lariat intron RNA or lariat intron–exon 2 RNAs. We have characterized the stereochemical specificity of the reverse of the second step of splicing and found it to be the opposite of that of the forward second step, in agreement with an S$_N$2 reaction mechanism.

So far the reverse splicing of group IIA introns has not been reported *in vitro.* However, there is a growing number *in vivo* observations (group

[48] M. Morl and C. Schmelzer, *Nucleic Acids Res.* **18,** 6545 (1990).
[49] C. Harris-Kerr, Ph.D. Dissertation, University of Pittsburgh (1992).

A

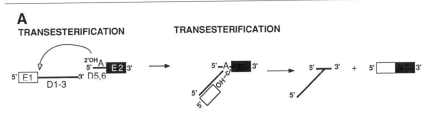

B

FIG. 4. Trans splicing. (A) Reactants and products of trans splicing of aI5γ RNA interrupted in domain 4.[39] (B) Reactants and products of a trans-splicing assay in which E1 is hydrolyzed from a model substrate containing just one nucleotide of the intron.[28]

II twintrons,[50] site-specific instabilities of fungal mtDNAs[51,52]) that appear to depend on reverse splicing reactions. There is considerable interest in those reactions, especially at ectopic sites, because they may illustrate ways in which group II introns may transpose to new sites. It was proposed some time ago that reverse splicing could have played a role in dispersing group I introns,[53] although, to date, there is no evidence comparable to that available for group II introns. It has been shown that a group IIB intron can catalyze the reverse of the second splicing reaction using a DNA strand as a substrate[54]; that reaction opens the possibility that introns might move directly into DNA, bypassing the proposed recombination step.

Reactions Catalyzed by Group II Intron Ribozymes *in Trans*

Some group II introns of plant organelle genomes trans-splice *in vivo*. One remarkable example is a group II intron of *Chlamydomonas* chloroplast DNA that appears to require three independent transcripts to assemble the active intron.[55] Jarrell *et al.* found that two inactive "half-molecules" of aI5γ made by interrupting the intron in domain 4 carry out all of the

[50] D. W. Copertino and R. B. Hallick, *Trends Biochem. Sci.* **18,** 467 (1993).

[51] M. W. Mueller, M. Allmaier, R. Eskes, and R. J. Schweyen, *Nature (London)* **366,** 174 (1993).

[52] C. H. Sellem, G. Lecellier, and L. Belcour, *Nature (London)* **366,** 176 (1993).

[53] T. R. Cech, *Int. Rev. Cytol.* **93,** 3 (1985).

[54] M. Morl, I. Niemer, and C. Schmelzer, *Cell (Cambridge, Mass.)* **70,** 803 (1992).

[55] R. A. Butow and P. S. Perlman, *Curr. Biol.* **1,** 331 (1991).

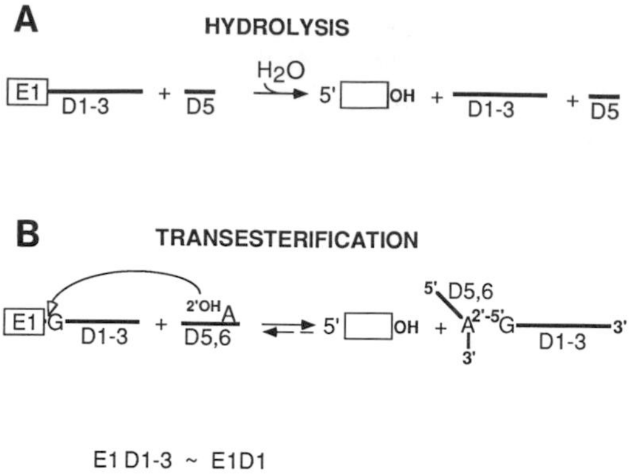

FIG. 5. Trans assays for step 1. (A) Reactants used in a standard trans assay for step 1 hydrolysis.[39] (B) Related reaction in which step 1 can be assayed by transesterification.

known reactions of the cis substrate when incubated together (see Fig. 4A).[39] The group IIA intron aI1 also trans-splices when interrupted in domain 4.[40] An important outcome of those early trans self-splicing experiments was the finding that the upstream half-molecule can be activated to carry out the hydrolytic release of the first exon by adding an RNA containing only the domain 5 portion of the downstream half-molecule (Fig. 5). That finding was the first example of an assay for the function of an individual intron domain. Since then, trans asays for several other intron segments have been devised, and a number of other bi- and even trimolecular reaction systems derived from the group II introns aI5γ and bI1 have been developed (see Figs. 5 and 6 for a summary). These trans assays allow kinetic analysis of individual reaction steps. Although reactions containing at least 0.5 M $(NH_4)_2SO_4$ are very suitable for studying cis reactions of aI5γ and bI1, that condition is not ideal for most trans reactions; instead, such experiments are usually carried out with 1 M NH_4Cl or KCl.

The multiple reaction pathways in which group II introns can engage, combined with an absolute requirement for only two domains, D1 and D5, for assembling the basic reaction center, determines a large number of combinations of intron fragments that can be studied in various trans reactions. Some of the reactions catalyzed by group II introns *in vitro* (DNA–RNA ligation, DNA cleavage) appear to have little relevance to the actual splicing reactions.[56] Nevertheless, they illustrate the catalytic proficiency of group II introns and open the possibility that several types of novel

[56] M. Morl, I. Niemer, and C. Schmelzer, *Cell (Cambridge, Mass.)* **70,** 803 (1992).

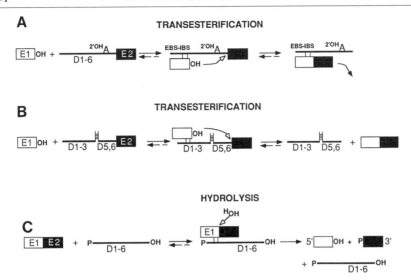

FIG. 6. Trans assays for step 2. (A, B) Trans assays for step 2 by transesterification. These same reactants can carry out step 2 by hydrolysis when incubated in KCl-containing buffer.[30,61] (C) Spliced exons reopening (SER) hydrolysis reaction, which appears to be related to step 2.[27,30]

reactions may occur at low levels *in vivo* and could influence the evolution of these introns.

A number of different reactions have been used to assay the first step of splicing (Fig. 5). In some cases, the RNAs are chosen so that only step 1 occurs; for example, by incubating a low concentration of labeled E1-D1-3 substrate with an excess of unlabeled D5 RNA in buffer containing 0.5–1.5 M NH_4Cl or KCl, hydrolysis of the 5′ splice junction occurs readily. To assay the first splicing reaction by transesterification, an unlabeled sample of the same substrate (≤ 1 nM) is reacted in buffer containing 1 M NH_4Cl with an excess of labeled D5,6 RNA, and the appearance of labeled Y-branched RNA is measured. Here branching depends on the salt used: if KCl is used instead of NH_4Cl, the step 1 reaction proceeds efficiently, but almost exclusively by hydrolysis. Finally, trans splicing in which both branching and second step splicing occur can be studied using E1-D1-3 and D5,6-E2 RNAs (Fig. 4A); in those cases, the course of the reaction can be studied by measuring the appearance of various products (or the disappearance of either or both precursors) depending on which RNAs are labeled (and how they are labeled.)

We have shown that these trans reactions at the 5′ splice site, whether by hydrolysis or transesterification, have the same stereochemical specificity as the first step transesterification or hydrolysis in cis splicing.[27] Two studies have analyzed this reaction enzymologically and have established that D5 binds with a relatively high affinity to D1 and that the reaction can occur with multiple turnover (with respect to D5).[57,58]

The first trans-splicing reaction ever reported for group II introns employed a substantially different E1-containing RNA, namely, E1 followed by just a small fragment of D1, as little as a single nucleotide.[28] In that system the second RNA was a transcript containing the entire intron and the second exon (see Fig. 4B). Those RNAs were reacted in low salt buffer and a slow "splicing" reaction was measured. Even though the cis-splicing substrate carries out step 1 exclusively by branching under that reaction condition, this trans reaction occurs without branching. Thus, step 1 hydrolysis can be promoted by features of the reacting RNAs and does not depend strictly on the presence of KCl. Altura et al. studied this reaction further and reported that exon 1 has to be connected to a D1 fragment of at least approximately 200 nucleotides (the exact number has yet to be determined) for these reactants to carry out the first step by transesterification.[59] Clearly, the original substrate, E1-G, lacks some feature needed for step 1 transesterification.

Several other studies have employed related E1 substrates containing a small fragment of D1 with similar results; in a variety of high salt buffers this RNA is only hydrolyzed.[60] We have studied the stereochemical selectivity of that reaction.[27] It is a mixed reaction, with some molecules apparently reacting by first step hydrolysis and others by a reaction related to the reverse of the second step. Because efficient second step splicing can occur following those mixed reactions leading to E1 release, it appears that two different pathways of splicing are possible. We have carried out an "exon swap" experiment in which the first step is E1 release by SER (the reverse of the second splicing reaction) and the second step is a normal (forward) second splicing reaction.[27] In that system, splicing appears to use just one of the two active sites of the group II intron (thus resembling group I introns which splice using a single active site).

Several trans assays for the second step of splicing have been developed (Fig. 6). Those reactions rely on the ability of free exon 1 to bind to D1 through the IBS–EBS base pairings. Incubation of labeled exon 1 with linear intron–exon 2 molecules in splicing buffer containing $(NH_4)_2SO_4$ results in second step exon ligation[30] when reacted in KCl, free 3' exon

[57] A. M. Pyle and J. B. Green, *Biochemistry* **33,** 2716 (1994).
[58] J. F. Franzen, M. Zhang, and C. L. Peebles, *Nucleic Acids Res.* **21,** 627 (1993).
[59] R. Altura, B. Rymond, B. Seraphin, and M. Rosbash, *Nucleic Acids Res.* **17,** 335 (1989).
[60] W. J. Michels, Jr., and A. M. Pyle, *Biochemistry* **34,** 2965 (1995).

RNA was the main product. In that study we reported that the intron–exon 2 RNA when incubated in NH₄Cl- or especially KCl-containing buffers without added exon 1 RNA actively cleaves itself at a site in the second exon following a sequence resembling IBS1; addition of exon 1 RNA appears to compete for that cryptic reaction leading to exon 2 release. In a variation of this assay, developed by Chanfreau and Jacquier, the active intron is formed by annealing D1-3 to D5,6E2 by means of GC clamps, and the reaction proceeds when exon 1 RNA is added in a reaction buffer containing 0.5 M (NH₄)₂SO₄.[61] This assay was used to analyze the roles of individual nucleotides and backbone phosphates of D5 and D6 in the second step.

The second step has also been assayed in reverse. Incubation of linear excised intron RNA with spliced exons RNA leads to the reverse of the second step resulting in exon 1 plus linear intron–second exon RNA as products (Fig. 3C). Because this reaction is reversible, it is not ideal for measuring the requirements for the second splicing step. A related reaction, that is irreversible and, so, is much more suitable for detailed analysis, is the spliced exons reopening (SER) reaction (Fig. 6C). We have determined the stereochemical requirements of the SER reaction and the reverse of the second step of splicing and found that both proceed with substrates containing an Rp-phosphorothioate at the splice junction.[27] If the SER reaction were homologous to the first step hydrolysis reaction then it would proceed with an Sp-phosphorothioate and would be blocked by Rp. We concluded that the SER reaction is not likely to be a model for the first reaction and is most likely the reverse of the second step, by hydrolysis.

Incubation of linear intron RNA—and here the exact ends of the RNA are not critical as they are in the reverse splicing assay—with spliced exons RNA in a buffer containing 100 mM Mg²⁺ plus 0.5–1.5 M KCl leads to the irreversible hydrolysis of the splice junction in the substrate RNA. This SER reaction is kinetically manageable and proceeds well in single turnover experiments with an excess of intron RNA. We have also analyzed the dependence of the reaction on domain 5 by carrying out a trimolecular reaction involving intron RNA lacking D5, D5 RNA, and substrate RNA.

Several other trans reactions have been reported. Deleting D3 from the intron appears to reduce the efficiency of SER that follows splicing in 0.5 M KCl; this defect is overcome by a higher salt concentration so that it is unlikely that D3 plays any direct or essential role in that reaction.[37] The addition of D3 RNA *in trans* can partly recover the reaction rate, this being a way to assay D3 function.[62] In this assay D3 RNA serves as an

[61] G. Chanfreau and A. Jacquier, *Science* **266,** 1383 (1994).
[62] M. Podar, S. D. Dib-Hajj, and P. S. Perlman, *RNA,* in press (1995).

activator with an apparent K_m of approximately 500 nM. Two postsplicing reactions that appear to be analogous to SER are known: in one reaction the intron RNA cleaves itself at a site in D2 that resembles IBS1, the so-called star cleavage reaction; and in another reaction IVS–E2 cleaves itself in E2. Also, Suchy and Schmelzer showed that deleting the C1 substructure of D1 of aI5γ blocks self-splicing and that splicing can be restored by adding a small RNA comprising C1 *in trans*.[63]

While trans reactions published to date involve RNAs from a given intron, it is obvious that these same approaches can be used for heterologous trans-splicing experiments. Some of these have been carried out, especially for D5 RNAs from various group II introns.[64] The general finding is that homologous trans reactions are invariably more active than are heterologous ones; however, none of these has yet been analyzed in sufficient detail to explain the apparent preference of a given intron ribozyme for its natural D5.

Acknowledgments

This research was supported by National Institutes of Health Grant GM31480 and Grant I-1211 from the Robert A. Welch Foundation. We thank Rick Padgett, Craig Peebles, and Jim Franzen for helpful comments.

[63] M. Suchy and C. Schmelzer, *J. Mol. Biol.* **222,** 179 (1991).
[64] S. C. Boulanger, C. L. Peebles, M. Zhang, J. Franzen, and P. S. Perlman, in preparation (1995).

[9] Genetic and Biochemical Approaches for Analysis of Mitochondrial RNase P from *Saccharomyces cerevisiae*

By Kathleen R. Groom, Yan Li Dang, Guo-Jian Gao,
Yan Chun Lou, Nancy C. Martin, Carol A. Wise,
and Michael J. Morales

Introduction

Yeast mitochondrial RNase P is the endonuclease that removes the 5' leader sequence from mitochondrial precursor transfer RNA (tRNA) molecules. Like the mitochondrial ribosome, cytochrome oxidase, cytochrome bc_1, and ATPase, it is a complex of dual genetic origin. Yeast mitochondrial RNase P is composed of both RNA and protein subunits coded by the mitochondrial and nuclear genomes, respectively.[1] In this methods review, we describe the *in vivo* and *in vitro* approaches we have taken to understand the nucleocytoplasmic interactions necessary for the biosynthesis of this dually derived complex of yeast mitochondria and to gain insight into RNase P structure and function.

Yeast Mitochondrial RNase P RNA

The gene coding for the RNA subunit was found by mapping the ability of petite deletion mutants to process tRNAs.[2] Biochemical and phylogenetic analysis of this gene in several yeasts demonstrated that it encodes an AU-rich RNA of varying size, yet all forms contain two short regions of striking sequence similarity to two regions highly conserved in all RNase P RNAs.[3] These regions participate in a long-range base pairing interaction to form a pseudoknot believed to compose part of the catalytic core of the enzyme.[4] This gene was the first yeast gene identified to affect mitochondrial RNase P activity and is named *RPM1* for RNase P mitochondrial.

In vitro studies of the RNA subunit of the enzyme have not been done, as conditions which allow Rpm1r to function alone *in vitro* have not been discovered.[5] Thus, in the absence of a successful protocol to reconstitute RNA and protein, the effect of RNA mutations *in vitro* cannot rigorously be assessed. Classic *in vivo* approaches to questions of structure and function would begin with *RPM1* mutants, but, despite repeated attempts, no mutants except those missing the gene have been isolated. An alternative approach for carrying out RNase P RNA structure–function studies has been presented by biolistic transformation of mitochondria and chloroplasts (reviewed elsewhere in this volume[6]). We reasoned that if an *RPM1* gene could be introduced by biolistic transformation into strains lacking mito-chondrial DNA and function, then DNA encoding mutant *RPM1* genes created *in vitro* could also be similarly introduced. Effects of such mutations on RNase P function could be determined by assessing the ability of the strain to process a reporter tRNA.

[1] M. J. Hollingsworth and N. C. Martin, *Mol. Cell. Biol.* **6,** 1058.

[2] K. Underbrink-Lyon, D. L. Miller, N. A. Ross, H. Fukuhara, and N. C. Martin, *Mol. Gen. Genet.* **191,** 512 (1983).

[3] C. A. Wise and N. C. Martin, *J. Biol. Chem.* **266,** 19154 (1991).

[4] E. S. Haas, D. P. Morse, J. W. Brown, F. J. Schmidt, and N. R. Pace, *Science* **254,** 853 (1991).

[5] C. A. Wise, E. E. R. Harris, M. J. Morales, H.-H. Shu, and N. C. Martin, *in* "Structure, Function and Biogenesis of Energy Transfer Systems" (E. Quagliariello, S. Papa, F. Palmieri, and C. Saccione, eds.), p. 215. Elsevier, Amsterdam, 1990.

[6] R. A. Butow, R. M. Henke, J. V. Moran, S. M. Belcher, and P. S. Perlman, this volume [24].

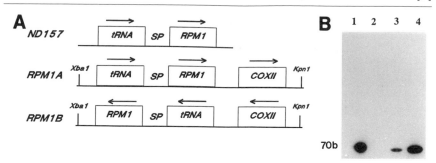

FIG. 1. (A) Mitochondrial DNA introduced into rho⁰ cells by biolistic transformation. The mitochondrial genome of the rho⁻ strain ND157 contains only *RPM1* and *tRNA^{Asp}* genes. Plasmids pMit *RPM1A* and *RPM1B* contain *Sau*3A fragments of ND157 cloned in either orientation into the *Bam*HI site of Bluescript M13 (−). *COXII* was introduced to provide a marker for mitochondrial transformation. (B) Northern analysis of whole-cell RNA from pMit *RPM1A* probed with a mitochondrial tRNA^{Asp}-specific probe. Lanes 1 and 4, *S. cerevisiae* wild-type ρ^+ control; lane 2, ρ^0 recipient; lane 3, ρ^0 cells transformed with pMit *RPM1A*. Identical results were obtained with pMit *RPM1B*.[7a]

Mitochondrial DNA from the rho⁻ strain ND157 (Fig. 1A) contains two genes, *RPM1* and a *tRNA^{Asp}* gene, repeated in tandem. A unique *Sau*3A site located at the 5′ end of the tRNA gene was used to linearize the petite genome and the resulting fragment cloned into the *Bam*HI site of Bluescript M13(−) (Stratagene, La Jolla, CA). The *COXII* gene encoding subunit 2 of cytochrome-*c* oxidase was subsequently introduced to provide a marker for mitochondrial transformation. The resulting plasmids pMit *RPM1A* and pMit *RPM1B* (Fig. 1A) were introduced into rho⁰ cells by biolistic transformation and mitochondrial transformants identified by their ability to complement the *coxII*-deficient tester strain TF145.[7] Northern analysis of whole-cell RNA extracted from positive transformants shows that mature tRNA is made (Fig. 1B). Like the original petite used to make these synthetic petite genomes, both mature Rpm1r and its precursors accumulate in these strains (data not shown). This technology has been applied to introduce two mutant *RPM1* genes created *in vitro*. One mutation eliminates, and the other greatly reduces, production of mature tRNA^{Asp}.[7a] Thus, this technology will allow a genetic approach to *in vivo* studies of structure–function and biogenesis of yeast mitochondrial RNase P RNA.

[7] T. D. Fox, J. C. Sanford, and T. W. McMullin, *Proc. Natl. Acad. Sci. U.S.A.* **85,** 7288 (1988).
[7a] P. Sulo, K. R. Groom, C. Wise, M. Steffen, and N. Martin, *Nucleic Acids Res.* **23,** 856 (1995).

Yeast Mitochondrial RNase P Protein Component(s)

In vitro, mitochondrial RNase P activity is protease as well as micrococcal nuclease sensitive,[1] with a buoyant density of 1.28 g/cm^3 in Cs$_2$SO$_4$[8] suggesting a ratio of protein to RNA higher than that found in the bacterial enzymes. As petite mutants are unable to make mitochondrial protein but many still have RNase P activity, any necessary proteins must be nuclear encoded. Initial attempts to identify genes coding for protein component(s) of yeast mitochondrial RNase P included the following: (1) screening existing collections of nuclear mutants for mutants deficient in 5' mitochondrial (mt) tRNA processing,[9] (2) attempting to use a yeast genomic library to complement an *Escherichia coli* mutant with a defective RNase P protein gene,[10] and (3) low stringency hybridization screening of a yeast genomic DNA library with the *E. coli* RNase P protein gene,[11] as described by Chen *et al.*[12] None of these approaches were fruitful, so a full-scale biochemical purification of the enzyme was undertaken.

Preparation of Precursor tRNA Substrate

A large-scale purification demanded a radiolabeled precursor tRNA substrate that could be synthesized in large amounts. Substrate was prepared from a plasmid containing the bacteriophage T7 promoter and a mitochondrial *tRNA$_f^{Met}$* gene modified to contain a *Bst*NI restriction enzyme site such that transcription of *Bst*NI-cut plasmid would make an RNA with a 3' CCA end.[8] Transcription reactions included 40 mM Tris-HCl, pH 7.5, 20 mM MgCl$_2$, 10 mM NaCl, 0.4 mM ATP, 0.4 mM CTP, 0.4 mM GTP, 0.4 mM UTP, plus 100 μCi of 800 Ci/mM [α-^{32}P] UTP or [α-^{32}P] ATP. Unincorporated nucleotides were removed with a Sephadex G-25 quick spin column (Boehringer Mannheim, Indianapolis, IN) and the RNA used directly. A schematic diagram of the plasmid used and the structure of the precursor is shown in Fig. 2.

Isolation of Mitochondria

Although the nuclear and mitochondrial RNase P enzymes of yeast are distinct, they do recognize and cleave similar substrate precursor tRNAs,

---

[8] M. J. Morales, C. A. Wise, M. J. Hollingsworth, and N. C. Martin, *Nucleic Acids Res.* **17,** 6865 (1991).

[9] A. Tzagoloff and C. L. Dieckmann, *Microbiol. Rev.* **54,** 211 (1990).

[10] P. Schedl and P. Primakoff, *Proc. Natl. Acad. Sci. U.S.A.* **70,** 2091 (1973).

[11] F. G. Hansen, E. B. Hansen, and T. Atlung, *Gene* **38,** 85 (1985).

[12] J.-Y. Chen, M. J. Hollingsworth, M. J. Morales, C. A. Wise, H. H. Shu, B. J. Clark, D. L. Caudle, H. P. Zassenhaus, and N. C. Martin, *NATO ASI Ser., Ser. H* **14,** p. 223 (1988).

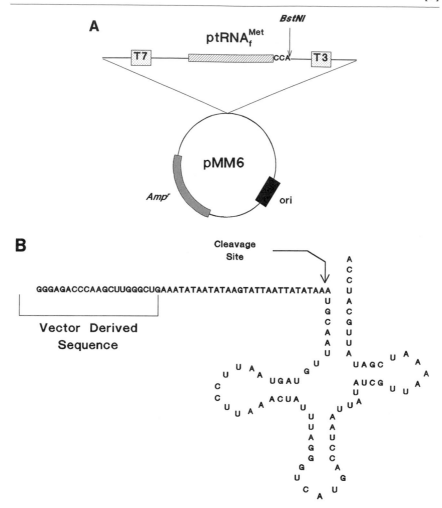

Fig. 2. (A) Plasmid pMM6 constructed for *in vitro* synthesis of precursor tRNA$_f^{Met}$ substrate. (B) Structure of the tRNA$_f^{Met}$ precursor. Arrow indicates the site of cleavage by RNase P.

including the precursor tRNA$_f^{Met}$. To assure purification of the mitochondrial enzyme, RNA extracted from isolated mitochondria was hybridized to probes specific to the nuclear[13] and mitochondrial RNase P RNAs and the relative radioactivity detected on Northern blots determined. Using

[13] J. Lee and D. Engelke, *Mol. Cell. Biol.* **9,** 2536 (1989).

the mitochondrial isolation protocol described below, nuclear RNase P RNA contamination was estimated at less than 5%.

Choice of strain was also important to the success of the purification. Nucleases in the extract made it difficult to assay RNase P activity because the substrate and products were degraded. Enzyme recovery in the initial experiments utilizing MH41-7B, the strain used for the earlier genetic studies, was disappointingly low, and the RNase P RNA was fragmented during purification. Although this fragmentation, interestingly enough, does not affect enzyme activity *in vitro,* its consequences on enzyme stability are not known. We reasoned that use of a nuclease-deficient strain might improve yield. One nuclease in *S. cerevisiae* mitochondria is the product of the nonessential *NUC1* gene.[14] Comparative RNase P enzyme assays using mitochondrial extracts prepared from *NUC1* and *nuc1* strains revealed no significant difference in stability of the RNase P RNA nor of RNase P activity. Thus, the product of the *NUC1* gene was not the nuclease responsible for the RNase P RNA fragmentation nor the nuclease that interfered with assays in crude extracts. We did, however, find use of the protease-deficient strain, CMY44, to result in greater enzyme recovery. CMY44 has mutations in the genes encoding proteinase B, carboxypeptidase Y, and PEP4p, a protein involved in general vacuolar function. Although these protease deficiencies may account, in part, for the improved yield, the fact that *pep4* mutants contain only 10% of the general ribonuclease of a wild-type cell[15] was perhaps also crucial. Although fragmentation of the RNase P RNA still occurred during purification, assaying for activity was easier because precursor substrate and product(s) were more stable during the assay.

The purification (summarized in Table I) begins with 42 liters of CMY44 cells grown in YPGal medium (1% (w/v) yeast extract, 1% (w/v) Bacto-peptone, 2% (w/v) galactose, and 0.2% (w/v) glucose) in batches of 3500 ml medium per 6-liter Erlenmeyer flask. Cells are grown at 30° with shaking at 200 rpm until reaching late log phase (OD_{640} of 3.5–5.0 or 2.5–3.5 × 10^8 cells/ml). The cells are harvested with a Millipore (Bedford, MA) Pellicon ultrafiltration apparatus (Model HPK40) fitted with an HVLP 0.45-μm cassette, washed once with 20 liters of water, and pelleted by centrifugation at 4400 g for 5 min at 4°. If mitochondrial isolation does not commence immediately, cells are then stored on ice at 4° overnight.

The 450–500 g (wet weight) of cells are converted to spheroplasts after

[14] E. Dake, T. J. Hofmann, S. McIntire, A. Hudson, and H. P. Zassenhaus, *J. Biol. Chem.* **263,** 7691 (1988).
[15] B. A. Hemmings, G. S. Zubenko, A. Hasilik, and E. W. Jones, *Proc. Natl. Acad. Sci. U.S.A.* **78,** 435 (1981).

TABLE I
PURIFICATION OF RNASE P

Sample	Protein (mg)	Activity (nmol/min)	Specific activity (nmol/min/mg)	Yield (%)	Purification (-fold)
S30	3033	173	0.057	100	1
DEAE	345	73.8	0.214	42.7	3.7
Heparin	37.4	30.5	0.816	17.6	14.3
Mono Q	10.4	20.6	1.98	11.9	34.7
Mono S	1.6	7.6	4.75	4.4	83.3
Glycerol	0.025	2.7	108	1.6	1894.7

suspension at 1 g cells/4 ml of 0.1 M Tris-HCl, pH 9.0, 2.5 mM dithiothreitol (DTT) and incubation at 30° for 20 min with gentle shaking. Following incubation, cells are washed once with distilled water and once with spheroplasting buffer (1.35 M sorbitol, 0.1 M EDTA). Cell wall digestion is accomplished by suspending cells in spheroplasting buffer containing 1.25 mg/ml yeast lytic enzyme (from *Arthrobacter luteus,* 70,000 units/mg; ICN, Costa Mesa, CA) at a ratio of 1 g cells/2 ml enzyme solution and incubating at 35° with gentle shaking. Digestion is monitored by measuring the decrease in optical density at 640 nm on osmotic lysis in water and is considered complete when values reach one-tenth of initial values (~2–3 hr).

Spheroplasts are pelleted by centrifugation at 6400 g for 5 min at 4° and suspended in 2 ml breakage buffer [0.6 M mannitol, 10 mM PIPES–NaOH, pH 6.7, 1.0 mM EDTA, and 0.3% (w/v) bovine serum albumin (BSA)] per gram cells. Spheroplasts are lysed by explosive decompression by two passages through a nitrogen bomb (Parr Instruments, Moline, IL) for 10 min at 300 psi and 4°. Lysed cells are diluted 2- to 3-fold with breakage buffer and large debris removed by centrifugation at 4400 g for 5 min. Mitochondria are recovered from this supernatant by centrifugation at 22,000 g for 12 min. The mitochondrial pellet is then suspended in 4 ml per gram cells of wash buffer (0.6 M mannitol, 1 mM PIPES–NaOH, pH 6.7, and 0.01% BSA) using a Dounce homogenizer (Wheaton Instruments, Millville, NJ), fitted with a loose-fitting pestle (size B). Large debris is removed from the suspension by centrifugation at 4400 g for 5 min, followed by centrifugation of the supernatant at 22,000 g for 12 min to pellet mitochondria. The pellet is suspended in 4 ml per gram cells of wash buffer and the mitochondria recovered by centrifugation at 22,000 g for 12 min at 4°. This washing procedure is repeated twice. Best yields are obtained if mitochondria are used immediately. If not used immediately, they are suspended in 0.1 ml wash buffer per gram original cells, frozen in liquid nitrogen, and stored at −70°.

RNase P Activity Assay

Enzyme is followed by assaying RNase P activity. Fractions are incubated in 50 mM Tris-HCl, pH 8.0, 10 mM MgCl$_2$, 30 mM NaCl, 200 μg/ml RNase-free BSA (Boehringer Mannheim), 0.25% (v/v) Tween 20 (Bio-Rad, Richmond, CA), with 4–40 nM *in vitro* transcribed, ^{32}P-labeled, precursor tRNA$_f^{Met}$ [10^4 to 10^6 disintegrations per minute (dpm) per assay] at a final monovalent cation concentration of less than 150 mM. If necessary, enzyme is diluted in 400 mM KCl, 25 mM HEPES–KOH, pH 7.9, 5 mM MgCl$_2$, 0.2 mM EDTA, 0.2 mM DTT, 10% (v/v) glycerol, 1% (v/v) Tween 20, and 200 μg/ml BSA prior to use. Incubation is for 5–20 min at 37°. Then 0.3 volume of 6 M urea, 50% (w/v) sucrose, 10 mM aurintricarboxylic acid, 0.01% (w/v) xylene cyanol, and 0.01% (w/v) bromphenol blue is added and the reaction products visualized by autoradiography after separation by electrophoresis on a 10% (w/v) acrylamide/6 M urea gel. For quantitation, the autoradiograph is used as a template for excision of radioactive RNAs, and the counts in precursor and product are determined by scintillation counting. One unit of enzyme is defined as the amount necessary to process 1 nmol precursor/min.

RNase P Enzyme Purification

The purification scheme is as originally published by Morales *et al.*[16] Freshly isolated or thawed frozen mitochondria are pretreated by suspension at a ratio of 1 g/ml of 1.0 M (NH$_4$)$_2$SO$_4$, 15 mM KCl, 25 mM HEPES–KOH, pH 7.9, 10 mM MgCl$_2$, 1.0 mM EDTA, 0.2 mM phenylmethylsulfonyl fluoride (PMSF), 0.2 mM benzamidine, 2.0 mM DTT, 1 μM leupeptin, and 1 μM pepstatin A. The mixture is stirred on ice for 30 min and the suspension subjected to centrifugation at 30,000 g for 30 min at 4°. The supernatant is discarded and the pellet suspended at a ratio of 1 g/5 ml of 1.0 M KCl, 25 mM HEPES–KOH, pH 7.9, 5 mM MgCl$_2$, 1.0 mM EDTA, 0.2 mM PMSF, 0.2 mM benzamidine, 2.0 mM DTT, 1 μM leupeptin, 1 μM pepstatin A, 10% (v/v) glycerol, and 1.2% (v/v) Tween 20. This suspension is stirred vigorously on ice for 30 min and subjected to centrifugation at 30,000 g for 30 min at 4°. The supernatant (S30), containing most of the mitochondrial RNase P, is dialyzed against 4 liters of 25 mM HEPES–KOH, pH 7.9, 5 mM MgCl$_2$, 0.2 mM EDTA, 0.2 mM PMSF, 0.2 mM benzamidine, 0.2 mM DTT, 10% (v/v) glycerol, and 1.0% (v/v) Tween 20 (buffer A). The buffer is changed every 45 min until the conductivity is below that of 0.15 M KCl in buffer A. The dialyzed S30 is applied at a rate of 7 ml/min to a 22 × 5 cm (450 ml) DE-52 cellulose column (Whatman, Clifton, NJ) previously

[16] M. J. Morales, Y. L. Dang, Y. C. Lou, P. Sulo, and N. C. Martin, *Proc. Natl. Acad. Sci. U.S.A.* **89,** 9875 (1992).

equilibrated in 0.10 M KCl plus buffer A. Once loaded, the column is washed with 500 ml of 0.10 M KCl plus buffer A at 7 ml/min. A further wash of 5400 ml of 0.15 M KCl plus buffer A at 8 ml/min is followed by elution of the enzyme by a step wash of 0.5 M KCl plus Buffer A at 3 ml/min (Fig. 3A).

Peak A_{280} absorbing fractions are pooled and diluted 1 to 2 with buffer A. This dilute DEAE fraction is then applied at a rate of 2.5 ml/min to an 18 × 2.5 cm (90 ml) heparin-Ultrogel (IBF, Columbia, MD) column equilibrated with 0.20 M KCl plus buffer A. The column is washed with 270 ml of 0.20 M KCl plus buffer A at 0.25 ml/min, followed by elution of the enzyme with a 900 ml linear gradient from 0.20 M KCl to 1.25 M KCl plus buffer A at a flow rate of 1.5 ml/min. RNase P elutes at about 0.75 M KCl (Fig. 3B).

The heparin pool is dialyzed against 4 liters of buffer A until the conductivity is less than that of 0.2 M KCl plus buffer A (~2 hr). The dialyzed pool is applied to an FPLC (fast protein liquid chromatography, Pharmacia, Piscataway, NJ) Mono Q HR 10/10 (1 × 10 cm, 8 ml) column equilibrated in 0.15 M KCl plus buffer A at 2.0 ml/min. The column is washed with 40 ml of 0.20 M KCl plus buffer A, and the enzyme is eluted with a 250 ml linear gradient from 0.20 M KCl to 0.75 M KCl plus buffer A at 0.25 ml/min. The enzyme elutes at approximately 0.35 M KCl (Fig. 3C).

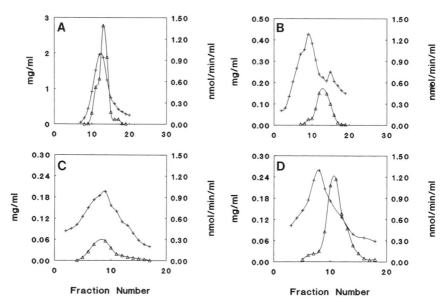

FIG. 3. Purification of RNase P. (A) DEAE-cellulose; (B) heparin-Ultrogel; (C) FPLC Mono Q; (D) FPLC Mono S. +, concentration of protein in mg/ml; Δ, RNase P activity expressed as nanomoles precursor processed per minute.

Active fractions from the Mono Q column are pooled and dialyzed against 2 liters buffer A for approximately 1 hr, until the conductivity is less than that of 0.2 M KCl plus buffer A. After reducing the conductivity of the dialyzate to that of 0.1 M KCl plus buffer A by dilution with buffer A, the sample is applied at 0.75 ml/min to an FPLC Mono S HR 5/5 (0.5 × 5 cm, 1.0 ml) column equilibrated in 0.1 M KCl plus buffer A. The column is washed with 5 ml of 0.1 M KCl plus buffer A and the enzyme eluted with a 20 ml linear gradient from 0.1 to 0.6 M KCl at a flow rate of 0.5 ml/min. RNase P elutes at 0.3 M KCl (Fig. 3D).

Glycerol gradients are formed by layering (from the bottom) 8, 9, 9, and 8 ml of 0.5 M KCl plus buffer A containing 35, 28, 22, and 15% (v/v) glycerol, respectively, in 1 × 3.5 inch Ultra-Clear (Beckman, Palo Alto, CA) tubes and left at 4° for 12–16 hr. The 5-ml pool from the Mono S column is concentrated in a Centricon 30 spin concentrator at 4300 g for 30 min and the concentrate layered in 0.9-ml aliquots onto 3 tubes containing the preformed glycerol gradients. Centrifugation is at 141,000 g for 30 hr in a Beckman SW28 rotor. Gradients are fractionated, from the bottom up, on an ISCO 185 gradient fractionator (ISCO, Inc., Lincoln, NE) at a flow rate of 3 ml/min using a Fluorinet-40 chase. Absorbance is monitored with an ISCO UA-5 absorbance detector and type 10 optical unit (Fig. 4).

Each column addition increases RNase P purification severalfold, with the most effective purification step being the final glycerol gradient fractionation (Table I). Figure 5 shows sodium dodecyl sulfate–polyacrylamide gel electrophoresis (SDS–PAGE) analysis of pooled fractions obtained from each purification step. A protein with a molecular mass of approximately 105 kDa was the most abundant protein in the glycerol gradient pool.

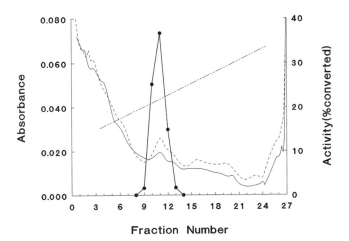

FIG. 4. Glycerol gradient fractionation of FPLC Mono S column fractions. ----, A_{254}; ——, A_{280}; —●—, activity; -----, glycerol.

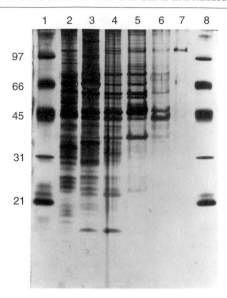

FIG. 5. Analysis by SDS-PAGE of column pools and glycerol gradient fractions. Lanes 1 and 8, molecular weight ($\times 10^{-3}$) standards; lane 2, 4.3 μg protein from mitochondrial extract; lane 3, 4.3 μg protein from DEAE column pool; lane 4, 2.2 μg from the heparin-Ultrogel column pool; lane 5, 2.2 μg protein from FPLC Mono Q column; lane 6, 0.8 μg protein from FPLC Mono S column; lane 7, protein from glycerol gradient fraction, amount unknown.

Proteins from active fractions of the glycerol gradient are precipitated by $CH_3OH/CHCl_3$, separated by Tris–Tricine–SDS–PAGE,[17] and transferred to an Immobilon polyvinylidene difluoride (PVDF) membrane (Millipore)[18] by electrophoresis at 90 V for 1 hr at 4°. Protein sequence information obtained from this membrane is used to design oligonucleotides and obtain the gene coding for the protein from a yeast genomic library.[19] Because it is the second gene required for RNase P activity (see below) it was named *RPM2*.

RPM2 Encodes a Protein Subunit of Yeast Mitochondrial RNase P

Biochemical evidence that *RPM2* codes for a protein subunit of yeast mitochondrial RNase P has been obtained by raising an antibody to Rpm2p and performing immunoprecipitation experiments.[19] DNA encoding amino acids 145–460 is cloned into a pATH expression vector and a TrpE-Rpm2p

[17] H. Schagger and G. von Jagow, *Anal. Biochem.* **166,** 368 (1987).
[18] S. W. Yuen, A. H. Chui, K. J. Wilson, and P. M. Yuan, *BioTechniques* **7,** 74 (1987).
[19] Y. L. Dang and N. C. Martin, *J. Biol. Chem.* **268,** 19791 (1993).

fusion protein produced for immunization of a rabbit. Immunoglobulin G (IgG) is separated from other serum proteins, including ribonucleases, by two passages through a protein G-Sepharose 4 column (MAbTrap G, Pharmacia LKB Biotechnology Inc.). Western blot analysis of mitochondrial proteins isolated from wild-type and *RPM2*-disrupted cells demonstrated a single protein of 105 kDa present only in wild-type cells. This antibody does not inhibit the enzyme activity, but an immunodepletion experiment demonstrated that it does immunoprecipitate RNase P activity. One hundred fifty micrograms of immune and preimmune IgG are bound to 50 μl protein A-agarose beads by incubation with rotation at 4° for 2 hr. Agarose particles with bound IgG are washed three times with 50 m*M* sodium phosphate, pH 7.0. Then 10 μl beads are mixed with 10 μl partially purified enzyme obtained from an active DEAE fraction and incubated with rotation for 2 hr at 4°. Following a 1-min centrifugation, pellet and supernatant are assayed for RNase P activity as previously described. Figure 6A demonstrates that RNase P activity remains in the supernatant of samples treated with preimmune sera but is quantitatively removed from the supernatant treated with immune IgG-protein A beads. Some, but not all, of the activity is recovered in the pellet of this immune precipitate, presumably because of difficulty in measuring activity in immunoprecipitated material.

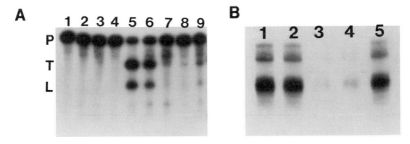

FIG. 6. (A) RNase P activity is precipitated by anti-Rpm2p immune but not preimmune serum. Substrate pre-tRNA$_f^{Met}$ was incubated at 37° for 5 min with buffer only (lane 1), protein A-agarose (lane 2), preimmune IgG (lane 3), immune IgG (lane 4), 1 μl DEAE fraction (lane 5), 2 μl supernatant with preimmune IgG (lane 6), 4 μl resuspended pellet with preimmune IgG (lane 7), 2 μl supernatant with immune IgG (lane 8), and 4 μl resuspended pellet with immune IgG (lane 9). P marks the precursor, T the tRNA product, and L the 5′ leader. (B) RNase P RNA is precipitated by immune but not preimmune serum. RNA extracted from supernatants and pellets of immunoprecipitates was separated on a 10% polyacrylamide gel, transferred to a Zeta-probe membrane, and probed with a RNase P RNA-specific probe. Lane 1, total RNA isolated from 50 μl starting material; lane 2, supernatant after immunodepletion with preimmune serum; lane 3, pellet after precipitation with preimmune IgG; lane 4, supernatant after immunodepletion with immune IgG; lane 5, pellet after precipitation with immune IgG. Reproduced from Ref. 17 with permission from The American Society for Biochemistry & Molecular Biology.

If the *RPM2* gene does indeed code for a protein subunit of yeast mitochondrial RNase P, then immunoprecipitates containing RNase P activity should also contain the RNase P RNA. Northern analysis of RNA isolated from the supernatants and pellets of the immunodepletion experiment described above and hybridized to an Rpm1r-specific probe clearly demonstrated a quantitative precipitation of Rpm1r by the anti-Rpm2p antibody (Fig. 6B). This result demonstrates that the 105-kDa protein and Rpm1r are together in a complex.

Genetic evidence that *RPM2* is required for RNase P activity is obtained by molecular genetic techniques. *RPM2* is disrupted by inserting a *Hpa*I fragment containing the *LEU2* gene into a unique *Hpa*I site located approximately two-thirds of the way into the coding region. The fragment containing the disrupted gene is isolated and introduced into haploid W3031A yeast cells by electroporation, and the transformants are selected by growth on Leu⁻ glucose medium. Strains with an *RPM2* disruption are respiratory deficient and do not retain a wild-type mitochondrial genome but instead become petite deletion mutants.[19]

To determine the effect of the disruption on mitochondrial tRNA processing, deletion mutants with this nuclear background and with the mitochondrial *RPM1* gene are selected. As strains without *RPM1* accumulate 5′ extended tRNA precursors, this gene needs to be present to assess the effect of the disruption on the 5′ maturation of mitochondrial tRNAs. Northern analysis of tRNA isolated from the disrupted strain retaining the mitochondrial *RPM1* and *tRNA_f^Met* genes demonstrated an accumulation of mitochondrial precursor tRNAs identical to that seen in a petite deletion mutant that does not contain *RPM1*.[19] The techniques of *in vitro* mutagenesis coupled with the ability to replace the endogenous genes in yeast will now allow studies of the effects of *RPM2* mutations on structure, function, and biogenesis of yeast mitochondrial RNase P.

Conclusion

Many questions about this enzyme and its biosynthesis remain despite the identification and current characterization of Rpm1r and Rpm2p. First, none of our experiments eliminate the possible presence of a nuclear encoded RNA in this enzyme. Second, although Rpm2p clearly fills all the criteria expected of a protein subunit of the enzyme and is the most abundant protein in near homogeneous preparations, it is possible that other proteins play either essential or stimulatory roles. The yield of activity from the purification reported here is very low, and proteins that facilitate activity but are not actually required could have been separated from Rpm1r and Rpm2p. Successful production of sufficient Rpm2p for *in vitro* reconstitution studies should enable us to determine if Rpm1r and Rpm2p are the only components necessary to support full activity *in vitro*. If they are, studies on the biochemistry and biogenesis of a well-defined yeast mitochondrial RNase P can be initiated. If not, genetic approaches to identify other proteins or RNAs should be possible by starting with the two genes that are now available. Immunoprecipitation of complexes or cross-linking between the known RNA and proteins offer other approaches to identify additional components that could be required for *in vitro* reconstitution or be associated with the enzyme *in vivo*.

Acknowledgment

This work was supported by National Institutes of Health Grant GM 27597.

[10] RNA Editing in Trypanosomatid Mitochondria

By LARRY SIMPSON, GEORGES C. FRECH, and DMITRI A. MASLOV

Introduction

The kinetoplastid protozoa, together with the euglenoids, represent one of the earliest branches of eukaryotes containing mitochondria.[1] There are two major subgroups within the Kinetoplastida: the bodonids–cryptobiids and the trypanosomatids. The former organisms are less investigated than the latter and are not discussed in this chapter. Members of the trypanosomatid genera *Trypanosoma* and *Leishmania* are the causal agents of several important diseases in humans and animals, including visceral and dermal leishmaniasis, South American Chagas disease, and African sleeping sickness. These "digenetic" species have a biphasic life cycle that involves both a vertebrate host and an invertebrate vector. Species which belong to the monogenetic genera *Crithidia*, *Blastocrithidia*, *Leptomonas*, and *Herpetomonas* parasitize only invertebrates. Most species can be grown axenically, and, in some cases, both stages of the life cycle can be maintained.[2]

The kinetoplastids have a single mitochondrion which contains a large mass of mitochondrial DNA situated in the region adjacent to the basal body of the flagellum. The region with the DNA is called the kinetoplast, and the DNA is called kinetoplast DNA or kDNA. The kDNA is organized into a giant network of approximately 10^4 catenated minicircles, 0.4 to 2.5

---

[1] L. Simpson and D. A. Maslov, *Curr. Opin. Genet. Dev.* **4,** 887 (1994).
[2] L. Simpson, *Int. Rev. Cytol.* **32,** 139 (1972).

kb in size, and 20–50 catenated maxicircles, 20 to 36 kb in size.[3] The maxicircle is the equivalent of the mitochondrial genome in other eukaryotic organisms, and encodes ribosomal RNAs as well as several components of the electron transport chain. The minicircles encode small 3′-oligouridylated transcripts known as guide RNAs (gRNAs),[4] which mediate a post-transcriptional process of RNA modification known as RNA editing.[5–12] The trypanosomatid type of RNA editing involves the insertion and, less frequently, deletion of uridylate (U) residues and affects coding regions of more than half of the maxicircle mRNA transcripts. Those genes, the transcripts of which are edited, are known as cryptogenes.[6] Depending on the cryptogene, editing creates initiation and/or termination codons, corrects frameshifts, and, in the case of pan-edited mRNAs, creates an entire reading frame leading in some cases to a doubling in size of the edited transcript. Editing proceeds in a 3′ to 5′ direction within an editing domain, and in most cases it terminates a few nucleotides upstream of the created or encoded translation initiation codon. It is assumed that only the mature fully edited mRNA is translationally competent; however, a direct proof that fully edited mRNAs are translated is still lacking.

The pattern of inserted and deleted U residues in the mature edited RNA is determined precisely by base-pairing interactions with gRNAs. Genes for gRNAs are present in both the maxicircle and the minicircle DNA molecules.[4,13,14] The gRNA contains at its 5′ end an anchor sequence that is complementary to the cognate mRNA just downstream of the first editing site. In pan-edited transcripts, multiple gRNAs mediate editing in a sequential manner: each downstream gRNA creates the anchor sequence for the adjacent upstream gRNA, thereby determining the overall 3′ to 5′ polarity of the editing cascade. The central portion of each gRNA contains

[3] L. Simpson, *Annu. Rev. Microbiol.* **41,** 363 (1987).

[4] B. Blum, N. Bakalara, and L. Simpson, *Cell (Cambridge, Mass.)* **60,** 189 (1990).

[5] R. Benne, J. Van den Burg, J. Brakenhoff, P. Sloof, J. Van Boom, and M. Tromp, *Cell (Cambridge, Mass.)* **46,** 819 (1986).

[6] L. Simpson and J. Shaw, *Cell (Cambridge, Mass.)* **57,** 355 (1989).

[7] K. Stuart, *Annu. Rev. Microbiol.* **45,** 327 (1991).

[8] L. Simpson, D. A. Maslov, and B. Blum, *in* "RNA Editing—The Alteration of Protein Coding Sequences of RNA" (R. Benne, ed.), p. 53. Ellis Horwood, New York, 1993.

[9] S. L. Hajduk, M. E. Harris, and V. W. Pollard, *FASEB J.* **7,** 54 (1993).

[10] K. Stuart, *in* "RNA Editing—The Alteration of Protein Coding Sequences of RNA" (R. Benne, ed.), p. 25. Ellis Horwood, New York, 1993.

[11] B. K. Adler and S. L. Hajduk, *Curr. Opin. Genet. Dev.* **4,** 316 (1994).

[12] R. Benne, *Eur. J. Biochem.* **221,** 9 (1994).

[13] N. R. Sturm and L. Simpson, *Cell (Cambridge, Mass.)* **61,** 879 (1990).

[14] V. W. Pollard, S. P. Rohrer, E. F. Michelotti, K. Hancock, and S. L. Hajduk, *Cell (Cambridge, Mass.)* **63,** 783 (1990).

the guiding nucleotides which, via base-pairing interactions with mRNA (including G/U base pairs), direct the insertions and deletions of U residues. At their 3′ ends, gRNAs contain heterogeneously sized oligo(U) tails that are added posttranscriptionally by a mitochondrial terminal uridylyltransferase (TUTase).[15,16]

Several models have been proposed to explain the mechanism of RNA editing. In the enzyme cascade model,[4] an endonuclease, a TUTase, and an RNA ligase sequentially interact during each editing cycle and the inserted U residues originate either from UTP or from the 3′ end of the gRNA.[17] In the transesterification model,[18,19] each editing cycle proceeds via two successive transesterification reactions, resulting in the transfer of a U from the 3′ end of the gRNA to the editing site, or, in the case of a deletion, from the editing site to the gRNA. Chimeric gRNA/mRNA molecules, which are predicted intermediates of this type of mechanism, have been visualized,[19] but it has not been established that these molecules represent true intermediates of the editing process.

Partially edited RNAs constitute a variable portion of the steady-state RNA. Frequently the junction region between the mature edited 3′ sequence and the preedited 5′ sequence exhibits unexpected editing patterns. These sequences have been attributed either to misediting caused by the mediation of noncognate gRNAs or to normal intermediates of the editing process.[20–23]

RNA editing has been studied extensively in three species to date: *Trypanosoma brucei*, *Leishmania tarentolae*, and *Crithidia fasciculata*. The authors have chosen to use *L. tarentolae*, which is a parasite of the gecko, as a model system to study the mechanism of RNA editing for the following reasons: (1) the cells are not pathogenic for humans; (2) the cells grow rapidly (division time 6–9 hr) to a high density (4×10^8 cells/ml) in an inexpensive medium; (3) cell fractionation protocols are well developed, and a large-scale isolation of the intact kinetoplast–mitochondria can be easily achieved; and (4) the informational portion of the mitochondrial genome is completely sequenced and virtually all gRNAs are identified. The major disadvantage of *L. tarentolae* as a model organism is that the

[15] B. Blum and L. Simpson, *Cell (Cambridge, Mass.)* **62,** 391 (1990).
[16] N. Bakalara, A. M. Simpson, and L. Simpson, *J. Biol. Chem.* **264,** 18679 (1989).
[17] B. Sollner-Webb, *Curr. Opin. Cell Biol.* **3,** 1056 (1991).
[18] T. R. Cech, *Cell (Cambridge, Mass.)* **64,** 667 (1991).
[19] B. Blum and L. Simpson, *Proc. Natl. Acad. Sci. U.S.A.* **89,** 11944 (1992).
[20] N. R. Sturm, D. A. Maslov, B. Blum, and L. Simpson, *Cell (Cambridge, Mass.)* **70,** 469 (1992).
[21] C. J. Decker and B. Sollner-Webb, *Cell (Cambridge, Mass.)* **61,** 1001 (1990).
[22] L. K. Read, R. A. Corell, and K. Stuart, *Nucleic Acids Res.* **20,** 2341 (1992).
[23] D. J. Koslowsky, G. J. Bhat, L. K. Read, and K. Stuart, *Cell (Cambridge, Mass.)* **67,** 537 (1991).

biology of the parasite within the natural host is poorly known, and an *in vitro* system which simulates the natural life cycle differentiation of the parasite is lacking.

This chapter describes a series of methods relevant to the study of RNA editing in *L. tarentolae*. However, with minor modifications, most procedures may be applied to the study of RNA editing in other trypanosomatid species.

Cell Culture

Leishmania tarentolae cells are grown in brain–heart infusion (BHI) medium (Difco Laboratories, Detroit, MI) supplemented with 10 μg/ml hemin. T-flasks may be used for small-scale cultures (3–10 ml), and 3.8-liter bottles, using a roller bottle culture apparatus maintained at approximately 16 rpm, can be used to grow cultures up to 1 liter. To avoid cultivation-induced changes, new cultures should be started from a frozen stock every few weeks. To freeze cells, aliquots of log-phase cultures (5×10^7 to 10^8 cells/ml) are transferred into sterile freezer vials, and an equal volume of 20% glycerol in BHI is added. The cells are slowly frozen overnight at $-80°$ and then transferred to liquid nitrogen for long-term storage.

Isolation of Kinetoplast DNA

The kDNA network has a sedimentation coefficient of approximately 4000 S and is relatively resistant to shear forces owing to its compact shape.[24,25] Network DNA can be isolated from a sheared total cell lysate by sedimentation through CsCl.[24,25] The maxicircle DNA, which represents approximately 5% of the kDNA and is also catenated into the network, can be isolated on the basis of its relatively higher AT content (84% A+T versus 55% A+T for minicircle DNA) after release from the network by digestion with a restriction enzyme that cuts only once or infrequently.[26] The complete sequence of the 23 kb maxicircle of *T. brucei* is known[27,28]

[24] L. Simpson and J. Berliner, *J. Protozool.* **21**, 382 (1974).
[25] L. Simpson and A. Simpson, *J. Protozool.* **21**, 774 (1974).
[26] L. Simpson, *Proc. Natl. Acad. Sci. U.S.A.* **76**, 1585 (1979).
[27] P. Sloof, A. De Haan, W. Eier, M. Van Iersel, E. Boel, H. Van Steeg, and R. Benne, *Mol. Biochem. Parasitol.* **56**, 289 (1992).
[28] P. J. Myler, D. Glick, J. E. Feagin, T. H. Morales, and K. D. Stuart, *Nucleic Acids Res.* **21**, 687 (1993).

and 21 kb of the 30-kb maxicircle genome of *L. tarentolae* has been sequenced.[29] The structural genes are clustered in approximately 17 kb, and the remainder, which is known as the divergent region, represents tandem repeats of varying complexity.[30]

The following protocol describes the isolation of kDNA from stationary phase cultures of *L. tarentolae*.

1. The cells are harvested by centrifugation (10 min at 2000–3000 g at 4°) or by filtration for large-scale cultures. A Pellicon transverse filter system (Millipore, Bedford, MA) is used to concentrate 5–20 liters of culture.

2. The cells are washed by resuspension in 50 volumes SET buffer (0.15 M NaCl, 0.1 M EDTA, 10 mM Tris-HCl, pH 7.5).

3. The cells are suspended at 1.2 × 10^9 cells/ml in SET buffer. Add pronase to 0.2 mg/ml and sarkosyl to 3% (w/v). Incubate at 60° for 1–3 hr.

4. The viscous lysate is passed through an 18-gauge syringe needle at 25 psi to shear nuclear DNA and reduce the viscosity. For small volumes (10–100 ml) use a 12-ml syringe and force the lysate through the needle by hand. For larger volumes, use a dispensing pressure vessel (Millipore XX67 OOP 05 or 10) with the pressure supplied by compressed air. Use a quick-release tube adaptor to attach the vessel to the tank with compressed air.

5. The lysate is centrifuged for 1.5 hr in an SW28 Beckman rotor at 22,000 rpm (Beckman Instruments, Columbia, MD). This pellets the network DNA.

6. Suspend each pellet in 1–2 ml TE buffer (10 mM Tris-HCl, 1 mM EDTA, pH 7.9) by shaking. Pour into a flask for suspension. Bring up the total volume with TE buffer to 6 ml per each 1–2 liters of original culture. Agitate 30 min to dissolve the network DNA.

7. Prepare CsCl step gradients in SW28 polyallomer tubes by slowly introducing 6 ml of the lower CsCl solution under 24 ml of the upper CsCl solution using a peristaltic pump. The upper CsCl solution contains 37.62 g CsCl in 62 ml TE buffer (refractive index at 25° 1.3705). The lower CsCl solution contains 29.2 g CsCl in 20 ml TE buffer plus 0.14 ml of 10 mg/ml ethidium bromide (refractive index at 25° 1.4040). Only the lower CsCl solution contains dye.

8. Layer 6 ml suspended crude kDNA solution on top of each gradient and centrifuge the gradients 15 min at 20,000 rpm at 20° in an SW28 rotor. The kDNA networks will sediment to the interface between the lower and

[29] V. de la Cruz, N. Neckelmann, and L. Simpson, *J. Biol. Chem.* **259**, 15136 (1984).
[30] M. Muhich, N. Neckelmann, and L. Simpson, *Nucleic Acids Res.* **13**, 3241 (1985).

upper CsCl solutions, whereas nuclear DNA, RNA, and protein will remain in the upper CsCl solution. If shearing of lysate was insufficient, some nuclear DNA will also sediment close to the interface.

9. Visualize the kDNA band at the interface with a UV source (302 nm). Collect the kDNA band by side puncture.

10. Remove the ethidium bromide by extracting twice with an equal volume of water-saturated *n*-butanol.

11. Dialyze the kDNA against 4 liters TE buffer overnight at 4°.

12. Concentrate the kDNA solution to approximately 400 μl by several extractions with *sec*-butanol. Use an equal volume of *sec*-butanol for each extraction, and the volume will be halved each time. Centrifuge 2 min in a clinical centrifuge to break the emulsion. Transfer to a microcentrifuge tube.

13. Extract the kDNA by vortexing with an equal volume of phenol–chloroform (1:1, v/v). Separate the phases by centrifugation at 10,000 g for 1 min. Excess centrifugation will result in loss of network DNA at the interface.

14. Remove the aqueous (upper) phase and remove traces of phenol by 2–4 extractions with an equal volume of water-saturated ether.

15. Add 0.1 volume of 2 M NaCl and 2 volumes ethanol. Incubate in dry ice–ethanol bath for 15 min or at $-20°$ overnight.

16. Recover the precipitated kDNA by centrifugation at 12,000 g for 15 min at 4° in a microcentrifuge.

17. Suspend the kDNA pellet in TE buffer at 1 mg/ml.

18. Check the integrity of the network DNA by diluting a sample 10-fold and adding DAPI (4′,6′-diamidino-2-phenylindole, Sigma, St. Louis, MO) to 1 μg/ml. Observe the stained kDNA using a UV microscope at 1000× magnification. The size and shape of the kDNA networks are distinctive for each kinetoplastid species. If shearing was too harsh, the networks will be fragmented. *Leishmania tarentolae* networks often break into half- or quarter-sized networks. *Crithidia fasciculata* networks are more resistant to shear forces. *Trypanosoma brucei* networks are smaller than those from *C. fasciculata* or *L. tarentolae*.

19. Store the kDNA in aliquots at $-20°$. Using the procedure described, the yield of kDNA is 0.5–1.0 mg/liter of culture. If desired, nuclear DNA can be isolated from the crude DNA preparation. Add 2 volumes of cold ($-20°$) ethanol and spool the nuclear DNA onto a glass rod. Dissolve the DNA in 10 ml TE buffer. Deproteinize by phenol extraction and recover the DNA by ethanol precipitation. This can be used directly for restriction enzyme digestion, or the contaminating RNA can be removed by digestion with RNase A or by ethidium bromide–CsCl isopycnic centrifugation.

Isolation of Maxicircle DNA

In all kinetoplastid species analyzed, maxicircle DNA has a relatively high A+T content as compared to minicircle DNA. This property can be exploited to allow the separation of maxicircle DNA on Hoechst 33258–CsCl density gradients.[26] The Hoechst dye binds preferentially to A+T-rich sequences, thereby decreasing the buoyant density. The maxicircle DNA from *L. tarentolae* is 30 kb in size and can be linearized at the single *Eco*RI site.

1. Dilute 1 mg kDNA into 2 ml *Eco*RI restriction endonuclease buffer. Add 1000 units *Eco*RI and incubate mixture 3 hr at 37°.

2. Monitor the extent of release of the 30-kb maxicircle DNA by running 10-μl samples on a 0.7% agarose gel.

3. Add TE buffer to 12 ml. Add 18.5 g CsCl.

4. Add 0.5 mg/ml Hoechst 33258 dye (Sigma) dropwise with mixing to approximately 1 μg dye per microgram DNA. Stop the addition if the solution becomes cloudy since this can result in the precipitation of the DNA.

5. Adjust density of the solution to a refractive index of 1.3950 at 25°.

6. Centrifuge in a Beckmann 50.2 or Ti60 rotor 48 hr at 40,000 rpm at 25°.

7. Visualize DNA bands with 302 nm UV illumination and recover the minor upper band. The lower band contains undigested kDNA networks and released minicircle DNA. Adjust the refractive index of the upper band to 1.3935, and centrifuge 6.5 ml per tube in a 50 rotor (Beckman) at 39,000 rpm for 48 hr at 25°. Recover the upper band. The second centrifugation is required to remove contaminating minicircle or kDNA.

8. Remove the dye by extraction with an equal volume of 2-propanol.

9. Dialyze the maxicircle DNA against TE buffer and concentrate by *sec*-butanol extraction.

10. Recover DNA by ethanol precipitation.

Isolation of Kinetoplast–Mitochondrial Fraction

Most kinetoplastid cells are generally quite resistant to rupture by shear forces in isotonic media. The method described here employs hypotonic conditions to allow efficient cell breakage by shear forces, thereby releasing a swollen kinetoplast–mitochondrion. Because of the complex, multilobular structure of the single mitochondrion, the isolation of intact mitochondria is most likely not possible. However, mitochondrial fragments apparently reseal effectively, and, in particular, the resealed portion containing the kinetoplast DNA can be isolated by its relatively high isopycnic buoyant density in Renografin density gradients.[31] There is no evidence that this portion of the mitochondrion differs enzymatically from the remainder of the mitochodrion.

1. Grow *L. tarentolae* cells to mid or late log phase ($1-2 \times 10^8$ cells/ml). Harvest the cells by centrifugation (4000 g for 10 min at 4°). Wash cells twice with a 50- to 100-fold excess of ice-cold 10 mM Tris-HCl, pH 8.0, 0.15 M NaCl, 0.1 M EDTA (SET).

2. Suspend the cells without clumps in cold 1 mM Tris-HCl (pH 7.9 at 4°), 1 mM EDTA (DTE). The volume in ml of DTE is determined by dividing the total cell number by 1.2×10^9. Monitor swelling of the cells by phase-contrast microscopy at a magnification of 400–1000×. If insufficient swelling has occurred, add more DTE. Leaving the cells on ice for 5–10 min is generally adequate for complete swelling.

3. Rupture the cells by passing through a 26-gauge needle connected to a dispensing pressure vessel (Millipore) driven by compressed air at 100 psi. Monitor the extent of breakage by phase-contrast microscopy. Small volumes can be processed by hand using a 1–10 ml syringe. Immediately add 0.125 volume of 60% (w/v) sucrose to the lysate. This causes the swollen kinetoplast–mitochondria to shrink into crescent-shaped refractile disks.

4. Centrifuge the lysate at 4° for 10 min at 16,000 g. Discard the supernatant and suspend the loosely packed pellet in cold 0.25 M sucrose, 20 mM Tris-HCl (pH 7.9), 3 mM MgCl$_2$ (STM), using one-sixth the volume of the lysate obtained in step 3. Add 0.005 volume of 2 mg/ml DNase I (Sigma) and incubate on ice 1 hr, in order to digest nuclear DNA, which would interfere with the fractionation procedures.

5. Add an equal volume of cold 8.56% sucrose, 10 mM Tris-HCl (pH 7.9), 2 mM EDTA (STE) and centrifuge at 4° for 10 min at 16,000 g. If the DNase digestion was successful, the pellet should now be well packed.

6. To the pellet add cold 76% Renografin (Squibb Diagnostics, Princeton, NJ: 66% diatrizoate meglumine, 10% diatrizoate sodium), 0.25 M sucrose, 0.1 mM EDTA (RSE). Use 4 ml of 76% RSE per liter of original culture. Vortex and let sit on ice for a few minutes to remove all air bubbles.

7. Layer 4–5 ml of the mixture underneath a 20–35% RSE gradient, using a syringe attached to an 18-gauge needle and polyethylene tubing. If the first drop of mixture floats rather than remaining at the bottom of the gradient, add more 76% RSE to increase the density. The gradients are prepared in 38-ml Beckman SW28 ultracentrifuge tubes by layering 16 ml of 20% Renografin, 8.65% sucrose, 20 mM Tris-HCl, pH 7.9, 0.1 mM EDTA, density 1.14 g/ml (20% RSTE), over 16 ml of 35% RSTE (density

[31] P. Braly, L. Simpson, and F. Kretzer, *J. Protozool.* **21,** 782 (1974).

1.26 g/ml). The tubes are frozen at $-20°$ and thawed overnight at $4°$ before used to establish the gradients.

8. Centrifuge the gradients at $4°$ for 2 hr at 24,000 rpm in a Beckman SW28 rotor. Visualize the kinetoplast–mitochondrial band at a density of approximately 1.22 g/ml by side illumination. Puncture the tube with a syringe needle, and collect the material.

9. Dilute the suspension with 2 volumes cold STE and centrifuge at $4°$ for 15 min at 16,000 g. Wash the mitochondrial fraction once with at least 50 pellet volumes cold STE.

10. For kRNA isolation, wash the pellet once with 50 volumes cold STM, and suspend the material in 10 mM Tris-HCl (pH 7.5), 10 mM MgCl$_2$, 5 mM NaCl (TMN), using 5 ml per liter of original culture.

11. For the preparation of mitochondrial extracts, the pellet from step 9 is suspended at a concentration of approximately 5 mg/ml in 20 mM HEPES–KOH (pH 7.5), 20 mM KCl, 1 mM EDTA, 9–20% (v/v) glycerol (K buffer). The suspension is either used directly or stored frozen at $-80°$.

Purification of Kinetoplast RNA

Isolation of intact kRNA from a purified kinetoplast–mitochondrial preparation is readily accomplished due to the low ribonuclease activity of this fraction.[32] More complex purification procedures using chaotropic salts are usually not necessary.

1. To the mitochondrial suspension in TMN add 20% sodium dodecyl sulfate (SDS) to a final concentration of 0.1% and leave the mixture on ice 5 min.

2. Perform an extraction with an equal volume of phenol–chloroform. Transfer the aqueous phase to a fresh tube, reextract the interface with water, and pool the aqueous phases.

3. Adjust the NaCl concentration to 0.2 M and add 2 volumes ethanol. Incubate at $-20°$ for at least 1 hr, and then recover the nucleic acids by centrifugation at $4°$ for 30 min at 12,000 g. Wash the pellet twice with 70% (v/v) ethanol.

4. Dissolve the nucleic acids in 1 ml TMN per each 2–3 liters original cell culture. Add 0.005 volume DNase I (RNase-free) and incubate at $37°$ for 30 min.

5. Extract with phenol–chloroform and precipitate the kRNA with ethanol as described in steps 2 and 3. Wash the pellet three times with 70% ethanol.

[32] L. Simpson and A. Simpson, *Cell* (*Cambridge, Mass.*) **14,** 169 (1978).

6. Dissolve the kRNA in water at a concentration of approximately 2 mg/ml. Store the RNA as frozen aliquots.

Cloning of Edited and Partially Edited mRNA Sequences

Preparations of steady-state kinetoplast RNA contain a mixture of preedited, partially edited, and fully edited transcripts in a ratio that varies for different cryptogenes as well as for different species and life cycle stages. The sequences of editing intermediates and fully edited mRNAs can be determined by reverse transcriptase (RT)-PCR (polymerase chain reaction) followed by cloning and sequencing.[33] Obtaining a sequence of fully edited mRNA may be a nontrivial task for some genes since most clones will contain partially edited or incorrectly edited (misedited) sequences. Often no single clone that corresponds to a fully edited mRNA can be obtained, and the mature editing pattern can only be deduced as a consensus edited sequence from the clones aligned according to the overall 3' to 5' progression of editing.[20,34] Because of this polarity, there will be more clones with 3' mature editing than with 5' editing, thus affecting the reliability of the consensus sequence at the 5' end.

The following strategy has been successfully employed to investigate the editing of many cryptogene-derived mRNAs. Putative cryptogenes are identified by defects in the reading frame or by a purine-rich region, which is typical of pan-editing.[35,36] The GCG (Genetics Computer Group, Madison, WI) programs Window and StatPlot can be used to locate (G+A)-rich regions in the DNA sequence. Synthesis of cDNA is performed using an oligo(dT) primer or a primer specific for the genomic sequence downstream of the preedited region. The latter may be preferable, since it results in a more specific amplification. In *L. tarentolae*, the 3'-most editing site is separated from the poly(A) tail by 20–40 nucleotides of unedited sequence. Therefore, if the 3' boundary of the putative cryptogene is well defined, an oligonucleotide complementary to the sequence just downstream of this boundary may be selected as a primer.

There are several options for choosing a 5' primer. For small cryptogenes, a sequence upstream of the preedited region may be chosen, thus yielding preedited, partially edited, and fully edited molecules in the same PCR. A stepwise approach may be more effective for large cryptogenes. Partially edited molecules are amplified with a 5' primer selected from an

---

[33] D. A. Maslov, N. R. Sturm, B. M. Niner, E. S. Gruszynski, M. Peris, and L. Simpson, *Mol. Cell. Biol.* **12,** 56 (1992).

[34] J. Abraham, J. Feagin, and K. Stuart, *Cell (Cambridge, Mass.)* **55,** 267 (1988).

[35] L. Simpson, N. Neckelmann, V. de la Cruz, A. Simpson, J. Feagin, D. Jasmer, and K. Stuart, *J. Biol. Chem.* **262,** 6182 (1987).

[36] J. E. Feagin, J. Abraham, and K. Stuart, *Cell (Cambridge, Mass.)* **53,** 413 (1988).

internal sequence of the purine-rich region. A consensus edited sequence for the 3' end of mRNA is deduced from these clones. Another PCR with an edited sequence-specific 3'-primer and an upstream genomic 5'-primer would extend the defined edited sequence further upstream. Information about the 5'-most editing sites is obtained by directly sequencing the 5' end of the mRNA or by taking a RACE (rapid amplification of cDNA ends)–PCR approach.

The overall procedure is as follows:

1. Anneal 10–50 pmol of the oligo(dT) primer or the downstream geno- mic primer to 2.5–5 μg kinetoplast or total cell RNA in 50 μl of 50 mM Tris-HCl (pH 8.3), 75 mM KCl, 3 mM MgCl$_2$ at 65° for 10 min, followed by incubation on ice for 10 min. Add 5 μl of 0.1 M dithiothreitol (DTT) and 5 μl dNTP solution (10 mM of each deoxynucleoside triphosphate). Add 2 μl of 200 units/μl SuperScript II RNase H$^-$ reverse transcriptase (GIBCO–BRL, Grand Island, NY). Incubate at 37° for 30 min. Add another aliquot of enzyme and incubate at 45° for another 30 min. Inactivate the enzyme by incubation at 95° for 5 min.

2. Set up a 50-μl PCR reaction containing 5 μl of the above mixture, 20 pmol of each primer, 20 mM Tris-HCl (pH 8.3), 1.5 mM MgCl$_2$, 25 mM KCl, 0.05% Tween 20, 0.1 mg/ml nuclease-free bovine serum albumin (BSA), 0.25 mM each dNTP, and 5 units Taq DNA polymerase. Denature at 95° for 5 min. Perform 5 PCR cycles at 95° for 30 sec, 45° for 30 sec, and 65° for 1 min. Perform 30 cycles at 95° for 30 sec, 50° for 30 sec, and 72° for 1 min.

3. Gel purification of the PCR product of the expected size may be required, as the reaction usually generates some spurious products, espe- cially if oligo(dT) was used to prime the cDNA synthesis. To improve the specificity, a single nucleotide-anchored 3' primer may be used for PCR. Because Taq DNA polymerase adds single, nonencoded A residues to the 3' ends, the purified PCR products can be cloned directly into vectors containing overhanging T residues (e.g., pT7Blue, Novagen, Madison, WI).

Construction of Guide RNA Libraries

Cloning of gRNAs is based on a RACE–PCR procedure that involves synthesis of cDNA, ligation of an anchor oligonucleotide, and PCR amplifi- cation[37] (Fig. 1). To monitor some critical steps of the procedure, cDNA synthesis is performed using 5'-labeled oligonucleotides. Because cloning of PCR products based on the single 3'-overhanging A residues is not very efficient, we have usually employed the CloneAmp system (GIBCO–BRL),

[37] O. H. Thiemann, D. A. Maslov, and L. Simpson, *EMBO J.* **13,** 5689 (1994).

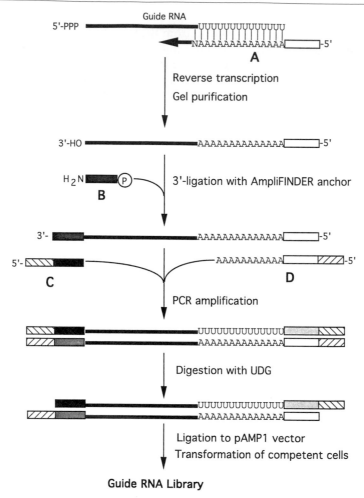

Guide RNA Library

FIG. 1. Construction of gRNA library. Thick black lines represent the encoded part of a guide RNA and the corresponding part of a cDNA. Open boxes in oligonucleotides A and D represent an adaptor sequence added for the purpose of increasing the melting temperature. The black box in oligonucleotide C represents a sequence complementary to the Ampli-FINDER anchor (oligonucleotide B). Note that both oligonucleotides A and B contain an *Eco*RI site (see text for the structure of the oligonucleotides) that can be used for cloning in an appropriate vector instead of the CloneAmp system. Hatched areas of oligonucleotides C and D represent sequences that contain deoxyuridines instead of thymidines.

which yields a large number of clones. The PCR primers are synthesized with 5'-end adaptors containing deoxyuridines (dU) instead of thymidines. Digestion of the PCR products with uracil DNA glycosylase generates protruding single stranded 5' ends suitable for efficient cloning in the pAMP1 vector. The method has been used successfully to obtain gRNA libraries from *T. cruzi*,[37a] *C. fasciculata*,[37b] *Trypanoplasma borreli* (S. Yasuhira, D. Maslov, and L. Simpson, unpublished results, 1995), as well as from two strains of *L. tarentolae*.[37]

1. Combine 1 μl of a mixture containing 10 pmol of each of the three oligonucleotides CGCGGATCC(A)$_{14}$C, CGCGGATCC(A)$_{14}$T, and CGCGGATCC(A)$_{14}$G (oligonucleotide A, Fig. 1), 2 μl of [γ-^{32}P]ATP (6000 Ci/mmol, 10 μCi/μl), 1 μl of 10× buffer (0.5 M Tris-HCl, pH 7.6, 0.1 M MgCl$_2$, 50 mM DTT, 1 mM spermidine, 1 mM EDTA), 1 μl of 10 units/ μl T4 polynucleotide kinase (GIBCO–BRL), and 5 μl double-distilled water. Incubate at 37° for 30 min. Purify the labeled oligonucleotides using NucTrap columns (Stratagene, La Jolla, CA). Add 2 volumes of 7.5 M ammonium acetate and precipitate the oligonucleotides with 3 volumes of ethanol. Suspend the pellet in 20 μl water.

2. Incubate 4 μl of the labeled oligonucleotides with 10–20 μg of kRNA or gel-purified guide RNA (usually less than 1 μg) in 38 μl of 50 mM Tris-HCl (pH 8.3), 75 mM KCl, 3 mM MgCl$_2$ at 65° for 5 min. Add 5 μl of 0.1 M DTT and 5 μl of dNTP solution (10 mM of each), and anneal for 10 min on ice. Add 2 μl of 200 units/μl SuperScript II RNase H$^-$ reverse transcriptase (GIBCO–BRL) and incubate at 16° for 30 min. Add another 1-μl aliquot of enzyme and incubate at 37° for 1 hr.

3. Hydrolyze the RNA by the addition of 50 μl of 0.4 M NaOH, 30 mM EDTA for 1 hr at 65°. Neutralize the solution by adding 12 μl of 3 M sodium acetate (pH 5.2). Precipitate the cDNA by adding 0.3 ml of ethanol. Suspend the pellet in 12 μl water.

4. Fractionate the cDNA on a 0.4 mm 8% (w/v) polyacrylamide–7.5 M urea sequencing gel. Visualize the cDNA bands by exposing the wet gel at 4°. This is a critical step of the procedure, since a failure to remove unreacted primers from the cDNA results in the addition of anchor oligonucleotides to the primers as well as to the cDNA, and causes a subsequent contamination of the library with clones containing anchored primers. Elute the cDNA overnight by diffusion. Precipitate the cDNA by the addition of 50 μg glycogen and 3 volumes ethanol. Suspend the final pellet in 12 μl water.

[37a] H. Avila and L. Simpson, *RNA*, in press (1995).
[37b] S. Yasuhira and L. Simpson, *RNA* **1**, 634 (1995).

5. Ligate the 3′-end-blocked anchor oligonucleotide B (Fig. 1), CAC-GAATTCACTATCGATTCTGGAACCTTCAGAGG (AmpliFINDER anchor, provided with 5′-AmpliFINDER RACE kit, Clontech Laboratories, Palo Alto, CA), by incubating 2.5 μl of the gel-purified cDNA, 2 μl of 2 pmol/μl oligonucleotide, 5 μl of 2× ligation buffer (provided by the manufacturer), and 0.5 μl of 20 units/μl T4 RNA ligase at room temperature for 16–24 hr. In an alternative approach, a homopolymer tail is added to the 3′ ends of the cDNA. The reaction is performed with terminal deoxynucleotidyl transferase (Boehringer Mannheim, Indianapolis, IN) by incubating 25 units enzyme and 1 μM dGTP for 30 min at 37°. If this method is chosen, the sequence of the PCR primers must be modified accordingly.

6. To PCR-amplify the anchor-ligated cDNA, combine 2 μl of the 10-fold diluted ligation product, 1 μl (200 pmol) of each of the two oligonucleotides (CAdU)$_4$GTTCCAGAATCGATAGTGAAT (oligonucleotide C, Fig. 1) and (CdUA)$_4$CGCGGATCC(A)$_{11}$ (oligonucleotide D, Fig. 1) in a 50-μl PCR reaction containing 50 mM KCl, 10 mM Tris-HCl (pH 8.3), 5 mM MgCl$_2$, 0.2 mM each dNTP, and 5 U Taq polymerase. Denature at 95° for 30 sec and amplify for 30 cycles at 94° for 5 sec, 60° for 10 sec, and 72° for 30 sec.

7. Analyze and, if desired, purify the cDNA further by using a 4% (w/v) NuSieve agarose gel or an 8% nondenaturing polyacrylamide gel. In addition to the expected PCR products (120–150 bp), a faster migrating product is usually observed, which results from amplification of carried-over cDNA synthesis primers.

8. To clone the cDNA, perform uracil DNA glycosylase treatment, ligation to pAMP1 vector (GIBCO–BRL), and transformation of DH5α Library Efficient competent cells (GIBCO–BRL).

Mitochondrial Extract Preparation and Fractionation by Glycerol Gradient Sedimentation

It is essential to first obtain crude kinetoplast–mitochondrial lysates, using the purified mitochondrial fraction obtained as described previously. This is most conveniently achieved by breaking the mitochondrial membranes either by sonication or by the addition of detergents. The following protocol describes the preparation of extracts containing Triton X-100,[16] and subsequent fractionation of the clarified extracts by glycerol gradient sedimentation.[38]

1. If necessary, thaw on ice a 0.2- to 0.3-ml aliquot of kinetoplast–

---

[38] M. Peris, G. C. Frech, A. M. Simpson, F. Bringaud, E. Byrne, A. Bakker, and L. Simpson, *EMBO J.* **13,** 1664 (1994).

mitochondrial fraction isolated as described above. Add 10% (v/v) Triton X-100 to a final concentration of 0.3% and mix gently.

2. Place the tube in an ice–water bath and homogenize the mixture for 15 sec using a motor-driven pellet pestle mixer (Kontes Glass Co., Vineland, NJ, Cat. No. 749520). This yields the TL (Triton lysate) extract.

3. Centrifuge the TL extract at 4° for 30 min at 12,000 g to obtain the clarified TS (Triton supernatant) extract, or at 4° for 1 hr at 100,000 g to obtain the S-100 extract. Store the extracts at $-80°$ or use them directly for glycerol gradient sedimentation and/or various assays as described below.

4. Prepare linear 10–30% glycerol gradients containing 20 mM HEPES–KOH (pH 7.5), 20 mM KCl, 1 mM EDTA, in 13.2 ml ultracentrifuge tubes (Beckman, Cat. No. 344059 or 331372), by using a standard gradient mixer or a gradient master (Biocomp Instruments, Fredericton, New Brunswick). Layer up to 0.2 ml TS or S-100 extract (containing a glycerol concentration of 9% or less) on top of a gradient. Concurrently, overlay equivalent gradients with S-value standards, such as alcohol dehydrogenase (7.6 S, Sigma), catalase (11.5 S, Pharmacia, Piscataway, NJ) thyroglobulin (19.3 S, Pharmacia), and *Escherichia coli* small (30 S) ribosomal subunits.

5. Centrifuge the gradients at 4° for 14 hr at 33,000 rpm in a Beckman SW-41 rotor.

6. Collect 16–20 fractions by using a density gradient fractionator (Instrumentation Specialties Co., Lincoln, NE). The gradient fractions may be used directly for various assays, or their volume and composition may first be modified by using Centricon centrifugal microconcentrators (Amicon, Danvers, MA) or microdialysis cassettes (Pierce, Rockford, IL).

Enzymatic Activities Possibly Involved in RNA Editing

Several enzymatic activities have been identified from mitochondrial extracts of *L. tarentolae* that are possibly involved in the editing of maxicircle transcripts. These include a terminal uridylyltransferase (TUTase) activity,[16] an RNA ligase activity,[16] and a site-specific endonuclease activity.[39,40] All of these activities can be solubilized with Triton X-100 and are detectable in the TL, TS, and S-100 extracts of purified kinetoplast–mitochondrial preparations.

An internal U-incorporation activity, which has several characteristics of an *in vitro* RNA editing activity, has also been identified in *L. tarentolae* mitochondrial extracts.[41] Uridine residues are incorporated into the preedited region of several synthetic mitochondrial transcripts. This activity is selectively inhibited by predigestion of the extract with micrococcal

[39] A. M. Simpson, N. Bakalara, and L. Simpson, *J. Biol. Chem.* **267**, 6782 (1992).
[40] M. Harris, C. Decker, B. Sollner-Webb, and S. Hajduk, *Mol. Cell. Biol.* **12**, 2591 (1992).
[41] G. C. Frech, N. Bakalara, L. Simpson, and A. M. Simpson, *EMBO J.* **14**, 178 (1995).

nuclease, suggesting the involvement of some type of endogenous RNA species. However, no direct evidence was obtained for the involvement of endogenous gRNAs, and exogenous synthetic gRNA was found to inhibit the internal U incorporation. This activity sediments in glycerol gradients at approximately 25 S and the ribonucleoprotein complexes which mediate this activity have operationally been termed the G complexes.[38]

Terminal Uridylyltransferase

The mitochondrial TUTase activity transfers UMP from UTP to the 3'-OH of RNA molecules. It is stimulated by at least one additional nucleoside triphosphate (NTP) but, unlike transcription, does not require all four NTPs to be present. Therefore, by omitting at least one nucleoside triphosphate from the reaction, the TUTase activity can be conveniently assayed in the absence of run-on transcription activity. The TUTase is relatively indiscriminate in terms of the RNA substrate, although certain RNA species, including tRNAs, are less efficiently labeled. The TUTase assay is as follows.[16]

1. Mix approximately 1 μg substrate RNA with mitochondrial extract in a 50-μl reaction containing 5 mM HEPES–KOH (pH 7.5), 60 mM KCl, 3 mM potassium phosphate (pH 7.5), 6 mM magnesium acetate, 20 mM DTT, 1 mM ATP, 1 mM GTP, and 10 μCi [α-^{32}P]UTP (800 Ci/mmol). Incubate the mixture at 27° for at least 30 min.

2. For a quantitative determination of TUTase activity, stop the reaction by adding an equal volume of 0.5 M sodium phosphate (pH 6.8), 0.5% (w/v) sodium pyrophosphate, 0.1% (w/v) SDS, and transfer aliquots onto DE-81 filter disks. Dry the disks before washing them three times in excess 0.5 M sodium phosphate (pH 6.8), 0.5% sodium pyrophosphate, and once in ethanol. Count the radioactivity retained on the dried disks in a liquid scintillation counter.

3. For an additional qualitative assessment of TUTase activity, an aliquot of the reaction mixture may be extracted with phenol–chloroform and the labeled RNAs analyzed by gel electrophoresis and subsequent autoradiography or PhosphorImager analysis.

RNA Ligase

The *L. tarentolae* mitochondrial RNA ligase activity[16] catalyzes the intra- or intermolecular ligation of a donor containing a 5'-phosphate with an acceptor containing a 3'-OH. The activity can be quantitatively assayed by addition of [^{32}P]pCp to the 3' termini of RNA molecules. However, not all RNA molecules are equally well modified by using this assay. If total kinetoplast RNA is used as a substrate, the tRNAs appear to incorporate most of the radioactivity.

1. Mix 0.5–2.5 μg kRNA with 10 μl mitochondrial extract in a 50-μl reaction containing 0.1 M HEPES–KOH (pH 7.9), 20 mM MgCl$_2$, 7 mM DTT, 0.2 mM ATP, 20% (v/v) dimethyl sulfoxide (DMSO), 1 unit RNase inhibitor, and 10 μCi [^{32}P]pCp (3000 Ci/mmol). Incubate the mixture at 27° for 2 hr, or at 4° overnight.

2. For a quantitative determination of ligase activity, transfer aliquots of the reaction onto DE-81 filter disks and process them as described above in step 2 of the TUTase reaction.

3. In addition, an aliquot of the reaction may be extracted with phenol–chloroform and then analyzed by gel electrophoresis and subsequent autoradiography or PhosphorImager analysis.

Cryptic RNase

A sequence- or structure-specific cryptic RNase activity can be detected in mitochondrial extracts (TL, TS, or S-100 extracts).[39] This cryptic RNase can be activated either by the addition of heparin (5 μg/ml) or by predigestion of the lysate with proteinase K. The RNA substrate used for this assay is a 280-nucleotide RNA synthesized by *in vitro* transcription from a recombinant plasmid using T7 RNA polymerase. The template DNA consists of the 22 nucleotide preedited region of the cytochrome *b* gene together with 56 nucleotides of 5′ flanking sequence, 129 nucleotides of 3′ flanking sequence, and 73 nucleotides of Bluescript vector sequence at the 5′ end. Cleavage by the cryptic RNase occurs at one major site and four minor sites within the preedited region.

1. Prepare uniformly ^{32}P-labeled synthetic RNA substrate [10^8 counts/min (cpm)/μg] by *in vitro* transcription from the recombinant plasmid or from PCR templates amplified with upstream primers containing the T7 phage RNA polymerase promoter.

2. Either add proteinase K (0.1 mg/ml final concentration) to the mitochondrial extract and incubate the extract 5 min at 37°, or add heparin (Sigma) to 5 μg/ml. Activations by protease predigestion and heparin are synergistic, and together result in better cleavage.

3. Mix approximately 10^4 cpm RNA substrate together with 10 μl activated mitochondrial extract (from step 2) in a 50-μl reaction containing 10 mM Tris-HCl (pH 7.5), 3 mM MgCl$_2$, 1 mM ATP, and 5 μg/ml heparin.

4. Incubate the mixture at 27° for 1 hr and extract with phenol–chloroform (1 : 1, v/v) followed by ethanol precipitation.

5. Analyze the cleavage products on an analytical polyacrylamide–7.5 M urea gel followed by autoradiography or PhosphorImager analysis.

Guide RNA–Messenger RNA Chimera-Forming Activity

To study the first step of the proposed transesterification model for RNA editing, cell-free systems were developed using mitochondrial extracts from both *T. brucei*[42,43] and *L. tarentolae*.[19] Chimera formation is monitored by following the covalent transfer of uniformly ^{32}P-labeled gRNA to a higher molecular weight nonradioactive test mRNA by gel electrophoresis. Both gRNA and mRNA are synthesized by *in vitro* transcription from recombinant plasmids or PCR-derived DNA templates. Sequence analysis of the reaction products by selective PCR amplification and cloning revealed gRNA/mRNA chimeric molecules as the most prominent products.[19] This *in vitro* system has been used to confirm the importance of the anchor sequence between the gRNA and mRNA.[19]

1. Suspend the mitochondrial fraction from 1 liter of cell culture in 2 ml ice-cold 10 mM HEPES–NaOH (pH 7.9), 0.5 mM EDTA. Keep the mixture on ice 10 min.

2. Disrupt the swollen mitochondria by sonication at 100 W for three 20-sec periods at 4° (Braunsonic No. 1510, microtip; Branson Ultrasonics, Danbury, CT).

3. Immediately add 2 ml freshly prepared 40 mM HEPES–NaOH (pH 7.9), 50% glycerol, 0.84 M NaCl, 1 mM DTT, 0.4 mM EDTA, 1 mM PMSF, and gently agitate the solution with a small magnetic stirrer for 30 min on ice.

4. Clarify the extract by centrifugation at 50,000 g for 30 min at 4°.

5. Concentrate the extract at 4° to approximately 0.1 ml by using two Centricon 10 centrifugal microconcentrators (Amicon). Exchange the buffer by adding twice to each microconcentrator 2 ml freshly prepared 20 mM HEPES–NaOH (pH 7.9), 20% glycerol, 0.1 M KCl, 0.5 mM DTT, 0.2 mM EDTA, 0.5 mM PMSF, and recentrifuging.

6. Quick-freeze the pooled extracts (~0.2 ml) in dry ice/ethanol. When samples are stored at −80°, chimera-forming activity is maintained for up to 3 months.

7. Anneal equimolar amounts of the test RNA molecules ([32P]UTP-labeled gRNA and cold mRNA) in 2.5 μl of 20 mM HEPES–NaOH (pH 7.9), 0.1 M KCl, 1 mM EDTA. Denature 3 min at 70° prior to annealing at 37° and 25° for 10 min each.

8. Add 8 μl of 8% (w/v) polyethylene glycol (PEG) 8000, 12.5 mM MgCl$_2$, 2.5 mM ATP, 1 unit/μl RNase inhibitor, and 15 μl thawed mitochon-

[42] M. E. Harris and S. L. Hajduk, *Cell* (*Cambridge, Mass.*) **68,** 1091 (1992).
[43] D. J. Koslowsky, H. U. Göringer, T. H. Morales, and K. Stuart, *Nature* (*London*) **356,** 807 (1992).

drial extract (from step 6). Chimera formation is observed on incubation for 15 to 120 min at 27°.

9. Stop the reaction by adding 0.1 ml of 0.25% *N*-laurylsarcosine, 25 m*M* EDTA, 0.25 mg/ml proteinase K, and incubating at 37° for 20 min. Extract the mixture with phenol–chloroform (1 : 1, v/v) and recover the RNA by ethanol precipitation.

10. Analyze the resulting RNA products on an analytical polyacrylamide–7.5 *M* urea gel followed by autoradiography or PhosphorImager analysis. RNA products may also be analyzed by sequence determination following reverse transcription, PCR amplification, and cloning. To amplify exogenous RNA specifically, a tag sequence may be added to the test RNA.

Internal Uridine-Incorporation Activity

Incubation of certain synthetic preedited mRNAs with mitochondrial extracts in the presence of [α-^{32}P]UTP leads to the internal incorporation of U residues, in addition to the addition of U residues to the 3′ termini caused by the TUTase activity.[41] This internal U-incorporation activity requires ATP and is stimulated by spermidine. In synthetic preedited transcripts derived from the cytochrome *b* mRNA and the NADH dehydrogenase subunit 7 mRNA, which are both edited *in vivo*, the preedited regions were identified as the predominant areas of *in vitro* internal U incorporation. To differentiate between internal U incorporation and 3′ addition of U residues, an assay is used (Fig. 2) that involves cleavage of the processed RNA by RNase H in conjunction with specific oligodeoxynucleotides and analysis of the heterogeneously sized cleavage products by gel electrophoresis followed by autoradiography or PhosphorImager analysis.

1. Prepare synthetic RNA substrates by *in vitro* transcription from recombinant, transcription-competent plasmids, or from PCR templates amplified with an upstream primer containing the T7 phage RNA polymerase promoter.

2. Mix approximately 1 μg synthetic transcript with mitochondrial extract in a 50-μl reaction containing 5 m*M* HEPES (pH 7.5), 60 m*M* KCl, 3 m*M* potassium phosphate (pH 7.5), 6 m*M* magnesium acetate, 20 m*M* DTT, 2 m*M* spermidine, 1 m*M* ATP, 1 m*M* GTP, 1 μ*M* unlabeled UTP, and 25 μCi [α-^{32}P]UTP (800 Ci/mmol). The optimal amount of mitochondrial extract should be determined by titration, since it varies with different mitochondrial preparations.

3. Incubate the mixture at 27° for 40–100 min and extract with phenol–chloroform.

4. Separate the RNA on a preparative polyacrylamide–7.5 *M* urea gel. Excise the full-length ethidium-stained RNA band (including material of

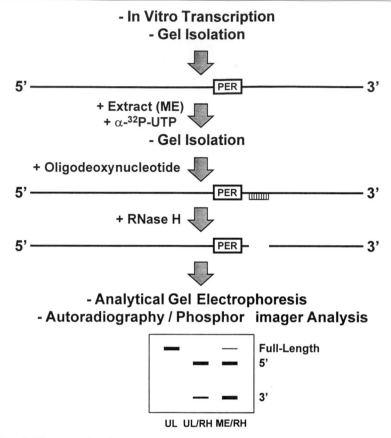

FIG. 2. Diagram of an *in vitro* assay to detect internal U incorporations into synthetic mRNA transcripts. ME, Mitochondrial extract; PER, preedited region; UL/RH, uniformly labeled RNA digested with RNase H; ME/RH, RNA labeled by incubation with mitochondrial extract and $[\alpha-^{32}P]UTP$, and subsequently digested with RNase H.

retarded mobility containing potential U incorporations) and elute the RNA in 0.5 M ammonium acetate, 10 mM magnesium acetate, 0.1% SDS, 0.1 mM EDTA, followed by phenol–chloroform extraction and ethanol precipitation.

5. Anneal the gel-purified RNA to 0.1–0.2 μg DNA oligomer, and incubate 1 hr at 37° in a 20-μl reaction containing 50 mM Tris (pH 7.5), 0.1 M KCl, 10 mM MgCl$_2$, 0.1 mM DTT, 10 μg/ml BSA, and 0.2–0.3 units RNase H (Pharmacia).

6. Visualize the digestion products containing incorporated label by autoradiography or PhosphorImager analysis of an analytical polyacrylamide–7.5 *M* urea gel.

Identification of Ribonucleoprotein Complexes Possibly Involved in RNA Editing

RNA processing reactions in eukaryotes occur within large ribonucleoprotein (RNP) complexes. This appears to also be the case with RNA editing in trypanosomatids.[38,44–47] Mitochondrial RNP complexes can be identified by sedimentation in glycerol gradients and by gel electrophoresis under native conditions. When *L. tarentolae* mitochondrial extract is incubated under TUTase conditions in the presence of [α-^{32}P]UTP and the mixture subsequently electrophoretically separated, several labeled complexes of different gel mobilities can be identified, which we have termed the T-complexes. These complexes sediment in a glycerol gradient at 10–13 S. All of these complexes contain RNA that is accessible to labeling by the TUTase, and some or all of them may be involved in RNA editing. The most intensely labeled T-complex (T-IV) was shown to contain gRNA, and mRNA fragments are found associated with all of the T-complexes. Another class of RNP complexes (G-complexes) comigrate with the internal U-incorporation activity at approximately 25 S.

Identification of [α-^{32}P]UTP-Labeled T-Complexes

1. Label TS mitochondrial extract by incubation for 40 min at 27° in 5 m*M* HEPES-KOH (pH 7.5), 60 m*M* KCl, 3 m*M* potassium phosphate (pH 7.5), 6 m*M* magnesium acetate, 20 m*M* DTT, 1 m*M* ATP, 1 m*M* GTP, and 0.2 mCi/ml [α-^{32}P]UTP. Note that CTP is omitted to avoid the occurrence of run-on transcription, which requires the presence of all four nucleoside triphosphates.

2. Pour a 4–16% polyacrylamide gradient gel containing 40 m*M* Tris–acetate (pH 8.0), 1 m*M* EDTA (1× TAE buffer), 0.1% (v/v) Tween 20, and a stabilizing gradient of 10–30% glycerol. Overlay with *n*-butanol and allow the gel to polymerize. After drainage of the *n*-butanol, add more 4%

[44] V. W. Pollard, M. E. Harris, and S. L. Hajduk, *EMBO J.* **11,** 4429 (1992).

[45] H. U. Göringer, D. J. Koslowsky, T. H. Morales, and K. Stuart, *Proc. Natl. Acad. Sci. U.S.A.* **91,** 1776 (1994).

[46] J. Köller, G. Nörskau, A. S. Paul, K. Stuart, and H. U. Göringer, *Nucleic Acids Res.* **22,** 1988 (1994).

[47] L. K. Read, H. U. Göringer, and K. Stuart, *Mol. Cell. Biol.* **14,** 2629 (1994).

gel solution and insert the comb 0.5 cm from the already polymerized gradient gel. Allow to polymerize.

3. Before loading the gel, add glycerol to the samples to a final concentration of at least 10%. The following molecular weight markers may be used: thyroglobulin (669 kDa, Pharmacia), ferritin (443 kDa, Pharmacia), and β-amylase (200 kDa, Sigma). Run the gel at 4° in 1× TAE buffer.

4. Stain the gel with rapid Coomassie stain (Diversified Biotech, Boston, MA) to identify the positions of the marker proteins. Visualize the labeled complexes by autoradiography or PhosphorImager analysis.

Detection of Specific T-Complexes Containing Terminal Uridylyltransferase Activity

Two specific T-complexes containing TUTase activity can be identified by performing a native gel *in situ* TUTase assay.[38] The endogenous RNAs in the remainder of the T-complexes presumably are labeled by interaction with these complexes.

1. Run an aliquot of unlabeled TS mitochondrial extract in a native gel as described above. Excise the gel lane (8 cm long) and transfer it to a 15-ml centrifuge tube.

2. Add 3 ml TUTase labeling mix containing 5 mM HEPES–KOH (pH 7.5), 60 mM KCl, 3 mM potassium phosphate (pH 7.5), 6 mM magnesium acetate, 20 mM DTT, 1 mM ATP, 1 mM GTP, and 0.2 mCi/ml [α-^{32}P]UTP. Incubate on a shaker 1 hr at room temperature.

3. Wash the gel slice with copious amounts of water to remove most of the unincorporated [α-^{32}P]UTP.

4. Realign the gel slice next to the original gel, which should contain a lane of endogenously labeled extract for comparison. Dry the gel under reduced pressure before autoradiography or PhosphorImager analysis.

Discussion

We have emphasized experimental protocols involving *L. tarentolae,* but similar protocols have been described for *T. brucei.* The major difference in RNA editing in *T. brucei* is the high gRNA complexity and the resulting presence of redundant gRNAs of different sequences overlapping in regions other than the anchor sequence.[48] In addition, editing in *T. brucei* appears to be regulated during the complex biphasic life cycle, which involves

---

[48] R. A. Corell, J. E. Feagin, G. R. Riley, T. Strickland, J. A. Guderian, P. J. Myler, and K. Stuart, *Nucleic Acids Res.* **21,** 4313 (1993).

periods of mitochondrial repression and derepression (see Ref. 10 for review). The mechanism of this transcript-specific regulation is not known.

A *T. brucei* mitochondrial TUTase,[44] an RNA ligase,[44,49] a preedited RNA site-specific ribonuclease,[40] and a gRNA/mRNA chimera-forming activity have been described.[42,43] In addition, mitochondrial RNP complexes containing these activities have been partially characterized, and a small set of proteins have been shown to interact with gRNAs.[44–47]

A breakthrough in the analysis of gRNA-mediated RNA editing was reported by Seiwert and Stuart.[50] They showed that a mitochondrial extract in the presence of exogenous synthetic gRNA could direct the deletion of U residues from the first editing site of the preedited ATPase subunit 6 (=MURF4) transcript and that this deletion activity was mediated by base pairing with guiding nucleotides in the gRNA. However, U additions were not observed in this system, suggesting that the mechanism for U deletions may differ from that for U additions.

It was shown that a trypanosomatid-like RNA editing also occurs in the cryptobid kinetoplastid *Trypanoplasma boreli*.[51,52] In this organism, however, the mitochondrial genome is not composed of a network of catenated mini- and maxicircles, but rather contains 80-kb circular molecules, which are the maxicircle equivalents, and 200-kb circular molecules with tandem repeats of gRNA-like genes. An analysis of this type of editing system is in progress.

It is clear that, except for the investigation of editing in the cryptobids and possibly in other lower kinetoplastids, the field has progressed past a purely descriptive stage and is entering a more biochemical stage. We hope that the procedures described in this chapter will aid in investigating the molecular and biochemical mechanisms involved in this type of RNA editing.

[49] T. White and P. Borst, *Nucleic Acids Res.* **15**, 3275 (1987).
[50] S. D. Seiwert and K. Stuart, *Science* **266**, 114 (1994).
[51] D. A. Maslov and L. Simpson, *Mol. Cell. Biol.* **14**, 8174 (1994).
[52] J. Lukes, G. J. Arts, J. Van den Burg, A. De Haan, F. Opperdoes, P. Sloof, and R. Benne, *EMBO J.* **13**, 5086 (1994).

[11] Purification of Mitochondrial DNA from Human Cell Cultures and Placenta

By CARLO AUSENDA and ANNE CHOMYN

Introduction

The need for mitochondrial DNA (mtDNA)-specific probes for the detection and analysis of mitochondrial RNA and mtDNA is usually satisfied by available cloned fragments of human mtDNA or by polymerase chain reaction (PCR)-amplified segments of mtDNA. On the other hand, there are instances in which it is desirable to use the total mitochondrial genome as a hybridization probe. Examples include the use of radiolabeled mtDNA to detect the overall pattern of the mitochondrial ribosomal and messenger RNAs on RNA blots[1] and to reveal duplication- and deletion-containing mtDNAs on Southern blots.[2] Thus, purified human mtDNA is very useful as a general hybridization probe, in addition to being a source of restriction enzyme fragments for subcloning. We describe here protocols for the purification of mtDNA from HeLa cells grown in suspension and from human placenta. The HeLa cell protocol can be applied to other cultured cells as well, including cells derived from other species. The yields of closed-circular mtDNA using these protocols are 1.2–2.4 μg/ml packed HeLa cells and 5–10 μg/100 g placenta.

Purification of Mitochondrial DNA from HeLa Cells

All operations should be carried out at 2°–4°, unless otherwise specified. The method of preparation of mitochondria by differential centrifugation is modified from Attardi et al.[3] The purification of mtDNA is based on the isolation of the closed-circular form, banded on a CsCl density gradient in the presence of ethidium bromide.[4]

Collection and Washing of Cells

Harvest cells from 2 liters or more of suspension culture in exponential phase by centrifugation in bottles for 10 min at 400 g_{av}. Suspend the cells

[1] A. Chomyn, A. Martinuzzi, M. Yoneda, A. Daga, O. Hurko, D. Johns, S. T. Lai, I. Nonaka, C. Angelini, and G. Attardi, *Proc. Natl. Acad. Sci. U.S.A.* **89,** 4221 (1992).

[2] D. R. Dunbar, P. A. Moonie, R. J. Swingler, D. Davidson, R. Roberts, and I. J. Holt, *Hum. Mol. Genet.* **2,** 1619 (1993).

[3] B. Attardi, B. Cravioto, and G. Attardi, *J. Mol. Biol.* **44,** 47 (1969).

[4] B. Hudson and J. Vinograd, *Nature (London)* **216,** 647 (1967).

in 10 volumes of 0.13 M NaCl, 5 mM KCl, 1 mM MgCl$_2$ (NKM), and centrifuge them in 50-ml graduated conical centrifuge tubes for 5 min at 165 g_{av} in a swinging-bucket rotor. Wash the cells again in NKM and then once more in Mg^{2+}-free isotonic buffer. We use 137 mM NaCl, 5 mM KCl, 0.7 mM Na$_2$HPO$_4$, 25 mM Tris-HCl, pH 7.4 at 25°. Note the volume of packed cells in each tube and discard the supernatant.

Cell Breakage and Differential Centrifugation

Working with one tube at a time, suspend the cell pellet in 6 volumes of 10 mM Tris-HCl, pH 6.7, 10 mM KCl, 0.1 mM EDTA. Allow the cells to swell in this hypotonic buffer for 2 min, then break the cells with a glass vessel–Teflon pestle (Potter–Elvehjem) homogenizer as follows: while the pestle is being rotated on its axis by a motor (~1600 rpm), slowly raise the vessel containing the cell suspension to bring the pestle to the bottom of the vessel. Sharply lower the vessel. One should hear a "pop." Repeat this operation 5 to 10 times until 60–70% of the cells have been broken. Monitor cell breakage with a phase-contrast microscope. This homogenizer is available from Arthur H. Thomas, (Swedesboro, NJ), and size C will accommodate up to 5 ml packed cells plus 30 ml hypotonic buffer. Immediately transfer the homogenate to a 50-ml centrifuge tube containing one-seventh volume of 2 M sucrose, 10 mM Tris-HCl, pH 6.7, 0.1 mM EDTA. Mix well and keep in ice until all the other cell pellets have been similarly processed. Centrifuge the tubes 3 min at 1160 g_{av} in a swinging-bucket rotor to pellet nuclei, unbroken cells, and large cytoplasmic debris.

Transfer the supernatants to fresh tubes, and centrifuge them 10 min at 8100 g_{av} to pellet the mitochondria. Add to each pellet one-half the homogenate volume of 250 mM sucrose, 10 mM Tris-HCl, pH 6.7, 0.1 mM EDTA, and suspend the mitochondrial fraction very well, breaking up any clumps so as to avoid loss of mitochondria in the next step. Centrifuge the tubes 2 min at 1100 g_{av} to pellet any remaining nuclei and large debris. Centrifuge the supernatants at 8100 g_{av} for 10 min. Suspend the pellets in 250 mM sucrose, 10 mM Tris-HCl, pH 7.4, 10 mM EDTA, using 0.5 to 1.0 ml/ml packed cells harvested. Pool the suspensions and proceed directly to the next step.

Lysis of Mitochondria and Equilibrium Density Gradient Centrifugation

Add sodium dodecyl sulfate (SDS) from a 20 or 25% (w/v) stock solution to a final concentration of 1.2%. Mix well by pipetting, and incubate at room temperature at least 10 min. Add one-sixth volume of 7 M CsCl, incubate in ice for 30 min, and centrifuge at 15,000 g_{av} for 15 min. Measure the volume of the supernatant, and add to it 0.741 g CsCl per milliliter solution. Transfer the solution to centrifuge tubes and keep it in ice 20 min. At this point, the procedure may be interrupted, if desired, and the solution stored at 4° overnight, or for longer periods at −20°.

Centrifuge the tubes 20 min at 30,000 g_{av} and transfer the solution with a Pasteur pipette to a cylinder, leaving behind the protein "skin." Measure the volume of the solution in a cylinder, cover the cylinder with aluminum foil, and add 1/50 volume of a 10 mg/ml ethidium bromide solution. Once the ethidium bromide has been added to the DNA, the DNA should be protected from light. Read the refractive index and adjust it to between 1.386 and 1.390 (ρ = 1.55 to 1.60 g/ml) by adding, as needed, solid CsCl, concentrated CsCl solution, or 10 mM Tris-HCl, pH 7.4, 10 mM EDTA. Transfer the final solution to one or more Quick-Seal polyallomer tubes for the Ty65 Beckman rotor (Beckman Instruments, Fullerton, CA), add paraffin oil to fill the tubes, if necessary, and centrifuge them in a Ty65 rotor at 38,000 rpm for 40–48 hr at 20°.

Collection of Mitochondrial DNA

Collection of the closed-circular DNA may be done at room temperature. Very carefully transport one tube at a time to a stand with a clamp in which the tube can be held securely. Wearing protective goggles or a full face shield that blocks ultraviolet light, illuminate the tube from the side for the minimum time with a long-wavelength ultraviolet light source. The longer the exposure to ultraviolet or near-ultraviolet light, the more the DNA will be damaged. There should be two main fluorescent bands in the upper half or middle portion of the gradient, a broad upper band containing nuclear DNA and nicked mtDNA and a thin lower one containing closed-circular DNA, including mtDNA. A broad, less strongly fluorescent band in the lower half of the tube contains polysaccharides. Insert a hypodermic needle into the top of the tube to allow intake of air. Affix a piece of Scotch tape to the side of the tube, puncture the tube through the Scotch tape, just below the lower band, with a hypodermic needle, bevel up, attached to a syringe, and withdraw the band material. To avoid the possibility of also collecting some nuclear DNA, it is advisable to cut away the very top of the tube and remove the upper part of the gradient, including the nuclear DNA band, with a Pasteur pipette or a syringe and needle, before beginning to collect the closed-circular DNA.

If very pure mtDNA is needed, the collected mtDNA is banded a second time in CsCl–ethidium bromide to remove any residual nuclear DNA. For this purpose, add enough CsCl–ethidium bromide solution (4.8 M CsCl, 200 μg/ml ethidium bromide, 10 mM Tris-HCl, pH 7.4, 10 mM EDTA) to the collected mtDNA to fill a Quick-Seal tube, adjust the density again to

1.55–1.60 g/ml, and centrifuge under the same conditions as in the first run. After the second run, again, two bands are seen in the gradient, the upper one being strongly reduced in amount as compared to the upper band after the first run. The lower band is then collected as described above.

In the following steps, continue to protect the mtDNA from light until all the ethidium bromide has been removed from the DNA solution. Extract the ethidium bromide from the DNA solution by shaking it with an equal volume of isoamyl alcohol. Incubate the mixture in ice for a few minutes to separate the phases. Remove the upper (alcohol) phase, repeat the extraction until no color is visible in the alcohol phase, and then repeat the extraction once more.

To remove the CsCl, dialyze the DNA solution at 4° against 1–2 liters of 10 mM Tris-HCl, pH 7.4, 1 mM EDTA, 0.1 M NaCl, with two to three changes for a total of 6 to 16 hr. Then, add NaCl to the DNA solution, to a final concentration of 0.2 M, and 2.5 volumes ethanol, to precipitate the DNA. To ensure good recovery of the mtDNA, chill the ethanol/DNA mixture overnight at −20° and then centrifuge 30 min at 12,000 g_{av}, at 4°. Remove the supernatant, wash the pellet with 70% (v/v) ethanol, and dissolve the mtDNA in a small volume of 10 mM Tris-HCl, pH 7.4–8.0, 1 mM EDTA.

Read the absorbance of the final preparation at 240, 260, and 280 nm to determine the spectrum. If the ratio A_{260}/A_{240} is close to 1.7, and the ratio A_{260}/A_{280} is close to 2, determine the mtDNA concentration from the absorbance at 260 nm. A 50 μg/ml solution of DNA will have an A_{260} equal to 1.0. If the above ratios differ significantly from the expected ones, pointing to contamination of the mtDNA with protein, the mtDNA should be purified further by one or two extractions with buffer-saturated phenol[5] and ethanol precipitation. The purity of the mtDNA preparation, as concerns the presence of residual nuclear DNA contaminants, should be checked by analyzing the final material by electrophoresis in an 0.8–1% agarose gel in Tris–borate/EDTA buffer or Tris–acetate/EDTA buffer.[5]

Preparation of Mitochondrial DNA from Human Placenta

Take the normal precautions that are to be used when working with large amounts of human blood and human tissues: wear a face shield, laboratory coat, and two pairs of gloves, preferably thicker than the usual disposable latex gloves. Because sharp instruments are used in this protocol, it is important to concentrate fully on the work at hand so as to avoid

---

[5] J. Sambrook, E. F. Fritsch, and T. Maniatis, "Molecular Cloning: A Laboratory Manual." Cold Spring Harbor Laboratory, Cold Spring Harbor, New York, 1989.

accidents. If a superficial cut or an exposure to blood should occur, allow the wound to bleed for a few minutes under running water, wash the exposed area or the wound thoroughly, and get medical attention.

Disruption of Tissue and Isolation of Mitochondria

The method of preparation of mitochondria from human placenta is modified from Hare *et al.*[6] and from Drouin.[7] Collect a placenta, within 60 min of delivery, from a local hospital. Try to ascertain that the placenta is free of viruses. Transport the placenta to the laboratory in ice.

The following operations should be carried out at 2°–4°, unless otherwise specified. Place the placenta in a large glass Pyrex baking dish in ice. Working in a cold room if possible, trim connective tissue from the placenta with a sharp kitchen knife, always cutting *away* from the hand that is holding the placenta in place. Cut the placenta into approximately 2-cm cubes, wrap the tissue in a square of cheesecloth, and rinse it three times in phosphate-buffered saline (PBS: 150 mM NaCl, 8.1 mM Na$_2$HPO$_4$, 1.5 mM KH$_2$PO$_4$, pH 7.65) in a beaker, 300–500 ml each time. Then rinse the tissue in approximately 200–300 ml homogenization buffer (0.25 M sucrose, 0.15 M KCl, 10 mM Tris-HCl, pH 7.5, 1 mM EDTA). Drain and weigh the nearly blood-free tissue in a clean, dry beaker.

Process the tissue for 60 sec with approximately 2 volumes homogenization buffer in a Waring or Hamilton Beach blender with an unbreakable container, at low speed. Transfer the disrupted tissue to centrifuge bottles, spin 10 min at 400 g_{av}, and collect the supernatant through several layers of cheesecloth lining a funnel, holding back the floating "cake," consisting largely of fat, with a glass rod. Centrifuge the supernatant in smaller centrifuge bottles or in tubes 7 min at 1000 g_{av} to pellet the red blood cells and any remaining debris. Centrifuge the supernatant 10 min at 8100 g_{av} to pellet the mitochondrial fraction, and discard the supernatant.

Gently disperse the mitochondria in one of two ways: (1) suspend the pellet by pipetting up and down in a small volume of STE (0.25 M sucrose, 10 mM Tris-HCl, pH 7.5, 1 mM EDTA) and then adjust the volume to about 45 ml, or (2) add STE to approximately 45 ml and homogenize the suspension, 20–25 ml at a time, very gently in a Potter–Elvehjem size C homogenizer with a loosely fitting, motor-driven pestle rotating at low speed. Layer the mitochondrial fraction over six 2-step sucrose gradients in SW27 polyallomer tubes, made by first depositing 15 ml of 1.7 M sucrose in 10 mM Tris-HCl, pH 7.5, 1 mM EDTA at the bottom of each tube and

[6] J. F. Hare, E. Ching, and G. Attardi, *Biochemistry* **19**, 2023 (1980).
[7] J. Drouin, *J. Mol. Biol.* **140**, 15 (1980).

then layering on top 15 ml of 1.0 M sucrose in 10 mM Tris-HCl, pH 7.5, 1 mM EDTA. After layering the mitochondrial fraction over the gradients, fill and balance the tubes with STE, if necessary.

Centrifuge at 22,000 rpm (63,000 g_{av}) for 30 min in an SW27 rotor. Mitochondria band between the 1.0 and 1.7 M sucrose layers. Collect the mitochondria with a large bore 5- or 10-ml pipette, and transfer each band (~10 ml) to a 50-ml centrifuge tube. Add 3–4 volumes STM (0.25 M sucrose, 10 mM Tris-HCl, pH 7.5, 2 mM magnesium acetate) to each tube (to introduce Mg^{2+} and to dilute the sucrose). Mix well, then centrifuge for 15 min at 20,000 g_{av} and discard the supernatant.

DNase Treatment

Suspend the pellets in STM, pool all the suspensions, and add STM to a final volume of 50 ml. Transfer the suspension to a 125-cm^3 Erlenmeyer flask, add 0.5 mg DNase I (Grade I, Boehringer Mannheim, Indianapolis, IN), and incubate at room temperature 30 min. This digestion eliminates much of the nuclear DNA contaminating the mitochondrial fraction. Add 2 ml of 0.5 M EDTA, mix well, and centrifuge the suspension 10 min at 20,000 g_{av}.

Suspend the pellets well in 250 M sucrose, 10 mM Tris-HCl, pH 7.4, 10 mM EDTA, using 1 ml for every 20 g original tissue. Continue with the protocol for the preparation of HeLa cell mtDNA described above, from the step entitled "Lysis of Mitochondria and Equilibrium Density Gradient Centrifugation."

Notes

1. The mitochondria should be intact up to the step of lysis with SDS. Therefore, we do not recommend the use of frozen mitochondria for the preparation of mtDNA.

2. For densities between 1.37 and 1.90 g/ml, the relationship between the density of a CsCl solution at 25°, $\rho^{25°}$, and its refractive index $\eta_D^{25°}$, is described by the following equation[8,9]:

$$\rho^{25°} = 10.8601\,\eta_D^{25°} - 13.4974 \qquad (1)$$

Other useful equations[8] for making or adjusting CsCl solutions are as follows:

$$\text{Weight \%} = 137.48 - 138.11(1/\rho^{25°}) \qquad (2)$$

[8] J. Vinograd, this series, Vol. 6, p. 854.
[9] R. Bruner and J. Vinograd, *Biochim. Biophys. Acta* **108**, 18 (1965).

$$v_w = v_c(\rho_c - \rho_f)/(\rho_f - 0.997) \tag{3}$$
$$v_{1,w} = (\rho_c - \rho_f)/(\rho_c - 0.997) \tag{4}$$

Equation (2) relates the weight percent [g CsCl/(g CsCl + g H_2O) × 100] of the CsCl solution to its density. Equation (3) indicates the volume of water or buffer, v_w, to be added to a volume, v_c, of concentrated solution of CsCl of density ρ_c to obtain the desired density ρ_f. Equation (4) can be used to calculate the volume of water, $v_{1,w}$, required to make 1 ml of the desired solution.

3. If too much DNA has been run in the CsCl gradient, and the bands representing nuclear DNA and closed-circular DNA are not resolved, then the entire DNA region of the gradient should be collected, divided among two to four tubes, and centrifuged in CsCl–ethidium bromide.

4. The closed-circular DNA band material derived from human cells includes, besides mtDNA, also nonmitochondrial closed-circular DNAs,[10] albeit in very low abundance when the DNA is prepared from isolated mitochondria. Because of this impurity, closed-circular band material should not be used as a probe to quantify mtDNA on dot or slot blots. A subcloned fragment of mtDNA should be used instead.

Acknowledgments

This work was supported by National Institutes of Health Grant GM-11726 to Giuseppe Attardi, a Muscular Dystrophy Association grant to Giuseppe Attardi and A. C., and an Italian Government Postdoctoral Fellowship and grant from Associazione Amici del Centro Dino Ferrari to C. A.

[10] R. Radloff, W. Bauer, and J. Vinograd, *Proc. Natl. Acad. Sci. U.S.A.* **57,** 1514 (1967).

[12] Mitochondrial DNA Transcription Initiation and Termination Using Mitochondrial Lysates from Cultured Human Cells

By PATRICIO FERNÁNDEZ-SILVA, VICENTE MICOL, and GIUSEPPE ATTARDI

Introduction

The complete nucleotide sequence of human and several other mammalian mitochondrial DNAs (mtDNAs) has been determined.[1-4] In human mitochondria, the main initiation sites for transcription of the heavy (H) and light (L) strands of the DNA have been identified *in vivo* and *in vitro*,[5-9] the discrete transcription products have been mapped,[10] and their structural and metabolic properties have been characterized.[11-15] By contrast, little is known about the enzymatic machinery involved in transcription in mammalian cells and the regulation of this process. One mitochondrial transcription factor, mtTFA, necessary for high levels of specific initiation of transcription at the L-strand promoter and rRNA-specific H-strand promoter, has been identified and cloned.[16,17] A different DNA-binding factor,

[1] S. Anderson, A. T. Bankier, B. G. Barrell, M. H. de Bruijn, A. R. Coulson, J. Drouin, I. C. Eperon, D. P. Nierlich, B. A. Roe, F. Sanger, P. H. Schreier, A. J. Smith, R. Staden, and I. G. Young, *Nature (London)* **290,** 457 (1981).

[2] M. J. Bibb, R. A. Van Etten, C. T. Wright, M. W. Walberg, and D. A. Clayton, *Cell (Cambridge, Mass.)* **26,** 167 (1981).

[3] S. Anderson, M. H. de Bruijn, A. R. Coulson, I. C. Eperon, F. Sanger, and I. G. Young, *J. Mol. Biol.* **156,** 683 (1982).

[4] M. N. Gadaleta, G. Pepe, G. De Candia, E. Quagliarello, E. Sbisà, and C. Saccone, *J. Mol. Evol.* **28,** 497 (1989).

[5] J. Montoya, T. Christianson, D. Levens, M. Rabinowitz, and G. Attardi, *Proc. Natl. Acad. Sci. U.S.A.* **79,** 7195 (1982).

[6] M. W. Walberg and D. A. Clayton, *J. Biol. Chem.* **258,** 1268 (1983).

[7] B. K. Yoza and D. F. Bogenhagen, *J. Biol. Chem.* **259,** 3909 (1984).

[8] D. J. Shuey and G. Attardi, *J. Biol. Chem.* **260,** 1952 (1985).

[9] D. D. Chang and D. A. Clayton, *Cell (Cambridge, Mass.)* **36,** 635 (1984).

[10] D. Ojala, C. Merkel, R. Gelfand, and G. Attardi, *Cell (Cambridge, Mass.)* **22,** 393 (1980).

[11] R. Gelfand and G. Attardi, *Mol. Cell. Biol.* **1,** 497 (1981).

[12] J. Montoya, G. Gaines, and G. Attardi, *Cell (Cambridge, Mass.)* **34,** 151 (1983).

[13] G. Gaines and G. Attardi, *J. Mol. Biol.* **172,** 451 (1984).

[14] G. Gaines and G. Attardi, *Mol. Cell. Biol.* **4,** 1605 (1984).

[15] G. Gaines and G. Attardi, *J. Biol. Chem.* **262,** 1907 (1987).

[16] R. P. Fischer and D. A. Clayton, *J. Biol. Chem.* **260,** 11330 (1985).

[17] M. A. Parisi and D. A. Clayton, *Science* **252,** 965 (1991).

mTERF (mitochondrial transcription termination factor), involved in termination of transcription at the 16 S rRNA/tRNA$^{Leu(UUR)}$ boundary, has been extensively characterized in our laboratory.[18,19] On the other hand, nothing is known about the mtRNA polymerase itself and the probable existence of other factors involved in transcription initiation at the downstream, whole H-strand specific promoter.[20]

Submitochondrial transcription systems have proved to be useful tools to study mtDNA transcription. These systems are easy to manipulate, and they have already provided valuable information about some aspects of mtDNA transcription, such as the energetic requirements for initiation and the sequences necessary for an effective and accurate initiation and termination of transcription.[9,21,22] Two different types of such "open" transcription systems have been developed. One of them utilizes partially purified protein components to promote transcription from mtDNA-derived templates[6,9,22]; the other one uses the S-100 of a mitochondrial lysate programmed by exogenous templates.[8,18,21,23] We have used the latter system extensively because of its efficiency and reproducibility, and particularly because it carries out several of the *in vivo* transcription activities, being at the same time accessible to external manipulation.

Using as a template a plasmid construct that allows the reproduction of the main events of *in vivo* transcription (Fig. 1), we have greatly improved the soluble transcription system previously developed in this laboratory.[8,21] In particular, the overall initiation and termination activities of the mitochondrial lysate have been increased about 8- and 15-fold, respectively, relative to the old protocol. Using this improved system, we have characterized in detail the effects on the transcription initiation and termination events of varying the protein and the template concentrations, the ATP level, the ionic strength, and the Ca^{2+} and Mg^{2+} concentrations. We have also used this system to characterize the termination-promoting activity of affinity-purified mTERF.[24] This transcription system is also expected to be very useful for the identification and purification of other factors active in mtDNA transcription.

In this chapter, we describe a modification of the soluble transcription

[18] B. Kruse, N. Narasimhan, and G. Attardi, *Cell* **58,** 391 (1989).
[19] A. Daga, V. Micol, D. Hess, R. Aebersold, and G. Attardi, *J. Biol. Chem.* **268,** 8123 (1993).
[20] A. Chomyn and G. Attardi, *in* "Molecular Mechanisms in Bioenergetics" (L. Ernster, ed.), p. 483. Elsevier, Amsterdam, 1992.
[21] N. Narasimhan and G. Attardi, *Proc. Natl. Acad. Sci. U.S.A.* **84,** 4078 (1987).
[22] R. P. Fischer, J. N. Topper, and D. A. Clayton, *Cell* **50,** 247 (1987).
[23] B. Kruse, N. N. Murdter, and G. Attardi, *Methods Mol. Biol.* **37** (*In Vitro* Transcription and Translation Protocols) (1995).
[24] V. Micol, P. Fernández-Silva, and G. Attardi, this volume [15].

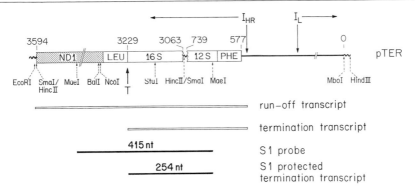

FIG. 1. Map of the clone pTER used for *in vitro* transcription. The map positions of the H-strand run-off and terminated transcripts, of the probe used in the S1 protection assays, and of the S1-protected terminated transcripts are shown. I_{HR}, Upstream initiation site for H-strand transcription; I_L, initiation site for L-strand transcription. Modified, with permission, from A. Daga, V. Micol, D. Hess, R. Aebersold, and G. Attardi, *J. Biol. Chem.* **268,** 8123 (1993).

system utilizing the 13,000 g supernatant (S-13) of a mitochondrial lysate from small-scale cell culture samples. Furthermore, we report an adaptation of the mobility shift assay for the semiquantitative detection of mTERF in the S-13 fraction from small amounts of cells, which allows an easy comparison of its relative content in different cell lines and under different growth conditions.

Procedures

Reagents and Solutions. All glassware should be baked at 180° at least 4 hr. Solutions should be prepared with diethyl pyrocarbonate (DEPC)-treated and autoclaved distilled water.

2 *M* Sucrose in autoclaved distilled water: sterilize by filtration and store frozen at −20°

0.5 *M* Dithiothreitol (DTT): dissolve in autoclaved distilled water and store in aliquots at −20° (add to buffers just before use)

0.5 *M* Phenylmethylsulfonyl fluoride (PMSF, Sigma, St. Louis, MO): dissolve in either ethanol or 2-propanol and store in aliquots at −20° (add to buffers just before use)

Polyoxyethylenesorbitan monolaurate (Tween 20, Sigma)

NKM buffer: 1 m*M* Tris-HCl, pH 7.4 at 25°, 0.13 *M* NaCl, 5 m*M* KCl, 7.5 m*M* MgCl$_2$

Homogenization buffer: 10 m*M* Tris-HCl, pH 6.7, 10 m*M* KCl, 0.15 m*M* MgCl$_2$

Mitochondria suspension buffer: 10 mM Tris-HCl, pH 6.7, 0.25 mM sucrose, 0.15 mM MgCl$_2$

Mitochondria lysis buffer: 25 mM HEPES–KOH, pH 7.6, 5 mM MgCl$_2$, 0.5 mM EDTA, 1 mM DTT, 1 mM PMSF, 10% glycerol v/v

TD buffer: 25 mM Tris-HCl, pH 7.4, 135 mM NaCl, 5 mM KCl, 0.5 mM Na$_2$HPO$_4$

Trypsin solution: 5 g/liter trypsin(1–300) (from porcine pancreas, ICN Biomedicals, Irvine, CA), 3.2 g/liter penicillin, 0.2 g/liter streptomycin, 0.003% phenol red w/v

TBE: 89 mM Tris base, 89 mM boric acid, 2 mM EDTA

Transcription buffer (2×): 20 mM Tris-HCl, pH 8.0, 20 mM MgCl$_2$, 2 mM DTT, 200 μg/ml bovine serum albumin (BSA), 20% (v/v) glycerol; sterilize by filtration and store at $-20°$

NTP mixture (25×): 25 mM ATP, 2.5 mM GTP and CTP, 0.25 mM UTP

Transcription stop buffer: 10 mM Tris base, 0.3 M sodium acetate, pH 7.0, 15 mM EDTA, 0.5% w/v SDS, final pH adjusted to 8.0 with HCl

0.5% SDS buffer: 10 mM Tris-HCl, pH 7.4, 0.2 M NaCl, 10 mM EDTA, 0.5% w/v sodium dodecyl sulfate (SDS)

T3 transcription buffer (5×): 200 mM Tris-HCl, pH 7.9, 30 mM MgCl$_2$, 10 mM spermidine, 50 mM NaCl

DNase buffer (5×): 100 mM Tris-HCl, pH 7.6, 50 mM CaCl$_2$, 50 mM MgCl$_2$

S1 hybridization buffer: 80% v/v formamide, 40 mM PIPES–HCl, pH 6.4, 380 mM NaCl, 0.5 mM EDTA; store at $-80°$

S1 digestion buffer (10×): 400 mM sodium acetate, 30 mM ZnCl$_2$, 2.5 M NaCl; adjust to pH 4.6 with HCl, filter, and store at $4°$

S1 stop buffer: 4 M ammonium acetate, 20 mM EDTA, 200 μg/ml yeast tRNA

Buffer C: 25 mM HEPES–KOH, pH 7.5, 100 mM KCl, 12.5 mM MgCl$_2$, 1 mM DTT, 20% (v/v) glycerol, 0.1% (v/v) Tween 20; store at $4°$

Urea–dye: 7 M urea, 0.01% w/v bromphenol blue, 0.01% w/v xylene cyanol in TBE; keep at $-20°$ and thaw at $65°$ before use

Tris–glycine buffer (5×): 250 mM Tris–base, 1.9 M glycine, 10 mM EDTA, pH ~8.5

Preparation of S-13 of Mitochondrial Lysate

An important difference in the preparation of the mitochondrial lysate with respect to the previously described protocol[8,21] is the use of Tween 20 instead of NP-40 (Nonidet P-40) in the lysis step. The substitution of NP-40 with Tween 20 dramatically increases the transcription initiation and termination activities of the lysate. [*Note:* The change in the detergent used to prepare the mitochondrial lysate was adopted after analyzing the effect, on the transcription initiation activities, of four different detergents (Triton X-100, sodium deoxycholate, NP-40, and Tween-20), Tween-20 being the one that gave the best results. The relative overall activities were in the ratio of 100 for Tween 20, 13 for NP-40, 5 for Triton X-100, and 3 for sodium deoxycholate. A detailed comparison between the NP-40 and Tween 20 lysates showed that the overall initiation and termination activities were, respectively, 8- and 15-fold higher when Tween 20 was used.]

The Tween 20 mitochondrial lysate containing the transcription machinery has been prepared from either suspension cultures or cultures grown on a solid substrate. We have used two different protocols, depending on the amount of starting material. The preparation of the S-100 of the mitochondrial lysate from a large-scale human cell suspension culture (e.g., HeLa cells grown in high-phosphate-containing Dulbecco's modified Eagle's medium supplemented with 5% v/v calf serum) is described in detail elsewhere in this volume.[24] Here, we report the preparation of the S-13 of the Tween 20 mitochondrial lysate from small-scale suspension cultures or from cell cultures grown on a solid substrate, which yield a packed cell volume ranging from 0.05 to 1.0 ml.

In the case of cells grown on a solid substrate, plates that are about 80–90% confluent are used. Each plate is washed once with 5 ml TD buffer and then trypsinized with 5 ml of a 1 : 10 dilution of trypsin solution in TD buffer containing 10 mM EDTA, prewarmed to 37°. The trypsin digestion is allowed to proceed for 5 min at room temperature, and then the cell suspension is collected with a Pasteur pipette and transferred to a conical tube containing 1 volume TD buffer with 5% calf serum, placed in ice. The cells (deriving from the trypsinized plates or from a small suspension culture) are pelleted at 500 g_{av} for 5 min and washed three times with NKM. Before the last centrifugation, the cells are transferred to a graduated Eppendorf tube(s), so that the approximate volume of packed cells can be estimated (from five plates containing a total of 10^7 cells one can expect a volume of packed cells around 100 μl). After removing the supernatant from the last washing step, the pelleted cells are placed at −20° for 15 min in order to weaken the cell membrane and facilitate the breakage in the homogenization step.

The samples are placed in ice, and the pellets are then suspended in a volume of homogenization buffer corresponding to 10 times the original volume of packed cells (vpc) and homogenized in a Thomas homogenizer (A. H. Thomas, Swedesboro, NJ) of appropriate size (0 for up to 1 ml homogenate, AA for up to 4 ml, and A for up to 10 ml) with 10 strong strokes, using a pestle rotating at approximately 1500 rpm. The degree of cell breakage obtained in this way is usually around 80–90%.

The homogenate is immediately poured into one or more Eppendorf tubes containing 2 M sucrose in an amount corresponding to one-sixth of the volume of the homogenate, mixed, and centrifuged at 1200 g to pellet nuclei, large debris, and unbroken cells. This centrifugation is repeated, and the last supernatant is transferred to a fresh Eppendorf tube(s), and centrifuged at approximately 13,000 g in a microcentrifuge for 1 min at 4°. The mitochondrial pellet(s) thus obtained is washed twice with 1 ml mitochondrial suspension buffer by centrifuging at approximately 13,000 g for 1 min, and finally suspended in 1 vpc mitochondria lysis buffer containing freshly added Tween 20 and KCl to final concentrations of 0.5% and 0.5 M, respectively. The mitochondria are lysed by pipetting up and down repeatedly until the suspension has become clear, and then by vortexing vigorously for 30 sec. The lysate is then kept on ice 5 min, and the vortexing step is repeated once in the same way. The lysate is then centrifuged for 45 min at approximately 13,000 g in the microcentrifuge at 4°, and the clear supernatant (S-13) is collected carefully, avoiding the fluffy layer. Aliquots of the S-13 are frozen in liquid N_2 and stored at $-80°$. The protein concentration of the S-13 preparation is determined by the method of Bradford.[25] The transcription activity of the S-13 is stable at $-80°$ for at least 6 months. Repeated freezing and thawing slowly reduce this activity.

In Vitro Transcription Assay

For the analysis of the transcription initiation and termination activities present in the S-13 of the mitochondrial lysate, we have chosen a construct (pTER)[17] that closely reproduces the known critical regions of the *in vivo* template, since it contains the whole promoter regions for L- and H-strand transcription and adjacent sequences, and the termination region located at the 16 S rRNA/tRNA$^{Leu(UUR)}$ boundary (Fig. 1).

In the standard transcription experiments, for each sample to be tested, the following components are mixed in an Eppendorf tube (placed in ice): 1 μl of the template DNA (pTER) at a concentration of 1 μg/μl, 2 μl of the 25× NTP mixture, 5 μl of the S-13 preparation to be analyzed, 25 μl of 2× transcription buffer, 1 μl of [α-^{32}P]UTP (400 Ci/mmol, 10 mCi/ml; Amersham, Arlington Heights, IL), and DEPC-treated autoclaved, distilled water to bring the volume to 50 μl. (*Note:* The protein concentration has been found to be 4–6 mg/ml in the S-13 preparation of the Tween 20 mitochondrial lysate from a small-scale preparation, to be compared to a concentration of 10–12 mg/ml in the S-100 of the Tween 20 lysate from a large-scale preparation. The concentrations of lysate and DNA template

---

[25] M. M. Bradford, *Anal. Biochem.* **72**, 248 (1976).

in the transcription assay may have to be adjusted in order to find the optimal ratio for each preparation.[21]) The mixtures are vortexed, briefly spun in an Eppendorf microcentrifuge, and incubated at 30° for 30 min. The reaction is stopped by adding to each sample 100 μl transcription stop buffer and 20 μg yeast tRNA, and the sample is extracted with 1 volume of phenol/CHCl$_3$/isoamyl alcohol (25:24:1, v/v). The nucleic acids are precipitated by adding 330 μl ethanol (prechilled at −20°) and keeping the mixtures at −80° for 30 min; then they are pelleted by centrifuging at full speed in a microfuge at 4° for 10 min. The pellet is dissolved in 150 μl 0.5% SDS buffer, extracted again with 1 volume of phenol/CHCl$_3$/isoamyl alcohol (25:24:1, v/v), and the nucleic acids are precipitated as described above. The final pellet is dissolved in DEPC-treated autoclaved distilled water, and the synthesized products are analyzed in a 5% polyacrylamide (acrylamide:bisacrylamide, 29:1)/7 M urea gel; alternatively, if so required, the transcripts are subjected to S1 protection analysis, as described below.

The pellet from each transcription reaction is dissolved in 100 μl of 1× DNase buffer, 10 units RNase-free DNase I is then added, and the mixture is incubated at room temperature for 20 min in order to destroy the template DNA. After phenol extraction and ethanol precipitation, the labeled RNA pellet is dissolved in DEPC-treated autoclaved distilled water, mixed with 0.2–0.4 μg of the unlabeled RNA probe (see below), and precipitated again with NaCl and ethanol. The new pellet is carefully suspended in 20 μl of S1 hybridization buffer by pipetting up and down repeatedly. Denaturing of the sample is achieved by heating it at 80° for 10 min, and hybridization is performed at 50° for 6 to 10 hr. After the hybridization, 200 μl of S1 digestion buffer containing 20 μg/ml denatured salmon sperm DNA and 250 to 400 units S1 nuclease (Boehringer Mannheim, Indianapolis, IN; 400 units/μl) are added to the sample, and the mixture is incubated at 41° for 30 to 45 min. Finally, 55 μl of S1 stop buffer is added to the sample, and the S1-resistant products are precipitated by adding 2.5 volumes ethanol and pelleted by centrifuging at approximately 13,000 g for 10 min at 4°C. The final pellets are suspended in the desired volume of DEPC-treated autoclaved distilled water, 1 volume of urea–dye is added, and, after denaturing at 80° for 10 min, the samples are loaded onto a 5% polyacrylamide (acrylamide:bisacrylamide, 29:1)/7 M urea gel in TBE and run in the same buffer for 2–3 hr at 400 V. After electrophoresis, the gel is washed twice for 10 min with autoclaved distilled water and vacuum-dried on a sheet of Bio-Rad (Richmond, CA) backing paper at 80° for 1 hr, before exposure for autoradiography.

To prepare the antisense unlabeled RNA to be used as a probe for S1 analysis of the transcripts, a 50-μl reaction mixture is set up containing the following: 1.5 μg plasmid pBSAND [containing the *Mae*I–*Mae*I fragment of pTER filled in and cloned into the *Sma*I site of plasmid pBS KS(+) (Promega, Madison, WI)] linearized with *Bam*HI, 10 μl of 5× T3 transcription buffer, 0.5 mM each of ATP, CTP, GTP, and UTP, 8 mM DTT, 40 units RNase inhibitor (RNasin, Promega; 40 units/μl), and 20 units T3 RNA polymerase (Promega; 20 units/μl). The mixture is incubated at 37° for 90 min, and then 20 units RNase-free DNase I (Boehringer Mannheim, 10 units/μl) is added and the mixture incubated for an additional 15 min at 37° in order to destroy the template DNA. After phenol extraction, the synthesized RNA is precipitated by adding ammonium acetate to a final concentration of 2 M and 2.5 volumes ethanol, then recovered by centrifugation at approximately 13,000 g for 10 min. The amount of probe obtained is determined by reading the A_{260}. A yield of 2 to 4 μg RNA per reaction is usually obtained.

The results of a typical transcription experiment are shown in Fig. 2. In lane 1 (Fig. 2), containing the labeled products, two main bands are visible, one corresponding to the H-strand runoff transcripts (H) and the other to the L-strand runoff transcripts (L). Lane 2 (Fig. 2), containing the S1-resistant products, shows two major bands corresponding to the

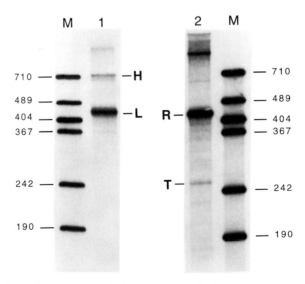

FIG. 2. Products of *in vitro* transcription assay using the S-13 of a Tween 20 mitochondrial lysate and *Eco*RI- and *HIN*dIII-digested pTER clone template. Lane 1 shows transcription products (H, H-strand runoff transcripts; L, L-strand runoff transcripts). Lane 2 shows S1-resistant products after hybridization of the *in vitro* products with the *Mae*I–*Mae*I unlabeled RNA probe (Fig. 1) (R, H-strand runoff transcripts; T, H-strand terminated transcripts).

protected fragments of the H-strand run-off (R) and terminated (T) transcripts.

Detection of Mitochondrial Transcription Termination Factor by Mobility Shift Assay

The protocol for the mobility shift assay of mTERF described elsewhere in this volume[24] was designed for the purpose of detecting and measuring the binding activity of mTERF purified by affinity chromatography.[20] In this chapter, we describe a modification of this procedure suitable for detecting the presence and estimating the relative abundance of mTERF in mitochondrial lysates derived from small-scale cell culture samples. The amount of probe per reaction has been increased in order to saturate the binding capacity of the factor; furthermore, in this modified protocol, the gel is run in the cold and is loaded with the current on to keep the dissociation of the DNA–protein complex to the minimum possible and to give sharp bands suitable for quantitation. By performing band-shift experiments with a constant amount of probe and increasing amounts of S-13 lysate, we were able to determine the conditions for a semiquantitative assay (Fig. 3).

The probe used for this assay is the double-stranded 44-mer fragment described elsewhere,[24] 3'-end-labeled by filling in with $[\alpha^{-32}P]dTTP$ and the Klenow fragment of *Escherichia coli* DNA polymerase I.[26] To set up the DNA-binding reactions, for each sample to be tested, a mixture is prepared, in an Eppendorf tube placed in ice, containing the following components: 0.1 pmol probe [labeled to a specific activity of ~200,000 counts/min (cpm)/pmol], 5 μg BSA, 5 μg poly(dI-dC) · (dI-dC), 10 μl buffer C containing 50 instead of 100 mM KCl, and the desired amount of the S-13 of the mitochondrial lysate to be tested (usually between 1 and 4 μl). The volume is then brought to 20 to 50 μl by adding buffer C lacking KCl, with the final KCl concentration being adjusted to 100 mM by the addition of 3 M KCl, if needed. The samples are vortexed, spun down briefly, and incubated at 25° for 20 min. After incubation, the samples are placed in ice and then loaded onto a 5% polyacrylamide : bisacrylamide (80 : 1) gel in Tris–glycine buffer containing 2.5% glycerol, which had been prerun for at least 2 hr at 200 V. After loading, the gel is run in the cold at 200–250 V for 4 hr, heat-dried at 80° for 2 hr, and exposed for autoradiography.

Figure 3 shows a calibration experiment carried out by incubating a constant amount of probe (0.1 pmol per reaction) with different amounts of the S-13 preparation of a mitochondrial lysate, corresponding to a range

[26] F. M. Ausubel, R. Brent, R. E. Kingston, D. D. Moore, J. G. Seidman, J. A. Smith, and K. Struhl, eds., "Current Protocols in Molecular Biology." Wiley(Interscience), New York, 1987.

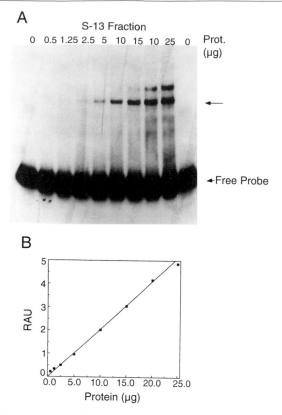

FIG. 3. Mobility shift assay calibration curve using the S-13 of a Tween 20 mitochondrial lysate. (A) A constant amount of the probe was incubated with different amounts of a sample of S-13, as described in the text. The specific retarded band that has been quantified in (B) is indicated by an arrow. (B) Quantification of the specific retarded band in the gel shown in (A) performed by Phosphor-Image detection and Image-Quant analysis (Molecular Dynamics).

of cell equivalents between 1×10^4 and 5×10^5. As the quantitation of the autoradiogram shows (Fig. 3b), there is a range of amounts of S-13/pmol probe in which the signal of the specific retarded band is proportional to the amount of lysate. (*Note:* For each sample to be analyzed by the semiquantitative band shift assay, we usually test two different amounts of the S-13 preparation of a Tween 20 mitochondrial lysate included in the linear portion of the calibration curve. Repeated experiments using different S-13 preparations from the same cell source have given consistent results in the relative abundance of the retarded band. The minimum amount of protein of an S-13 preparation that gives quantifiable results under our experimental conditions corresponds to $\sim 5 \times 10^4$ cells.)

Acknowledgments

This work was supported by National Institutes of Health Grant GM-11726 (to G. A.) and scholarships from the Fulbright/Spanish Ministry of Education and Science Visiting Scholar Program (to V. M.) and from Caltech (Gosney Fellowship, to P. F.-S.).

[13] Mapping Promoters in Displacement-Loop Region of Vertebrate Mitochondrial DNA

By GERALD S. SHADEL and DAVID A. CLAYTON

Introduction

In all vertebrates examined to date, mitochondrial (mt) DNA exists and can be isolated as a small covalently closed-circular molecule that has a promoter-containing regulatory locus called the displacement-loop (D-loop) region (Fig. 1). It is in the D-loop region where leading-strand DNA replication and the majority of mitochondrial gene transcription initiates. This chapter outlines a general strategy for locating the major promoters in the D-loop region of the mitochondrial genome of vertebrate organisms. We propose two approaches for the isolation of a D-loop region-containing DNA fragment that utilize conserved nucleotide sequences that occur within or near the D-loop region of vertebrates. Once in hand, promoters can be precisely located on this template using a crude mtRNA polymerase preparation to assay promoter activity in an *in vitro* transcription reaction. Rather than presenting the conventional methods for mapping the 5' ends of primary transcripts and defining boundaries of promoters in detail, those aspects unique to analyzing mitochondrial promoters are emphasized. We refer to the published methods that were used to locate D-loop region promoters of human[1] and mouse[2,3] mtDNAs as models for this type of analysis.

Isolation of Mitochondria and Mitochondrial DNA from Vertebrate Cells

Many of the procedures outlined in this chapter require the isolation of mitochondria and mtDNA from vertebrate cells. Methods for the purifi-

---

[1] D. D. Chang and D. A. Clayton, *Cell (Cambridge, Mass.)* **36,** 635 (1984).
[2] D. D. Chang and D. A. Clayton, *Mol. Cell. Biol.* **6,** 3253 (1986).
[3] D. D. Chang and D. A. Clayton, *Mol. Cell. Biol.* **6,** 3262 (1986).

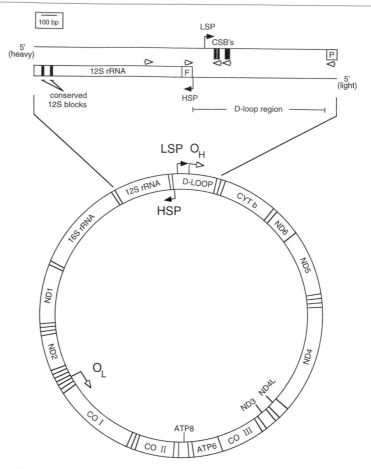

FIG. 1. The mouse mitochondrial genome is shown as an example of a typical vertebrate. The circular mouse mtDNA molecule is presented with the region between the 12 S rRNA and tRNA^Pro genes expanded to emphasize important elements in the D-loop region of the genome. In the circular diagram, tRNAs are depicted as open boxes. The rRNA and protein coding genes are labeled. Also indicated are the H-strand promoter (HSP), L-strand promoter (LSP), and the origins of H-strand (O_H) and L-strand (O_L) DNA replication. The promoters are indicated by the filled arrows and the origins by open arrows. In the expanded portion of the figure, the tRNA^Phe and tRNA^Pro genes are indicated by the boxes F and P, respectively. The black boxes in the D-loop region are the conserved sequence blocks (CSBs). The black boxes in the 12 S rRNA gene are conserved sequences present in mitochondrial 12 S rRNAs, E. coli 16 S rRNA, and yeast 18 S rRNA. The transcription initiation sites are indicated by bent arrows. The heavy and light mtDNA strands are indicated. The open arrowheads indicate the orientation and location of putative primers designed to amplify a D-loop region-containing fragment by polymerase chain reaction (PCR).

cation of intact mitochondria by centrifugation through a sucrose step gradient and isolation of mitochondrial nucleic acids have been documented in a previous volume in this series.[4] Mitochondrial DNA isolated by this method can be used to generate a library of mtDNA fragments and is a suitable template for amplification by polymerase chain reaction (PCR).

Isolation of Vertebrate D-Loop Regions by Polymerase
    Chain Reaction

In all vertebrates examined to date, the D-loop region of the mitochondrial genome is flanked by two tRNA genes (Fig. 1). The tRNA[Phe] gene is universally located downstream of the major heavy (H)-strand promoter (HSP), and the tRNA[Pro] gene is located downstream of the major light (L)-strand promoter (LSP) (except for birds, which have tRNA[Gln] in this position). Also, the 12 S rRNA gene is located immediately downstream of tRNA[Phe]. In addition, there are conserved sequence blocks (CSBs) within the D-loop regions of vertebrates that are most commonly downstream of a LSP.[5] These sequences are associated with transition points from RNA to DNA synthesis during initiation of H-strand DNA replication. The nucleotide sequences of the mitochondrial tRNAs and rRNAs are the most highly conserved mitochondrial coding sequences in vertebrates, and CSB I and CSB II are sufficiently conserved to allow the rational design of either homologous or degenerate primers for use in a PCR amplification. Published alignments of these mitochondrial sequences that will aid in the design of these primers include the alignment of the tRNAs from *Xenopus,* human, mouse, and bovine by Roe *et al.*[6]; alignment of the 12 S rRNAs and CSB I from *Xenopus,* chicken, and human by Desjardins and Morais[7]; alignment of tRNAs and 12 S rRNAs of human and bovine by Anderson *et al.*[8]; and alignment of the CSBs from human, mouse, and rat by Walberg and Clayton.[5] In addition, a similar PCR protocol for the amplification of the human D-loop region has been described which may also prove helpful in the design of the precise PCR strategy.[9]

Procedure. Design a set of primers based on conserved regions of

[4] D. P. Tapper, R. A. Van Etten, and D. A. Clayton, this series, Vol. 97, p. 426.

[5] M. W. Walberg and D. A. Clayton, *Nucleic Acids Res.* **9,** 5411 (1981).

[6] B. A. Roe, D. Ma, R. K. Wilson, and J. F. H. Wong, *J. Biol. Chem.* **260,** 9759 (1985).

[7] P. Desjardins and R. Morais, *J. Mol. Biol.* **212,** 599 (1990).

[8] S. Anderson, M. H. L. DeBruijn, A. R. Coulson, I. C. Eperon, F. Sanger, and I. G. Young, *J. Mol. Biol.* **156,** 683 (1982).

[9] C. Orrego and M. C. King, *in* "PCR Protocols: A Guide to Methods and Applications" (M. A. Innis, D. H. Gelfand, J. J. Sninsky, and T. J. White, eds.), p. 418. Academic Press, San Diego, 1990.

tRNA^{Phe}, tRNA^{Pro} (or tRNA^{Gln} for birds), 12 S rRNA, CSB I, and CSB II. Primers should correspond to the mtDNA H strand for tRNA^{Phe} and 12 S rRNA and the mtDNA L strand for tRNA^{Pro} (tRNA^{Gln}), CSB I, and CSB II (Fig. 1). Amplify a D-loop region-containing DNA fragment by PCR using either purified vertebrate mtDNA (or a library of cloned mtDNA fragments) as a template and the following pairs of primers:

Pair 1: primer 1, tRNA^{Phe}; primer 2, tRNA^{Pro} (or tRNA^{Gln} for birds)
Pair 2: primer 1, 12 S rRNA; primer 2, tRNA^{Pro} (or tRNA^{Gln} for birds)
Pair 3: primer 1, tRNA^{Phe}; primer 2, CSB I
Pair 4: primer 1, 12 S rRNA; primer 2, CSB I
Pair 5: primer 1, tRNA^{Phe}; primer 2, CSB II
Pair 6: primer 1, 12 S rRNA; primer 2, CSB II

The optimal PCR conditions must be determined empirically. Any PCR products obtained should be cloned and the nucleotide sequence determined to ensure that the amplified fragment is similar to known vertebrate D-loop regions. As with the sequence of any PCR product, several independent clones should be analyzed to ensure that sequencing errors were not introduced by the PCR protocol.

Probing Vertebrate Mitochondrial Genomic Library with
    12 S rRNA Probe

The following hybridization strategy is meant to serve as an alternative to the PCR strategy above. This procedure requires obtaining or constructing a library of mtDNA from the vertebrate of interest. Within vertebrate 12 S rRNA genes are many significant stretches of nucleotide identity that can be used to design homologous probes for the gene. Also, comparison of the genes for the human mitochondrial 12 S rRNA, the *Escherichia coli* 16 S rRNA, and the yeast 18 S rRNA revealed two significantly conserved regions near the 3' end of the gene[10] (Fig. 1).

Procedure. Design oligonucleotide probes based on sequences that are conserved between the mitochondrial 12 S rRNA genes from known vertebrates. Also, design two oligonucleotides based on the two sequences that are highly conserved between vertebrate mitochondrial 12 S rRNAs and the small rRNA subunits of *E. coli* and yeast. Use these oligonucleotides to probe for the 12 S rRNA gene in purified mtDNA by Southern analysis. A positive signal on genomic mtDNA indicates which probe will be sufficient to screen a bacterial library of the vertebrate mtDNA fragments. Because of the conserved location of the 12 S rRNA gene in mtDNA

---

[10] I. C. Eperon, S. Anderson, and D. P. Neirlich, *Nature (London)* **286,** 460 (1980).

(downstream of the HSP after tRNA^{Phe}) (Fig. 1), this probe would serve to identify a DNA fragment in a mtDNA library that contained sequences near the D-loop region. Isolate a fragment that hybridizes to this probe and that has sequences at least 1000 base pairs on each side of the 12 S rRNA. Such a fragment would likely contain all of the D-loop region promoters.

Isolation of Mitochondrial RNA Polymerase from Vertebrate Cells

For the purpose of mapping D-loop region promoters it is necessary to isolate a mtRNA polymerase fraction that is competent for specific transcription initiation. It is not necessary to purify the enzyme highly; in fact, extensive purification is not encouraged for this purpose because it would greatly increase the chance that the mtRNA polymerase will be purified away from protein factors required for high levels of specific transcription initiation. The procedure we recommend is adapted from that used to isolate human mtRNA polymerase.[11] A mtRNA polymerase fraction isolated in this manner should be competent for transcription of all D-loop region promoters.

Solutions and Reagents

Wash buffer: 20 mM Tris hydrochloride (pH 8.0), 0.2 mM EDTA, 1 mM dithiothreitol (DTT), 1 mM phenylmethylsulfonyl fluoride (PMSF), 0.25 M sucrose, 15% (v/v) glycerol

Buffer A: 10 mM Tris hydrochloride (pH 8.0), 0.1 mM EDTA, 1 mM DTT, 7.5% (v/v) glycerol, 0.5 mM PMSF

Buffer B–0.1 M KCl: buffer A without PMSF but with 0.1 M KCl

Buffer B–0.3 M KCl: buffer A without PMSF but with 0.3 M KCl

Storage buffer: 10 mM Tris hydrochloride (pH 8.0), 0.1 mM EDTA, 1 mM DTT, 50% (v/v) glycerol, 0.5 mM PMSF

Procedure. To prepare a soluble extract (S-130) of vertebrate mitochondrial enzymes, mitochondria from approximately 1×10^{10} to 3×10^{10} cells are isolated by sucrose step gradient centrifugation as described.[4] The pelleted mitochondria from this procedure are washed once with 300 ml wash buffer and centrifuged in a JA-20 rotor (15 min, 18,000 g_{av}, 4°). The washed mitochondria are then suspended in 30 ml lysis buffer (same as wash buffer, except that sucrose is omitted). The volume of the suspension is then adjusted to 53 ml with deionized water, and the mitochondria are lysed by addition of 1.5 ml of 20% (v/v) Triton X-100 (final concentration 0.5%) and homogenized in a glass Dounce homogenizer. The KCl concen-

---

[11] R. P. Fisher and D. A. Clayton, *Mol. Cell. Biol.* **8,** 3496 (1988).

tration is then brought to 0.35 M by the addition of 5.25 ml of 4 M KCl, and a second homogenization is done. The lysate is then centrifuged 60 min at 45,000 rpm (130,000 g_{av}) in a Ty75Ti rotor (Beckman Instruments, Palo Alto, CA), and the supernatant (S-130) is removed.

Next, the S-130 extract is subjected to chromatography on DEAE-Sephacel. Dilute the S-130 3.5-fold with buffer A to reduce the KCl concentration to 0.1 M and load the material directly onto a 60-ml DEAE-Sephacel column that has been equilibrated in buffer B–0.1 M KCl. Wash the column with buffer B–0.1 M KCl until a constant A_{280} of the effluent is achieved. Next, wash the column with buffer B–0.3 M KCl and collect 1-ml fractions. To identify fractions containing mtRNA polymerase, nonspecific RNA polymerase activity is assayed using poly(dA-dT) as a template and quantitated by a filter-binding assay.[12] A standard reaction contains 10 mM Tris hydrochloride (pH 8.0), 10 mM MgCl$_2$, 100 μg/ml bovine serum albumin (BSA) (RNase-free), 1 mM DTT, 400 μM ATP, 150 μM CTP and GTP, 1 μM UTP (unlabeled), 0.2 μM [α-^{32}P]UTP (DuPont NEN, Boston, MA), and 10 μg/ml poly(dA-dT). Portions of the fractions from the DEAE-Sephacel column are diluted or dialyzed in buffer A to reduce the salt concentration of the sample (vertebrate mtRNA polymerases are very sensitive to salt concentration) and then assayed in a total volume of 25 μl with the standard reaction components. Transcription reactions are incubated at 28° for 30 min, precipitated in 2 ml trichloroacetic acid and 50 μM sodium pyrophosphate, and then filtered through Whatman (Clifton, NJ) GF/B paper. After several washes, the filters are dried and radioactive RNA bound to the filter paper is counted by liquid scintillation. Fractions containing the peak of nonspecific RNA polymerase activity are pooled and dialyzed against storage buffer.

Alternative Procedure. Originally, a heparin-Sepharose chromatography step was used instead of the DEAE-Sephacel step during the isolation of human mtRNA polymerase.[12,13] We refer to this procedure as a logical alternative to the DEAE-Sephacel step should it prove unsuccessful.

Specific Initiation of Transcription from D-Loop Region Promoters

Once a pool of active vertebrate mtRNA polymerase has been obtained, it can be tested for the ability to initiate transcription from the D-loop region promoters. For the initial attempts at assaying specific transcription products, start with a linearized D-loop region template that would be predicted to give a transcript from a LSP (a LSP is likely to be the strongest

---

[12] R. P. Fisher and D. A. Clayton, *J. Biol. Chem.* **260**, 11330 (1985).
[13] M. W. Walberg and D. A. Clayton, *J. Biol. Chem.* **258**, 1268 (1983).

promoter). Once a LSP-specific product is obtained, transcription from the other strand of mtDNA can be assayed. For any vertebrate, one should expect to observe at least one transcript initiating from each strand of the mtDNA. A typical result from the assay of human mtRNA polymerase on a linear D-loop region template containing both the LSP and HSP is shown in Fig. 2. The run-off transcription products originating from the LSP and

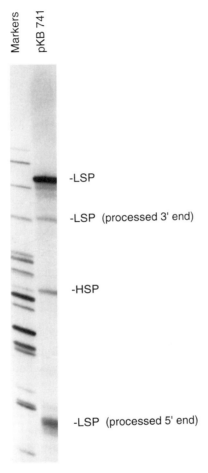

FIG. 2. A typical *in vitro* transcription reaction using human mtRNA polymerase on a D-loop region-containing template harboring both the HSP and LSP. Run-off transcripts from the HSP and LSP are indicated. Additional RNA species resulting from RNA processing activities in the crude mtRNA polymerase fraction are indicated. As discussed in the text, one should expect at least one transcription initiation event originating from each strand of mtDNA in the vertebrate D-loop region. Reproduced with permission from J. N. Topper and D. A. Clayton, *Mol. Cell. Biol.* **9,** 1200 (1989).

HSP are clearly seen. The *in vitro* transcription procedure below was originally described by Walberg and Clayton[13] and modified by Fisher and Clayton.[12]

Transcription Procedure for Specific Initiation from D-Loop Promoters. The total reaction volume is 25 μl and contains, in addition to the added mtRNA polymerase fraction purified as described above, 10 mM Tris-HCl (pH 8.0), 10 mM MgCl$_2$, 1 mM DTT, 0.1 mg/ml RNase-free BSA, 400 μM ATP, 150 μM CTP, 150 μM GTP, 0.2 μM UTP, 0.2 μM [α-^{32}P]UTP (DuPont NEN), and 10 μg/ml linearized D-loop region DNA template. The reactions are incubated 30 min at 28°, stopped by the addition of 50 μl of 0.5 M sodium acetate (pH 5.2), and extracted with 75 μl phenol (pH 5.2, equilibrated with sodium acetate). The RNA in the aqueous phase is precipitated in the presence of 1 μg tRNA by the adding of 200 μl ethanol and incubating the sample on dry ice for 30 min. The RNA pellet is recovered by centrifugation (14,000 rpm, 15 min, 4°) in a microcentrifuge, washed with 70% (v/v) ethanol and dried. Radiolabeled RNA products are separated by electrophoresis through a 6% polyacrylamide, 8 M urea, 1× TBE gel and visualized by autoradiography.

Mapping of RNA 5′ Ends and Delimiting Boundaries of Promoter Elements

Because there is nothing special about mtRNA that would make analyses of general promoter elements by mapping different from methods used for analyzing any promoter, we do not describe these methods in detail. Again, we refer to published methods used to map the 5′ ends of primary transcripts and the 5′ and 3′ boundaries of the HSP and LSP of human and mouse mtDNA as models for such analyses.[1-3] In short, the RNA 5′ ends are mapped in either an S1 nuclease protection assay or primer extension by reverse transcriptase using the *in vitro*-generated RNA as a substrate. The 5′ and 3′ boundaries of the promoter elements are determined by generating a series of deleted D-loop region templates that encroach each mapped RNA 5′ end from upstream (for the 5′ boundary) or downstream (for the 3′ boundary). These deletion series are then assayed using the specific *in vitro* transcription assay. Those deletions that no longer support specific initiation delimit the boundaries of the promoter element. The boundaries can be defined to the nucleotide level if so desired.

Special Considerations

Although the occurrence of a D loop is an apparently universal feature of vertebrate mtDNA, the size of D-loop regions, the structure of the

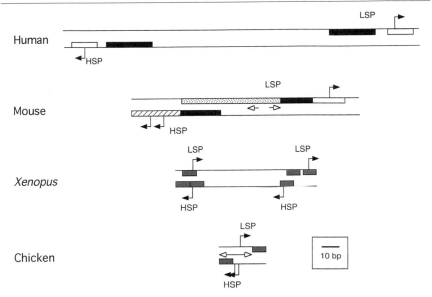

FIG. 3. Comparison of the promoter elements from four vertebrates. Transcription initiation sites are indicated by bent arrows. The L-strand promoters (LSP) and H-strand promoters (HSP) are labeled. The mtDNA H strand is drawn on top (5' to 3', left to right) in all four cases. In the chicken and *Xenopus* drawings, the gray boxes indicate close or perfect matches to a conserved octanucleotide sequence associated with these promoters. In the human and mouse drawings, the black boxes indicate binding sites for mtTFA, and white boxes indicate sequences surrounding the transcription initiation site that are required for transcription initiation. The hatched box represents the region surrounding the mouse HSP that affects transcription initiation efficiency. However, these sequences can be replaced with non-mtDNA sequences and transcription initiation is still observed. The speckled box represents additional upstream sequences that influence transcription from the mouse LSP. Open bidirectional arrows represent inverted repeat sequences.

promoter elements they contain, and the number of specific transcription initiation sites vary considerably (Fig. 3). Several complicating factors exist that affect the analysis of vertebrate mitochondrial promoters. The first is promoter bidirectionality. Chicken and frog mitochondrial promoters are bidirectional as a result of overlapping or nearly overlapping divergently oriented promoter elements,[14,15] whereas human mitochondrial promoters are primarily unidirectional with some bidirectional character.[16] Thus, promoters of vertebrates should be tested thoroughly for bidirectional tran-

[14] D. L'Abbé, J. Duhaime, B. F. Lang, and R. Morais, *J. Biol. Chem.* **266,** 10844 (1991).
[15] D. F. Bogenhagen and M. F. Romanelli, *Mol. Cell. Biol.* **8,** 2917 (1988).
[16] D. D. Chang, J. E. Hixon, and D. A. Clayton, *Mol. Cell. Biol.* **6,** 294 (1986).

scription properties. The second complication is cis-acting DNA elements. The promoters of human and mouse are more complex than other vertebrates owing to the presence of upstream sequences that influence promoter activity (Fig. 3). These upstream sequences in most cases are the specific binding site for the major transcription factor mtTFA (mitochondrial transcription factor A). Thus, vertebrate promoters must also be tested for cis-acting DNA elements that influence promoter activity. A third complication is the occurrence of multiple transcription initiation sites in some systems. For example, two transcription initiation sites are observed from the mouse and chicken HSP (Fig. 3). It is difficult to determine the significance of multiple initiation sites *in vitro,* and care should be taken in interpreting these data in the absence of matching 5'-end-mapping data from RNA synthesized *in vivo.* Finally, mitochondrial transcripts are the targets for RNA processing events. Analysis of transcription products using crude mtRNA polymerase extracts can be complicated by contaminating RNA processing activities. As a result, the 5' ends of the RNA species must be analyzed to determine if they correspond to an actual site of transcription initiation or result from nucleolytic activity. An example of this phenomenon can be seen in Parisi and Clayton,[17] where an additional RNA species is detected along with the full-length RNA transcripts generated in an *in vitro* transcription assay of the human D-loop region. This shorter LSP RNA species (300 nucleotides) is generated by an endonucleolytic activity that cleaves the full-length LSP transcript (416 nucleotides). The 5' and 3' processed portions of the human LSP transcript are also shown in Fig. 2.

Acknowledgments

This research was supported by Grant GM33088-24 from the National Institute of General Medical Sciences. G. S. S. is supported by the Cancer Research Fund of the Damon Runyon-Walter Winchell Foundation, Fellowship DRG-1208.

[17] M. A. Parisi and D. A. Clayton, *Science* **252,** 965 (1991).

[14] Isolation and Characterization of Vertebrate Mitochondrial Transcription Factor A Homologs

By GERALD S. SHADEL and DAVID A. CLAYTON

Introduction

The D-loop region of mammalian mitochondrial (mt) DNA is a key regulatory locus for both mtDNA transcription and replication. In this region are two major promoters, the heavy (H)-strand promoter (HSP) and the light (L)-strand promoter (LSP), that are responsible for transcription of both strands of mtDNA.[1] Transcripts from the LSP are required not only for gene expression, but also for priming H-strand DNA replication. Binding sites for the mitochondrial transcription factor A (mtTFA) are located at various points throughout the D-loop region of mammals, most notably at sites located upstream of the transcription initiation region of the HSP and LSP (Fig. 1). Binding to these upstream elements is necessary for efficient transcription from both promoters. As a result, the mtTFA molecule plays a crucial role in mtDNA transcription and, most likely, in controlling mtDNA replication through its interaction at the LSP. In addition, this protein has a high affinity for nonspecific DNA sequences and binds specifically at other regions of mtDNA, suggesting that it may have multiple critical functions in mtDNA metabolism. It remains unclear if the primary role in transcription initiation in mammalian mitochondria is maintained in other vertebrates. This is certainly an important question given the remarkably different promoter arrangements found in other vertebrates and the less critical role for mtTFA in transcription initiation in yeast mitochondria.[2]

The genes for mtTFA homologs have been isolated from human[3] (h-mtTFA), mouse[4] (m-mtTFA), and the yeast *Saccharomyces cerevisiae*[5] (sc-mtTFA), and all contain two tandem repeats of the high mobility group (HMG)-box DNA-binding domain (Fig. 1). In addition, a putative mtTFA homolog has been isolated from *Xenopus* that has an amino-terminal sequence that matches a portion of the HMG-box domain.[6] Thus, it appears

[1] D. A. Clayton, *Annu. Rev. Cell Biol.* **7,** 453 (1991).
[2] G. S. Shadel and D. A. Clayton, *J. Biol. Chem.* **268,** 16083 (1993).
[3] M. A. Parisi and D. A. Clayton, *Science* **252,** 965 (1991).
[4] N.-G. Larsson, J. D. Garman, and D. A. Clayton, manuscript in preparation.
[5] J. F. X. Diffley and B. Stillman, *Proc. Natl. Acad. Sci. U.S.A.* **88,** 7864 (1991).
[6] R. Ghrir, B. Mignotte, and M. Guéride, *Biochemie* **73,** 615 (1991).

A

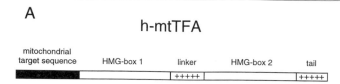

h-mtTFA

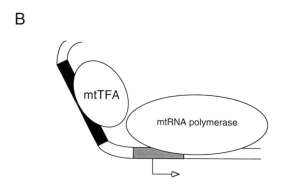

B

FIG. 1. (A) Schematic representation of human mtTFA. The h-mtTFA molecule comprises five functional regions: a mitochondrial targeting sequence (black rectangle), two HMG-box DNA-binding domains (open rectangles), a linker region (rectangle with plus signs) separating the two HMG-box domains, and a C-terminal tail (rectangle with plus signs). The plus (+) signs in the linker and tail region denote the clustering of basic charges in these regions. (B) Interaction of mtTFA at a mitochondrial promoter. The upstream mtTFA-binding site is the black rectangle, and sequences overlapping the transcription initiation site (bent arrow) that are required for transcription initiation are denoted by the gray rectangle. The mtTFA molecule (small oval) is bound at the upstream element of the promoter and is shown bending the DNA. The mtRNA polymerase complex (large oval) is shown interacting with the mtTFA–promoter complex.

that this protein is ubiquitous in mitochondria, which suggests that its role in mitochondrial transcription and mtDNA metabolism is a critical one. This chapter presents a strategy for the isolation of a cDNA for vertebrate mtTFA homologs and subsequent characterization of transcriptional properties. The basic procedures are summarized in Fig. 2.

Cloning cDNA for Vertebrate Mitochondrial Transcription Factor A Homologs by Nucleotide Sequence Homology

It has been possible to isolate a cDNA for mtTFA from a mouse (m-mtTFA) cDNA library using a labeled DNA probe generated from the

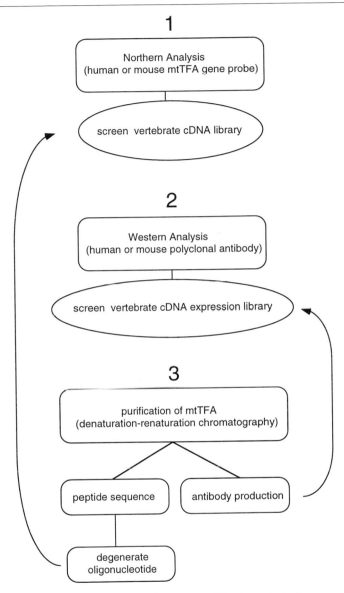

FIG. 2. Outline of mtTFA cDNA cloning strategy. The three strategies are numbered in the order that we recommend they be attempted. The end products of the final strategy are referred back (arrows) to the stage in the earlier strategies that can be used to isolate the cDNA.

h-mtTFA gene.[4] Therefore, we suggest that the most logical first step in an attempt to isolate a cDNA from a vertebrate closely related to humans or mice is to use a probe generated from either the human or mouse mtTFA gene to screen for the homologous gene by Northern analysis. A positive result from the Northern analysis would allow screening of a cDNA library of the vertebrate of interest. It should be noted that probes generated from the sc-mtTFA gene *(ABF2)* do not hybridize to either the human or mouse sequences by Southern analysis, suggesting that this strategy may not be useful for vertebrates evolutionarily distant from humans or mice.

Cloning cDNA for Vertebrate Mitochondrial Transcription Factor A by Cross-Immunoreactivity with mtTFA Antibodies

As with the human gene probe described above, polyclonal antibodies raised to h-mtTFA react with m-mtTFA in a Western analysis. Therefore, if a positive result is not obtained using the human or mouse gene probe by Northern analysis, cross-reactivity to the human and (soon to be available) mouse antibody should be checked by Western analysis. If sufficient cross-reactivity is observed, then a bacterial expression library can be screened with the appropriate antibody. Again, it is noted that the human antibody does not react with sc-mtTFA in a Western analysis, suggesting that this strategy may not be applicable to all vertebrates.

Biochemical Purification of Mitochondrial Transcription Factor A Homologs

If cloning of the vertebrate mtTFA homolog is not successful using the homologous gene probes or available antibodies, then biochemical isolation is necessary. Here we suggest the approach that was used to isolate the cDNA for h-mtTFA.[3] The mtTFA homolog is purified by denaturation–renaturation chromatography and then used to obtain a partial amino acid sequence of the amino terminus or an internal peptide of the protein. This information can then be used to generate a degenerate oligonucleotide probe specific for the mtTFA gene of interest. Alternatively, the purified protein can be used to generate an antibody specific for the mtTFA homolog desired and then used to isolate the cDNA by the method outlined above.

Purification by Denaturation–Renaturation Chromatography

A general purification method for a DNA-binding protein has been developed that allows the easy isolation of substantial amounts of highly

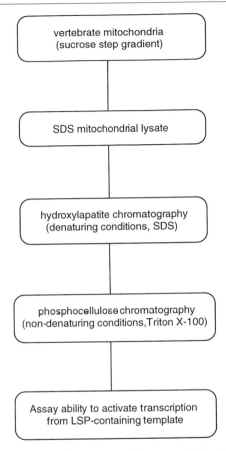

FIG. 3. Flowchart of denaturation–renaturation protocol for mtTFA purification.

purified mtTFA from mitochondria.[7] A flowchart of the isolation procedure is presented in Fig. 3. This method has proved effective in the purification of both human and yeast mtTFA and should allow the easy isolation of any mtTFA homolog. The procedure takes advantage of the fact that mtTFA is an abundant, perhaps the most abundant, DNA-binding protein in mitochondria and is readily renatured to its active form after complete denaturation. Mitochondria are isolated by normal means and lysed in the presence of boiling sodium dodecyl sulfate (SDS) to immediately denature all proteins in the extract. The denatured proteins are then applied to a hydroxylapatite column in the presence of SDS (denaturing conditions).

[7] R. P. Fisher, T. Lisowsky, G. A. M. Breen, and D. A. Clayton, *J. Biol. Chem.* **266,** 9153 (1991).

Denatured proteins eluted from the column are then renatured in the presence of excess Triton X-100 and assayed for DNA-binding and transcriptional activation properties. The pool of mtTFA activity is subjected to chromatography on a final phosphocellulose column to yield highly purified mtTFA. An SDS–polyacrylamide gel of protein fractions from the various stages of the purification of h-mtTFA by this method is shown in Fig. 4.

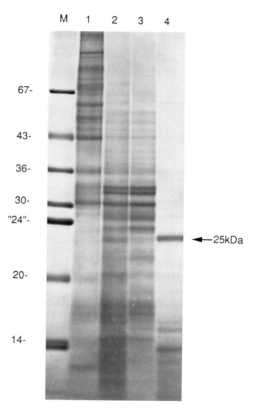

FIG. 4. Analysis by SDS–polyacrylamide gel electrophoresis of proteins from various steps of the denaturation–renaturation purification of h-mtTFA. Lane M contains molecular weight standards; lane 1, crude SDS mitochondrial lysate (40 μg); lane 2, pooled fraction of h-mtTFA DNA-binding activity from the hydroxylapatite column (sample loaded onto the phosphocellulose column, 40 μg); lane 3, 0.1 M NaCl flow-through from the phosphocellulose column (40 μg); lane 4, pool of h-mtTFA from the phosphocellulose column (20 μg). The h-mtTFA protein is indicated by the arrow at an apparent molecular mass of 25 kDa. Reproduced with permission from R. P. Fisher, T. Lisowsky, G. A. M. Breen, and D. A. Clayton, *J. Biol. Chem.* **266**, 9153 (1991).

Solutions

Wash buffer: 0.1 *M* sodium phosphate, pH 6.8, 0.25 *M* sucrose, 15% (v/v) glycerol, 1 m*M* dithiothreitol (DTT)

SDS-lysis buffer: 0.1 *M* sodium phosphate, 2% (w/v) SDS, 20 m*M* DTT, 0.5 m*M* phenylmethylsulfonyl fluoride (PMSF)

Lysis dilution buffer: 0.1 *M* sodium phosphate, 1 m*M* DTT, 0.5 m*M* PMSF

Renaturation buffer 1: 2% (v/v) Triton X-100, 9 m*M* Tris-Cl, pH 8.0, 45 m*M* NaCl, 0.1 m*M* EDTA, 45% (v/v) glycerol, 1 m*M* DTT

Renaturation buffer 2: 4% (v/v) Triton X-100, 20 m*M* Tris-Cl, pH 8.0, 0.2 m*M* EDTA, 2 m*M* DTT, 0.25 m*M* PMSF, 15% (v/v) glycerol

Buffer A: 10 m*M* Tris-Cl, pH 8.0, 0.1 m*M* EDTA, 1 m*M* DTT, 7.5% (v/v) glycerol, 0.1% (v/v) Triton X-100, 0.1 *M* NaCl

Storage buffer: 10 m*M* Tris-Cl, pH 8.0, 50 m*M* NaCl, 0.1 m*M* EDTA, 5 m*M* DTT, 50% (v/v) glycerol

Procedure. Isolate mitochondria by the sucrose step-gradient procedure as described.[8] Wash the purified mitochondria once with ice-cold wash buffer and concentrate by centrifugation (15 min at 15,000 rpm in a Beckman JA-20 rotor at 4°). The pelleted mitochondria are then suspended in approximately 50 ml boiling SDS-lysis buffer. The actual volume of SDS-lysis buffer used is determined by calculating the amount of protein present in the mitochondrial fraction and adding enough buffer so that an approximately 10-fold mass excess of SDS to protein is achieved. The mitochondria are gently mixed until a clear solution is obtained, then boiled for 5 min and diluted 10-fold with lysis dilution buffer. Load a sample of the lysate (~100 mg total protein) onto a 60-ml hydroxylapatite (Bio-Rad, Richmond, CA) column equilibrated in 0.1 *M* sodium phosphate, 0.1% SDS, 1 m*M* DTT. Wash the column extensively with lysis dilution buffer containing 0.1% SDS. Proteins are then eluted with a linear gradient of 0.1 to 0.5 *M* sodium phosphate in 0.1% SDS, 1 m*M* DTT. Fractions are screened for mtTFA activity by diluting a sample from each fraction 10-fold with renaturation buffer 1 and assaying for DNA-binding activity using the gel retardation method as described.[7]

Fractions containing mtTFA activity are pooled and diluted 3-fold with ice-cold renaturation buffer 2 and loaded onto a 25-ml phosphocellulose (Whatman, Clifton, NJ) column equilibrated in buffer A at a flow rate of 20 ml/hr. Wash the column extensively with buffer A, and then elute the column with a linear gradient of NaCl, 0.1 to 1.0 *M*, in buffer A. Assay fractions directly for DNA binding by gel retardation as before. For refer-

[8] G. S. Shadel and D. A. Clayton, this volume [13].

ence, h-mtTFA elutes from this column between 0.65 and 0.70 M NaCl. Pool the fractions containing the peak of DNA-binding activity, dialyze against storage buffer, and keep at $-20°$.

Assaying Mitochondrial Transcription Factor A-Dependent Transcription Activity

The most definitive assay for mtTFA, at least in mammals, is its ability to activate transcription from the D-loop region promoters. In human and mouse, mtTFA is absolutely required for high levels of specific transcription initiation. Elsewhere in this volume[8] we have described a strategy for identifying the D-loop region promoters from vertebrates using a crude mtRNA polymerase preparation. To assay mtTFA effectively it is necessary to generate a mtRNA polymerase fraction that is dependent on mtTFA. This requires a second chromatography step on phosphocellulose that separates mtTFA from mtRNA polymerase.[9,10] We suggest developing this assay using a LSP-containing template from vertebrates, since this is the most efficient mtTFA-dependent promoter of human and mouse mtDNA. A mtTFA-dependent mtRNA polymerase preparation is then used to assay for the transcriptional activation capacity of mtTFA prepared by denaturation–renaturation chromatography.

Solutions

P11 wash buffer: 10 mM Tris hydrochloride (pH 8.0), 0.1 mM EDTA, 1 mM DTT, 7.5% (v/v) glycerol, 0.3 M KCl

P11 elution buffer: 10 mM Tris hydrochloride (pH 8.0), 0.1 mM EDTA, 1 mM DTT, 7.5% (v/v) glycerol

Procedure for Isolation of mtTFA-Dependent Mitochondrial RNA Polymerase. The following procedure was originally described by Fisher and Clayton.[9,10] Follow the procedures for the isolation and assay of crude mtRNA polymerase from vertebrate cells described elsewhere in this volume.[8] The peak fractions from the DEAE-Sephacel column are pooled and loaded directly onto a 25-ml column of phosphocellulose P11 (Whatman) equilibrated in P11 wash buffer. Wash the column extensively with 0.3 M KCl (P11 wash buffer). Then elute proteins with a linear gradient of KCl (0.3–1.0 M) in 200 ml of P11 elution buffer at a flow rate of 10 ml/hr and collect fractions of approximately 4 ml. The fractions are then assayed for nonspecific RNA polymerase activity using poly(dA-dT) as a template. For a reference point, human mtRNA polymerase activity is eluted at

[9] R. P. Fisher and D. A. Clayton, *J. Biol. Chem.* **260,** 11330 (1985).
[10] R. P. Fisher and D. A. Clayton, *Mol. Cell. Biol.* **8,** 3496 (1988).

approximately 0.5 M KCl, whereas h-mtTFA stimulatory activity elutes at approximately 0.7 M KCl. Fractions containing the peak of nonspecific RNA polymerase activity are used to assay for specific initiation on a linearized LSP-containing template as described,[8] except purified mtTFA prepared by the denaturation–renaturation purification strategy above is added to the reaction.

Verification of Potential Mitochondrial Transcription Factor A cDNA

We have been successful in expressing both the human and yeast mtTFA proteins in *Escherichia coli* using the pT7-7 system.[11] The recombinant proteins have DNA-binding and transcriptional activation capabilities that are indistinguishable from the proteins purified from mitochondria. After overexpression, we recommend a simple procedure for the isolation of the recombinant mtTFA from polyacrylamide gels and subsequent renaturation.[12] This method has proved effective in the isolation of both h-mtTFA and sc-mtTFA. The recombinant protein isolated by this method can be used in the mtTFA-dependent transcription reaction to demonstrate that the isolated cDNA indeed encodes mtTFA.

Special Considerations

The mtTFA protein serves a critical function in transcription initiation in mammalian mitochondria and probably plays a role in controlling mtDNA replication through its interaction at the LSP. It is important to note that these capacities have only been demonstrated for the closely related human and mouse systems. Thus, it remains an open question as to what degree mtTFA influences transcription initiation and DNA replication in other vertebrates. The fact that mtTFA can vary in its ability to effect transcription initiation is seen in the yeast system where sc-mtTFA only stimulates transcription initiation a fewfold. Nonetheless, sc-mtTFA is required for mtDNA maintenance *in vivo*. Therefore, it is possible that the mitochondrial promoters in other vertebrates are not dependent on mtTFA to the same degree as in humans and mice. This would complicate our cDNA verification strategy above since it is based on the transcriptional activation properties of mtTFA. In this case, it would be necessary to attempt to characterize potential specific DNA-binding features of mtTFA in the

---

[11] S. Tabor, *in* "Current Protocols in Molecular Biology" (F. A. Ausubel, R. Brent, R. E. Kingston, D. D. Moore, J. A. Deidman, and K. Struhl, eds.), p. 16.2.1. Greene and Wiley (Interscience), New York, 1990.

[12] D. Hagar and R. R. Burgess, *Anal. Biochem.* **109,** 76 (1980).

D-loop region as a measure of potential function in mtDNA metabolism. It has been shown that h-mtTFA interacts specifically with other sites in the D-loop region besides the promoters, including a site near conserved sequence block (CSB) I that potentially is involved in the transition from RNA to DNA during initiation of mtDNA replication. Interaction of mtTFA at key sites in the D-loop region, or elsewhere in mtDNA, may point to other potential functions of this protein in mtDNA metabolism.

Acknowledgments

This research was supported by Grant GM33088-24 from the National Institute of General Medical Sciences. G.S.S. is supported by the Cancer Research Fund of the Damon Runyon-Walter Winchell Foundation, Fellowship DRG-1208.

[15] Isolation and Assay of Mitochondrial Transcription Termination Factor from Human Cells

By VICENTE MICOL, PATRICIO FERNÁNDEZ-SILVA, and GIUSEPPE ATTARDI

Introduction

In vivo studies[1,2] and *in organello* transcription experiments[3-5] have indicated that transcriptional mechanisms play an important role in the differential expression of mitochondrial genes in human cells. The initiation sites for transcription of the two strands of human mitochondrial DNA (mtDNA) are located in a small gene-free region near the origin of replication of the heavy (H) strand, called the D-loop region.[6-8] From the single start site for the light (L) strand transcription originate giant polycistronic transcripts that contain the sequences of the mRNA for

[1] R. Gelfand and G. Attardi, *Mol. Cell. Biol.* **1**, 497 (1981).
[2] J. Montoya, G. Gaines, and G. Attardi, *Cell (Cambridge, Mass.)* **34**, 151 (1983).
[3] G. Gaines and G. Attardi, *J. Mol. Biol.* **172**, 451 (1984).
[4] G. Gaines and G. Attardi, *Mol. Cell. Biol.* **4**, 1605 (1984).
[5] G. Gaines and G. Attardi, *J. Biol. Chem.* **262**, 1907 (1987).
[6] J. Montoya, T. Christianson, D. Levens, M. Rabinowitz, and G. Attardi, *Proc. Natl. Acad. Sci. U.S.A.* **79**, 7195 (1982).
[7] D. D. Chang and D. A. Clayton, *Cell (Cambridge, Mass.)* **36**, 635 (1984).
[8] D. F. Bogenhagen, E. F. Applegate, and B. K. Yoza, *Cell (Cambridge, Mass.)* **36**, 1105 (1984).

subunit 6 of NADH dehydrogenase and of eight tRNAs.[9] On the other hand, there are two closely located initiation sites for the transcription of the H strand, which differ in activity.[2,6–8,10] Transcription initiating at the upstream site is responsible for the synthesis of the bulk of the rRNAs and of the tRNA^{Phe} and tRNA^{Val}, and it terminates at the 3' end of the 16 S rRNA gene.[2] Transcription initiating at the less active downstream site results in the synthesis of giant polycistronic molecules that are precursors of all the mRNAs and most of the tRNAs encoded in the H strand.[2] A large body of evidence indicates that this organization of H-strand transcription, with two overlapping transcription units, underlies the mechanism by which the rRNA species are synthesized at a rate 15–50 times higher than the mRNAs for the proteins encoded in the H strand.[1] In this mechanism, besides the different rate of initiation of the two transcription units, a central role is played by the control of termination at the 3' end of the 16 S rRNA gene.

A DNA-binding protein(s) that protects a 28-bp region immediately adjacent and downstream of the 3' end of the 16 S rRNA gene has been identified and purified by sequence-specific DNA affinity chromatography from a HeLa cell mitochondrial lysate.[11] A specific termination-promoting activity copurified with the DNA-binding protein(s).[11] The fractions containing this activity have been characterized in terms of protein composition and the functional properties of the protein components.[12] Three polypeptides closely related in sequence, two of 34 kDa molecular mass and one of 31 kDa, have been shown to be associated with the specific DNA-binding and footprinting activity of the factor. The transcription termination activity has been found to be associated with the two 34-kDa polypeptides, which thus appear to constitute the mitochondrial transcription termination factor (mTERF). An interesting observation concerning this factor is that an A to G transition in mtDNA, which is associated with the syndrome of mitochondrial myopathy, encephalopathy, lactic acidosis, and stroke-like episodes (MELAS) in humans, is located in the middle of the mTERF protected region and decreases the binding affinity of mTERF for its target sequence.[13,14]

Several methods based on DNA affinity chromatography have been

[9] G. Attardi, *BioEssays* **5,** 34 (1986).
[10] A. Chomyn and G. Attardi, *in* "Molecular Mechanisms in Bioenergetics" (L. Ernster, ed.), p. 483. Elsevier, Amsterdam, 1992.
[11] B. Kruse, N. Narasimhan, and G. Attardi, *Cell (Cambridge, Mass.)* **58,** 391 (1989).
[12] A. Daga, V. Micol, D. Hess, R. Aebersold, and G. Attardi, *J. Biol. Chem.* **268,** 8123 (1993).
[13] J. F. Hess, M. A. Parisi, J. L. Bennett, and D. A. Clayton, *Nature (London)* **351,** 236 (1991).
[14] A. Chomyn, A. Martinuzzi, M. Yoneda, A. Daga, O. Hurko, D. Johns, S. T. Lai, I. Nonaka, C. Angelini, and G. Attardi, *Proc. Natl. Acad. Sci. U.S.A.* **89,** 4221 (1992).

described for the isolation of DNA-binding proteins.[15–20] We have chosen an approach involving the use of complementary multimeric oligodeoxynucleotides covalently coupled to agarose beads[20] for the isolation of the mitochondrial transcription termination factor. In particular, this low abundance DNA-binding protein(s) has been purified to homogeneity from the S-100 preparation of a mitochondrial lysate from HeLa cells by sequential chromatography through a heparin-agarose column and an oligodeoxynucleotide affinity column, followed by sodium dodecyl sulfate (SDS)–polyacrylamide gel electrophoresis (PAGE). This procedure can be a model for the purification of other low abundance mtDNA-binding proteins.

All procedures described below are performed at 4°, unless otherwise specified.

Procedure

Reagents and Solutions. All glassware should be baked at 180° for at least 4 hr. Solutions should be prepared with diethyl pyrocarbonate (DEPC)-treated and autoclaved distilled water.

> 2 M Sucrose in autoclaved, distilled water: sterilize by filtration and store frozen at −20° in aliquots
> 0.5 M Dithiothreitol (DTT): dissolve in autoclaved distilled water and store in aliquots at −20° (add to buffers just before use)
> 0.5 M Phenylmethylsulfonyl fluoride (PMSF, Sigma, St. Louis, MO): dissolve in either ethanol or 2-propanol and store in aliquots at −20° (add to buffers just before use)
> 1 mM Pepstatin A (Sigma): dissolve in methanol, and store at −20° (add to buffers just before use)
> Polyoxyethylenesorbitan monolaureate (Tween 20, Sigma)
> Poly(dI-dC) · (dI-dC) (Pharmacia, Uppsala, Sweden): dissolve to a final concentration of 1 mg/ml in TE (see below) containing 100 mM NaCl, heat to 90°, and slowly cool down to room temperature; store at −20°
> TE buffer: 10 mM Tris-HCl, pH 7.6 at 25°, 1 mM EDTA

[15] B. Alberts and G. Herrick, this series, Vol. 21, p. 198.
[16] D. J. Arndt-Jovin, T. M. Jovin, W. Bohr, A. M. Frischauf, and M. Marquandt, *Eur. J. Biochem.* **54,** 411 (1975).
[17] L. A. Chodosh, R. W. Carthew, and P. A. Sharp, *Mol. Cell. Biol.* **6,** 4723 (1986).
[18] C. Wu, S. Wilson, B. Walker, I. David, T. Paisley, V. Zimarino, and H. Ueda, *Science* **238,** 1247 (1987).
[19] R. Blanks and L. W. McLaughlin, *Nucleic Acids Res.* **16,** 10283 (1988).
[20] J. T. Kadonaga and R. Tjian, *Proc. Natl. Acad. Sci. U.S.A.* **83,** 5889 (1986).

NKM buffer: 1 mM Tris-HCl, pH 7.4, 0.13 M NaCl, 5 mM KCl, 7.5 mM MgCl$_2$

Homogenization buffer: 10 mM Tris-HCl, pH 6.7, 10 mM KCl, 0.15 mM MgCl$_2$

Mitochondria suspension buffer: 10 mM Tris-HCl, pH 6.7, 0.15 mM MgCl$_2$, 0.25 mM sucrose

Mitochondria lysis buffer: 25 mM HEPES–KOH, pH 7.6, 5 mM MgCl$_2$, 0.5 mM EDTA, 1 mM DTT, 1 mM PMSF, 10% (v/v) glycerol

Buffer A: 25 mM HEPES–KOH, pH 7.6, 100 mM KCl, 5 mM MgCl$_2$, 0.5 mM EDTA, 1 mM DTT, 0.2 mM PMSF, 0.1 μM pepstatin A, 10% (v/v) glycerol, 0.1% v/v Tween 20

Buffer B: 25 mM HEPES–KOH, pH 7.8, 12.5 mM MgCl$_2$, 1 mM DTT, 0.2 mM PMSF, 0.1 μM pepstatin A, 20% (v/v) glycerol, 0.1% (v/v) Tween 20

Buffer C: 25 mM HEPES–KOH, pH 7.5, 100 mM KCl, 12.5 mM MgCl$_2$, 1 mM DTT, 20% (v/v) glycerol, 0.1% (v/v) Tween 20

Buffer D: 10 mM Tris-HCl, pH 7.5, 0.3 M NaCl, 10 mM EDTA

Buffer E (10×): 500 mM Tris-HCl, pH 7.4, 100 mM MgCl$_2$, 1 mM DTT, 500 μg/ml bovine serum albumin (BSA)

Calf intestinal alkaline phosphatase (CIP) buffer (10×): 500 mM Tris-HCl, pH 9.0, 10 mM MgCl$_2$, 1 mM ZnCl$_2$, 10 mM spermidine

Polynucleotide kinase (PNK) buffer (10×): 500 mM Tris HCl, pH 7.6, 100 mM MgCl$_2$, 50 mM DTT, 1 mM spermidine, 1 mM EDTA

DNase stop solution: 20 mM EDTA, pH 8, 0.2 M NaCl, 1% w/v SDS, 250 μg/ml carrier RNA (total yeast RNA that has been exhaustively extracted with phenol and subsequently dialyzed against TE)

Transfer buffer: 25 mM Tris base, 192 mM glycine, 20% (v/v) methanol, 0.05% w/v SDS; the pH of this buffer is ~8.0

Transcription buffer (2×): 20 mM Tris-HCl, pH 8.0, 20 mM MgCl$_2$, 2 mM DTT, 200 μg/ml BSA, 20% (v/v) glycerol

25× NTP mixture: 25 mM ATP, 2.5 mM each GTP and CTP, 0.25 mM UTP, prepared from 100 mM stock solutions neutralized with NaOH

Transcription stop buffer: 10 mM Tris base, 0.3 M sodium acetate, pH 7.0, 15 mM EDTA, 0.5% w/v SDS, final pH adjusted to 8.0 with HCl

0.5% SDS buffer: 10 mM Tris-HCl, pH 7.4, 0.2 M NaCl, 10 mM EDTA, 0.5% w/v SDS

DNase buffer (5×): 100 mM Tris-HCl, pH 7.6, 50 mM CaCl$_2$, 50 mM MgCl$_2$

S1 hybridization buffer: 40 mM PIPES–HCl, pH 6.4, 380 mM NaCl, 0.5 mM EDTA, 80% (v/v) formamide; store at $-80°$

S1 digestion buffer (10×): 400 mM sodium acetate, 30 mM ZnCl$_2$, 2.5 M NaCl; adjust to pH 4.6 with HCl, filter, and store at 4°

S1 stop buffer: 4 M ammonium acetate, 20 mM EDTA, 200 μg/ml yeast tRNA

TBE (1×): 89 mM Tris base, 89 mM boric acid, 2 mM EDTA

Tris–acetate buffer: 40 mM Tris base, 20 mM sodium acetate, 1 mM EDTA; adjust to pH 7.4 with glacial acetic acid

Formamide–dye: 95% (v/v) formamide, 20 mM EDTA, 0.05% w/v bromphenol blue, 0.05% w/v xylene cyanol FF

Urea–dye: 7 M urea, 0.01% bromphenol blue, 0.01% xylene cyanol in TBE; keep at −20° and thaw at 65° before use

Ficoll–dye: 30% w/v Ficoll (Sigma), 0.05% w/v bromphenol blue, 0.05% w/v xylene cyanol in TE

Preparation of DNA Affinity Column

The optimal length of the oligodeoxynucleotides recommended to be used for the preparation of a DNA affinity resin ranges between 14 and 51 nucleotides.[20] We have successfully used two 44-mers corresponding to partially complementary sequences of the H and L strand of human mtDNA in the 16 S rRNA/tRNA$^{Leu(UUR)}$ boundary region, as indicated below in the numbering system of the Cambridge sequence[21]:

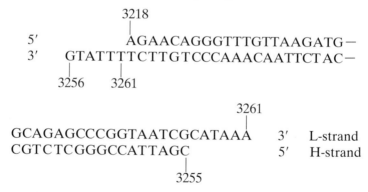

The two sequences contain the protected regions previously identified by the footprinting assays[11] and, when annealed, exhibit complementary 3'-end-protruding hexanucleotide stretches. The two oligodeoxynucleotides are purified on a 15% polyacrylamide–8.3 M urea gel, as described elsewhere.[22] 5'-Phosphorylation, annealing, and ligation of the oligodeoxynucleotides is carried out essentially as previously described,[20] the main

---

[21] S. Anderson, A. T. Bankier, B. G. Barrell, M. H. de Bruijn, A. R. Coulson, J. Drouin, I. C. Eperon, D. P. Nierlich, B. A. Roe, F. Sanger, P. H. Schreier, A. J. Smith, R. Staden, and I. G. Young, *Nature (London)* **290**, 457 (1981).

[22] J. Normanly, R. C. Ogden, S. J. Horvath, and J. Abelson, *Nature (London)* **341**, 213 (1986).

difference being the inversion in order of the 5'-phosphorylation and annealing. A sample of the ligation mixture should be run on a 1% agarose gel in parallel with size markers. The ligation products should cover a size range up to more than 2000 bp, with the majority being between 400 and 1000 bp in size.

Coupling of the ligated oligodeoxynucleotides to CNBr-modified Sepharose CL-2B (Pharmacia) is carried out in 10 mM potassium phosphate buffer (pH 8.0) in a final volume of 10 ml, essentially as described elsewhere.[16] The DNA-Sepharose product, after the coupling reaction, is washed extensively with autoclaved distilled water and then incubated with 2 ml of 1 M ethanolamine, adjusted to pH 8.0 with HCl, overnight at room temperature on a rotatory wheel to inactivate the cyanate groups on the Sepharose. Thereafter, the resin is washed with 100 ml buffer D. The resin is stored at 4° in buffer D with 0.02% NaN_3.

Protein Purification

HeLa cells for the preparation of the mitochondrial lysate are grown in suspension to late exponential phase in high phosphate-containing Dulbecco's modified Eagle's medium supplemented with 5% v/v calf serum. Large-scale cultures are prepared in sets of three 5-liter balloons containing 3 liters medium each. High yields of cells (up to ~2 × 10^6 cells/ml) are obtained by an improved method of growth that involves flushing over the medium, at 2 liters/min, a filtered mixture of 5% CO_2/95% air (v/v) in order to keep the pH of the medium constant.[22a]

The cells are collected by centrifugation in large bottles at approximately 370 g_{av} for 7 min. The supernatant is decanted, and cells are suspended in 10 volumes of packed cells (vpc) of NKM buffer and distributed among 50-ml conical tubes. This washing step is repeated twice more, and the cells are finally suspended in 6 vpc homogenization buffer. The cells in each tube are mixed gently using a large orifice pipette, transferred to a glass homogenizer (A. H. Thomas, Swedesboro, NJ, size C), and kept 2 min in ice. The homogenization is performed using 6–8 strokes with a motor-driven Teflon pestle, and the cells are examined under a phase-contrast microscope to check the degree of cell breakage. We found the percentage of cell breakage to be a critical parameter, with the optimum being around 60%. The homogenate is immediately poured into a conical centrifuge tube containing 1 vpc of 2 M sucrose solution and mixed gently. The unbroken cells, nuclei, and large debris are pelleted at approximately 1200 g_{av} for 3 min, and the supernatant is centrifuged once more under the same condi-

[22a] J. F. Hare, E. Ching, and G. Attardi, *Biochemistry* **19**, 2023 (1980).

tions, again discarding the pellet. The mitochondrial fraction is pelleted at 7000 g_{av} for 10 min and suspended in 3 vpc mitochondria suspension buffer. The contaminating nuclei and large debris are pelleted once more at approximately 1200 g_{av} for 3 min, and the supernatant is centrifuged at 9500 g_{av} for 10 min to obtain the final mitochondrial pellet.

The mitochondrial fraction is suspended in one-third vpc lysis buffer and poured into a Thomas homogenizer (size B). The suspension is homogenized with a motor-driven Teflon pestle by five slow strokes; then Tween 20 and KCl are added to final concentrations of 0.5% and 0.5 M, respectively, and the mixture is vortexed and allowed to stand in ice 5 to 10 min. (*Note:* See elsewhere in this volume[23] for discussion of the experimental evidence which led to the choice of Tween 20 as the detergent that allowed the highest transcription initiation and termination activities in the lysates.) The homogenization is repeated 10 times, and the final mitochondrial lysate is spun at 39,000 rpm (100,000 g_{av}) for 60 min in a TY65 Beckman rotor at 4°. The clear supernatant is carefully collected with a sterile Pasteur pipette, avoiding the fluffy layer over the pellet, to yield the final S-100 fraction of the mitochondrial lysate. This S-100 preparation is either frozen in liquid nitrogen and stored at −80° or immediately treated as described below. The protein concentration of the samples is determined by the Bradford method.[24]

A typical purification experiment involving mTERF is carried out utilizing the S-100 fraction of the mitochondrial lysate prepared from approximately 200 g packed cells (deriving from ten to twelve 5-liter balloons, and yielding 65–70 ml S-100 containing 750–1000 mg total protein). The lysate is dialyzed 2 hr against a 1-liter volume (four changes) of buffer A and then centrifuged at 10,000 g_{av} for 15 min (the conductivity of the S-100 preparation, after dialysis, should be around that corresponding to 0.18 M KCl). The supernatant, containing 650–850 mg total protein at a concentration of about 10–12 mg/ml, is applied, at a 0.5 ml/min flow rate, onto a heparin-agarose column,[25] previously equilibrated with buffer A, adjusting the ratio of protein to heparin-agarose to around 40 mg/ml bed volume. The column is washed with 3 column volumes buffer A, and bound components are sequentially eluted with 3 volumes each of buffer A containing 0.3, 0.5, and 0.8 M KCl. Fractions (4 ml in size) are collected, adjusted to 20% glycerol, and frozen under liquid N_2.

The fractions of the 0.5 and 0.8 M KCl eluate giving conductivity measurements corresponding to between 450 and 650 mM KCl are pooled and

[23] P. Fernández-Silva, V. Micol, and G. Attardi, this volume [12].
[24] M. M. Bradford, *Anal. Biochem.* **72**, 248 (1976).
[25] B. L. Davidson, T. Leighton, and J. C. Rabinowitz, *J. Biol. Chem.* **254**, 9220 (1979).

diluted with buffer B until the conductivity reaches a value corresponding to 175 mM KCl. After addition of poly(dI-dC) · (dI-dC) to 8 μg/mg protein, the solution is incubated 20 min on ice and then applied, at a flow rate of 0.28 ml/min, onto two 0.9-ml DNA affinity columns equilibrated with buffer B containing 150 mM KCl. It is important to keep a ratio of 40–50 mg protein/ml packed oligonucleotide affinity resin (prepared as described above), in order to avoid saturation of the binding capacity. The columns are washed with 10 to 15 column volumes buffer B containing 150 mM KCl, and bound proteins are eluted with the same buffer containing 1.0 M KCl. The eluate is diluted 5-fold with buffer B lacking KCl (to give a final conductivity corresponding to ~220 mM KCl), incubated with 2 μg poly(dI-dC) · (dI-dC)/mg protein for 15 min in ice, and reapplied onto one of the previously used DNA affinity columns, preequilibrated with buffer B containing 200 mM KCl. The column is then washed with 10 to 15 column volumes of the last mentioned buffer, and bound components are eluted with 1-ml portions of buffer B containing 0.3, 0.4, 0.5, 0.6, 0.8, and 1 M KCl or, alternatively, with a 12-ml linear 0.2–0.8 M KCl-containing buffer B gradient, collecting 0.5 ml fractions.

The fractions are tested for DNA binding activity or transcription termination-promoting activity, or analyzed by SDS–PAGE, either directly or after being concentrated 10× and dialyzed using Amicon (Danvers, MA) Centricon-10 microconcentrators.

Assay of Purified Protein

DNase I Footprinting Assay. The probe for the DNase I protection assay that we use is the *Stu*I (position 3148)-*Nco*I (position 3332) fragment of mtDNA ^{32}P-labeled at the 5' end of the H strand at the *Nco*I site. To prepare this probe, we utilize the pTER plasmid,[11,12] a clone containing the human mtDNA *Mbo* fragment between positions 1 and 739 and the fragment between positions 3063 and 3594 joined together through a linker (Fig. 1), constructed as previously described.[11] In brief, 25 μg pTER plasmid is digested with the restriction enzyme *Nco*I, and then the DNA is phenol-extracted, ethanol-precipitated, and dissolved in 45 μl TE. Dephosphorylation is performed by mixing the whole digested plasmid (~20 mol terminal phosphate) with 25 units calf intestinal alkaline phosphatase (Boehringer-Mannheim, Indianapolis, IN) and the appropriate amount of 10× CIP buffer, and incubating the mixture at 37° for 2 hr. The sample is diluted to 200 μl final volume, extracted once with phenol/CHCl$_3$/isoamyl alcohol (25 : 24 : 1, v/v), once with CHCl$_3$/isoamyl alcohol (24 : 1, v/v), precipitated with ethanol, and finally dissolved in TE at a concentration of 1 μg/ml.

The kinasing reaction is performed using 10 μg dephosphorylated DNA,

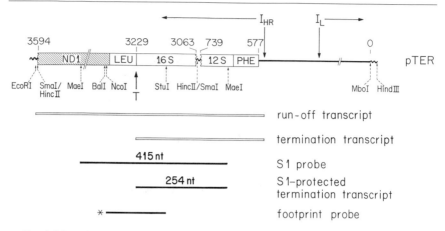

FIG. 1. Map of the transcription termination clone pTER and map positions of the *in vitro* transcripts and probes used in the S1 protection assays and in the footprinting assays. I_{HR}, Upstream initiation site for H-strand transcription; I_L, initiation site for L-strand transcription. Reproduced, with permission, from A. Daga, V. Micol, D. Hess, R. Aebersold, and G. Attardi, *J. Biol. Chem.* **268**, 8123 (1993).

in a total reaction volume of 50 μl, with 5 μl of 10× PNK buffer, 2 μl [γ-^{32}P]rATP (>5000 Ci/mmol, ~30 pmol/μl, >150 mCi/ml; Amersham, Arlington Heights, IL), corresponding to a 7.5-fold molar excess of ATP relative to 5'-DNA termini, and 1 μl T4 polynucleotide kinase (10,000 units/ml; New England BioLabs, Beverly, MA). The mixture is incubated at 37° for 1 hr, and the reaction is stopped by heating at 70° for 15 min to inactivate the enzyme. Two precipitations with 2 M (final concentration) ammonium acetate and 2.5 volumes ethanol are performed in order to eliminate the free ATP. The final DNA pellet is washed with 1 ml of 70% (v/v) ethanol and dissolved in 10 μl TE.

The second restriction digestion to obtain the probe is performed using a large excess of the enzyme *Stu*I (10 units/μg DNA), to ensure that the reaction goes to completion, in a total volume of 50 μl for 3 hr at 37°. After the digestion, 5 μl Ficoll–dye is added, and the mixture is run in a horizontal 1% agarose gel in Tris–acetate buffer. The tank of the electrophoresis apparatus should not be filled with running buffer above the level of the gel surface to avoid sample dilution and contamination of the buffer. The 184-bp DNA fragment originating after the *Stu*I digestion, 5'-end-labeled at the H strand, is excised from the agarose gel and purified using the QIAEX agarose gel extraction system (Qiagen Inc., Chatsworth, CA). The probe is extracted in a final volume of 100 μl TE and the radioactivity incorporated determined. Assuming an efficiency of 5'-end-labeling of about 30%, the expected specific activity is around 3000 counts/min (cpm)/fmol; in several experiments, the final radioactivity recovered ranged between 50,000 and 70,000 cpm/μl.

For the footprinting assay (see Fig. 2), amounts of 1 to 20 μl of the purified protein fractions and appropriate amounts of buffer C needed to reach a final volume of 25 μl after the subsequent additions are introduced into Eppendorf tubes (previously chilled in ice). Then the probe [0.5–2 ng (5–20 fmol)] is added to each tube, and the samples are mixed by flicking the tubes and incubated in ice 15 min. The final KCl concentration is adjusted to 0.1 M by adding 3 M KCl, if necessary. [*Note:* No competitor DNA should be used for footprinting assays with pure or nearly pure protein fractions, since it can inhibit the specific binding of mTERF.[26] On the other hand, when testing the S-100 fraction of a mitochondrial lysate or partially purified protein fractions, a nonspecific competitor DNA, for example, poly(dI-dC) · (dI-dC), in the amount of 1 μg/ml, should be used. The concentration of DNase I to be used in this case is also different; typically, we use 4 μl DNase I diluted to 200 μg/ml when testing samples of S-100 mitochondrial lysate.]

After the incubation, the samples are removed from the ice and left to stand at room temperature 1 min. Then 50 μl of a 10 mM MgCl$_2$, 5 mM CaCl$_2$ solution is added, and DNase I digestion is performed. Typically, we use 4 μl DNase I (Worthington, Freehold, NJ; DPPF grade, dissolved in autoclaved distilled water at a concentration of 10 mg/ml and frozen at $-20°$ in 10-μl aliquots) diluted to 200 μg/ml for nonpurified proteins, or to 2–6 μg/ml for purified proteins. The reaction is allowed to take place for exactly 1 min; it is then stopped by adding 90 μl DNase I stop solution and vortexing. The samples are purified by extraction with 200 μl phenol/CHCl$_3$/isoamyl alcohol (25:24:1, v/v), and the DNA is precipitated with 2.5 volumes cold ethanol. The pellets are washed once with 70% (v/v) ethanol, dried in a desiccator, and dissolved in 4–10 μl formamide–dye. The samples are boiled for 3 min and then run on a standard 0.4-mm-thick, 6% polyacrylamide (acrylamide:bisacrylamide, 19:1)/7.7 M urea sequencing gel in TBE. The gel is washed with 12% v/v methanol, 10% v/v acetic acid in autoclaved distilled H$_2$O for 15 min, heat-dried, and exposed.

Southwestern Blotting. The Southwestern blotting experiments are carried out essentially as described by Mangalam *et al.,*[27] with a few modifica-

[26] J. T. Kadonaga, this series, Vol. 208, p. 10.
[27] H. J. Mangalam, V. R. Albert, H. A. Ingraham, R. Kapiloff, L. Wilson, C. Nelson, H. Elsholtz, and M. G. Rosenfeld, *Genes Dev.* **3,** 946 (1989).

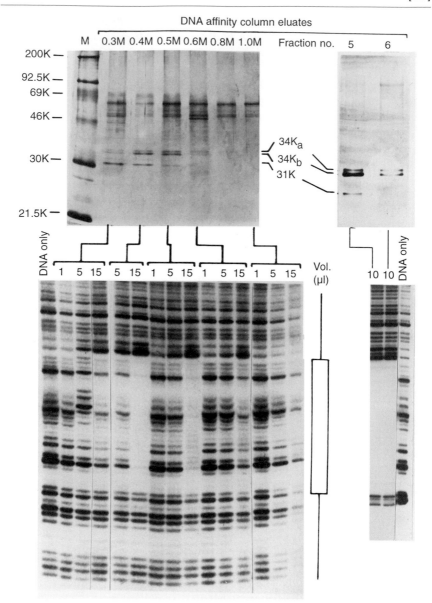

FIG. 2. Identification of the proteins associated with the footprinting activity of affinity-purified mTERF fractions. *Top left:* Silver-stained patterns of proteins eluted from an affinity column with 0.3, 0.4, 0.5, 0.6, 0.8, and 1.0 *M* KCl steps and analyzed by SDS–PAGE. *Top right:* Protein patterns of the two peak fractions eluted from an affinity column with a linear KCl concentration gradient from 0.2 to 0.8 *M*. *Bottom:* DNase I footprinting patterns obtained

tions. After separation of the affinity-purified protein fractions on a 12% SDS–polyacrylamide gel[28] using Acrylaide (FMC Bioproducts, Rockland, ME) as a cross-linking agent, the proteins are electrotransferred onto a nitrocellulose filter at 50 V for 1 hr at 4° in transfer buffer, using a Bio-Rad (Hercules, CA) Trans-Blot cell. The filter is then incubated in the presence of buffer C containing 6 M guanidine hydrochloride for 10 min. The filter is washed two times with buffer C for 10 min each and incubated in the above-mentioned buffer containing 3% nonfat dry milk for 10 min, and, finally, the washing step with buffer C is repeated once more.

The probe used is a double-stranded oligodeoxynucleotide with the following sequence:

The probe is labeled by incubating, in a final volume of 50 μl, the following 5 μl of 10× buffer E, 1 μg double-stranded oligodeoxynucleotide, 0.2 mM each dATP, dCTP, and dGTP, 10 μl [α-^{32}P]dTTP (3000 Ci/mmol, 10 mCi/ml; Amersham), and 10 units of Klenow fragment of *Escherichia coli* DNA polymerase I (New England Biolabs; 5 units/μl).[29] The reaction is allowed to proceed for 1 hr at room temperature, and the labeled probe is precipitated by adding 20 μg yeast tRNA as a carrier, ammonium acetate to a final concentration of 2 M, and 2.5 volumes ethanol. The specific activity of the probe obtained in several experiments, using the conditions described above, was found to be around 2000 cpm/fmol.

The binding reaction is performed in the minimum volume of buffer B (lacking PMSF and pepstatin A and containing 50 mM KCl) that is sufficient to cover the filter. The hybridization solution contains approximately 10^6 cpm/ml of the labeled oligodeoxynucleotide and a 2000-fold excess of salmon sperm DNA, utilized as a nonspecific competitor DNA. The incubation with the probe is performed overnight at 4° under gentle shaking. After the incubation, the filter is washed at room temperature three times,

[28] U. K. Laemmli, *Nature (London)* **227**, 680 (1970).
[29] F. M. Ausubel, R. Brent, R. E. Kingston, D. D. Moore, J. G. Seidman, J. A. Smith, and K. Struhl, eds., "Current Protocols in Molecular Biology." Wiley(Interscience), New York, 1987.

with the indicated volumes of the corresponding fractions. Reproduced, with permission, from A. Daga, V. Micol, D. Hess, R. Aebersold, and G. Attardi, *J. Biol. Chem.* **268**, 8123 (1993).

10 min each, with buffer C, to reduce the radioactivity background. The filter is then exposed wet for autoradiography, to check the effectiveness of the washing step, and finally dried at room temperature and exposed (see Fig. 3).

Transcription Termination Activity. To test the termination-promoting activity of the purified protein fractions, *in vitro* transcription reactions are carried out using different amounts of such fractions. For a detailed description of the *in vitro* transcription system see elsewhere in this volume.[23]

The following components are introduced into Eppendorf tubes (placed in ice): 1 μl template DNA (pTER) at a concentration of 1 μg/μl, 2 μl of 25× NTP mixture, 5 μl of the S-100 fraction of HeLa cell mitochondrial lysate, 25 μl of 2× transcription buffer, 1 μl [α-^{32}P]UTP (400 Ci/mmol, 10 mCi/ml; Amersham), the desired volume of the fraction to be tested (usually, between 1 and 15 μl), and DEPC-treated autoclaved distilled water to bring the volume to 50 μl. After the components are mixed by vortexing, the mixture is briefly spun in an Eppendorf microfuge and then incubated

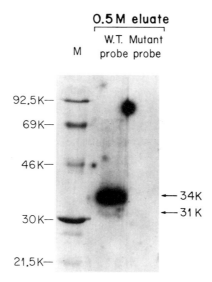

FIG. 3. Southwestern blotting of affinity-purified mTERF eluted with 0.5 *M* KCl. The specific DNA-binding affinity of the individual polypeptides was assessed by probing the nitrocellulose filter with either wild-type (W.T.) labeled double-stranded oligodeoxynucleotide or an equivalent probe carrying a 4-bp deletion within the mTERF binding site. Adapted, with permission, from A. Daga, V. Micol, D. Hess, R. Aebersold, and G. Attardi, *J. Biol. Chem.* **268,** 8123 (1993).

at 30° for 30 min. The reactions are stopped by adding 100 μl transcription stop buffer and 20 μg yeast tRNA. The nucleic acids are extracted with 1 volume of phenol/CHCl$_3$/isoamyl alcohol (25:24:1, v/v), precipitated by adding 330 μl ethanol prechilled at $-20°$ and keeping the samples at $-80°$ for 30 min, and then pelleted by centrifuging at full speed in a microcentrifuge at 4° for 10 min. The pellets are dissolved in 150 μl of 0.5% SDS buffer and extracted again with 1 volume of phenol/CHCl$_3$/isoamyl alcohol (25:24:1, v/v), and the nucleic acids are then precipitated as described above. The pellets can be dissolved in DEPC-treated autoclaved distilled water, and the synthesized products can be analyzed by PAGE; alternatively, if S1 protection analysis of the transcripts is required, the pellets are treated as described below.

Each pellet is dissolved in 100 μl DNase buffer, 10 units RNase-free DNase I (Boehringer-Mannheim; 10 units/μl) is added, and the mixture is incubated at room temperature 20 min. After phenol extraction and precipitation, the RNA pellet is dissolved in DEPC-treated autoclaved distilled water, mixed with the unlabeled MaeI–MaeI RNA probe (Fig. 1; the preparation of this probe is described in detail elsewhere in this volume[23]), and precipitated again with ethanol in the presence of 0.2 M NaCl. The pellet is carefully dissolved in 20 μl S1 hybridization buffer by pipetting up and down repeatedly. Denaturing of the sample is achieved by heating at 80° for 10 min, and hybridization is performed at 50° for 6 to 10 hr. After the hybridization, 200 μl S1 digestion buffer containing 20 μg/ml heat-denatured salmon sperm DNA and 250 to 400 units S1 nuclease (Boehringer-Mannheim; 400 units/μl) is added to each sample, and the mixtures are incubated at 41° for 30 to 45 min. Finally, 55 μl S1 stop buffer is added to each sample, and the S1-resistant products are precipitated with 2.5 volumes ethanol and collected by centrifugation 10 min at 4°. The final pellets are suspended in the desired volume of DEPC-treated autoclaved distilled water, 1 volume urea–dye is added, and the samples are denatured at 80° for 10 min. The samples are then loaded onto a 5% polyacrylamide/7 M urea gel in TBE and run for 2–3 hr at 400 V. After the running, the gel is washed twice with autoclaved distilled water, vacuum-dried on a sheet of Bio-Rad paper at 80° for 1 hr, and then exposed for autoradiography (see Fig. 4).

Mobility Shift Assay

The protocol for the mobility shift assay of the mTERF purified fractions is the one described by Ausubel et al.,[29] with some modifications. The probe used for this assay is the double-stranded 44-mer fragment used for the

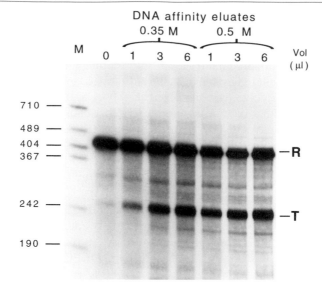

FIG. 4. Transcription termination promoting activity of affinity-purified mTERF fractions. The S1-resistant products of transcription reactions carried out in the absence or in the presence of the indicated volumes of 10× concentrated and dialyzed affinity column eluates are shown. R, H-strand run-off transcripts; T, H-strand terminated transcripts.

Southwestern blotting experiment. This DNA fragment is 3'-end-labeled by filling in with $[\alpha\text{-}^{32}P]TTP$ and the Klenow fragment of *E. coli* DNA polymerase I, as described above under Southwestern blotting.

The desired amounts of the protein fractions to be tested (usually between 1 and 10 μl) are mixed, in Eppendorf tubes placed in ice, with 10 fmol probe (~2000 cpm/fmol), 5 μg BSA, 1 μg poly(dI-dC) · (dI-dC), and buffer C containing 50 mM instead of 100 mM KCl. The final KCl concentration is adjusted to 100 mM with 3 M KCl, and the volume is brought to 20 μl by adding buffer C lacking KCl. The samples are vortexed, spun down briefly, and incubated in ice for 30 min. After incubation, 0.1 volume Ficoll–dye is added to each sample, and the samples are loaded onto a 5% polyacrylamide : bisacrylamide (49.4 : 0.6, v/v) gel in 0.5× TBE buffer. The gel is run at 12 V/cm at room temperature, until the bromphenol blue has migrated about 20 cm, heat-dried at 80° for 2 hr, and exposed for autoradiography.

Results

After analyzing the various affinity column fractions by SDS–PAGE and silver staining, the pattern of polypeptides displaying DNA binding activity, as shown by DNase I footprinting, gel retardation, and Southwestern blot assays, was found to exhibit approximately equivalent amounts of two 34-kDa polypeptides and a smaller amount (10–20% of the total 34-kDa components) of a 31-kDa polypeptide, which showed a lower affinity for the recognition sequence than the 34-kDa components. On the other hand, the transcription termination activity was found to be associated exclusively with the two 34-kDa polypeptides.[12]

Acknowledgments

This work was supported by National Institutes of Health Grant GM-11726 (to G. A.) and scholarships from the Fulbright/Spanish Ministry of Education and Science Visiting Scholar Program (to V. M.) and from Caltech (Gosney Fellowship, to P. F.-S.).

[16] Isolation of Mitochondrial tRNAs from Human Cells

By Michael P. King

Introduction

Mammalian mitochondrial tRNAs possess many unique sequence, structural, and functional properties. All mammalian mitochondrial DNAs (mtDNAs) code for 22 tRNAs, which are sufficient for the translation of the mtDNA-encoded mRNAs, because a single tRNA is able to read all codons of a four-codon family.[1,2] In contrast to prokaryotic, eukaryotic nuclear, and even chloroplast tRNAs, whose structures are remarkably conserved,[3] there are numerous sequence and structural alterations found among mammalian mitochondrial tRNAs. These include changes in highly conserved nucleotides or nucleotide base modifications, variations in the sizes of stem and loop structures, and alterations in conserved tertiary interactions (summarized by Wolstenholme[4]). Variations in the primary

[1] B. G. Barrell, S. Anderson, A. T. Bankier, M. H. L. de Bruijn, E. Chen, A. R. Coulson, J. Drouin, I. C. Eperon, D. P. Nierlich, B. A. Roe, F. Sanger, P. H. Schreier, A. J. H. Smith, R. Staden, and I. G. Young, *Proc. Natl. Acad. Sci. U.S.A.* **77,** 3164 (1980).
[2] S. Anderson, A. T. Bankier, B. G. Barrell, M. H. L. de Bruijn, A. R. Coulson, J. Drouin, I. C. Eperon, D. P. Nierlich, B. A. Roe, F. Sanger, P. H. Schreier, A. J. H. Smith, R. Staden, and I. G. Young, *Nature (London)* **290,** 457 (1981).
[3] M. Sprinzl, T. Hartmann, J. Weber, J. Blank, and R. Zeidler, *Nucleic Acids Res.* **17**(Suppl.), r1 (1989).
[4] D. R. Wolstenholme, *Int. Rev. Cytol.* **141,** 173 (1992).

structures of human mitochondrial tRNAs have been implicated in numerous human diseases.[5] Therefore, it may be desirable to isolate mammalian mitochondrial tRNAs in pure form for a variety of studies, including direct RNA sequence analysis, analysis of base modifications, structural studies, and functional studies.

Procedures have been described for the isolation of single tRNA species based on a variety of fractionation methods. Initially, procedures were based on the isolation of total nucleic acids from mitochondrial fractions of cell and tissue homogenates. The tRNA fraction, isolated by DEAE-cellulose, was subsequently further fractionated by electrophoresis through successive polyacrylamide gels.[6,7] Such analyses required partial sequence analysis of the tRNAs in the gel fractions in order to identify the tRNA of interest. Other procedures were based on successive fractionations of the mitochondrial tRNAs by column chromatography. The tRNAs of interest were identified in column fractions either by assay of radioactive amino acid accepting activity with its cognate aminoacyl-tRNA synthetase[8] or by hybridization with radioactively labeled tRNA-specific probes.[9] Alternatively, a specific tRNA of interest was directly selected from a crude fraction of mitochondrial nucleic acids by hybrid selection with an oligonucleotide complementary to a portion of the tRNA.[10] Here we describe a procedure for cultured mammalian cells that allows the isolation of a highly purified total mitochondrial tRNA preparation. This procedure has been used previously to investigate the steady-state levels and metabolic properties of tRNAs isolated from HeLa cell mitochondria.[11]

Rationale

A procedure previously developed for the isolation of mitochondrial nucleic acids from mammalian cells in pure form depended on the use of micrococcal nuclease to degrade the extramitochondrial nucleic acids.[12] This procedure was unsuitable for the isolation of mitochondrial tRNAs

[5] D. C. Wallace, *Proc. Natl. Acad. Sci. U.S.A.* **91,** 8739 (1994).
[6] M. H. L. de Bruijn, P. H. Schreier, I. C. Eperon, and B. G. Barrell, *Nucleic Acids Res.* **8,** 5213 (1980).
[7] K. Randerath, H. P. Agrawal, and E. Randerath, *Biochem. Biophys. Res. Commun.* **100,** 732 (1981).
[8] T. Ueda, T. Ohta, and K. Watanabe, *J. Biochem.* (*Tokyo*) **98,** 1275 (1985).
[9] Y. Kumazawa, T. Yokogawa, E. Hasegawa, K.-I. Miura, and K. Watanabe, *J. Biol. Chem.* **264,** 13005 (1989).
[10] H. Tsurui, Y. Kumazawa, R. Sanokawa, Y. Watanabe, T. Kuroda, A. Wada, K. Watanabe, and T. Shirai, *Anal. Biochem.* **221,** 166 (1994).
[11] M. P. King and G. Attardi, *J. Biol. Chem.* **268,** 10228 (1993).
[12] G. Attardi and J. Montoya, this series, Vol. 97, 435 (1983).

because of the large amounts of small nucleic acid fragments formed by the nuclease. An alternative procedure for the isolation of mitochondrial tRNAs involved gentle breakage of cells by homogenization and treatment of the crude mitochondrial fraction with high concentrations of ethylenediaminetetraacetic acid (EDTA) to disrupt the ribosomes of the rough endoplasmic reticulum (which unavoidably contaminates mitochondria[13]) and thus to minimize any cytoplasmic tRNA contamination, followed by two cycles of sucrose gradient centrifugation of the extracted RNA. In a previous study that utilized this procedure to isolate HeLa cell mitochondrial tRNAs for aminoacylation analysis,[14] less than 10% of the tRNAs isolated were found to represent cytoplasmic tRNA species. In the protocol described here, a similar approach has been developed and has been used to obtain from the twice EDTA-washed mitochondrial fraction an RNA fraction that is subsequently size-fractionated by sucrose gradient sedimentation and polyacrylamide gel electrophoresis.

Procedure

The tRNAs are typically prepared from HeLa S3 cells grown in spinner culture. Preparations are initiated from 2–3 g (wet weight) cells. All operations are conducted at 4°. After centrifugation to pellet the cells, the cells are washed twice in 0.13 M NaCl, 5 mM KCl, and 1 mM MgSO$_4$, then once in 25 mM Tris-HCl, pH 7.4, 0.13 M NaCl, 5 mM KCl, and 0.7 mM Na$_2$HPO$_4$.

The washed cells are suspended in a minimum of 6 cell volumes of 10 mM Tris-HCl, pH 6.7, 10 mM KCl, and 0.1 mM EDTA. After 2 min to allow cell swelling, the cells are homogenized with a motor-driven Teflon pestle (~1600 rpm) in a glass homogenizer (size C; Thomas Scientific, Swedesboro, NJ) until about 60% of the cells are broken. Breakage of cells should be monitored visually by light microscopy. Homogenization is conducted to maximize breakage of the cells and fragmentation of the cytoplasm, but breakage of the nuclei should be minimized. In particular, the nuclei should maintain a round appearance, and shearing or fragmentation should not be permitted because this releases nuclear nucleic acids. It has previously been shown that, under these conditions, less than 1% of [3H]thymidine-labeled DNA is found in the cytoplasmic fraction.[13] Typically, 12–18 quick strokes are required to achieve this level of breakage. Proper fit of the Teflon pestle to the glass homogenizer is important for

[13] B. Attardi, B. Cravioto, and G. Attardi, *J. Mol. Biol.* **44**, 47 (1969).
[14] D. C. Lynch and G. Attardi, *J. Mol. Biol.* **102**, 125 (1976).

correct breakage of the cells. It may be necessary to try several combinations of pestles and homogenizers to obtain the desired fit and to obtain cell breakage meeting the required parameters.

We have also had success with a rotor-stator based cell homogenization system. In particular, the Tissue-Tearor (Biospec Products, Bartlesville, OK) is used for cell homogenization, and is found to provide excellent breakage of cells, with minimal damage to nuclei. With this model of homogenizer, swollen cells are homogenized at a speed of 2.3 for 25 to 35 sec. Again, homogenization should be monitored by light microscopy in order to verify the level and quality of breakage.

The homogenate is brought to a final concentration of 0.25 M sucrose with a 2 M sucrose, 10 mM Tris-HCl, pH 7.0, solution. After thorough mixing of the homogenate, it is centrifuged at 1100 g for 3 min to remove unbroken cells, nuclei, and large pieces of cytoplasm. Typically, this centrifugation is conducted in 50-ml conical centrifuge tubes using an IEC (Needham Heights, MA) Model 269 swinging-bucket rotor. The supernatant from this centrifugation is removed, added to a new centrifuge tube, and spun at 1000 g for 2 min. The supernatant is then spun at 12,000 g for 10 min in 50-ml polypropylene centrifuge tubes in a Sorvall SS-34 fixed-angle rotor. The resulting pellet is resuspended well in at least 10 cell volumes of 0.25 M sucrose, 10 mM Tris-HCl, pH 7.0, and the suspension is centrifuged at 1100 g for 2 min to remove any residual nuclei or large cell fragments. The supernatant is brought to 40 mM EDTA with a stock solution of 0.4 M EDTA, pH 7.0, incubated on ice for 20 min, then centrifuged at 12,000 g for 10 min in the SS-34 rotor. The pellet, resuspended in 10 cell volumes of 10 mM Tris-HCl, pH 7.0, 0.25 M sucrose, 10 mM EDTA, is incubated on ice for 15 min, then centrifuged at 12,000 g for 10 min. The pellet, resuspended in 1 cell volume of 10 mM Tris-HCl, pH 7.4, 0.15 M NaCl, 1 mM EDTA, 100 μg pronase/ml, is incubated on ice 5 min, then brought to 1% (w/v) sodium dodecyl sulfate (SDS) and incubated an additional 30 min at room temperature. The solution is extracted twice with phenol, mixed with an equal volume of chloroform and isoamyl alcohol (25:1, v/v), which has been preequilibrated with 10 mM Tris-HCl, pH 7.4, 0.1 M NaCl, 1 mM EDTA, 0.5% SDS. The nucleic acids are then ethanol-precipitated, dried, and dissolved in 10 mM Tris-HCl, pH 7.4, 0.15 M NaCl, 1 mM EDTA, 0.5% SDS.

The mitochondrial RNAs are then size-fractionated by sedimentation through a sucrose gradient. The nucleic acids are layered on 15–30% sucrose gradients prepared in the resuspension buffer. Approximately 5–6 A_{260} units is layered on each 35-ml gradient and centrifuged in an SW27 rotor at 95,000 g for 12 to 20 hr. The gradients are fractionated from the top, and the fractions containing the tRNAs are identified by electrophoresis of a portion of the fractions through 5% polyacrylamide–7 M urea gels. The fractions containing tRNAs are pooled and ethanol-precipitated. Because the nuclear DNA and larger RNA species will be found in the lower portions of the gradients, it is important to fractionate beginning with the top of the gradients. Fractionating from the bottom results in contamination of the later fractions by the large amounts of high molecular weight nucleic acids found in the lower portions of the gradients. If one includes a subsequent purification step such as polyacrylamide gel fractionation (see below), it may be sufficient to pool the top 10 ml of each gradient prior to ethanol precipitation. This material may be used directly for polyacrylamide gel analysis. For hybridization studies, it is recommended that an additional purification step, as described below, be undertaken.

In particular, for further purification, the same material pooled from the top 10 ml of each gradient is electrophoresed through a 5% polyacrylamide–7 M urea gel. The left lane of Fig. 1 shows an autoradiogram of the low molecular weight RNAs long-term ^{32}P-labeled *in vivo,* after fractionation on a sucrose gradient and electrophoresis through a 5% polyacrylamide–7 M urea gel. A group of closely migrating species in the lower portion of the gel corresponds to the mitochondrial tRNAs. A pronounced 5 S rRNA band and a few minor species migrating more slowly than the 5 S rRNA are also evident. These minor species presumably represent contaminating small nuclear RNAs. There is no evidence of any significant degradation of higher molecular weight RNAs. A sample of a similar preparation of low molecular weight RNAs is electrophoresed through a 20% polyacrylamide–7 M urea gel, and the right lane of Fig. 1 shows an autoradiogram of a portion of this gel. In the lower part of this lane, one can see a group of closely migrating bands representing the mitochondrial tRNA species, expected to have sizes ranging from 62 to 78 nucleotides including the noncoded CCA. Near the top of this lane is a small amount of cytoplasmic 5 S rRNA migrating as a doublet band. Cytoplasmic tRNAs are generally larger than the mitochondrial tRNAs[3] and would migrate in an area of the gel extending from a position well above the mitochondrial tRNAs to the area including the top several mitochondrial tRNA species. Bands were not detected in the portion of the gel immediately above the mitochondrial tRNAs, pointing to the substantial purity of the mitochondrial tRNA preparation.

The tRNA region of the 5% polyacrylamide gels is typically visualized by ethidium bromide staining, and the region indicated in Fig. 1 is excised from the gel. The nucleic acids in the gel fragments are isolated by the crush and soak method. The gel pieces are chopped into fine fragments with a sterile razor blade or scalpel and are incubated with shaking overnight at 37° in sufficient buffer (10 mM Tris-HCl, pH 7.5, 0.15 M NaCl, 1 mM

5% 20%

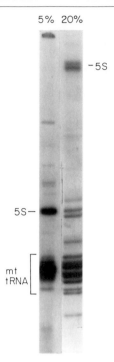

FIG. 1. Autoradiograms of long-term labeled mitochondrial tRNAs electrophoresed through 5 or 20% polyacrylamide–7 M urea gels. Low molecular weight RNAs, long-term labeled with ortho[^{32}P]phosphate, were size-selected by sucrose gradient centrifugation and electrophoresed through either 5 or 20% polyacrylamide–7 M urea gels. The indicated region of the 5% polyacrylamide gel was excised, and the RNA was eluted and subsequently used for hybridization analysis. 5S, Cytoplasmic 5 S rRNA. Reprinted with permission from M. P. King and G. Attardi, *J. Biol. Chem.* **268,** 10228 (1993).

EDTA, 0.5% SDS) to cover the gel fragments. The buffer is removed, and the nucleic acids and other eluted material are ethanol-precipitated. Typically, greater than 90% of the nucleic acids are eluted by this procedure. The majority of any residual material can be isolated with a second round of elution.

If necessary, the residual polyacrylamide and other contaminants are removed by dissolving the precipitated nucleic acids in 10 mM Tris-HCl, pH 7.5, 0.2 M NaCl, 1 mM EDTA and passing them through a 0.5-ml Sephadex A-25 column. After thoroughly washing the column (usually with 10 column volumes of buffer), the tRNAs are eluted with 10 mM Tris-HCl, pH 7.5, 1 M NaCl, 1 mM EDTA, 1% SDS at 60°.

Purity of Mitochondrial tRNAs

The substantial purity of the mitochondrial tRNAs isolated by these procedures is indicated by the results of DNA excess or RNA excess hybridization of *in vivo* labeled tRNAs with mtDNA. As shown in Fig. 2a, the results of a typical experiment in which an excess of H-strand or L-strand mtDNA was annealed with the ^{32}P-labeled mitochondrial tRNA preparation indicate that more than 70% of the labeled material hybridized with

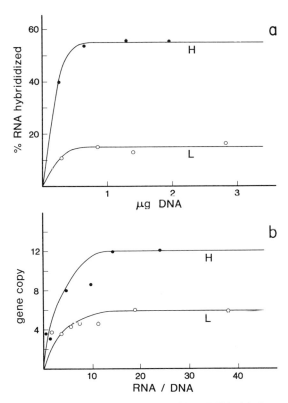

FIG. 2. Hybridization of mitochondrial tRNAs with mtDNA. (a) An excess of H-strand (H) or L-strand (L) mtDNA was annealed with a constant amount of purified tRNAs, labeled *in vivo* with ortho[^{32}P]phosphate for 48 hr. The percentage of the tRNA samples hybridized with each strand at varying DNA concentrations is plotted. (b) An excess of long-term (48 hr) ^{32}P-labeled tRNAs were annealed with a constant amount of H-strand (H) or L-strand (L) mtDNA. The relative gene copy number per DNA molecule, at various ratios of tRNAs to mtDNA (weight to weight), is plotted. Reprinted with permission from M. P. King and G. Attardi, *J. Biol. Chem.* **268,** 10228 (1993).

the mtDNA. Most of the remaining 30% probably represents mitochondrial tRNAs for which the hybridization conditions were not optimal. The tRNAs transcribed from the H strand have on the average only a 35% G+C content, with four having less than 30% G+C, whereas L-strand-encoded tRNAs average a 43% G+C content. However, the hybridization conditions utilized in the present work (68° in 0.4 M salt) were close to optimal for RNAs of higher G+C content (~44%). These conditions were chosen to favor the melting of the secondary structure of the tRNAs, but they may have been too stringent for the hybridization of the tRNAs with the lowest G+C content with the mtDNA.

The results of a representative experiment of mitochondrial tRNA excess hybridizations with H and L strands of mtDNA are shown in Fig. 2b. These data indicate that there is little, if any, contamination of the tRNAs by degradation products of other mtDNA-coded RNAs. The absence of any significant contamination of the tRNAs by degradation products of other mtDNA-coded RNAs, which is indicated by the results discussed above, is further supported by the lack of hybridization of the tRNA preparations with M13 clones of mtDNA not containing a tRNA gene.

Two-Dimensional Separation of Mitochondrial tRNAs

Figure 3 shows an autoradiogram of a two-dimensional separation of the mitochondrial tRNAs long-term labeled with ortho[^{32}P]phosphate. The tRNAs were separated in the first-dimension by electrophoresis through a 20% polyacrylamide–7 M urea gel and in the second-dimension by electrophoresis through a 20% polyacrylamide–3 M urea gel. This system of electrophoresis is based on that described by de Bruijn et al.[6] Greater than 20 distinct species are resolved in this pattern.

The mitochondrial tRNAs are prepared for electrophoresis through the first-dimension 40-cm polyacrylamide–urea gel by dissolving a portion of the mitochondrial tRNAs in 10 μl of 10 mM Tris-HCl, pH 7.5, 1 mM EDTA. Urea crystals are added to this solution until it is saturated with urea. One microliter of a solution of 0.5% xylene cyanol, 0.5% bromphenol blue in distilled water is added to the tRNAs, and the material is heated to 68° for 5 min prior to electrophoresis through a 0.4-mm-thick, 40-cm-long, 20% polyacrylamide (19 : 1 acrylamide: bisacrylamide, w/w)–7 M urea gel, using 0.5× Tris–borate buffer (1×: 90 mM Tris, 90 mM boric acid, 2.5 mM EDTA, pH 8.3), for 22 to 26 hr at 1800 V. The gel is prerun for 1 hr to increase the temperature of the gel to approximately 45°. Typically, the lane width of the first dimension is less than 5 mm.

The tRNA portion of the material separated in the first dimension is identified by autoradiography and excised from the gel. If the nucleic acids

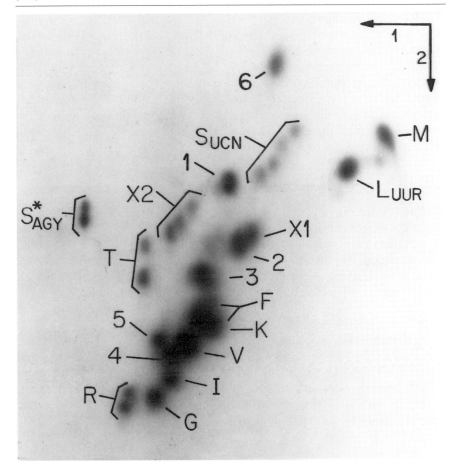

FIG. 3. Two-dimensional polyacrylamide gel fractionation of purified mitochondrial tRNAs. Autoradiogram of long-term (24 hr) ^{32}P-labeled tRNAs separated by two-dimensional electrophoresis. The mitochondrial tRNAs, indicated by the single-letter amino acid code, were identified by hybridizing individual eluted tRNA species with M13 clones of mtDNA. Species X1 and X2 are either tRNA^{Ala}, tRNA^{Asn}, tRNA^{Cys}, or tRNA^{Tyr}. tRNA$^{Ser(AGY)}$, marked with an asterisk, was tentatively identified on the basis of its unique migrational properties. Species 1 through 6 are presumably mitochondrial tRNAs but have not been identified. Reprinted with permission from M. P. King and G. Attardi, *J. Biol. Chem.* **268**, 10228 (1993).

of interest are not radioactively labeled, radioactively labeled markers can be run in parallel to identify the region of the gel containing the tRNAs. The second-dimension 0.8-mm-thick, 40-cm-long polyacrylamide gel is prepared by pouring 20% polyacrylamide–3 *M* urea in 0.5× Tris–borate buffer to within approximately 5 cm from the top of the gel electrophoresis plates.

After the gel is allowed to polymerize, the excised portion of the first-dimension gel is placed along the top of the plates of the second dimension, and the plates are clamped to apply pressure in order to hold the gel slice in place. The gel slice is subsequently held in place with a 5% polyacrylamide–7 M urea solution. Electrophoresis in the second dimension is carried out at 1200 V for approximately 16 to 22 hr at 4°. The time of electrophoresis varies according to the temperature of the polyacrylamide gel during electrophoresis. Following electrophoresis, the gel is covered with Saran wrap and exposed directly to Kodak (Rochester, NY) XAR film.

Isolation of tRNAs from Other Sources

These procedures have been developed for the isolation of mitochondrial tRNAs from HeLa cells. We have used the procedures for larger scale preparations from cultured cells, but care must be taken to maintain the recommended proportions of buffer to cells, not only for homogenization, but also for washing of the mitochondrial fraction. Inadequate washing of mitochondria, or washing of mitochondria in too small a volume, may result in significant contamination of the mitochondrial tRNAs by nuclear and cytoplasmic nucleic acids.

In theory, these procedures could be adapted for other cell types or for tissues. Cell homogenization or disruption is probably the crucial step, and it is worth emphasizing that the process is greatly aided by preswelling of cells in low ionic strength buffer. It is strongly advised that the operation be monitored visually to ensure minimal contamination by nuclear nucleic acids. In this way, mitochondrial tRNAs can readily be isolated in highly purified form.

Acknowledgments

This work was supported by Grants GM11726 (to Giuseppe Attardi) and EY10085 (to M. P. K.) from the National Institutes of Health. M. P. K. is an Investigator of the American Heart Association, New York City Affiliate.

[17] Analysis of Aminoacylation of Human Mitochondrial tRNAs

By José Antonio Enríquez and Giuseppe Attardi

Introduction

The increasing interest in the study of human diseases associated with mitochondrial DNA (mtDNA) mutations has attracted the attention of investigators to the functional analysis of mitochondrial tRNAs. Among the mitochondrial genes whose mutations have been associated with human diseases, the tRNA genes appear to be most frequently affected.[1] In this chapter, we describe several experimental approaches aimed at investigating the aminoacylation capacity of mitochondrial tRNAs.

The analysis of human mitochondrial tRNAs has been hampered so far by the presence in the cell of a large excess of cytosolic tRNAs ($\sim$130:1 in HeLa cells[2]). Furthermore, because all the described tRNA mutations associated with human diseases are present in heteroplasmic form in all tissues and primary cell culture preparations, it is more difficult to obtain highly purified mutant mitochondrial tRNAs for structural and functional analysis. In addition, mammalian mitochondrial tRNAs generally exhibit a noncanonical structure,[3] thereby preventing extrapolation of observations made in bacterial and cytoplasmic systems concerning the influence of a given mutation on tRNA functions. The unavailability of direct genetic manipulation approaches and the poor molecular characterization of the mammalian mitochondrial translation apparatus have also contributed to the slow progress in understanding the molecular basis of the mitochondrial tRNA defects.

The development of the mtDNA-less (ρ^0) cell mitochondria-mediated transformation approach[4] has opened the way to the isolation of cell lines almost completely homoplasmic for a given mitochondrial tRNA mutation[5,6] and to the isolation of reasonable amounts of mutant mitochondrial tRNAs.[7] However, in the study of mitochondrial tRNA aminoacylation,

[1] D. C. Wallace, *Annu. Rev. Biochem.* **61,** 1175 (1992).
[2] M. P. King and G. Attardi, *J. Biol. Chem.* **268,** 10228 (1993).
[3] Anderson *et al., Nature (London)* **290,** 427 (1981).
[4] M. P. King and G. Attardi, *Science* **246,** 500 (1989).
[5] A. Chomyn, *et al., Mol. Cell. Biol.* **11,** 2236 (1991).
[6] A. Chomyn, *et al., Proc. Natl. Acad. Sci. U.S.A.* **89,** 4221 (1992).
[7] J. A. Enriquez, A. Chomyn, and G. Attardi, *Nat. Genet.* **10,** 47 (1995).

other difficulties are encountered. Because the link between amino acid and tRNA in the charged tRNA is extremely labile, special conditions have to be used for purification of aminoacyl-tRNA complexes. Furthermore, it is known that some cytoplasmic tRNAs can also be aminoacylated by mitochondrial enzymes in *in vitro* assays.[8] Therefore, a relative small contamination of a mitochondrial tRNA preparation by cytoplasmic tRNAs can complicate the interpretation of the results, if these are based solely on radioactive labeling measurements. Electrophoretic fractionation of the tRNAs and autoradiography are therefore necessary. However, standard electrophoretic conditions for nucleic acids do not preserve the aminoacyl-tRNA complexes, and, again, special conditions for electrophoresis of the latter have to be used. Finally, some mitochondrial tRNAs have a size very similar to that of their cytoplasmic homologous species, and electrophoresis does not permit their clear separation and independent quantification. In such cases, RNA transfer hybridization has to be used to quantify the relative amount of a given mitochondrial tRNA and, whenever possible (see below), to estimate the relative proportions of the aminoacylated and nonaminoacylated species. The above considerations indicate that different experimental approaches or combinations of approaches have to be used to investigate the aminoacylation of mitochondrial tRNAs. The choice of the appropriate approach(es) will depend on the particular tRNA under investigation. Below, specific examples are described.

Cell Lines and Media

The pT1 human cell line, carrying in nearly homoplasmic form the A to G transition at position 8344 in the mitochondrial tRNA^{Lys} gene, which is associated with myoclonus epilepsy associated with ragged red fibers (MERRF syndrome),[9] and the pT3 cell line, carrying in homoplasmic form the wild-type version of the tRNA^{Lys} gene, are isolated by transfer into human mtDNA-less $\rho^0 206$ cells of mitochondria from myoblasts of a MERRF patient.[5] The transformants are grown in Dulbecco's modified Eagle's medium (DMEM), supplemented with 10% v/v dialyzed fetal calf serum (FBS) and 100 μg/ml bromodeoxyuridine (BrdU). The parental line of $\rho^0 206$, 143B.TK^{-4}, is grown in DMEM with 5% FBS and 100 μg/ml BrdU. HeLa S3 cells are grown in suspension as previously described.[10]

[8] D. C. Lynch and G. Attardi, *J. Mol. Biol.* **102,** 125 (1976).
[9] J. M. Shoffner, *et al., Cell (Cambridge, Mass.)* **61,** 931 (1990).
[10] F. Amaldi and G. Attardi, *J. Mol. Biol.* **33,** 737 (1968).

Mitochondria Isolation

Two different methods are used for mitochondria isolation depending on the objective of the analysis. Thus, when isolation of organelles is performed to obtain highly purified total mitochondrial tRNA, the method that we define as the standard method is chosen. On the other hand, when mitochondria preparations are destined to be used for *in organello* aminoacylation experiments or for purification of *in vivo* preexisting aminoacyl-tRNAs, a much faster procedure for mitochondria isolation is utilized.

Standard Method

In the standard method,[11] exponentially growing cells ($\sim 3.0 \times 10^8$ to $\sim 3.0 \times 10^9$ cells) are harvested, washed two times in 6 volumes of the packed cell volume of 1 mM Tris-HCl (pH 7.0 at 25°), 0.13 M NaCl, 5 mM KCl, 7.5 mM MgCl$_2$, and then broken in a Thomas (Swedesboro, NJ) homogenizer of appropriate size with a motor-driven Teflon pestle in 10 mM Tris-HCl (pH 6.7), 10 mM KCl, 0.15 mM MgCl$_2$. The homogenate is brought to 0.25 M sucrose, and spun at 4° for 3 min at 1600 g_{av} to pellet unbroken cells, large debris, and nuclei. The supernatant is recentrifuged under the same conditions, and the final supernatant is centrifuged for 10 min at 8100 g_{av} at 4° to obtain the mitochondrial pellet. (*Note:* Centrifugation at 5000 g_{av} yields a mitochondrial fraction less contaminated by rough endoplasmic reticulum,[12] and it is therefore preferred whenever the amount of available material is not limiting.) To eliminate the maximum amount of contaminating cytoplasmic tRNAs associated with rough endoplasmic reticulum,[11] the mitochondrial pellet is suspended in 2 to 5 ml of 10 mM Tris-HCl (pH 7.1), 30 mM EDTA, 0.25 M sucrose, kept in ice for 10 min, and then centrifuged at 11,000 g_{av} for 10 min at 4°. The pellet is rinsed with 0.5 ml of 10 mM Tris-HCl (pH 7.1), 1.5 mM EDTA, 0.25 M sucrose, and suspended in 8 ml of the same medium. The suspension is immediately centrifuged at 11,000 g_{av} for 10 min, 4°, to obtain the final mitochondrial fraction.

Rapid Method

Mitochondria are isolated from different cell lines by a modification of a previously described method,[7,13] as follows. All fractionation processes are performed at 4°. Exponentially growing cells are harvested, washed

[11] B. Attardi, B. Cravioto, and G. Attardi, *J. Mol. Biol.* **44,** 47 (1969).
[12] M. Lederman and G. Attardi, *Biochem. Biophys. Res. Commun.* **40,** 1492 (1970).
[13] G. Gaines and G. Attardi, *Mol. Cell. Biol.* **4,** 1605 (1984).

two times with 1 mM Tris-HCl (pH 7.0), 0.13 M NaCl, 5 mM KCl, 7.5 mM MgCl$_2$, and broken in one-half of the packed cell volume of 3.5 mM Tris-HCl (pH 7.8), 2 mM NaCl 0.5 mM MgCl$_2$, by using a Thomas homogenizer with a motor-driven Teflon pestle. The homogenate is immediately mixed with one-ninth of the cell volume of 0.35 M Tris-HCl (pH 7.8), 0.2 M NaCl, 50 mM MgCl$_2$, and spun for 3 min at 1600 g_{av} to pellet unbroken cells, large debris, and nuclei. The supernatant is recentrifuged under the same conditions. The final supernatant is partitioned among Eppendorf tubes and spun at approximately 13,000 g in an Eppendorf microcentrifuge for 1 min. The mitochondrial pellets are washed serially with 35 mM Tris-HCl (pH 7.8), 20 mM NaCl, 5 mM MgCl$_2$, then twice with 10 mM Tris-HCl (pH 7.4), 1 mM EDTA, 0.32 M sucrose.

Isolation of Mitochondrial Nucleic Acids or Total Mitochondrial or Cytosolic tRNA Complement

Nonacid Conditions

Highly purified mitochondrial tRNA preparations exhibiting low contamination by cytoplasmic tRNA are obtained by phenol–chloroform extraction of nucleic acids from twice EDTA-washed mitochondria, followed by sucrose gradient fractionation and polyacrylamide gel electrophoresis, as described below[2] (see also Ref. 14 for details). These preparations are suitable for quantification and size analysis of single mitochondrial tRNA species, for determination of the proportion of mutant tRNA relative to wild-type tRNA, and for *in vitro* aminoacylation experiments.

The mitochondrial fraction, isolated by the standard protocol (see above), is suspended in 10 mM Tris-HCl (pH 7.4), 0.15 M NaCl, 1 mM EDTA, and, after addition of proteinase K and sodium dodecyl sulfate (SDS) to 200 μg/μl and 2.5%, respectively, incubated for 10 min at 37°. Total mitochondrial nucleic acids are extracted with an equal volume of phenol–chloroform–isoamyl alcohol (25:25:1, v/v), preequilibrated with 10 mM Tris-HCl (pH 7.4), 0.15 M NaCl, 1 mM EDTA, and 0.5% SDS, and then precipitated with ethanol.[15] To eliminate most of the mtDNA and large RNAs, the nucleic acid pellet is dissolved in 10 mM Tris-HCl (pH 7.4), 0.15 M NaCl, 1 mM EDTA, and 0.5% SDS, layered on 15 to 30% w/v sucrose gradients (5–6 A_{260} units per gradient) in the same buffer, and centrifuged in a Beckman SW-27 rotor at 95,000 g_{av} for about 20 hr at room temperature. The top 10 ml of each gradient (of 30 ml) are pooled

[14] M. P. King, this volume [16].
[15] G. Attardi and J. Montoya, this series, Vol. 97, 435.

and ethanol-precipitated. Further purification of total mitochondrial tRNA is achieved by running the fraction obtained from the sucrose gradients through a 20-cm-long, 1-mm-thick, 5% polyacrylamide/7 M urea gel in TBE (89 mM Tris base, 89 mM boric acid, 2 mM EDTA). After the run, the gel is stained with ethidium bromide, and the broad group of bands corresponding to the tRNAs is excised, cut into small pieces, and the tRNAs eluted by incubation for 6 to 8 hr in 10 mM Tris-HCl (pH 7.4), 0.15 M NaCl, 1 mM EDTA, and 0.2% SDS. The tRNAs are then extracted with phenol–chloroform–isoamyl alcohol, ethanol-precipitated, dissolved in sterile water, divided into aliquots, and stored at $-80°$. When the initial amount of cells is small (1 or 2 ml packed cells), the sucrose gradient centrifugation step can be omitted and mitochondrial tRNAs can be purified directly by preparative electrophoresis.

The amount of tRNA obtained is estimated by spectrophotometry and/ or by electrophoresis in 5% polyacrylamide/7 M urea minigels in TBE in parallel with tRNA samples of known concentration. The yield is approximately 80 ng total mitochondrial tRNA per 10^6 cells.

Cytosolic tRNAs from HeLa cells are purified from the post-microsomal supernatant (obtained by centrifugation of the homogenate at 30,000 g_{av} for 15 min) following the protocol described above.

Acid Conditions

In the experiments where aminoacyl-tRNA complexes, synthesized *in vitro* or *in organello* or preexisting *in vivo*, have to be preserved, a method of RNA extraction under acid conditions is used.[7,8] The mitochondrial fraction, purified by the rapid method, is suspended in 10 mM Tris-HCl (pH 7.4), 0.15 NaCl, 1 mM EDTA, and, after addition of proteinase K and SDS to 200 $\mu g/\mu l$ and 2.5%, respectively, incubated 10 min at $37°$. The total nucleic acids from mitochondria or the tRNAs from the *in vitro* aminoacylation assays are extracted with an equal volume of phenol, pre-equilibrated with 0.1 M sodium acetate (pH 4.7), 0.5 M NaCl, and 1% SDS, and precipitated with ethanol. The samples are then dissolved in 50 mM acetic acid (pH 4.7), 0.25 M NaCl, and 0.5% SDS, and precipitated with ethanol twice. Samples are dissolved in sterile water, divided into aliquots, and stored at $-80°$.

Electrophoresis of tRNAs

Standard Conditions

The highly purified tRNA preparations or the total mitochondrial RNA are electrophoresed, for quantification purposes, through a 10% polyacrylamide/7 M urea gel in TBE, or, for size determination, through a 6.5% polyacrylamide/7 M urea sequencing gel in TBE (in both cases after heating the sample at 90° for 5 min). The gels are then electroblotted onto a Zeta-probe membrane (Bio-Rad, Hercules, CA) for hybridization analysis with specific tRNA probes (see below). Aminoacyl-tRNAs synthesized *in vitro* or *in organello,* using [^{35}S]methionine or a ^{3}H-labeled amino acid, can be detected after electrophoresis in polyacrylamide/urea gels in TBE. However, aminoacyl-tRNA complexes are progressively decomposed during the electrophoresis owing to the alkaline pH. Electrophoresis under such conditions is, therefore, not recommended.

Conditions for Preserving Aminoacylation

In the experiments in which the mitochondrial or cytosolic aminoacyl-lated tRNAs synthesized *in vitro* or the mitochondrial aminoacylated tRNAs synthesized in isolated organelles or preexisting *in vivo* (see below) have to be analyzed quantitatively, the tRNAs or the total mitochondrial RNA isolated under acid conditions are electrophoresed at 4° under acid conditions.[7,16] In particular, the samples are mixed with 1.5 volumes loading buffer [0.1 M sodium acetate buffer (pH 5.0), 7 M urea, 0.05% bromphenol blue, and 0.05% xylene cyanol] and fractionated on a 20-cm-long, 1-mm-thick 6.5% polyacrylamide (acrylamide : bisacrylamide, 29 : 1, w/w)/7 M urea gel in 0.1 M sodium acetate buffer (pH 5.0). (*Note:* Electrophoresis under the same conditions, but at pH 6 or pH 4, gives results similar to those obtained at pH 5.) Electrophoresis is carried out at 100–200 V in a cold room (4°) for 20–24 hr, with the buffer being continuously recirculated. The gels are then either treated with Amplify (Amersham, Arlington Heights, IL), dried, and exposed for fluorography, or electroblotted onto a Zeta-probe membrane for hybridization with specific tRNA probes, as detailed below.

Quantification of Single Nucleotide Mutations in tRNA

Disease-causing mtDNA mutations are frequently present in a cell in heteroplasmic form, that is, wild-type and mutant copies of mtDNA coexist in the cell. Although it is important to determine the proportion of mutant mtDNA in the cells under investigation, this may not reflect the proportion of mutant gene product, owing to differences in expression of the mutant and wild-type genes. Therefore, in some investigations, it may be necessary to determine the amount of the mutant product. An approach is described below for quantifying point mutations in tRNAs, as applied to the case of

---

[16] U. Varshney, C.-P. Lee, and U. L. RajBhandary, *J. Biol. Chem.* **266,** 24712 (1991).

the MERRF mutation at position 8344 in the mitochondrial tRNA^{Lys} gene.[9] By polymerase chain reaction (PCR) amplification, using a mismatched primer, of a mtDNA segment encompassing the mutation in total cell DNA samples and restriction enzyme digestion, it was previously determined that the pT1 cell line contained at least 98% mutant mtDNA, whereas the pT3 cell line contained the wild-type mtDNA in homoplasmic form.[7] The approach described below involves sequence-dependent termination of primer extension.

For this purpose, a greater than 10-fold excess of ^{32}P 5′-end-labeled oligonucleotide complementary to the 3′-end-proximal 19 nucleotides of the mitochondrial tRNA^{Lys} is mixed with samples of total mitochondrial tRNA purified from the mutant (pT1) or wild-type (pT3) cell line, and containing approximately equal amounts of mitochondrial tRNA^{Lys}, in M-MLV (Moloney murine leukemia virus) reverse transcriptase reaction buffer (BRL, Gaithersburg, MD), containing 50 μM dGTP, 50 μM dTTP, and 1 mM ddCTP, in a volume of 18 μl. The mixtures are heated to 95° for 1 min and cooled immediately in ice–water. After addition of 200 U of M-MLV reverse transcriptase (to a final volume of 20 μl), the samples are incubated 20 min at 37° and then extracted with phenol–chloroform. Equal portions of the two samples are run in a 50-cm-long 20% polyacrylamide (acrylamide : bisacrylamide, 19 : 1, v/v)/7 M urea sequencing gel in TBE, which is then dried and exposed for fluorography. As shown in Fig. 1, using the protocol described above, no mutant tRNA was detected in the pT3 sample, and no wild-type tRNA was observed in the pT1 sample. These data are in substantial agreement with the determined proportion of mutant mtDNA, indicating that pT1 cells contain essentially pure mutant tRNA^{Lys} and that pT3 cells contain essentially pure wild-type tRNA^{Lys}.

Quantification of Individual Mitochondrial tRNAs and Determination of Their Charged and Noncharged Proportions *in Vivo*

The relative total amount of a given mitochondrial tRNA species in different samples and the relative proportions of its charged and uncharged forms, whenever they can be separated electrophoretically, are determined by RNA transfer hybridization analysis with specific probes,[7] using either total mitochondrial tRNAs or total mitochondrial RNA isolated and electrophoresed under nonacid or acid conditions, depending on the purpose of the analysis, as described above.

Electroblotting of the gel onto a Zeta-probe membrane is performed in a Bio-Rad Trans-Blot cell (20 × 15 cm) in TAE buffer (40 mM Tris, 20 mM sodium acetate, pH 7.4, 1 mM EDTA) at 15 V for 30 min and then at 25 V for 2.5 hr. As a probe for the detection of a given tRNA, either a

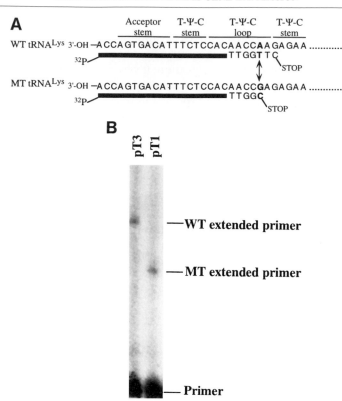

FIG. 1. Quantification of the MERRF tRNA^Lys mutation in the tRNA from the pT1 and pT3 cell lines by sequence-dependent termination of primer extension. (A) Schematic representation of the experiment. The solid bars represent the oligodeoxynucleotide primers. (B) Fluorogram, after electrophoresis through a sequencing gel, of the ^{32}P 5'-end-labeled primer extension products obtained from samples of total mitochondrial tRNA purified from the pT1 and pT3 cell lines. WT, Wild type; MT, mutant. See text for details.

riboprobe synthesized in the presence of [α-^{32}P]UTP from the T7, T3, or Sp6 promoter of an appropriate plasmid or a ^{32}P 5'-end-labeled tRNA-specific oligonucleotide is used. [*Note:* Verification of the specificity of the probe used for hybridization with a given mitochondrial tRNA is strongly recommended, to avoid measuring cross-hybridization with its cytoplasmic counterpart(s).] The hybridization is carried out in a medium containing 5× SSC (SSC is standard saline citrate; 1× SSC is 150 mM NaCl, 15 mM trisodium citrate, pH 7.0), 50% formamide, 5× Denhardt's solution [1× Denhardt's solution is 0.02% Ficoll, 0.02% polyvinylpyrrolidone, 0.02% (w/v) bovine serum albumin (BSA)], 0.1% (w/v) SDS, and 200 μg/ml salmon sperm DNA, for 16 to 20 hr at 42°. After hybridization, the blot is washed two times for 30 min in 2× SSC, 0.1% SDS at 50°, and two times for 30 min in 0.2× SSC, 0.1% SDS at room temperature. Quantification of radioactivity in each band is made either by densitometric analysis of the autoradiogram, using an LKB laser instrument, or by Phosphor-Image detection and ImageQuant analysis (Molecular Dynamics, Sunnyvale, CA).

Figure 2 shows the autoradiogram of an RNA transfer hybridization blot of total mitochondrial RNA from 143B.TK⁻ and PT3 cells, extracted and electrophoresed under acid conditions, using a specific probe for mito-chondrial tRNA^{Lys}. The experiment illustrates the clear separation of the charged and noncharged tRNA^{Lys} species. The electrophoretic conditions described above have also proved to be effective in separating the charged and uncharged forms of nine other tRNAs (see below). It is interesting that charged and uncharged tRNAs specific for amino acids other than lysine can also be separated under the conditions described above.

Isolation of Human Mitochondrial Aminoacyl-tRNA Synthetase Fraction

To perform *in vitro* aminoacylation assays, a crude aminoacyl-tRNA synthetase fraction is prepared from HeLa cell mitochondria.[8] For this purpose, a mitochondrial fraction is isolated from approximately 20 ml packed cells by the standard method, except that 10 mM 2-mercaptoethanol is added to all buffers. The mitochondrial pellets are resuspended in one-half of the original packed cell volume of 0.1 M Tris-HCl (pH 7.8), 10 mM KCl, 10 mM MgCl$_2$, 10 mM 2-mercaptoethanol. The suspension is sonicated at 0° to approximate clarity with eight to ten 15-sec bursts from a Branson (Danbury, CT) Sonifier at maximum power, and then centrifuged 90 min

143B pT3

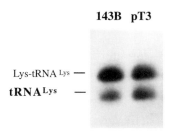

Lys-tRNA ^{Lys} —

tRNA^{Lys} —

FIG. 2. *In vivo* aminoacylation assays. Samples of total mitochondrial RNA purified under acid conditions from 143B and pT3 cells were run in an acid polyacrylamide gel, electroblotted onto a Zeta-probe membrane, and hybridized with a ^{32}P 5′-end-labeled oligodeoxynucleotide probe specific for mitochondrial tRNA^{Lys}. See text for details.

at 40,000 rpm in a fixed-angle Beckman 65 rotor. The supernatant is loaded onto a Sephadex G-100 column (2 × 40 cm) equilibrated with the sonication buffer. The first approximately 20 ml following the void volume is concentrated to approximately 1.4 ml by ultrafiltration with Centricon-30 concentrators (Amicon, Danvers, MA). The retained material is mixed with an equal volume of glycerol, and, after the addition of 2-mercaptoethanol to a final concentration of 20 mM, stored at −20°. The protein concentration of the final preparation ranges between 1.5 and 2.5 mg/ml. The aminoacyl-tRNA synthetase preparation remains active for several months.

In Vitro Aminoacylation Assay

In vitro aminoacylation[8] of mitochondrial or cytosolic tRNAs is carried out using a crude aminoacyl-tRNA synthetase fraction, prepared from HeLa cell mitochondria as described above. The aminoacylation reaction is performed in 50 μl of medium containing 80 mM Tris-HCl (pH 7.8), 10 mM MgCl$_2$, 2 mM ATP, 0.1 mM CTP, 100 μM of each of the nonradioactive 19 amino acids, approximately 1 μg purified tRNA, about 25 μg protein of the aminoacyl-tRNA synthetase preparation, and 75 μCi of either a ^{3}H-labeled amino acid or [^{35}S]methionine (1200 Ci/mmol; Amersham). After 30 min at 37°, the aminoacylated tRNAs are purified under acid conditions and then ethanol-precipitated three times. The alcohol-precipitable ^{3}H or ^{35}S radioactivity is routinely determined by scintillation counting to provide an estimate of aminoacylation, and the reaction products are then analyzed by electrophoresis under acid conditions, as described above. A critical factor in the experiments aimed at quantifying the *in vitro* aminoacylation capacity of a given mitochondrial tRNA species is that the electrophoretic conditions used are adequate to separate the labeled aminoacylated mito-chondrial tRNA species from possible contaminanting cytosolic counter-part(s). For this purpose, it is essential to run in the gel, in parallel with the mitochondrial tRNA sample aminoacylated *in vitro* with a given amino acid, a sample of cytosolic tRNA aminoacylated with the same amino acid. If necessary, the electrophoretic conditions should be modified so as to achieve the optimum resolution.

As shown in Fig. 3A (lanes 2 and 3), electrophoresis under acid conditions described above gave a good separation between the bands corresponding to the [^{35}S]methionyl-tRNA^{Met} synthesized *in vitro* from the cytosolic and mitochondrial tRNA preparations from HeLa cells. As expected, after RNA transfer hybridization, only the latter band gave a signal with the mitochondrial tRNA^{Met}-specific probe (Fig. 3A, lanes 5 and 6). It should be noted that the practically complete absence of contamination of the mitochondrial sample by *in vitro* synthesized cytosolic methionyl-tRNA^{Met}

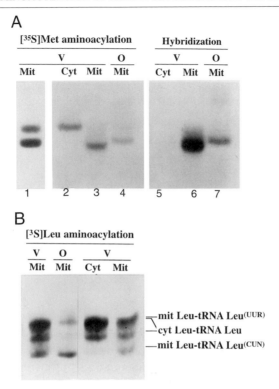

FIG. 3. *In vitro* (V) and *in organello* (O) aminoacylation products analyzed by acid poly-acrylamide gel electrophoresis. (A) Lane 1 shows the fluorogram, after electrophoresis under acid conditions, of the products of an *in vitro* aminoacylation assay of highly purified total mitochondrial tRNA from HeLa cells, carried out using [^{35}S]methionine. Lanes 2 to 4 show the autoradiogram, on Hyperfilm-βmax (Amersham), of *in organello* synthesized [^{35}S]methionyl-tRNA^{Met} from HeLa cells (Mit) and *in vitro* synthesized mitochondrial (Mit) and cytosolic (Cyt) [^{35}S]methionyl-tRNA^{Met} from the same cells, fractionated in an acid polyacrylamide gel and electroblotted onto a Zeta-probe membrane. Lanes 5 to 7 show results when the blot in the left-hand panel (lanes 2 to 4) was hybridized with a ^{32}P 5'-end-labeled mitochondrial tRNA^{Met}-specific oligodeoxynucleotide probe, and then subjected to fluorography using a screen intensifier. (B) Fluorogram, after electrophoresis under acid conditions, of the products of *in vitro* aminoacylation assays carried out, using [^{3}H]leucine, on highly purified total mitochondrial (Mit) or cytosolic (Cyt) tRNA from HeLa cells, and of the aminoacylation products synthesized in the presence of [^{3}H]leucine in isolated mitochondria from HeLa cells (Mit).

is not a constant observation. In some experiments, a band of variable intensity corresponding to the cytoplasmic contaminant could be identified in the mitochondrial sample (see, e.g., Fig. 3A, lane 1). Both the degree of purification of the mitochondrial tRNA sample and the age of the mito-chondrial synthetase preparation appeared to be responsible for this variability. In any case, the clear resolution of the mitochondrial aminoacylated tRNA^{Met} from its cytoplasmic counterpart(s) made it always possible to quantify the *in vitro* aminoacylation capacity of the organelle-specific tRNA^{Met}.

Another mitochondrial tRNA species whose *in vitro* aminoacylated form appears to be well resolved from the cytosolic aminoacylated counterpart(s) is the tRNA$^{Leu(CUN)}$. As shown in Fig. 3B (lanes 1 and 4), *in vitro* aminoacylation with [^{3}H]leucine of the mitochondrial tRNA sample produced three distinct bands, of which the fastest moving was the least abundant. In contrast, the *in vitro* aminoacylated cytosolic tRNA sample revealed only two bands corresponding in mobility to the two slowest bands of the mitochondrial sample. RNA transfer hybridization with specific probes (data not shown) indicated that the fastest moving band in the mitochondrial sample represented the aminoacylated derivative of the tRNA$^{Leu(CUN)}$, whereas the slowest moving band contained, in addition to the expected cytosolic contaminant, the aminoacylated derivative of the other mitochondrial tRNA^{Leu}, namely, tRNA$^{Leu(UUR)}$ (see below).

In organello Aminoacylation Assay

For *in organello* aminoacylation, a crude mitochondrial fraction is isolated by the rapid method (see above), and the mitochondrial pellets are then suspended in the appropriate incubation buffer and spun down at approximately 13,000 g for 1 min in an Eppendorf microcentrifuge. Incubation of the isolated mitochondria is carried out under conditions previously described for *in organello* RNA[17,18] or DNA[19] synthesis. In particular, the isolated organelles (~1 mg protein) are incubated, in 1.5-ml Eppendorf tubes, in 0.5 ml buffer containing 10 mM Tris-HCl, pH 7.4, 100 mM KCl, 5 mM MgCl$_2$, 10 mM K$_2$HPO$_4$, 50 μM EDTA, 1 mM ADP, 10 mM glutamate, 2.5 mM malate, 25 mM sucrose, 75 mM sorbitol, 1 mg/ml BSA, a mixture of all amino acids (except the labeled one) at a final concentration of 10 μM each, and 75 μCi of either a ^{3}H-labeled amino acid or [^{35}S]methionine (1200 Ci/mmol). Incubation is carried out at 37° for 15 min in a rotary shaker (12 rpm), and analysis of aminoacylation is performed as detailed in the previous section. Although in the *in organello* assay described above aminoacylation of cytosolic tRNAs contaminating the mitochondrial fraction is almost negligible (Fig. 3), it is advisable to run in the gel, in parallel with the mitochondrial *in organello* aminoacylated tRNA sample, a sample

[17] J. A. Enriquez, M. J. López-Pérez, and J. Montoya, *FEBS Lett.* **280,** 32 (1991).
[18] J. A. Enriquez, A. Pérez-Martos, M. J. López-Pérez, and J. Montoya, this volume [6].
[19] J. A. Enriquez, J. Ramos, A. Pérez-Martos, M. J. López-Pérez, and J. Montoya, *Nucleic Acids Res.* **22,** 1861 (1994).

of cytosolic tRNAs charged *in vitro* with the same amino acid, in order to verify the separation of the mitochondrial and cytosolic tRNAs.

As shown in Fig. 3A (lane 4), *in organello* synthesized [^{35}S]methionyl-tRNA^{Met} from HeLa cells, fractionated through a 6.5% acid polyacrylamide/8 M urea gel and then electroblotted onto a Zeta-probe membrane, exhibited a band which migrated slightly behind the band of *in vitro* aminoacylated tRNA^{Met} (Fig. 3A, lane 3). It is possible that the slight difference in migration of the *in organello* synthesized aminoacylated tRNA^{Met} is due to its formylation. It is known that, under the conditions of *in vitro* aminoacylation used here, no significant formylation of the human mitochondrial methionyl-tRNA^{Met} occurs.[8] Furthermore, it has been previously reported that, in a bacterial system, the glutaminyl-tRNA complex migrates when formylated slightly more slowly than the nonformylated complex, under electrophoretic conditions similar to those used here.[16] As expected, the band of the *in organello* synthesized aminoacylated tRNA^{Met} hybridized with the mitochondrial tRNA^{Met}-specific probe (Fig. 3A, lane 7). In spite of its slower migration, relative to the *in vitro* aminoacylated tRNA^{Met}, the *in organello* aminoacylated tRNA^{Met} migrated clearly faster than the *in vitro* synthesized cytosolic methionyl-tRNA^{Met}. This finding indicates the substantial absence of contamination of the *in organello* aminoacylated tRNA^{Met} by its cytosolic counterpart(s).

Another *in organello* aminoacylated tRNA, namely, the aminoacylated tRNA$^{Leu(CUN)}$, was shown to be free or nearly free of contamination by cytosolic aminoacylated tRNA^{Leu}. As illustrated in Fig. 3B (lane 2), aminoacylation with [^{3}H]leucine of isolated HeLa cell mitochondria produced a major band corresponding in migration to the aminoacylated mitochondrial tRNA$^{Leu(CUN)}$ and a minor band migrating like the aminoacylated mitochondrial tRNA$^{Leu(UUR)}$, as identified by hybridization of the bands, after RNA transfer, with specific probes (data not shown). The absence or near absence of the faster moving cytosolic aminoacylated tRNA^{Leu} band suggests that the upper band in the *in organello* aminoacylated tRNA sample represents pure or near-pure aminoacylated mitochondrial tRNA$^{Leu(UUR)}$. These observations confirm the absence of a significant contamination by cytoplasmic components of the *in organello* aminoacylated tRNAs, which had been previously observed for tRNA^{Met} (Fig. 3A, lane 4).

Conclusions

As pointed out in the introduction, analysis of the aminoacylation capacity of human mitochondrial tRNAs and how it may be affected by mutations is still laborious, and no single experimental protocol is available. A specific strategy has to be designed for each tRNA, by choosing the experimental approach or combination of approaches most suitable in each case. One has to be aware of the multiple possible causes of misinterpretation, and the specific examples described in this chapter have illustrated the advantages and limitations of each approach.

Whenever applicable, quantification of a given tRNA and determination of its aminoacylated proportion *in vivo* represent the approach of choice. Thus far, analysis of the *in vivo* aminoacylation capacity per cell has been successfully applied by us to $tRNA^{Lys}$, $tRNA^{Leu(UUR)}$, $tRNA^{Leu(CUN)}$, $tRNA^{Gly}$, $tRNA^{Val}$, $tRNA^{Ile}$, $tRNA^{Ser(AGY)}$, $tRNA^{His}$, $tRNA^{Arg}$, and $tRNA^{Trp}$, and it is expected that it should be applicable to other tRNAs as well. However, we have been unable to separate the charged and uncharged forms of mitochondrial $tRNA^{Met}$, $tRNA^{Glu}$, $tRNA^{Asp}$, and $tRNA^{Phe}$.

As concerns *in vitro* aminoacylation, despite the use of a procedure that yields highly purified mitochondrial tRNA preparations, a minor and variable proportion of cytoplasmic tRNA contaminants may remain in these preparations. Because of this and because of the fact that some mitochondrial aminoacyl-tRNAs have electrophoretic mobilities very similar to those of their cytoplasmic counterparts, the *in vitro* aminoacylation approach is suitable for only a subset of the mitochondrial tRNAs. The *in organello* aminoacylation assays have the advantage over the *in vitro* assays that the conditions of aminoacylation, in particular the ratio of enzyme to substrate, are close to the *in vivo* situation. Furthermore, any cytoplasmic tRNA contaminating the mitochondrial fraction would be expected not to be accessible to the mitochondrial synthetases, and, therefore, not to be acylated, nor to compete with the acylation of mitochondrial tRNAs. In support of this conclusion, we have seen no significant labeling of cytoplasmic tRNA contaminants in the *in organello* aminoacylation experiments when [^{35}S]methionine or [^{3}H]leucine were used to label the aminoacylated complexes.

Acknowledgments

This work was supported by National Institutes of Health Grant GM-11726 (to G. A.) and an P.F.P.I. Fellowship from the Spanish Ministry of Education (to J. A. E.).

[18] *In Vivo* Labeling and Analysis of Human Mitochondrial Translation Products

By ANNE CHOMYN

Introduction

Since the subject of mitochondrial biogenesis was last covered in this series in Volume 97,[1] the elucidation of the genetic content of the human mitochondrial genome has been brought to completion. In particular, the functional assignment of all the previously unidentified reading frames[2] has been made. Key in the identification of the genes encoding the seven subunits of the respiratory NADH dehydrogenase[3,4] was the ability to label *in vivo* selectively the mitochondrial translation products, using inhibitors of extramitochondrial protein synthesis (cycloheximide[5] and emetine[6]), along with the development of high-resolution polyacrylamide gel electrophoresis systems able to separate all the mitochondrial DNA (mtDNA)-encoded polypeptides.[7]

The analysis of mitochondrial translation products has proved to be useful for the study of mtDNA mutations causing diseases in humans or isolated in cultured cell systems. Among the mutants analyzed in this way, there are some that exhibit a decreased overall protein synthesis rate,[8–11] some that synthesize additional, abnormal polypeptides, due to premature

---

[1] S. Fleischer and B. Fleischer (eds.), this series, Vol. 97.

[2] S. Anderson, A. T. Bankier, B. G. Barrell, M. H. L. de Bruijn, A. R. Coulson, J. Drouin, I. C. Eperon, D. P. Nierlich, B. A. Roe, F. Sanger, P. H. Schreier, A. J. H. Smith, R. Staden, and I. G. Young, *Nature (London)* **290**, 457 (1981).

[3] A. Chomyn, P. Mariottini, M. W. J. Cleeter, C. I. Ragan, A. Matsuno-Yagi, Y. Hatefi, R. F. Doolittle, and G. Attardi, *Nature (London)* **314**, 592 (1985).

[4] A. Chomyn, M. W. J. Cleeter, C. I. Ragan, M. Riley, R. F. Doolittle, and G. Attardi, *Science* **234**, 614 (1986).

[5] A. Linnane, in "Biochemical Aspects of the Biogenesis of Mitochondria" (E. C. Slater, M. Tager, S. Papa, and E. Quagliariello, eds.), p. 333. Adriatica Editrice, Bari, 1968.

[6] S. Perlman and S. Penman, *Biochem. Biophys. Res. Commun.* **40**, 941 (1970).

[7] G. Attardi and E. Ching, this series, Vol. 56, p. 66.

[8] A. Chomyn, G. Meola, N. Bresolin, S. T. Lai, G. Scarlato, and G. Attardi, *Mol. Cell. Biol.* **11**, 2236 (1991).

[9] J. A. Enriquez, A. Chomyn, and G. Attardi, *Nat. Genet.* **10**, 47 (1995).

[10] M. P. King, Y. Koga, M. Davidson, and E. A. Schon, *Mol. Cell. Biol.* **12**, 480 (1992).

[11] A. Chomyn, A. Martinuzzi, M. Yoneda, A. Daga, O. Hurko, D. Johns, S. T. Lai, I. Nonaka, C. Angelini, and G. Attardi, *Proc. Natl. Acad. Sci. U.S.A.* **89**, 4221 (1992).

termination[8,9] or the presence of deletions,[12] and some that do not synthesize one or another given polypeptide due to frameshift mutations.[13,14] Because of the simplicity of the mammalian mitochondrial genetic system, with only 13 protein-coding genes (and a minimal complete set of tRNA and rRNA genes), and because of the high signal-to-noise ratio in the labeled translation products, *in vivo* labeling and analysis of the mitochondrial translation products provide a very powerful tool for the dissection of the molecular pathogenetic mechanisms of mtDNA mutations affecting the translation apparatus.[9] This chapter describes methods utilized in our laboratory for the study of mitochondrial protein synthesis *in vivo* in human cells in culture. These methods can easily be applied to other mammalian cells.

Experimental Procedures

Choice of Cell Type

Labeling of mitochondrial translation products is most easily done on a small scale in cultures of anchorage-dependent cells. Labeling on a small scale is appropriate for the detection in mutant cell lines of defects in mitochondrial protein synthesis or of aberrant translation products, and for studies of kinetics of labeling and/or stability of mitochondrial translation products. Usually, whole cell lysates are suitable for gel analysis, but, when necessary, a subcellular fraction enriched in mitochondria can be isolated.

For relatively large preparations of mitochondria (10 mg or more of mitochondrial protein) from cells labeled *in vivo* in their mitochondrial translation products, we usually utilize suspension cultures of HeLa cells. Not only is it easier to grow and collect the large number of cells required, but also a mitochondrial fraction is isolated more easily from such large-scale preparations than from small-scale preparations. A mitochondrial fraction is preferable to a whole-cell lysate for immunoprecipitation experiments or for purification of individual radiolabeled mitochondrial translation products.

Pulse Labeling of Mitochondrial Translation Products in
*   Anchorage-Dependent Cells*

Sufficient radioactively labeled material to run in several lanes of analytical gels is obtained from labeling one 6-cm or 10-cm petri dish of subconfluent cells. The cells are used within 3 days of plating, or receive a fluid

[12] J.-I. Hayashi, S. Ohta, A. Kikuchi, M. Takemitsu, Y.-I. Goto, and I. Nonaka, *Proc. Natl. Acad. Sci. U.S.A.* **88,** 10614 (1991).
[13] G. Hofhaus and G. Attardi, *EMBO J.* **12,** 3043 (1993).
[14] G. Hofhaus and G. Attardi, *Mol. Cell. Biol.* **15,** 964 (1995).

change within 24 hr of use. We grow anchorage-dependent cells in Dulbecco's modified Eagle's medium (DMEM), supplemented with 5 or 10% (v/v) dialyzed or nondialyzed bovine calf or fetal bovine serum, depending on the cell type. The cells are grown and labeled at 37° in a 5% (v/v) CO_2 atmosphere.

The media used for washing and labeling cells in the following procedures are prewarmed to 37°. The growth medium is removed from the cells; the cells are then rinsed with methionine-free medium (without serum) and incubated in the same medium 5 min at 37°. After the rinse and incubation steps have been repeated, the medium is removed, and 2–3 or 4–5 ml methionine-free medium supplemented with 10% (v/v) dialyzed serum, containing 100 μg/ml emetine (emetine hydrochloride, Sigma, St. Louis, MO), is added to the 6- or 10-cm petri dish, respectively. After 6 min of incubation at 37°, 0.1 to 1 mCi [^{35}S]methionine (1000 Ci/mmol; Amersham, Arlington Heights, IL) is added to the plate, and the cells are incubated at 37° for the desired time. For a qualitative analysis of mitochondrial translation products, the cells are labeled for 1 to 2 hr; for an analysis of protein synthesis rates, they are labeled for 15 to 30 min.

At the end of the pulse, the radioactive medium is removed, and the plate is rinsed once with complete medium (i.e., containing methionine) and twice with a buffered isotonic solution, like TD [25 mM Tris-HCl, pH 7.4–7.5 (25°), 137 mM NaCl, 10 mM KCl, 0.7 mM Na_2HPO_4]. The cells are harvested by trypsinization, the suspension is added to an equal volume of medium plus 10% calf serum, and the cells are centrifuged at 115 g_{av} in a tabletop International Equipment Co. (Needham Heights, MA) clinical centrifuge for 5 min. The cells are washed once in about 10 ml TD, suspended in 1 ml TD, transferred to a microcentrifuge tube, and centrifuged for 5 min at approximately 100 g_{av}. After washing the cells again in 1 ml TD, they are finally suspended in 50 μl TD to which phenylmethylsulfonyl fluoride (PMSF) has been freshly added to 1 mM. The cells are then stored at or below 70° until needed.

When a mitochondrial fraction has to be isolated, the protocol described above is scaled up to labeling 5 to 10 10-cm petri dishes. After labeling and collecting the cells by trypsinization, the mitochondrial fraction is isolated as detailed below.

Pulse Labeling of Mitochondrial Translation Products in Suspension Cultures of Cells

HeLa cells of the S3 clonal strain are grown in a spinner in high-phosphate-containing Dulbecco's modified Eagle's medium (DMEMP),[15]

---

[15] L. Levintow and J. E. Darnell, *J. Biol. Chem.* **235,** 70 (1960).

supplemented with 5% (v/v) bovine calf serum. The labeling procedure is a modified version of one described in an earlier volume of this series.[7]

The cells to be labeled are collected, under sterile conditions and at room temperature, from an 800-ml spinner culture in logarithmic phase of growth by centrifugation for 7 to 10 min at 170 g_{av}. One to two milliliters of packed cells are suspended in 25 to 50 volumes of methionine-free DMEMP, supplemented with 5% (v/v) dialyzed bovine calf serum. The cells are pelleted by centrifugation for 5 min at 115 g_{av}, and this washing step is repeated. The cells are suspended in 80 to 160 ml of the same medium at 37° in a 200-ml balloon flask that has had a side arm attached to it, and that contains a small magnetic stirring bar. The final cell concentration is 1.5–2 $\times$ 10^6 cells/ml, or about 1 g cells per 80 ml of suspension. Emetine is added from a 10 mg/ml stock solution in water to a final concentration of 100 μg/ml. The flask is stoppered and placed on a magnetic stirrer in a 37° room. After 5 to 15 min at 37°, [^{35}S]methionine ($\sim$1000 Ci/mmol; Amersham) is added to a final concentration of 15 to 60 μCi/ml. The vessel is stoppered, and the cells are incubated at 37° under stirring.

Since the medium contains sodium bicarbonate, we flush a 5% CO_2/95% air (v/v) mixture over the medium at the rate of 2 liters/min, to maintain the correct pH of the medium.[16] The CO_2/air mixture is introduced into the flask through the side-arm opening of the flask, which is closed with a rubber stopper pierced with two syringe needles. One of these is connected via two pieces of latex tubing joined by a Pasteur pipette, plugged with loosely packed cotton, to the supply of CO_2/air mixture. The other syringe needle is connected via latex tubing to another Pasteur pipette, unplugged, the end of which is submerged in a flask of water, which serves as a trap for radioactivity. The stopper, latex tubing, and Pasteur pipette assembly is sterilized by autoclaving prior to use.

The labeling is continued for the desired length of time. A 2-hr pulse is usually employed to label translation products for immunoprecipitation experiments or for purification purposes. The cells are harvested by simply pouring the cell suspension into centrifuge bottles or tubes on ice. For very short labeling pulses (30 min or less), as is done for determining labeling kinetics, the cells are chilled rapidly by pouring the cell suspension into a flask or beaker immersed in a rock salt/ice mixture and containing an equal volume of frozen, crushed NKM (0.13 M NaCl, 5 mM KCl, 1 mM MgCl$_2$). The mixture is swirled until the frozen NKM has melted, and the container is kept in ice until the cells can be centrifuged. If mitochondria are not to be isolated, the cells are washed three times with NKM, resuspended in a

[16] J. F. Hare, E. Ching, and G. Attardi, *Biochemistry* **19,** 2023 (1980).

small volume of NKM or TD, 1 mM PMSF, divided into aliquots, and stored at or below $-70°$ until needed.

Pulse–Chase Labeling of Mitochondrial Translation Products

For the purpose of detecting mitochondrial translation products in immunoprecipitated respiratory chain complexes, the period of labeling with [^{35}S]methionine in the presence of an inhibitor of extramitochondrial protein synthesis must be followed by a chase, that is, a period of continued growth in the presence of unlabeled methionine and in the absence of any protein synthesis inhibitor. Furthermore, the cells to be labeled are incubated with chloramphenicol (CAP) prior to labeling, in order to stabilize the mitochondrial translation products synthesized after removal of the drug.[17] The rationale for this is that the CAP pretreatment, by inhibiting mitochondrial protein synthesis (reversibly), allows the accumulation within the organelles of a pool of cytosolically synthesized components of the respiratory chain complexes, which will facilitate the incorporation into these complexes of labeled mitochondrial translation products synthesized after removal of the CAP. Pretreatment with CAP also increases the amount of radioactive precursor incorporated into at least some mitochondrial translation products during the labeling period.[17,18] The CAP is removed by the washes with methionine-free medium done prior to the start of the labeling.

Chloramphenicol is added to the cell cultures to 40 μg/ml 22–24 hr prior to the start of labeling with [^{35}S]methionine. The labeling is carried out as described above in the section on pulse labeling, except that cycloheximide (Calbiochem, La Jolla, CA) is substituted, weight by weight, for emetine. In contrast to the inhibition of protein synthesis by emetine,[19] that by cycloheximide is reversible.[20,21] After labeling for 2 hr, unlabeled methionine is added to the medium to a final concentration equivalent to its concentration in normal growth medium (DMEM or DMEMP). Subsequently, if the cells are growing on plates, they are washed twice with DMEM; then, DMEM supplemented with serum is added, and the cultures are grown for an additional 16 to 18 hr. If the cells are in suspension culture, they are collected by centrifugation after the addition of unlabeled methionine, washed once in 20 to 40 volumes DMEMP supplemented with

[17] P. Costantino and G. Attardi, *J. Biol. Chem.* **252**, 1702 (1977).
[18] P. Mariottini, A. Chomyn, R. F. Doolittle, and G. Attardi, *J. Biol. Chem.* **261**, 3355 (1986).
[19] A. P. Grollman, *Proc. Natl. Acad. Sci. U.S.A.* **56**, 1867 (1966).
[20] H. L. Ennis and M. Lubin, *Science* **146**, 1474 (1964).
[21] B. Colombo, L. Felicetti, and C. Baglioni, *Biochim. Biophys. Acta* **119**, 109 (1966).

serum, resuspended in 1 liter of the same medium, and grown for an additional 16 to 18 hr. The cells are harvested as described above.

Notes on Labeling Procedures

1. It is essential that the medium to be used for labeling be free of methionine. For this reason, to supplement the medium used in the labeling reactions, we use bovine calf serum or fetal bovine serum which has been dialyzed extensively, that is, for 10 hr, with hourly changes, against 25 mM Tris-HCl, pH 7.4, 137 mM NaCl, 10 mM KCl, and then filter-sterilized.

2. It is important that the solutions and media used in the labeling protocols be sterile, especially in experiments with long labeling periods and in pulse–chase experiments.

3. The preincubation with emetine or cycloheximide should not exceed 10 to 15 min, as longer preincubations will reduce the amount of [^{35}S]methionine incorporated.[17]

4. When the labeling time is short, that is, no longer than 45 min, the washes prior to the labeling are carried out in a 37° room to avoid cooling of the cell cultures, which would reduce the rate of labeling.

5. Expre^{35}S^{35}S (DuPont/New England Nuclear, Boston, MA), a mixture of [^{35}S]methionine and [^{35}S]cysteine, is an economical alternative to [^{35}S]methionine for *in vivo* labeling studies. The labeling obtained with Expre^{35}S^{35}S is similar to that obtained with [^{35}S]methionine.

Isolation of Mitochondrial Fraction

A mitochondria-rich fraction is isolated by differential centrifugation as described previously,[22] with some modifications as concerns the buffers used. The following operations are done at 0°–4°. After the first centrifugation of the cells, they are washed three times with 15 to 30 volumes NKM, collecting the cells each time by centrifugation for 5 min at 150 g_{av}. These washes are carried out in graduated centrifuge tubes, so that the volume of the final cell pellet can be determined. The supernatant is removed, and the cells are suspended in 6 volumes TKM [10 mM Tris-HCl, pH 6.7 (25°), 10 mM KCl, 1.5 × 10^{-4} M MgCl$_2$]. The cells are allowed to swell in this hypotonic buffer in ice for 2 or 3 min. The cells are broken in a Potter–Elvehjem homogenizer (glass vessel and motor-driven Teflon pestle; A. C. Thomas Scientific, Swedesboro, NJ) as described elsewhere in this volume.[23] Cell disruption is carried out until 60–80% of the cells are broken (more breakage tends in fact to result in nuclear breakage, with release of frag-

---

[22] B. Attardi, B. Cravioto, and G. Attardi, *J. Mol. Biol.* **44,** 47 (1969).
[23] C. Ausenda and A. Chomyn, this volume [11].

ments that will cofractionate with the mitochondria). Five to twelve strokes of the homogenizer are usually sufficient to achieve this extent of breakage of the cells, as monitored by phase-contrast microscopy. As soon as the cells have been sufficiently broken, the homogenate is transferred to a tube containing a volume of 2 M sucrose in TKM equivalent to the original volume of packed cells. The suspension is mixed well and centrifuged 3 min at 1000 g_{av}, to pellet the nuclei, unbroken cells, and large cytoplasmic debris.

The supernatant is transferred to another centrifuge tube, and the mitochondria are pelleted by centrifugation for 10 min at 7700 g_{av}. The mitochondrial fraction is then suspended in one-half the homogenate volume of 0.25 M sucrose, 10 mM Tris-HCl, pH 6.7, 1.5 × 10^{-4} M MgCl$_2$. Suspension is done thoroughly, starting with a small volume of buffer and using a small-bore pipette, such as a 2-ml glass tissue culture pipette, for dispersing clumps. Any residual nuclei or large cytoplasmic debris are removed by centrifugation for 3 min at 750 g_{av}. The mitochondria are pelleted again by centrifugation for 10 min at 7700 g_{av}, and the final mitochondrial pellet is suspended in a small volume of 0.25 M sucrose, 10 mM Tris-HCl, pH 6.7 (250–600 μl for every milliliter original packed cell volume). This subcellular fraction is highly enriched for mitochondria but also contains rough endoplasmic reticulum and other membrane components.[22] The suspension is divided into aliquots in microcentrifuge tubes and stored at or below −70°.

Notes on Isolation of Mitochondrial Fraction from Labeled Cells

1. When breaking cells with a Potter–Elvehjem homogenizer, it is important to choose the correct homogenizer size. For volumes of packed cells less than or equal to 0.7 ml, size A is used. For volumes of packed cells between 0.7 and 2.0 ml, size B is used. Using a homogenizer that is smaller than indicated for a given volume greatly increases the chance that the glass vessel will break during the cell disruption.

2. Because the cells are radioactive, special care is taken to minimize splashing during cell disruption with the Potter–Elvehjem homogenizer. We extend the height of the vessel by 2 inches by wrapping a wide sheet of Parafilm around the neck. The motor that drives the pestle is placed in a box constructed of Lucite, with openings at the sides for access to the pestle. The cell disruption is thus carried out inside the box, confining any possible splashing of homogenate.

Determination of Specific Activity of Labeled Samples

To determine the concentration of incorporated [^{35}S]methionine, 20 μl of 0.5% sodium dodecyl sulfate (SDS) and 50 μl of 2 mg/ml bovine serum albumin are added to duplicate 2-μl samples of mitochondria or cells, well suspended just prior to sampling. After mixing the samples with 2 ml of 15% (w/v) trichloroacetic acid (TCA) containing 1 mM methionine, and incubating them on ice 30 to 60 min, the TCA-precipitated protein is collected on Millipore filters under suction. The filters are washed under suction three times with 5% TCA (8 to 10 ml each time) and, finally, with a small volume of ethanol. The filters are allowed to dry in scintillation vials, they are covered with 5–10 ml of a suitable scintillation fluid, and the radioactivity is then determined by scintillation counting.

To determine the protein concentration by the Bradford assay,[24] duplicate 4-μl samples of mitochondria or cell suspensions are mixed in 10-ml glass test tubes with 16 μl of 0.5% (w/v) SDS (Bio-Rad, Hercules, CA) and 76 μl distilled water. Samples for a standard curve are prepared using 10 to 80 μl of 1 mg/ml bovine γ-globulins in phosphate-buffered saline (150 mM NaCl, 8.1 mM Na$_2$HPO$_4$, 1.5 mM KH$_2$PO$_4$), 16 μl of 0.5% (w/v) SDS, and sufficient water to bring the samples to a final volume of 96 μl. Then, 5 ml Bio-Rad Protein Assay Reagent (5-fold diluted Dye Reagent Concentrate) is added to each tube, and the tube is vortexed immediately. The absorbance at 595 nm is read within 1 hr.

Electrophoretic Resolution of ^{35}S-Labeled Mitochondrial Translation Products

The 13 mammalian mitochondrial translation products are best resolved in a 15 to 20% exponential gradient polyacrylamide gel,[8,25] similar to that described by Douglas et al.[26] This gel system gives sharper bands than the urea–SDS gel system[3] or the linear gradient polyacrylamide–SDS gel system.[4]

Gel Apparatus. The apparatus for gel electrophoresis is basically the slab gel and running stand described by Studier.[27] The final size of the separating gel is 18 × 13.6 × 0.1 cm. A 2-cm high stacking gel is attached to the 18-cm-tall separating gel. We use 1-mm-thick combs, with 8 to 12 evenly spaced teeth.

We pour the gel in a casting stand, which has a base consisting of a Lucite platform into which a trough has been milled; the glass plate–spacer assembly is set into the trough. Affixed to the base toward one end is a vertical metal rod with a clamp, which holds the glass plates in a vertical

[24] M. M. Bradford, *Anal. Biochem.* **72,** 248 (1976).
[25] A. Chomyn and S. S.-A. T. Lai, *in* "Structure, Function and Biogenesis of Energy Transfer Systems" (E. Quagliariello, S. Papa, F. Palmieri, and C. Saccone, eds.), p. 179. Elsevier, Amsterdam, 1990.
[26] M. Douglas, D. Finkelstein, and R. A. Butow, this series, Vol. 56, p. 58.
[27] F. W. Studier, *J. Mol. Biol.* **79,** 237 (1973).

position. For electrophoresis, the glass plate sandwich containing the gel is removed from the casting stand and mounted onto a vertical running stand that accommodates 500 ml buffer in both the upper and lower chambers.

Gel Solutions. We use the SDS buffer system of Laemmli,[28] slightly modified.

4× Separating buffer: 1.5 M Tris-HCl, pH 8.8, 0.4% SDS, 8 mM EDTA
4× Stacking buffer: 0.5 M Tris-HCl, pH 6.8, 0.4% SDS, 8 mM EDTA
Acrylamide–bisacrylamide stock solution: 30% (w/v) acrylamide, 0.8% (w/v) bisacrylamide; filter through Whatman (Fairfield, NJ) No. 1 paper, if any particles are visible, and store in the dark at 4°
10% (w/w) Ammonium persulfate (freshly made every day or two)
Electrode buffer: 3 g Tris base, 14.4 g glycine, 5 ml of 20% (w/v) SDS per liter; adjust to pH 8.3–8.45 with HCl
5× Sample buffer: 2.5 ml of 4× stacking buffer, 2.5 ml of 20% SDS, 0.25 ml of 0.4 M EDTA, 0.50 ml 2-mercaptoethanol, 4.25 ml glycerol, 5 mg bromphenol blue; store at 4°

Pouring the Gel. The gel is poured the day before it is used. First, the spacers are sealed in the glass plates with 1% molten agarose. The glass plate "sandwich" is then clamped into the casting stand, and the following base gel mixture is poured into the trough:

9 ml acrylamide–bisacrylamide stock solution
4.5 ml of 4× separating buffer
4.33 ml distilled water
20 μl TEMED (N,N,N',N'-tetramethylethylenediamine)
150 μl of 10% ammonium persulfate

The base gel solution comes up between the glass plates to a height of approximately 1 cm and polymerizes in about 5 min. The lower edge of the glass plate sandwich is now sealed. If buffer has been extruded from the base gel, it is removed by tilting the casting stand and absorbing the liquid with a long, narrow piece of Whatman 3MM filter paper inserted between the glass plates. Next, 24 ml (the separating gel volume) of the 15% acrylamide gel mixture and 12 ml (half the separating gel volume) of the 20% acrylamide gel mixture are prepared in 50-ml Erlenmeyer flasks according to the recipes in Table I.

A magnetic stirring bar is introduced into each flask, the flasks are placed on a two-place or four-place magnetic stirring plate or on two plates of identical height, and the mixtures are stirred briefly before starting the pouring. The acrylamide gel mixtures need not be degassed, if stirring has been done at a moderate speed. A two-channel or multichannel peristaltic

[28] U. K. Laemmli, *Nature* (*London*) **227**, 680 (1970).

TABLE I
PREPARATION OF ACRYLAMIDE GEL SOLUTIONS

Component	15% Acrylamide	20% Acrylamide
Acrylamide–bisacrylamide stock solution	12 ml	8 ml
4× Separating buffer	6 ml	3 ml
Distilled water	5.9 ml	1 ml
TEMED	12 μl	5 μl

pump is placed in position between the magnetic stirring plate(s) and the glass plates in their casting stand. A length of pump tubing is led from the mixing flask (20% gel solution), through the pump, to the center of the top opening in the glass plate assembly. An equal length of pump tubing of identical diameter (or flow rate) is led from the 15% gel solution, through the pump, to the mixing flask. The ends of the pump tubing that are inserted into the flasks are fitted with disposable capillary pipettes. With the magnetic bars rotating at slow to moderate speed, 35 μl of 10% ammonium persulfate is added to the 20% gel solution in the mixing flask, and 70 μl of 10% ammonium persulfate is added to the 15% gel solution. The peristaltic pump is started, keeping the magnetic stirrers on (stirring the 15% gel solution helps to prevent premature polymerization). As the solution is pumped out of the mixing flask into the space between the glass plates, the 15% gel solution is being pumped into the mixing flask at the same rate, thus keeping the volume in the mixing flask constant, and thereby creating an exponential gradient of polyacrylamide between the glass plates. The solutions are pumped, at a rate of about 3.5 ml/min, until the gel solution between the glass plates comes up to a level about 3 cm from the top of the shorter plate. This leaves enough space for a 2 cm-high stacking gel. The 15% gel solution flask is empty at this point. An overlay of 2-propanol is now applied onto the separating gel, and the pump tubing is rinsed with water. After the gel has been allowed to polymerize for at least 1 hr, the 2-propanol is rinsed away with distilled water and replaced with 1× separating buffer. The gel is covered with Saran wrap and a plastic bag, and stored at 4° overnight.

The following day, the gel is allowed to come to room temperature, and, within 1 hr of running the gel, the stacking gel is mixed according to the following recipe:

    1.2 ml acrylamide–bisacrylamide stock solution
    2.0 ml of 4× stacking buffer
    4.7 ml distilled water

20 μl TEMED

75 μl of 10% ammonium persulfate

The 1× separating buffer overlay is removed completely from the separating gel, the stacking gel is poured, and the comb is inserted. The gel polymerizes within 20 min. The comb is removed, and the gel is mounted in the running stand. Electrode buffer is added to the upper and lower chambers.

Sample Preparation. Sample preparation is done before pouring the stacking gel or while waiting for it to polymerize. Between 1 and 150 μg protein are loaded per well. All the samples to be run on one gel are adjusted to the same final sample volume, which should be as small as possible, that is, 40 μl or less. The samples are prepared by adding one-fifth the final volume of 5× sample buffer to each portion of mitochondria or whole-cell suspension and bringing the sample to the final volume with distilled water. All the samples are mixed and kept at room temperature until they are loaded on the gel, up to 2 hr. The samples are not boiled, as aggregation of some of the translation products can occur at high temperature.[7] Whole-cell lysates are likely to be very viscous, owing to the nuclear DNA present. To reduce viscosity and facilitate loading the sample, the sample is vortexed vigorously for 1 min or more. The samples are loaded using a Pipetman and yellow tip or gel loading tip, or a 50-μl Hamilton syringe with a piece of intramedic tubing attached, which is of sufficient length to allow samples to enter only the tubing. The tips and intramedic tubing are changed between different samples. Any empty wells that are adjacent to wells containing samples receive 1× sample buffer in the same final volume.

The gel is run at constant current, 20 mA, for 30 min, and then at 30 mA until 1 hr after the bromphenol blue has run off the gel. The gel is run with two cooling fans, one in front and one in back. At the end of the run, the gel is fixed in 30% methanol, 10% acetic acid, 60 min to overnight, with one change. If left overnight, the gel is kept in its fixing solution at 4°. The gel is then dried for autoradiography or treated for fluorography.

Fluorography. In our experience, for fluorography of gels containing ^{35}S-labeled proteins, the method of Bonner and Laskey,[29] described below, gives the sharpest bands on the exposed film. Once the procedure for fluorography has begun, it is carried through to completion on the same day, as the shorter polypeptides tend to be lost during long incubations of the gel in dimethyl sulfoxide (DMSO).

The water in the gel is replaced with DMSO by two 200 ml washes in a glass tray, 30 min each, under gentle agitation. After the gel has been

[29] W. M. Bonner and R. A. Laskey, *Eur. J. Biochem.* **46,** 83 (1974).

transferred into DMSO, it is handled wearing chemical-resistant gloves, because DMSO penetrates the skin and latex gloves. The gel is transferred to 4 volumes (i.e., 110–120 ml) of 22% (w/v) 2,5-diphenyloxazole (PPO) in DMSO and agitated vigorously on a rotary shaker for 2 hr. The gel is then transferred to a tray containing a large volume of water and agitated slowly for 1 hr, with frequent changes. During this step, the PPO precipitates in the gel (the gel turns white), and the DMSO is washed away. At this stage, the gel is ready to be dried on a standard gel dryer (with heat and vacuum). If the DMSO has not been washed away thoroughly, the gel will be difficult to dry. The dried gel is exposed to sensitized (preflashed)[30] X-ray film (Kodak, Rochester, NY; X-OMAT-AR) at $-70°$ for one to several days. The film is developed according to the manufacturer's instructions.

Because PPO is expensive, one may want to recover it from the PPO/DMSO solution. The PPO is simply precipitated from solution by adding several volumes of water and collected by passing the slurry through a Whatman No. 1 filter in a Buchner funnel. The PPO is washed extensively with water, and then spread out in a dish in a fume hood to dry. When completely dry, the PPO can be used to prepare fresh PPO/DMSO solution.

Notes on Gel Electrophoresis of Mitochondrial Translation Products

1. Human cytochrome-*c* oxidase subunit II (CO II) displays an anomalous migration behavior on SDS gels, that is, its mobility is determined not only by its size, but also by the purity, or brand, of SDS and by the electrophoresis buffer composition. The supplier of SDS used in the preparation of the buffers for polyacrylamide gel electrophoresis in our laboratory is Sigma (Cat. No. L-4509). It is possible to use other brands or types of SDS, but the user should be aware that the migration of CO II relative to H^+-ATPase subunit 6 (ATPase 6) may vary with the brand of SDS. Under our conditions of electrophoresis, CO II migrates slower than ATPase 6. We have observed CO II migrating faster than ATPase 6 when a rather impure preparation of SDS was used. Another factor that influences the migration of CO II relative to ATPase 6 is the pH adjustment of the electrode buffer. Prior to any pH adjustment, the pH of the electrode buffer is, in our hands, approximately pH 8.6. Adjustment with HCl to reduce the pH by 0.1–0.2 units results in a pattern in which CO II migrates between ATPase 6 and cytochrome-*c* oxidase subunit III (CO III). Adding more HCl causes the CO II to migrate slower, closer to CO III. Adding less HCl causes CO II to migrate faster, closer to or together with ATPase 6.

2. The concentration of SDS in the sample to be loaded on the gel

[30] R. A. Laskey and A. D. Mills, *Eur. J. Biochem.* **56,** 335 (1975).

should not exceed 1% (w/v). Higher SDS concentrations will result in a diffuse appearance of the low molecular weight polypeptides.

Results and Discussion

The incorporation of [^{35}S]methionine into HeLa cell mitochondrial translation products labeled in suspension culture in methionine-free DMEMP supplemented with 5% (v/v) dialyzed bovine calf serum, in the presence of emetine, is linear for 30 min. Some of the polypeptides continue to be labeled at a linear rate for longer times.[31] Nevertheless, ^{35}S-labeled mitochondrial translation products continue to accumulate (non-linearly) for up to 4 hr.[17] Clearly, for comparing protein synthesis rates in different samples or among the different polypeptides, labeling times within the range that gives linear incorporation in the cell cultures under investigation should be used.

When testing new labeling conditions, or when preparing labeled mitochondrial translation products for immunoprecipitation experiments, it is important to determine the specific activity, that is, the number of disintegrations per minute per microgram of (dpm) protein. A 30-min pulse labeling of HeLa cells with 25 μCi/ml [^{35}S]methionine in the presence of emetine should yield a mitochondrial fraction with a specific activity of about 700 dpm/μg protein. A 2-hr pulse under the same conditions should yield a mitochondrial fraction with a specific activity of about 1600 dpm/μg protein. The specific activity of whole cells labeled in the presence of emetine is 2- to 5-fold lower than the specific activity of the mitochondrial fraction. The specific activity of mitochondrial translation products labeled in anchorage-dependent HeLa cells is somewhat lower than the specific activity obtained from HeLa cell suspension cultures labeled with the same concentration of [^{35}S]methionine. Cultures of nontransformed cells tend to give considerably lower specific activities of the mitochondrial translation products than established cell lines. When a higher specific activity of the labeled proteins is desired, a longer starvation period in methionine-free medium or a higher concentration of [^{35}S]methionine in the labeling medium should be tried.

Figure 1 displays the pattern of human mitochondrial translation products separated on a 15–20% exponential polyacrylamide–SDS gel. Note that, in the pulse and chase sample, which was derived from cells that had been pretreated with CAP for 1 day, the labeling of H$^+$-ATPase subunit 6 and H$^+$-ATPase subunit 8 relative to the other polypeptides is markedly enhanced. This effect is probably due to the CAP pretreatment.[17,18] Note also that ND3 in HeLa cells (ND3') exhibits an abnormal mobility, migrat-

---

[31] A. Chomyn, unpublished observations (1987).

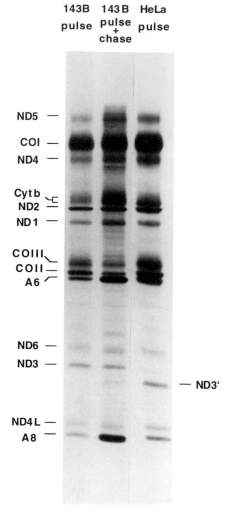

FIG. 1. ^{35}S-Labeled human mitochondrial translation products separated by electrophoresis on a 15–20% exponential gradient polyacrylamide–SDS gel. The human osteosarcoma-derived cell line 143B.TK$^-$ was labeled with ^{35}S]methionine for 1 hr in the presence of emetine (143B pulse) or was treated for 23 hr with chloramphenicol, then labeled with ^{35}S]methionine in the presence of cycloheximide for 2.25 hr, and grown further in the presence of unlabeled methionine and in the absence of inhibitors for 16.5 hr (143B pulse + chase). HeLa cells grown in suspension were labeled with ^{35}S]methionine for 2 hr in the presence of emetine (HeLa pulse). Lane 1 shows the pattern of a whole-cell lysate; lanes 2 and 3 show those of mitochondrial lysates. ND1, ND2, ND3, ND4, ND4L, ND5, and ND6 refer to subunits 1, 2, 3, 4, 4L, 5, and 6 of the rotenone-sensitive NADH dehydrogenase. CO I, CO II, and CO III denote subunits I, II, and III of cytochrome-c oxidase. A6 and A8 are subunits 6 and 8 of the H$^+$-ATPase. Between 15,000 and 20,000 counts/min (cpm) were run in each lane. The gel was impregnated with PPO as described in the text, dried, and exposed for fluorography for 4 days.

ing faster than the wild-type ND3, seen in 143B cells. This polymorphism, which has been described previously, is due to a point mutation in the ND3 gene.[32]

Acknowledgment

This work was supported by National Institutes of Health Grant GM-11726 to Giuseppe Attardi.

[32] N. A. Oliver, B. D. Greenberg, and D. C. Wallace, *J. Biol. Chem.* **258,** 5834 (1983).

[19] Mitochondrial Protein Synthesis in Rat Brain Synaptosomes

By Paola Loguercio Polosa and Giuseppe Attardi

Introduction

Synaptosomes, a subcellular fraction unique to the central nervous system, have proved to be an excellent model system for studies on the molecular mechanisms underlying presynaptic phenomena, since synaptosomes appear to retain the general organization and chemical composition of the original nerve terminals. Mitochondria of presynaptic endings are of particular interest both from the point of view of biogenesis and from that of synaptic function. In fact, the segregation of mitochondria in the nerve terminals creates for the cell the problem of maintaining a continuous supply of nuclear-coded components from the cell body to the peripheral organelles, with significant implications from the point of view of regulation of gene expression. Furthermore, understanding of the energetic metabolism at the nerve endings is expected to increase our knowledge of the mechanism of synaptic transmission.

This chapter describes the preparation of rat brain synaptosomes that are able to support *in vitro* mitochondrial protein synthesis, and the analysis of [35S]methionine-labeled translation products by sodium dodecyl sulfate–polyacrylamide gel electrophoresis (SDS–PAGE) combined with fluorography.

Isolation of Synaptosomes from Rat Brain

Principles. The conditions utilized to prepare synaptosomes and the duration of the procedure can greatly affect the viability of this subcellular fraction. In the original description of isolation of synaptosomes from rat brain, sucrose density gradients were used.[1] This procedure, however, caused loss of the functional integrity of the nerve endings and gave relatively low yields. A fractionation method utilizing an isosmotic Percoll/sucrose gradient was introduced by Nagy and Delgado-Escueta.[2] This method allows the rapid isolation of reasonably pure, biosynthetically competent synaptosomes. We have adapted their method for the purpose of isolation of synaptosomes that are able to support mitochondrial protein synthesis, introducing slight modifications involving a milder homogenization and a shorter preparation time.

Reagents and Solutions. All solutions are filter-sterilized and stored frozen.

> 2.5 M Sucrose in distilled water
> Medium I: 10% (w/w) sucrose in 5 mM Tris-HCl (pH 6.7, at 25°), 0.1 mM EDTA (Na$^+$ salt)
> Medium II: 0.25 M sucrose in 5 mM Tris-HCl (pH 7.0), 0.1 mM EDTA (Na$^+$ salt)
> Percoll (polyvinylpyrrolidone-coated colloidal silica, Pharmacia, Piscataway, NJ) solutions (8.5, 10, and 16%, and stock solution of isosmotic Percoll) are prepared just prior to use, and kept refrigerated. The stock solution of isosmotic Percoll (SIP) is made by addition of 9 volumes of the original sterile Percoll solution (designated as 100%, v/v) to 1 volume of 2.5 M sucrose in distilled water (the final concentration of Percoll is 90%, v/v). The 2.5 M sucrose solution is added dropwise to the 100% Percoll solution under constant stirring, to ensure a homogeneous mixing of the components. The other diluted Percoll solutions (8.5, 10, and 16%, v/v) are prepared by dropwise addition of medium II to the appropriate volume of SIP, under constant stirring. Typically, to prepare four gradients, 15 ml of SIP and 20 ml of each diluted Percoll solution are made, and the pH is adjusted to pH 7.0 with dilute HCl.
> SIP (15 ml): 1.5 ml of 2.5 M sucrose is added to 13.5 ml of 100% Percoll
> 8.5% Percoll solution (20 ml): 18.1 ml medium II is added to 1.9 ml SIP
> 10% Percoll solution (20 ml): 17.8 ml medium II is added to 2.2 ml SIP
> 16% Percoll solution (20 ml): 16.5 ml medium II is added to 3.5 ml SIP

Procedure. In the experiments described here, Sprague-Dawley male rats and Fisher 344 male rats have been used. One or two rats (or more

[1] E. G. Gray and V. P. Whittaker, *J. Anat.* **96,** 79 (1962).
[2] A. Nagy and A. V. Delgado-Escueta, *J. Neurochem.* **43,** 1114 (1984).

animals, if less than 1 month old) are sacrificed by decapitation for each preparation. The skull of each animal is rapidly opened, and the cerebral hemispheres are removed without the olfactory bulbs. The cortices are aseptically dissected from the brain, weighed, and finely chopped with a pair of scissors, while being frequently washed with ice-cold medium I to remove blood. The dissection is carried out rapidly on ice, and all solutions to be used in subsequent manipulations of the tissue are ice-cold.

The suspension is homogenized manually in medium I (10–15 ml per gram original wet tissue) with an all-glass Dounce homogenizer [15 or 40 ml capacity, pestle type A (tight fitting), Wheaton, Millville, NJ], using five up-and-down strokes. The crude homogenate is centrifuged at 1300 g_{av} for 4 min to sediment nuclei and unbroken cells. The low speed supernatant is carefully withdrawn with a large bore pipette, leaving 1 cm above the pellet, and is spun at 12,000 g_{av} for 5 min to pellet the membrane fraction, which contains the presynaptic endings. The pellet is carefully suspended in 1 ml of medium I per gram original wet tissue, by pipetting up-and-down through a large bore pipette. The membrane suspension is then gently diluted with 4 volumes of the 8.5% Percoll solution (final concentration of Percoll is 6.8%), and 3.5 ml of this dilution is layered onto a freshly prepared two-step Percoll density gradient.

Discontinuous density gradients are prepared in 1.5 × 7.5 cm polyallomer centrifuge tubes by layering 4 ml of 10% Percoll solution over 4 ml of 16% Percoll solution, using a peristaltic pump tubing with a needle attached to one end and connected to a pump. The gradients are kept in ice until use. The tubes are centrifuged at 20,500 g for 20 min at 4° in a Beckman Ty65 fixed-angle rotor. The centrifugation time includes the acceleration, but not the deceleration time; no brake is used during deceleration. Three diffuse bands of membrane material are separated in the gradient, as shown in Fig. 1a. As characterized in the original paper,[2] the uppermost two bands consist mainly of myelin and large membrane fragments and some unsealed synaptosomes (Fig. 1a, region M), whereas the band appearing at the interface of the 10 and 16% Percoll layers (Fig. 1a, region S) is the most enriched in well-preserved synaptosomes, with little evidence of free mitochondria. A loose pellet at the bottom of the gradient is represented mainly by free mitochondria, with only a small amount of synaptosomes. [35S]Methionine labeling experiments have confirmed the presence in the S fraction of labeled polypeptides exhibiting a pattern corresponding closely to that of the mitochondrial translation products of the rat R2 fibroblast line (Fig. 1b, lanes S and R2); in contrast, the M fraction lacks any such polypeptides, indicating the absence of a significant amount of intact synaptosomes in this fraction (Fig. 1b, lane M). Identification of the

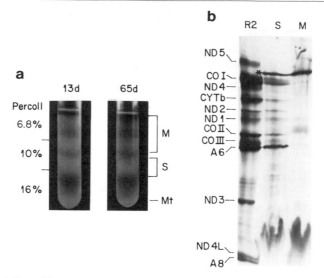

FIG. 1. (a) Percoll/sucrose gradients, used for purifying synaptosomes from 13- and 65-day-old rats, after the 20,500 g_{av} centrifugation. M indicates the two diffuse bands in the upper region of the gradient, containing mainly myelin and large membrane fragments. S is the 10%/16% Percoll interphase band consisting of synaptosomes. Mt is the mitochondrial pellet. (b) Analysis by SDS–polyacrylamide gradient gel electrophoresis of translation products from bands S and M of the Percoll/sucrose gradient, labeled with [^{35}S]methionine for 30 min at 30° in the presence of 100 μg/ml emetine. The mitochondrial protein labeling pattern from rat R2 cells exposed to [^{35}S]methionine for 2 hr is shown for comparison. COI, COII, and COIII denote subunits I, II, and III of cytochrome-c oxidase; ND1, ND2, ND3, ND4, ND4L, and ND5, subunits of the respiratory chain NADH dehydrogenase; CYTb, apocytochrome b; A6 and A8, H$^+$-ATPase subunits 6 and 8.

various rat mitochondrial translation products has been made by comparison of the R2 pattern with the HeLa cell pattern, and by immunoprecipitation experiments.[3]

The M region of the gradient is drawn off with a Pasteur pipette and discarded; the synaptosome diffuse band is gently collected with a tissue culture Pasteur pipette and immediately used without further purification. There is no need to remove Percoll from the purified fraction, since it is substantially inert. Less than 1 hr passes between sacrificing the animal and starting the protein labeling experiment.

[3] G. Attardi, A. Chomyn, and P. Loguercio Polosa, in "Advances in Myochemistry" (G. Benzi, ed.), Vol. 2, p. 55. Libbey Eurotext, London and Paris, 1989.

Protein Labeling

Samples (1–1.5 ml) of the synaptosomal fraction recovered from the Percoll/sucrose gradients are each gently added to 2 ml prewarmed methionine-free Dulbecco's modified Eagle's medium (DMEM), supplemented with 5% (v/v) extensively dialyzed fetal bovine serum (FBS). Subsequent incubation of the samples is carried out in stoppered 25-ml Erlenmeyer flasks in a 37° water bath under gentle rotatory shaking. After 3 min of incubation, the cytoplasmic protein synthesis inhibitor cycloheximide is added at 100 μg/ml. Emetine, another inhibitor of cytosolic protein synthesis, at 100 μg/ml, was found to inhibit considerably the labeling of all mitochondrial translation products (see below). To investigate the effects of the mitochondrial protein synthesis inhibitor chloramphenicol, this is added to the reaction mixture, at 100 μg/ml, 1 min after the addition of cycloheximide. Five minutes later, the labeling reaction is initiated by the addition of 1 mCi of [^{35}S]methionine (1100 Ci/mmol; Amersham, Arlington Heights, IL). The reaction is allowed to proceed 20 or 30 min and is then terminated by the addition of 5 volumes ice-cold DMEM plus 5% FBS. The labeled synaptosomes are collected by centrifuging the reaction mixture at 12,000 g_{av} for 10 min. The supernatant is discarded, and the pellet is washed twice by suspending it in medium II and centrifuging it at the same speed. Finally, the labeled pellet is suspended in a small volume (300 μl) of the same medium in the presence of 5 mM of the protease inhibitor phenylmethylsulfonyl fluoride. Protein concentration is measured by the Bradford method,[4] and the samples are divided into aliquots and stored at −70°.

For analysis of the mitochondrial translation products by polyacrylamide gel electrophoresis, the samples of synaptosome suspension are treated with ethanol in order to remove material giving rise to a smear, presumably containing lipids, in the low molecular weight region of the electrophoretic pattern (Fig. 1b). The ethanol treatment is carried out as follows: 120–150 μg protein suspension is diluted to 40 μl with medium II, then 360 μl of cold 100% ethanol is added (final ethanol concentration 90%). The sample is mixed and incubated on ice 10 min. After centrifugation at 12,000 g for 10 min, the pellet is dried under nitrogen and dissolved in an appropriate volume of Laemmli sample buffer.[5] The mitochondrial translation products are then subjected to gel electrophoresis on 15–20% exponential polyacryl-

[4] M. Bradford, *Anal. Biochem.* **72,** 248 (1976).
[5] U. K. Laemmli, *Nature (London)* **227,** 688 (1970).

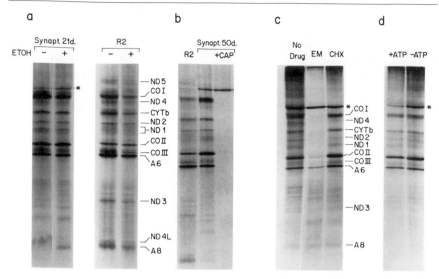

FIG. 2. Characterization of *in vitro* protein synthesis in the rat brain synaptosomal fraction. (a) Electrophoretic patterns of 30-min [^{35}S]methionine-labeled proteins of the synaptosomal fraction from a 21-day-old rat. Mitochondrial protein translation products labeled for 2 hr from the R2 cell line are shown for comparison. Equivalent samples (120 μg protein) of each fraction were electrophoresed directly or after treatment with 90% (v/v) ethanol. (b) Effect of chloramphenicol (CAP) at 100 μg/ml on the protein labeling profile, after a 20-min incubation with [^{35}S]methionine, of a 50-day-old rat synaptosomal fraction. (c) Comparison of protein synthesis in synaptosomes from a 5-day-old rat, after a 20-min [^{35}S]methionine labeling, in the absence of inhibitors of cytoplasmic protein synthesis (no drug) or in the presence of 100 μg/ml emetine (EM) or 100 μg/ml cycloheximide (CHX). (d) Effects of 1 mM ATP on the protein labeling pattern, after a 20-min pulse, of the synaptosomal fraction from a 5-day-old rat. Synaptosomal samples in b, c, and d were treated with 90% (v/v) ethanol prior to electrophoresis. Reproduced, with permission, from P. Loguercio Polosa and G. Attardi, *J. Biol. Chem.* **266,** 10011 (1991).

amide gradient gel in the presence of SDS.[6,7] The gel is then prepared for fluorography as reported.[8]

Characterization of *in Vitro* Mitochondrial Protein Synthesis in Rat Brain Synaptosomes

The smeary material at the bottom of the gel, which obscures the ND4L and ATPase 8 polypeptides (Fig. 1b) is completely removed by ethanol

[6] A. Chomyn and S. Lai, *in* "Structure, Function and Biogenesis of Energy Transfer Systems" (E. Quagliariello, S. Papa, F. Palmieri, and C. Saccone, eds.), p. 179. Elsevier, Amsterdam, 1990.

[7] A. Chomyn, this volume [18].

[8] W. M. Bonner and R. A. Laskey, *Eur. J. Biochem.* **46,** 83 (1974).

treatment, as shown in Fig. 2a.[9] A similar treatment applied to the mitochondrial fraction from 2-hr [^{35}S]methionine-labeled R2 cells has no significant effect on the pattern of the mitochondrial translation products (Fig. 2a). A pronounced labeled band moving somewhat more slowly than COI (marked with an asterisk in Fig. 2a) has been consistently observed in the protein labeling patterns from both the synaptosomal and the membrane fractions (Figs. 1b and 2). The labeling of this band is resistant to cytosolic (Fig. 2a–d) as well as to mitochondrial protein synthesis inhibitors (Fig. 2b), suggesting that its presence is not due to protein synthesis, but rather reflects some end-labeling event. Figure 2c shows that the pattern of mitochondrial translation products is clearly recognizable when the labeling is carried out in the absence of inhibitors of cytoplasmic translation, indicating a low level of extramitochondrial protein synthesis in the synaptosomal fraction. This argues in favor of the absence or the presence of only a low amount of cytoplasmic polysomes in rat brain synaptosomes, pointing at the same time to a very small contamination of this fraction by cytoplasmic fragments from neuronal or glial cells. Also, in contrast to the situation observed in HeLa cells,[10] it appears that emetine considerably inhibits mitochondrial protein synthesis, whereas cycloheximide has no effect. Accordingly, cycloheximide should be used as cytosolic protein synthesis inhibitor with isolated synaptosomes (Fig. 2c). Figure 2d shows that addition of 1 mM ATP to the incubation medium has no effect on the labeling profile of synaptosome translation products. This argues against any significant contamination of the synaptosomes by free mitochondria (deriving from broken neuronal bodies or glial cells). In fact, it had previously been shown that protein synthesis in isolated HeLa cell mitochondria depends on the presence in the medium of ATP or, alternatively, of ADP, phosphate, and a respiratory substrate.[11]

A distinctive feature of the newly synthesized translation products in synaptosomes is the apparent absence of labeling of the ND5 subunit, one of the mtDNA-encoded subunits of NADH dehydrogenase. Using the conditions of rapid isolation and mitochondrial translation analysis of synaptosomes described in this chapter, the rate of mitochondrial protein synthesis has been measured in rat brain synaptosomes during postnatal development and maturation of the animal.[9]

Acknowledgment

This work was supported by National Institutes of Health Grant GM-11726 to G. A.

[9] P. Loguercio Polosa and G. Attardi, *J. Biol. Chem.* **266,** 10011 (1991).
[10] P. Costantino and G. Attardi, *J. Biol. Chem.* **252,** 1702 (1977).
[11] M. Lederman and G. Attardi, *Biochem. Biophys. Res. Commun.* **40,** 1942 (1970).

[20] Immunoprecipitation of Human Mitochondrial Translation Products with Peptide-Specific Antibodies

By HERMAN A. C. M. BENTLAGE and ANNE CHOMYN

Introduction

Antibodies directed against a short synthetic peptide sequence predicted from a given gene sequence were shown by Walter *et al.*[1] to be able to recognize a full-length protein containing that short peptide sequence. This approach has been applied extensively for the identification of the proteins encoded in cloned genes.[1] A successful application has been in the identification of the mitochondrial translation products[2-6] corresponding to the reading frames[7] of human mitochondrial DNA (mtDNA). Peptide-specific antibodies specific for individual mtDNA-encoded polypeptides have also been used to immunoprecipitate entire respiratory complexes containing them,[8-10] as well as in the molecular analysis of defects associated with mitochondrial diseases.[11-13]

This chapter focuses on the use of peptide-specific antibodies in immu-

---

[1] G. Walter, K.-H. Scheidtmann, A. Carbone, A. P. Laudano, and R. F. Doolittle, *Proc. Natl. Acad. Sci. U.S.A.* **77,** 5197 (1980).

[2] R. F. Doolittle "Of URFs and ORFs: A Primer on How to Analyze Derived Amino Acid Sequences." University Science Books, Mill Valley, California, 1986.

[3] P. Mariottini, A. Chomyn, G. Attardi, D. Trovato, D. D. Strong, and R. F. Doolittle, *Cell (Cambridge, Mass.)* **32,** 1269 (1983).

[4] A. Chomyn, P. Mariottini, N. Gonzalez-Cadavid, G. Attardi, D. D. Strong, D. Trovato, M. Riley, and R. F. Doolittle, *Proc. Natl. Acad. Sci. U.S.A.* **80,** 5535 (1983).

[5] P. Mariottini, A. Chomyn, M. Riley, B. Cottrell, R. F. Doolittle, and G. Attardi, *Proc. Natl. Acad. Sci. U.S.A.* **83,** 1563 (1986).

[6] A. Chomyn, M. W. J. Cleeter, C. I. Ragan, M. Riley, R. F. Doolittle, and G. Attardi, *Science* **234,** 614 (1986).

[7] S. Anderson, A. T. Bankier, B. G. Barrell, M. H. L. de Bruijn, A. R. Coulson, J. Drouin, I. C. Eperon, D. P. Nierlich, B. A. Roe, F. Sanger, P. H. Schreier, A. J. H. Smith, R. Stadin, and I. G. Young, *Nature (London)* **290,** 457 (1981).

[8] A. Chomyn, P. Mariottini, M. W. J. Cleeter, C. I. Ragan, A. Matsumoto-Yagi, Y. Hatefi, R. F. Doolittle, and G. Attardi, *Nature (London)* **314,** 592 (1985).

[9] P. Mariottini and A. Chomyn, this series, Vol. 260 [13].

[10] P. Mariottini, A. Chomyn, R. F. Doolittle, and G. Attardi, *J. Biol. Chem.* **261,** 3355 (1986).

[11] H. A. C. M. Bentlage, A. J. M. Janssen, A. Chomyn, G. Attardi, J. E. Walker, H. Schägger, R. C. A. Sengers, and F. J. M. Trijbels, *Biochim. Biophys. Acta* **1234,** 63 (1995).

[12] J. A. Enriquez, A. Chomyn, and G. Attardi, *Nat. Genet.* **10,** 47 (1995).

[13] H. J. Tritschler, E. Bonilla, A. Lombes, F. Andreetta, S. Servidei, B. Schneyder, A. F. Miranda, E. A. Schon, B. Kadenbach, and S. DiMauro, *Neurology* **41,** 300 (1991).

noprecipitation experiments involving mtDNA-encoded polypeptides. However, before going into the details of these experiments, we summarize the steps that were carried out in producing the peptide-specific antibodies.

Selection and Synthesis of Peptides

The peptides used for the studies mentioned above on mtDNA-encoded polypeptides were 7 to 18 amino acids long (Table I). The length depended on the presence and location of amino acids utilizable for coupling the peptide to a carrier protein and was limited by the presence of amino acids that would present difficulties for peptide synthesis, such as proline near the COOH terminus and cysteine. Sometimes the peptide was lengthened to increase the distance between proline and the COOH terminus.

Whenever possible, the COOH-terminal peptide was chosen for synthesis, because it was considered to be likely to be exposed in the complete polypeptide chain. In addition, the attachment of its NH_2 terminus to a carrier protein is easily done.[2] The second choice was the NH_2-terminal peptide, which in the intact protein might also be easily accessible to antibodies. In some cases, internal peptides were tried. Among these, some (ND1-I1, ND1-I2, ND6-I1, and ND6-I2) were chosen from among the hydrophilic portions of the protein, as determined by a Kyte–Doolittle hydropathy plot.[14] Other peptides (ND4-I1, ND4-I2, ND4-I3, ND4-I4, ND4-I5, ND5-I, and ND6-I3) were chosen on the basis of hydrophilicity, surface probability, flexibility, high probability of β turns, and low α-helix or β-sheet propensity[15] using the computer programs PEPPLOT or PLOTS-TRUCTURE.[16] When no amino acid suitable for coupling was present in the selected peptide sequence, such an amino acid was added at the NH_2 terminus of COOH-terminal peptides and at the COOH terminus of NH_2-terminal peptides, and at either or both ends of internal peptides (Table I). Alternatively, the activatable S-acetylthioacetyl (SATA) group was attached to the NH_2 terminus.[17]

For the determination of the peptide–carrier ratio, a radioactive amino acid was included in the synthesized peptide,[2,3] or the peptide was extended with L-norleucine at the NH_2 terminus.[17] The peptides were synthesized

[14] J. Kyte and R. F. Doolittle, *J. Mol. Biol.* **157,** 105 (1982).

[15] M. H. V. Van Regenmortel, J. P. Briand, S. Muller, and S. Plaué, *in* "Laboratory Techniques in Biochemistry and Molecular Biology" (R. H. Burdon and P. H. Van Knippenberg, eds.), p. 1. Elsevier, Amsterdam, 1988.

[16] J. Devereux, P. Haeberli, and O. Smithies, *Nucleic Acids Res.* **12,** 387 (1984).

[17] W. J. G. Schielen, M. Voskuilen, G. I. Tesser, and W. Nieuwenhuizen, *Proc. Natl. Acad. Sci. U.S.A.* **86,** 8951 (1989).

TABLE I
ABILITY TO IMMUNOPRECIPITATE POLYPEPTIDE[a]

Peptide	Sequence	Attachment	Carrier molecule	Number of peptides per carrier	RAIP[h]
COII-N[3 b,c]	MAHAAQVGLQ(E)[d]	EDC[e]	BSA	15	+++
COII-C[3]	KIFEMGPVFTL	GLUT	BSA	19	+++
A8-N[3]	MPQLNTTV(Y)	BDB	BSA	33	+
A8-C[3]	(K)SLHSLPPQS	GLUT	BSA	5	+++
A6-C[4]	(K)VSLYLHDNT	GLUT	BSA	21	+++
ND1-I1[2] (3397–3420)	AcMQLRKGPN(Y)	BDB	BSA	30[f]	−
ND1-I2[2] (3475–3498)	(Y)TKEPLKPA(Y)	BDB	BSA	28[f]	−
ND1-C[4]	(K)PITISSIPPQT	GLUT	BSA	26	+++
ND3-C1[4]	KGLDWTE	GLUT	BSA	35	+++
ND3-C2[4]	(Y)LQKGLDWTE	BDB	BSA	28	+++
Cytb-N[2]	MTPMRKINP(E)	EDC	BSA	37[f]	−
ND2-N[5]	INPLAQPV(E)	EDC	BSA	21	++
ND4-N[5]	MLKLIVP(E)	EDC	BSA	15	−
ND4-C[5]	(K)PDIITGFSS	GLUT	BSA	18	++
ND4L-N[5]	MPLIYNM(E)	EDC	BSA	12	−
ND4L-C[5]	(K)HNLNLLQ-X	GLUT	BSA	24	++
ND5-N[5]	MTMHTTM(AE)	EDC	BSA	16	−
ND5-C[5]	(K)TLLLIT	GLUT	BSA	32	−
ND6-N[6]	MMYALF(E)	EDC	BSA	16	−
ND6-C[6]	(K)IVIEIARGN	GLUT	BSA	33	−
ND6-I1[6] (14,422–14,454)	AIEEYPEAWGS	GLUT	BSA	24	+
ND6-I2[6] (14,593–14,622)	AcGFSSKPSPIY	BDB	BSA	25	−
ND4-I1[11] (11,000–11,041)	*QRHLSSEPLSRKKL[g]	Sulfo-MBS	OA	5	−
ND4-I2[11] (11,162–11,194)	*RWGNQPERLNA	Sulfo-MBS	OA	11	−
ND4-I3[11] (11,587–11,684)	*LRQTDLKSL	Sulfo-MBS	OA	8	−
ND4-I4[11] (11,744–11,776)	*LANSNYERTHS	MHS	BSA	18	−
ND4-I5[11] (12,017–12,064)	*THHINNMKPSFTRENT	Sulfo-MBS	OA	4	+++
ND4-C[11] ND5-I1[11] (12,397–12,441)	*PDIITGFSS	Sulfo-MBS	OA	3	+++
ND6-I3[11] (14,242–14,292)	*TTLVNPNKKNSYPHY	MHS	BSA	13	+
	*EGEGSGFIREDPIGAGA	MHS	BSA	27	++

[a] The first 22 peptides were synthesized and antisera to them were produced by Dennis Trovato, Donna D. Strong, Marcia Riley, Barbara Cottrell, and Russell F. Doolittle, Department of Chemistry, University of California at San Diego, La Jolla, California. The last eight peptides were

by Merrifield solid-phase procedures.[18,19] In most cases, the amino acid composition of the synthesized peptides was checked by amino acid analysis after acid hydrolysis. Furthermore, the homogeneity and correct charge were in some cases determined by paper electrophoresis.[2-6] The homogeneity was also checked in some cases by high-performance liquid chromatography.[3,11] Mass spectroscopy was also used to confirm the correct molecular weight of the peptide.[11] Checks such as these were done (1) to ensure that the correct amino acids had been used in the synthesis and (2) to ensure that blocking groups, such as those used on acidic side chains, had been completely removed.[2]

Coupling of Peptides to Carrier Proteins

Peptides were coupled to a carrier protein [bovine serum albumin (BSA) or myoglobin, or ovalbumin] with one of several chemical linkers. Table I lists the coupling reagents that were used for the various peptides. An easy linker to use[2] is glutaraldehyde, which links amino groups in the peptide to amino groups on the carrier. The method using water-soluble carbodii-

[18] R. B. Merrifield, *J. Am. Chem. Soc.* **85,** 2149 (1964).

[19] J. M. Stewart and J. D. Young, "Solid Phase Peptide Synthesis." Freeman, San Francisco, 1969.

synthesized by G. I. Tesser, Laboratory of Organic Chemistry, University of Nijmegen, The Netherlands.

[b] N, C, and I denote NH_2-terminal, COOH-terminal, and internal portions of the polypeptide.

[c] COII, Cytochrome-c oxidase subunit II; A8, H^+-ATPase subunit 8, originally known as A6L[3]; A6, H^+-ATPase subunit 6; ND1, ND2, ND3, ND4, ND4L, ND5, and ND6, subunits of the rotenone-sensitive NADH dehydrogenase. The reading frames for the ND polypeptides were originally designated URF1, URF2, URF3, etc.,[7] and were renamed subsequent to the functional identification of the respective translation products.[6,8]

[d] Amino acids in parentheses were not present in the protein sequence but were included to facilitate attachment to the carrier molecule. Ac, Acetyl; X, α-aminobutyric acid.

[e] BDB, Bisdiazobenzidine; BSA, bovine serum albumin; EDC, water-soluble carbodiimide; GLUT, glutaraldehyde; MHS, 6-maleimidohexanoic acid N-hydroxysuccinimide ester; OA, ovalbumin; Sulfo-MBS, 6-maleimidobenzoic acid N-hydroxysulfosuccinimide ester.

[f] R. F. Doolittle, personal communication (1981, 1983).

[g] The asterisk (*) indicates the extension of the peptide at its NH_2 terminus with S-acetylthioacetyl and norleucine.

[h] Relative ability of the antiserum raised against the peptide to immunoprecipitate the corresponding polypeptide. −, no immunoprecipitate formed; +, ++, +++, relatively low, intermediate, and high efficiency of specific immunoprecipitation, respectively, as estimated by comparison of autoradiogram band intensities of the polypeptide in the immunoprecipitate and in the starting material, total mitochondrial lysate.

mide links carboxyl groups on the peptide to amino groups on the carrier protein.[2] Bisdiazobenzidine forms bridges between phenolic rings, thus linking tyrosines to tyrosines.[2] S-Acetylthioacetyl (SATA) generates a sulfhydryl group on the amino terminus of the peptide after deprotection, which can be used for coupling to a maleimidated carrier molecule.[17] 6-Maleimidohexanoic acid N-hydroxysuccinimide ester (MHS; Pierce, Rockford, IL) and 6-maleimidobenzoic acid N-hydroxysulfosuccinimide ester (Sulfo-MBS; Pierce) are used to maleimidate the carrier molecules and also serve as a linker between the peptide and the carrier molecules.[11]

Some thought should be given to the carrier protein to be used, because it is likely that antibodies to the carrier protein, as well as to the peptide, will be produced by the rabbit.[3] These carrier protein-specific antibodies can be dealt with in one of two ways, as discussed below in the section on immunization.

Determination of Peptide–Carrier Ratio

The average number of peptides attached to a molecule of carrier protein was determined in one of two ways: either a small amount of radioactive amino acid was incorporated in the peptide sequence, and then the radioactivity of the peptide–carrier conjugate after dialysis was determined, or the peptides were extended with L-norleucine (Nle) at the NH_2 terminus, and then the ratio of Nle to the other amino acids was determined by amino acid analysis of the hydrolyzed conjugate.[17] Tricine–sodium dodecyl sulfate–polyacrylamide gel electrophoresis (SDS-PAGE)[20] was sometimes used to get a rough idea of the peptide–carrier ratio in the conjugates and an indication of the extent of cross-linking between carrier molecules.

Immunization and Antibody Preparation

All antisera were raised in rabbits. Two rabbits were injected with a given antigen at 2- to 3-week intervals. Each rabbit received the peptide–carrier protein conjugate mixed in a 1 : 1 (v/v) ratio with the Freund's adjuvant system [complete, FCA (Difco, Laboratories, Detroit, MI), for initial injections, or incomplete, FIA, for booster injections]. The Ribi adjuvant system (Ribi ImmunoChem Research, Inc., Hamilton, MT) yielded only low levels of peptide-specific antibodies.[11] One of the protocols for the administration of adjuvant preparations was as follows: antigen conjugate-FCA (1 mg conjugate, total volume 2 ml) was injected intradermally at 20 sites on the back of the animal; antigen conjugate-FIA (0.5-mg conjugate, total volume 1 ml) was

[20] H. Schägger and G. von Jagow, *Anal. Biochem.* **166,** 368 (1987).

injected intramuscularly and subcutaneously. Blood samples were taken before the first injection (preimmune serum) and either 2 weeks after or immediately before the booster injections. The sera were prepared and then stored at $-20°$. In some cases, γ-globulins were isolated from the serum by precipitation with 33% saturated ammonium sulfate[9] or by passing the serum over a protein-A Sepharose column (Pharmacia/LKB Biotechnology, Piscataway, NJ). In earlier experiments, we routinely removed the carrier protein-specific antibodies from the γ-globulin fraction by passing the latter over an affinity column to which the carrier protein alone had been covalently linked. The flow-through thus contained peptide-specific antibodies devoid of carrier-specific antibodies. This purification step has since been found to be unnecessary for immunoprecipitation experiments, as the same carrier protein is never used in experiments done with the antibodies. For example, the peptides were coupled in most instances to BSA (Table I). In immunoprecipitation protocols using the antibodies generated to those peptides, ovalbumin, rather than BSA, is used as the blocking agent. Thus, any BSA-specific antibodies that may be present in the antibody preparation do not form immunoprecipitates.

Enzyme-Linked Immunosorbent Assay

The presence of peptide-specific antibodies was detected by double immunodiffusion against peptide coupled to a different carrier[3] or by solid-phase radioimmunoassay with antibodies after they had been passed over a carrier protein-affinity column twice.[3] Titers were determined by enzyme-linked immunosorbent assay (ELISA),[21] using the peptides coupled with a different linker to a different carrier protein as coating antigens, in order to exclude reactions of no interest.

Protocol for Labeling and Immunoprecipitation of Mitochondrial Translation Products

Pulse Labeling with [35S]Methionine

To label specifically mitochondrial translation products, an inhibitor of cytosolic protein synthesis, such as emetine or cycloheximide, is used. Emetine is superior for suppressing extramitochondrial protein synthesis, but it should be noted that its action is irreversible. We have usually labeled cells with at least 25 μCi/ml [35S]methionine for 2 hr in methionine-free medium in the presence of 100 μg emetine/ml. For this labeling protocol

[21] H. J. Geerligs, W. J. Weijer, W. Bloemhoff, G. W. Welling, and S. Welling-Wester, *J. Immunol. Methods* **106,** 239 (1988).

we do not carry out a chase. The details of the labeling protocol are given elsewhere in this volume[22] (see also note 1 below).

To obtain cleaner immunoprecipitates, it is advisable to isolate the mitochondrial fraction as described.[22] The mitochondria can be stored frozen at $-70°$ until needed. Alternatively, whole-cell lysates can be used in the immunoprecipitation protocol; however, these lysates will have lower specific activity than the mitochondrial fraction and may give less clean immunoprecipitates. The protein concentration of an SDS lysate of a sample and the concentration of incorporated radioactivity of the isolated mitochondria or of the whole-cell preparation should be determined.[22] The specific activity of a mitochondria preparation should preferably be at least 1000 disintegrations/min (dpm)/μg, and that of a whole-cell preparation should be at least 500 dpm/μg. The amount of protein used in an immunoprecipitation reaction is 30 to 100 μg.

Immunoprecipitation

Stock Solutions

1 *M* Tris-HCl, pH 8.0 (25°)
1 *M* Phenylmethylsulfonyl fluoride (PMSF) in dimethyl sulfoxide (DMSO) or 0.25 *M* PMSF in ethanol (PMSF is very toxic; therefore, the solution of PMSF in DMSO should never be made in amounts greater than 1 ml and should be used with care, because, if the solution is spilled on the skin, the DMSO, and presumably the PMSF dissolved in it, will be absorbed through the skin)
20% (w/v) Sodium dodecyl sulfate (SDS)
10 mg/ml Ovalbumin in water (use another protein if ovalbumin was used as the carrier molecule for immunization)
Freshly made 10% (w/v) Triton X-100
10× NET: 1.5 *M* NaCl, 0.5 *M* Tris base, 50 m*M* EDTA, 0.2% (w/v) NaN$_3$; adjust to pH 9.0 with hydrochloric acid (see note 2 below)

Procedure. Within 24 h of use, prewash the immunoadsorbent, formaldehyde-fixed cells of *Staphylococcus aureus,* strain Cowan I (Zysorbin, Zymed Laboratories, San Francisco, CA), with approximately 10 volumes of 1× NET, 0.5% Triton X-100. Incubate the well-suspended immunoadsorbent in this buffer at room temperature for 15 min. Centrifuge the suspension (at 10,000 g for 8 min in a Sorvall rotor, or in a microcentrifuge at 12,800 g for 3 min), and suspend the pellet in the same buffer. Centrifuge the suspension immediately and suspend the pellet in 1× NET. Centrifuge the

[22] A. Chomyn, this volume [18].

suspension once more and suspend the final pellet in 1× NET containing 1 mg/ml ovalbumin, 1 mM PMSF, 1 mM methionine, at the original suspension concentration, that is, so as to make a 10% (v/v) suspension. Keep the suspension at 4° until needed.

Lyse the mitochondria or cell suspension in 1% SDS (w/v) at 1.4 mg protein/ml. We carry out this step by diluting the preparation, for example, 40 μg protein per reaction, to 2.8 mg protein/ml with water or with 0.25 M sucrose, 10 mM Tris-HCl, pH 6.7, and then adding an equal volume, in this case, 14.3 μl, of 10 mM Tris-HCl, pH 8.0, 2% SDS, 1 mM PMSF. Incubate the lysate at 37° for 15 to 60 min, and then add 9 volumes of a mixture consisting of 1 ml of 10× NET, 1 ml of 10 mg/ml ovalbumin, 7 ml distilled water, and 10 μl of 1 M PMSF at room temperature. Mix and add 120 μl $S.$ $aureus$ suspension (prewashed). Agitate at room temperature 30 to 60 min. Pellet the immunoadsorbent, and add to the supernatant 100 μg of the γ-globulin fraction of the antiserum or 10 μl of the antiserum. Incubate the lysate–antibody mixture overnight at 4° with gentle mixing. Bring samples to room temperature and add 120 μl fixed $S.$ $aureus$ suspension (prewashed). Incubate at room temperature with agitation for 1 hr. Centrifuge the suspension to collect the $S.$ $aureus$ (2 min in a microcentrifuge at 12,800 g) and wash the pellet three times at room temperature, twice with 1 ml of 1× NET, 1 mg/ml ovalbumin, 0.1% SDS, 1 mM PMSF, and 2 mM methionine, and a third time with 1 ml of 10 mM Tris-HCl, pH 8.0. Remove the last supernatant completely (carry out an extra, brief spin to collect the last bit of supernatant) and add to the pellet 60 μl of 5 mM Tris-HCl, pH 8.0, 1% SDS, 1 mM PMSF. Suspend the pellet well and incubate at 37° for 30 min. Centrifuge the suspension in the microcentrifuge for 4 min (12,800 g) and transfer the supernatant (the eluate) to a fresh tube. Store at −70°, or analyze 30–40 μl immediately by SDS–gel electrophoresis. To prepare the sample for electrophoresis on a Laemmli[23] gel, add one-fourth volume of 5× sample buffer which contains only 1% SDS. An SDS concentration greater than 1% in the sample will cause low molecular weight polypeptides to run as diffuse bands.[22] The conditions for gel electrophoresis and fluorography are given elsewhere in this volume.[22]

Control immunoprecipitation reactions should always be done, to distinguish specific immunoprecipitates from background precipitates. An essential control reaction is one performed, preferably, with preimmune serum or, otherwise, with normal serum in the place of the peptide-specific antiserum. Another useful control is an antibody reaction carried out in the presence of 5 to 20 μg of the corresponding synthetic peptide.

[23] U. Laemmli, $Nature$ ($London$) **227**, 680 (1970).

Notes

1. Pretreatment of the cells with chloramphenicol at 40 μg/ml in complete medium for 22 hr prior to labeling will increase the amount of label incorporated into at least some of the mitochondrial translation products during a 2-hr pulse.[10,22]

2. We have found that, for most of the peptide-specific antibodies directed against mitochondrial translation products, more of the respective polypeptide is immunoprecipitated in a high pH buffer (pH 9.0 at 25°), as compared to a pH 7.4 buffer, and fewer nonspecific polypeptides contaminate the immunoprecipitate. There are three exceptions: H^+-ATPase subunit 6, H^+-ATPase 8, and ND2 are more efficiently precipitated by the A6-C, A8-C, and ND2-N antibodies at pH 7.4.

3. Purification of the peptide-specific antibodies by affinity chromatography with a column to which the specific peptide is covalently linked does not lead to improved immunoprecipitation of the respective polypeptide.

4. The immunoadsorbent protein A-Sepharose, in our hands, does not substitute well for formaldehyde-fixed *S. aureus.*

5. Triton X-100, which is often used in conjunction with SDS in immunoprecipitation experiments, is actually inhibitory for the immunochemical reactions described here.[3]

Results

The ability of all the peptide-specific antibodies tested to immunoprecipitate human mitochondrial translation products is listed in Table I. Also included are the peptide sequence used, the attachment procedure, the carrier molecule, and the peptide–carrier ratio. It can be seen that in general either the COOH- and/or NH$_2$-terminal peptides gave good results. The C-terminal peptides were more often successful than their N-terminal counterparts. Only in the cases of ND5 and ND6 did these antibodies not precipitate their respective polypeptide. On the other hand, antibodies against predicted internal epitopes were successful in precipitating the ND5 and ND6 polypeptides. It must be stated, however, that the prediction of internal epitopes was successful in only one of five and two of three cases for ND4 and ND6, respectively. Only one internal peptide sequence of ND5 was tried, which yielded weakly reactive antibodies. Two internal peptides of ND1 failed to yield antibodies capable of precipitating ND1. There was no correlation between the peptide–carrier ratio and the efficiency of the antiserum in immunoprecipitating the respective polypeptide.

Other Uses of Peptide-Specific Antibodies

Peptide-specific antibodies against COII-C, ND1-C, ND3-C, and ND4L-C (Table I) have been used in Western blots of HeLa cell mitochondria (COII-C)[10] and bovine (ND1-C, ND3-C, and ND4L-C) heart mito-chondria.[24] Similarly, antibodies against ND4-I5, ND4-C, ND5-I, and ND6-I3 (Table I) have been used successfully in Western blots of human skeletal muscle mitochondria, both from control individuals and from mitochondrial myopathy patients with various respiratory chain enzyme deficiencies.[11,25] The specificity of the reaction can be determined by a preincubation of the antibodies with the free peptide. In addition, antibodies against an internal peptide of *Neurospora crassa* ND1[26] that has 12 of 14 amino acids identical with the homologous human sequence reacted in this assay.[25] The enhanced chemiluminescence (ECL) kit (Amersham, Arlington Heights, IL) is very sensitive for detection of weakly bound antibodies.

Peptide-specific antibodies to human mitochondrial translation products have also been used successfully for immunohistochemistry. Specifically, COII-C antibodies (Table I) have been used to detect the presence or absence of the corresponding polypeptides in skeletal muscle sections.[13]

A word of caution should be given here regarding the use of peptide-specific antibodies for immunohistochemistry. As the peptides can be quite short, there is a possibility that the peptide-specific antibodies could cross-react with a noncorresponding polypeptide(s) containing a similar peptide sequence. In fact, such a case has been reported.[3] Antibodies specific for a COOH-terminal peptide of the mitochondrial translation product H+-ATPase subunit 8 cross-reacted with ND1. The peptide sequences were

A8-C: (K) S L H S L P P Q S

ND1-C: (K) P I T I S S I P P Q T

Thus, four of the last eight amino acids in H+-ATPase subunit 8 were identical to four amino acids in the COOH-terminal octapeptide of ND1, and, furthermore, the two polypeptides had similar amino acids in three of the remaining four positions in the COOH-terminal octapeptide. Similarly, antibodies directed against a COOH-terminal peptide of ND1 cross-reacted with H+-ATPase subunit 8.[4] Thus, any peptide-specific antibodies that will be used to detect the presence of a protein by a method that does not reveal the size of the protein should first be characterized at the molecular level, by immunoprecipitation or Western blots, ideally using the same source of material that will be used in the immunohistochemistry, for example, whole cells, muscle preparations, or subcellular fractions.

[24] F. G. P. Earley, S. D. Patel, C. I. Ragan, and G. Attardi, *FEBS Lett.* **219,** 108 (1987).

[25] H. A. C. M. Bentlage, R. de Coo, H. ter Laak, R. Sengers, F. Trijbels, W. Ruitenbeek, W. Schlote, K. Pfeiffer, S. Gencic, G. von Jagow, and H. Schägger, *Eur. J. Biochem.* **227,** 909 (1995).

[26] R. Zauner, J. Christner, G. Jung, U. Borchart, W. Machleidt, A. Videira, and S. Werner, *Eur. J. Biochem.* **150,** 447 (1985).

Acknowledgments

This research was supported by grants from the Prinses Beatrix Fonds and The Netherlands Organization for Scientific Research (NWO) to H. A. C. M. B. and by National Institutes of Health Grant GM-11726 to Giuseppe Attardi.

[21] Genetic Strategies for Identification of Mitochondrial Translation Factors in *Saccharomyces cerevisiae*

By THOMAS D. FOX

Introduction

The mechanisms of mitochondrial (mt) translation initiation are poorly understood. To a large extent, this is because translation systems of mitochondria have resisted detailed *in vitro* analysis: although isolated intact organelles will carry out protein synthesis,[1] no extract of mitochondria with reproducible mRNA-dependent translational activity has been described. For example, whereas yeast mitochondrial ribosomes will carry out the elongation steps of translation on synthetic polyribonucleotides, they will not initiate correctly in the systems used to date.[2] Although initiation factor 2[3] and several elongation factors have been biochemically assayed and purified from bovine mitochondrial extracts by virtue of their ability to function in *Escherichia coli*-derived *in vitro* systems,[4] this approach has limitations for the study of processes specific to mitochondria such as translational control of organellar genes.

In the absence of *in vitro* systems, genetic analysis of mitochondrial translation in the yeast *Saccharomyces cerevisiae* has provided what little insight we have into this important level of organellar gene expression. To date, it has allowed the demonstration of *in vivo* functional roles for some general translation factors homologous to those of other biological systems as well as the demonstration of the existence of important regulatory proteins for which no homologs in other systems have, as yet, been found. This chapter summarizes general strategies employed to study yeast mitochondrial translation genetically. Detailed descriptions of procedures for the isolation of mutants, genetic manipulation of yeast, and phenotypic

[1] E. E. McKee and R. O. Poyton, *J. Biol. Chem.* **259,** 9320 (1984).

[2] P. J. T. Dekker, B. Papadopoulou, and L. A. Grivell, *Curr. Genet.* **23,** 22 (1993).

[3] H. X. Liao and L. L. Spremulli, *J. Biol. Chem.* **265,** 13618 (1990).

[4] C. J. Schwartzbach, M. Farwell, H.-X. Liao, and L. L. Spremulli, this volume [23].

analysis of mitochondrial functions are presented elsewhere in this volume[5,6] and in Volume 194 (Guide to Yeast Genetics and Molecular Biology) of this series.[7-12] Although the specific strategies discussed here are directly applicable only to *S. cerevisiae* at the present time, it is worth considering how they might be extended to other organisms.

Factors Required for Translation of All Mitochondrial mRNAs

Saccharomyces cerevisiae and a few other yeasts are peculiar in that spontaneous mutants bearing large deletions of mtDNA and totally lacking mitochondrial protein synthesis, termed *rho⁻* (*ρ⁻*) or cytoplasmic *petite*, arise at very high frequency. Thus, any screen among collections of nonrespiring yeast mutants for strains with lesions in specific genes (nuclear or mitochondrial) must be set up to eliminate the frequent *rho⁻* mutants. Although there are several ways to achieve this goal,[6,12] the *rho⁻* problem creates a special difficulty for the isolation of mutants defective in genes required generally for mitochondrial protein synthesis: for reasons that are not clear, yeast cells totally lacking mitochondrial translation cannot maintain their wild-type mtDNA and rapidly become *rho⁻*.[13] Thus a null mutation in a nuclear gene coding a mitochondrial translation factor, that completely disrupted protein synthesis, would cause the cells bearing that mutation to become *rho⁻*. Such mutants would be indistinguishable in a rapid screen from the spontaneous *rho⁻* mutants lacking nuclear mutations and thus extremely difficult to isolate.

Despite these difficulties, several important mitochondrial translation factor genes have been identified by screening collections of nonrespiring mutants.[14] In several cases, wild-type DNA complementing the mutations was cloned[10] and revealed by DNA sequence analysis to encode homologs of bacterial translation factors. In these cases, the originally isolated nuclear mutations caused only partial loss of mitochondrial translation: mitochondrial gene expression was diminished to the point that the cells failed to

[5] R. A. Butow, R. M. Henke, J. V. Moran, S. M. Belcher, and P. S. Perlman, this volume [24].

[6] R. E. Gray, R. H. P. Law, R. J. Devenish, and P. Nagley, this volume [33].

[7] J. C. Schneider and L. Guarente, this series, Vol. 194, p. 373.

[8] M. P. Yaffe, this series, Vol. 194, p. 627.

[9] R. Rothstein, this series, Vol. 194, p. 281.

[10] M. D. Rose and J. R. Broach, this series, Vol. 194, p. 195.

[11] C. W. Lawrence, this series, Vol. 194, p. 273.

[12] T. D. Fox, L. S. Folley, J. J. Mulero, T. W. McMullin, P. E. Thorsness, L. O. Hedin, and M. C. Costanzo, this series, Vol. 194, p. 149.

[13] A. M. Myers, L. K. Pape, and A. Tzagoloff, *EMBO J.* **4**, 2087 (1985).

[14] A. Tzagoloff and C. L. Dieckmann, *Microbiol. Rev.* **54**, 211 (1990).

respire normally, but not to the point where the stability of wild-type mtDNA was completely compromised. Once the genes identified by such "leaky" mutations were isolated, their sequences could be used to artificially create corresponding null mutations.[9] The null mutations completely blocked protein synthesis and led to loss of wild-type mtDNA (conversion to *rho⁻*). Genes falling into this category include those coding for several mitochondrial tRNA synthetases,[15] homologs of the bacterial initiation factor 2 and elongation factor G,[16] a release factor,[17] and at least one gene of as yet unknown function.[18] Curiously, in several cases the original leaky mutations, affecting general translation system components, had gene-specific effects on mitochondrial gene expression.

It seems very likely that many yeast genes required for mitochondrial translation are not represented in collections of unconditional nonrespiring mutants, as leaky mutations reducing translation to levels that allow phenotypic detection without loss of mtDNA are probably rare. One possible way around this problem would be to isolate temperature-sensitive nonrespiratory mutants that become *rho⁻* during incubation at the nonpermissive temperature. Although collections of such mutants have been obtained,[19,20] they will, unfortunately, also fail to identify many genes of interest that cannot easily mutate to produce the desired conditional phenotype.

One obvious approach to the identification of yeast mitochondrial homologs of bacterial translation factors that bypasses these genetic difficulties is the use of the bacterial genes as hybridization probes for the screening of clone banks. However, it seems that this approach has only been successfully used once, for the isolation of the nuclear gene encoding mitochondrial elongation factor Tu.[21] An extension of this approach will be to search in the sequence of yeast genomic DNA[22,23] for homologs of translation factors identified in other systems.

Although ribosomes are not translation factors per se, they are intimately involved with them. The fact that mitochondrial ribosomes can be purified has allowed the use of reverse genetics for the identification of

[15] L. K. Pape, T. J. Koerner, and A. Tzagoloff, *J. Biol. Chem.* **260**, 15362 (1985).

[16] A. Vambutas, S. J. Ackerman, and A. Tzagoloff, *Eur. J. Biochem.* **201**, 643 (1991).

[17] H. J. Pel, C. Maat, M. Rep, and L. A. Grivell, *Nucleic Acids Res.* **20**, 6339 (1992).

[18] J. J. Mulero, J. K. Rosenthal, and T. D. Fox, *Curr. Genet.* **25**, 299 (1994).

[19] A. Genga, L. Bianchi, and F. Foury, *J. Biol. Chem.* **261**, 9328 (1986).

[20] D. M. Mueller, T. K. Biswas, J. Backer, J. C. Edwards, M. Rabinowitz, and G. S. Getz, *Curr. Genet.* **11**, 359 (1987).

[21] S. Nagata, Y. Tsunetsugu-Yokota, A. Naito, and Y. Kaziro, *Proc. Natl. Acad. Sci. U.S.A.* **80**, 6192 (1983).

[22] B. Dujon, *et al., Nature* (*London*) **369**, 371 (1994).

[23] S. G. Oliver, *et al., Nature* (*London*) **357**, 38 (1992).

nuclear genes coding ribosomal proteins. One successful strategy has been to use antibodies directed against ribosomal proteins to screen expression libraries for corresponding clones. This has allowed the isolation of genes absolutely required for mitochondrial protein synthesis[24] as well as genes whose function is only marginally required for translation.[25] A second strategy depending on purified ribosomes is the use of amino acid sequence data to direct the synthesis of oligonucleotide probes for the corresponding genes.[26] If *in vitro* assays for yeast translation factors can be developed and used to guide purification, as in the case of bovine factors,[4] then these reverse genetic strategies could be applied more generally.

Once a yeast gene coding for a protein of interest is in hand, there are several genetic strategies for the identification of genes encoding proteins that functionally (possibly physically) interact. Thus, starting with a known translation factor gene it should be possible to identify other important components that may have escaped detection in standard mutant hunts. In addition, the pattern of *in vivo* functional interactions can often provide insights into mechanism. One purely genetic strategy relies on the possibility that a mutation disrupting a protein–protein interaction can sometimes be suppressed by a second mutation in an interacting partner.[27] Similarly, a mutation that disrupts a functional or physical interaction can sometimes be suppressed by increasing the intracellular concentration of an interacting partner.[28]

Another strategy is based on the observation that a protein complex or a functional pathway may work with either of two mutant components but fail to function when both components are mutated at the same time: in this case one observes a strongly enhanced phenotype (synthetic defect) in the double mutant strain compared to either single mutant.[27] Finally, a genetic test for protein–protein interactions termed the "two-hybrid system"[29] takes advantage of the fact that transcriptional activators such as the yeast GAL4 protein can activate transcription even when their DNA binding domains and transcriptional activation domains are held together by noncovalent interactions. Nuclearly targeted hybrid proteins, consisting of each domain fused to other proteins of interest, can be coexpressed in yeast cells. If the proteins of interest interact physically, then the transcriptional activator is reconstituted *in vivo* and a reporter gene or selectable marker is expressed, revealing the interaction. Of course the two-hybrid

[24] K. Fearon and T. L. Mason, *Mol. Cell. Biol.* **8,** 3636 (1988).
[25] J. A. Partaledis and T. L. Mason, *Mol. Cell. Biol.* **8,** 3647 (1988).
[26] L. Grohmann, H.-R. Graack, and M. Kitakawa, *Eur. J. Biochem.* **183,** 155 (1989).
[27] T. C. Huffaker, M. A. Hoyt, and D. Botstein, *Annu. Rev. Genet.* **21,** 259 (1987).
[28] J. Rine, this series, Vol. 194, p. 239.
[29] S. Fields and R. Sternglanz, *Trends Genet.* **10,** 286 (1994).

system can be used to identify interacting proteins coded by genes of any species. All of these strategies have complications and pitfalls, a critical discussion of which is beyond the scope of this chapter. However, they have all proved to be useful for the analysis of yeast mitochondrial translation.[30,31]

mRNA-Specific Translational Activators

Among collections of nonrespiring mutants that maintain wild-type mtDNA, one finds strains that have defects in the ability to express single mitochondrial genes.[32] Several of these mutations have been shown to lie in genes required to activate translation of specific mitochondrial mRNAs. The precise mechanism by which these translational activators work remains unclear in the absence of *in vitro* studies. In the case of three nuclear genes required to activate *COX3* translation, genetic analysis, combined with submitochondrial localization of the gene products, indicates that the three activator proteins are in a complex bound to the inner membrane that mediates an interaction between the *COX3* 5′-untranslated leader and the small subunit of mitochondrial ribosomes.[31,33] Existing information on the translation of four other mitochondrial mRNAs is consistent with this picture.[34] It seems clear that many translational activator genes remain to be identified and that further study of their interactions with other components is necessary to understand their mechanisms of action.

To date there is no published biochemical assay for these mRNA-specific translational activators. The genetic evidence strongly suggests that at least some of them are RNA-binding proteins.[31,35,36] However, their genes are expressed at very low levels,[37,38] and their products have not been detected by analysis of mitochondrial RNA-binding proteins.[39,40] All of the activator genes identified so far have been detected in screens of nuclear nonrespiratory mutants (*Pet*⁻) for strains lacking individual mitochondrial translation products. The most general current procedure for

---

[30] P. Haffter, T. W. McMullin, and T. D. Fox, *Genetics* **127**, 319 (1991).
[31] N. G. Brown, M. C. Costanzo, and T. D. Fox, *Mol. Cell. Biol.* **14**, 1045 (1994).
[32] M. C. Costanzo and T. D. Fox, *Annu. Rev. Genet.* **24**, 91 (1990).
[33] T. W. McMullin and T. D. Fox, *J. Biol. Chem.* **268**, 11737 (1993).
[34] H. J. Pel and L. A. Grivell, *Mol. Biol. Rep.* **19**, 183 (1994).
[35] M. C. Costanzo and T. D. Fox, *Mol. Cell. Biol.* **13**, 4806 (1993).
[36] J. J. Mulero and T. D. Fox, *Mol. Biol. Cell* **4**, 1327 (1993).
[37] D. L. Marykwas and T. D. Fox, *Mol. Cell. Biol.* **9**, 484 (1989).
[38] T. D. Fox and Z. Shen, *in* "Protein Synthesis and Targeting in Yeast" (J. E. G. McCarthy and M. F. Tuite, eds.), p. 157. Springer-Verlag, Berlin, 1992.
[39] S. D. J. Elzinga, A. L. Bednarz, K. van Oosterum, P. J. T. Dekker, and L. A. Grivell, *Nucleic Acids Res.* **21**, 5328 (1993).
[40] P. J. T. Dekker, B. Papadopoulou, and L. A. Grivell, *Biochimie* **73**, 1487 (1991).

such screens is to examine each mutant for the pattern of mitochondrial translation products labeled *in vivo* in the presence of the cytoplasmic protein synthesis inhibitor cycloheximide[12,41] (see also Ref. 42). However, in the future, reporter genes placed into the mitochondrial genome should greatly simplify the task of identifying nuclear mutations with specific effects on mitochondrial gene expression.

The absence of a single mitochondrial translation product (or a subset thereof) could result from defects in transcription, mRNA processing, mRNA stability, translation, posttranslational modification, or protein stability. Indeed, mutants defective at several of these levels have been found.[32] The next step in identification of translational activator mutants is to examine the levels of mRNA coding for the missing translation product. Typically, this mRNA will be present in easily detectable amounts, although the levels may be lower than in the wild type.[43] A decreased steady-state level of mRNA may indicate that RNA degradation is more rapid in the absence of translation. Indeed, the observation of reduced, but detectable, mRNA levels can be taken as evidence that the block in gene expression is not occurring at a posttranslational step. In the case of a mutation in a nuclear gene required for mRNA stability per se, the steady-state mRNA level would be reduced as dramatically as that of the translation product itself.[44]

A complication arises in the phenotype of mutants defective for translational activators of the mitochondrial *COB* and *COX1* mRNAs, whose primary transcripts contain introns. The mature mRNAs for these genes are produced by a series of splicing steps, several of which require the activity of intron-encoded proteins termed maturases.[45] Because the translation of at least some of the maturases depends on translation of the first exon, a nuclear mutation blocking translation of these mRNAs will also block the production of mature mRNA and lead to the accumulation of unspliced precursors.[46] Thus, such a mutation can be easily confused with a mutation that directly affects mRNA processing. A relatively simple tool for distinguishing whether a nuclear mutation that causes accumulation of *COB* and/or *COX1* precursor transcripts is primarily defective in translation or mRNA splicing has been developed by constructing a yeast mitochondrial genome that lacks all known introns.[47] This mitochondrial ge-

[41] M. Douglas and R. A. Butow, *Proc. Natl. Acad. Sci. U.S.A.* **73**, 1083 (1976).
[42] A. Chomyn, this volume [18].
[43] C. G. Poutre and T. D. Fox, *Genetics* **115**, 637 (1987).
[44] C. L. Dieckmann, T. J. Koerner, and A. Tzagoloff, *J. Biol. Chem.* **259**, 4722 (1984).
[45] L. A. Grivell, *Eur. J. Biochem.* **182**, 477 (1989).
[46] G. Rödel, A. Körte, and F. Kaudewitz, *Curr. Genet.* **9**, 641 (1985).
[47] G. Séraphin, A. Boulet, M. Simon, and G. Faye, *Proc. Natl. Acad. Sci. U.S.A.* **84**, 6810 (1987).

nome can be introduced into the nuclear mutant background and the resulting phenotype determined. If the nuclear mutation directly causes a defect in splicing, it should be suppressed by the presence of this mitochondrial genome. However, if the nuclear mutation causes a defect in translation initiation (or any other step in functional gene expression), then the mutation will not be suppressed and resulting strain will fail to respire.[48]

A nuclear mutation that causes a posttranscriptional block in expression of a specific mitochondrial gene could define a function required either for translation or for some posttranslational step. In the former case, the target of that function would most likely be the mRNA, whereas in the latter case it would be the mitochondrially encoded protein itself. One strategy to distinguish these possibilities is to select mitochondrially inherited mutations that suppress the original nuclear mutation.[49] For example, a nuclear mutation *(pet494)* blocking the translation of the *COX3* mRNA was suppressed (bypassed) by mitochondrial DNA rearrangements that generated chimeric genes carried on *rho⁻* mtDNAs. The chimeric genes encoded *COX3* mRNAs with the 5' leaders of other mitochondrial mRNAs.[50] Thus, substitution of a new 5' leader on the *COX3* mRNA allowed translation in the absence of *PET494* function. As these chimeric mRNAs encoded the same polypeptide chain as the wild-type *COX3* mRNA, these suppressors virtually ruled out posttranslational functions for the *PET494* nuclear gene. Taken together with the fact that the nuclear mutation had no effect on *COX3* mRNA steady-state levels,[49] they provided very strong evidence for the role of *PET494* in activating translation.

The selection of such "mitochondrial revertants" of nuclear mutations begins with the plating of large numbers of mutant cells on nonfermentable medium. Generally it is necessary to determine whether respiring revertant clones carry suppressor mutations in the nuclear or mitochondrial genomes. These possibilities can be distinguished by growing each revertant in medium containing ethidium bromide to remove mtDNA.[12] The resulting *rho⁰* (ρ^0) derivatives can then be crossed back to a tester strain carrying the original nuclear mutation, examining the respiratory phenotype of the resulting diploids to see whether the suppressor mutation has also been eliminated: failure of a diploid to respire suggests a mitochondrial suppressor. [A recessive nuclear suppressor would produce the same result: these two possibilities can be distinguished by also crossing the original *rho⁺* (ρ^+) revertant to the tester strain.]

---

[48] E. Decoster, M. Simon, D. Hatat, and G. Faye, *Mol. Gen. Genet.* **224,** 111 (1990).
[49] P. P. Müller, M. K. Reif, S. Zonghou, C. Sengstag, T. L. Mason, and T. D. Fox, *J. Mol. Biol.* **175,** 431 (1984).
[50] M. C. Costanzo and T. D. Fox, *Mol. Cell. Biol.* **6,** 3694 (1986).

A characteristic property of strains carrying chimeric mitochondrial genes that suppress nuclear mutations is that their ability to respire is highly unstable during mitotic growth (and during meiosis).[49,50] The respiring cells contain both wild-type mtDNA and the *rho*⁻ mtDNA encoding the chimeric mRNA in a heteroplasmic state. Because different mtDNAs present in the same cell segregate rapidly from one another during mitosis in *S. cerevisiae*,[51] most of the cells contain either the wild-type mtDNA alone or the *rho*⁻ mtDNA alone, and therefore cannot give rise to respiring colonies. This is true even for cultures grown under selective conditions on nonfermentable carbon sources. This mitotic instability can be easily revealed by streaking (and restreaking) revertant clones to plates containing nonfermentable carbon sources: whereas a stable revertant produces uniformly growing subclones, an unstable revertant will produce a few growing subclones surrounded by a large number of nonrespiring microcolonies. (In one reported case, a chimeric gene, that suppressed the instability of the *COB* mRNA in a *cpb1* mutant, was integrated into *rho*⁺ mtDNA by an illegitimate recombination process producing a more stable strain.[52])

Demonstrating that a candidate mutation affects an mRNA-specific translational activator by selection of mitochondrial bypass suppressors is effective, but it suffers from several technical drawbacks. First, the desired mitochondrial gene rearrangements are typically rare, so that, if the nuclear mutation itself is capable of reverting, considerable effort must be expended in screening for revertants carrying mitochondrial suppressors. In the case of the *COX1* mRNA-specific activator *MSS51,* no such chimeric mitochondrial genes could be obtained, probably because of the rarity of crossovers that would produce them.[48] The second drawback is that the selected mitochondrial gene rearrangements frequently change not only the 5′ leader of the mRNA but also the amino terminus of the encoded protein,[43,49,50] leading to ambiguity in the identification of the target of the nuclear gene function (see above). These problems can be circumvented by constructing chimeric genes *in vitro* that will encode appropriate chimeric mRNAs. Plasmids carrying the chimeric genes are transformed[5] into mitochondria of a *rho*⁰ host[53] bearing the nuclear mutation in question, generating synthetic *rho*⁻ strains. To test whether the constructed chimeric genes suppress the nuclear mutation, the synthetic *rho*⁻ strains are mated to a *rho*⁺ strain that also carries the nuclear mutation.[54] This generates diploid zygotes, homozygous

[51] B. Dujon, *in* "The Molecular Biology of the Yeast *Saccharomyces:* Life Cycle and Inheritance" (J. N. Strathern, E. W. Jones, and J. R. Broach, eds.), p. 505. Cold Spring Harbor Laboratory, Cold Spring Harbor, New York, 1981.

[52] C. L. Dieckmann and T. M. Mittelmeier, *Curr. Genet.* **12,** 391 (1987).

[53] T. D. Fox, J. C. Sanford, and T. W. McMullin, *Proc. Natl. Acad. Sci. U.S.A.* **85,** 7288 (1988).

[54] J. J. Mulero and T. D. Fox, *Genetics* **133,** 509 (1993).

for the nuclear mutation, that contain both rho^+ mtDNA and the plasmid bearing the chimeric gene. If the nuclear mutation inactivates an mRNA-specific translational activator that functions through the 5' leader, then the mRNA coded by the chimeric gene will be translated, leading to mitotically unstable respiratory growth. Such a test could, for example, be applied in the case of the *MSS51* gene: would the precise fusion of the 5' leaders of the *COX2* or *COX3* mRNAs to the *COX1* coding sequence suppress an *mss51* mutation?

To demonstrate that the target of a particular translational activator resides solely within a given 5' leader, it is necessary to place that 5' leader on a different mRNA and show that the activator now works on the new mRNA. This experiment has been carried out using chimeric genes selected genetically as bypass suppressors of translational activator mutations,[55,56] and using chimeric genes constructed *in vitro*.[54] In the absence of reporter genes, this strategy requires a nuclear mutation in a known translational activator (A) that normally works on the 5' leader of a mitochondrial mRNA (A'). One then asks whether the function of the second putative translational activator (B) is necessary to translate a chimeric mRNA bearing the B' leader and the A' coding sequence when the translational activator A is defective. The plasmid bearing the B' leader–A' coding sequence chimeric gene is transformed into a rho^0 strain with nuclear mutations in both the A and B genes. The resulting synthetic rho^- is first mated to a rho^+ strain with a single nuclear mutation in the A gene, to determine whether the chimeric mRNA bypasses the need for A function in expression of the A' mitochondrial gene.

If the chimeric mRNA bypasses the need for A function, then the synthetic rho^- is next mated to a rho^+ strain with mutations in both the A and B genes: if B acts in the B' leader then there should be no expression of the A' coding sequences in this situation. Unfortunately this cannot be tested at the level of respiratory growth because the mutation in B prevents respiration in any case by preventing expression of B'. Furthermore, the heteroplasmic diploids generated by this mating cannot be propagated in the absence of selection for respiratory growth. Therefore, the requirement for B function in translation of the B' leader–A' coding sequence chimeric mRNA must be monitored by transiently labeling mitochondrial translation products in the freshly mated zygotes.[57] If the mutation in B blocks translation of the B' leader–A' coding sequence chimeric mRNA, then the site of action of B has been mapped to the B' leader.

[55] G. Rödel and T. D. Fox, *Mol. Gen. Genet.* **206,** 45 (1987).
[56] M. C. Costanzo and T. D. Fox, *Proc. Natl. Acad. Sci. U.S.A.* **85,** 2677 (1988).
[57] R. L. Strausberg and R. A. Butow, *Proc. Natl. Acad. Sci. U.S.A.* **74,** 2715 (1977).

As noted above for general translation factors, once one has a particular gene in hand one can search for other genes encoding interacting products by a variety of genetic strategies. Such studies on mRNA-specific activators can demonstrate functional interactions between specific activator proteins[31] as well as interactions with general components of the translation system.[30] Furthermore, once the mRNA target of a translational activator has been located, that target can be subjected to mutagenesis *in vitro* and the mutated forms returned to mitochondria. Selection of respiring revertants from the resulting mitochondrial mutants can demonstrate *in vivo* functional interactions between the mRNA and the translational activators, as well as lead to the identification of new genes that play a role in the translational control of mitochondrial gene expression.[35,36]

Acknowledgment

Research in the author's laboratory was supported by a grant (GM29362) from the National Institutes of Health.

[22] Bovine Mitochondrial Ribosomes

By Thomas W. O'Brien and Nancy D. Denslow

Introduction

The bovine mitochondrial ribosome is being developed as a model system for studying the structure and function of mammalian mitochondrial ribosomes. These 55 S ribosomes from mammalian mitochondria resemble bacterial and eukaryotic cytoplasmic ribosomes in general functional properties, but in terms of composition, fine structure, and physicochemical properties they differ unexpectedly from both of those kinds of ribosomes, as well as from other kinds of mitochondrial ribosomes. Products of two genomes, mammalian mitochondrial ribosomes are considered to be members of the prokaryotic class, because they have more homologies with bacterial ribosomes than with eukaryotic cytoplasmic ribosomes. Nevertheless, they contain scarcely more than half the RNA of bacterial ribosomes, the bulk of their mass being contributed by a large number of proteins. The unusual properties of these ribosomes raise questions about their relation to other kinds of ribosomes, and their large number of proteins raises questions about their functional and structural organization, and also about the identity of individual mitoribosomal proteins that are homologous to proteins in other ribosomes.

Preparation of Mitochondrial Ribosomes from Bovine Liver

Mammalian mitochondria contain relatively few ribosomes,[1] necessitating the development of procedures for the large-scale preparation of ribosomes in order to obtain the amounts required for structural and functional studies. Mitochondria are isolated from 2- or 4-kg batches of bovine liver by differential centrifugation, using digitonin treatment to reduce the amount of contaminating cytoplasmic ribosomes. Ribosomes released by detergent treatment of mitochondria are recovered by differential centrifugation over sucrose cushions. Ribosomes in large volumes may be concentrated by batch adsorption/elution using DEAE-cellulose, in combination with ultrafiltration, before recovery by high speed centrifugation. Highest yields of ribosomal subunits are obtained by lysing mitochondria under ribosome dissociating conditions, in solutions containing higher salt concentrations (300 mM KCl). However, such ribosomal subunits will be contaminated with variable amounts of cosedimenting particles. Ribosomal subunits of higher purity are routinely obtained by first isolating 55 S monoribosomes, using lysis and sucrose gradient conditions that stabilize monoribosomes, followed by separation of the 28 S- and 39 S-derived subunits in a second sucrose gradient under ribosome dissociating conditions. Derived subunits are recovered from regions of the second gradient essentially free of materials originally cosedimenting with the 55 S monoribosomes.[1,2]

Isolation of Mitochondria

Solutions

Mitochondrial isolation medium: 0.34 M sucrose (commercial grade), 1 mM EDTA, 5 mM Tris-HCl, pH 7.5

4× Mitochondrial freezing buffer: 160 mM KCl, 60 mM $MgCl_2$, 20 mM 2-mercaptoethanol, 0.2 mM EDTA, 0.2 mM spermine, 0.2 mM spermidine, 40 mM Tris, pH 7.5

Digitonin solution: Mitochondrial isolation medium containing digitonin, 500 μg/ml

Procedure. Fresh bovine liver is obtained and chilled as quickly as possible, by immersion of liver slices (1–2 cm thick) in a slurry of crushed ice in mitochondrial isolation medium. Care should be taken to avoid larger vessels and connective tissue that will interfere with homogenization of the

---

[1] T. W. O'Brien, *J. Biol. Chem.* **245,** 3409 (1971).

[2] D. E. Matthews, R. A. Hessler, N. D. Denslow, J. S. Edwards, and T. W. O'Brien, *J. Biol. Chem.* **257,** 8788 (1982).

liver. All subsequent operations are at $0°-4°$, maintaining containers holding the ground liver, mitochondrial suspensions, and centrifuge tubes in crushed ice. Centrifuge tubes are transported to and from the centrifuge immersed in crushed ice. Grind the liver in 2-kg batches, using a meat grinder with a fine sieving screen. Strain the ground liver through fiberglass screening (1–2 mm grid) while stirring with 1 liter ice-cold isolation medium. The retentate is transferred in batches to a Waring blender, with equal volumes of isolation medium, where it is further reduced by running the blender for a few brief pulses at moderate speed. After each pulse, allow several seconds for the larger liver particles to settle before the next pulse. Connective tissue is removed by filtration of the small liver particles through the screen. For each 2-kg batch of ground liver, add isolation medium to bring the volume to 6 liters, in preparation for homogenization using a tissue disrupter such as the Tekmar SD-45K unit with a G-454 generator (Tekmar Co., Cincinnati, OH). Operation of the Tekmar SD-45K unit for 1 min at maximum speed (10,000 rpm) is sufficient to disrupt more than 90% of the liver cells. Cell disruption should be monitored by microscopy and terminated when approximately 90% complete to avoid damage to the mitochondria. (Excessive homogenization, especially using devices of the Waring blender variety, yields ribosomes with reduced RNA content.)

Cells, cell debris, and nuclei are removed by centrifugation of the homogenate at 1000 g_{av} for 10 min [3000 rpm in Beckman (Beckman Instr., Fullerton, CA) JA-10 or Sorvall (Newtown, CT) GS-3 rotors (each rotor has a 3-liter capacity)]. Average centrifugal force (g_{av}) is used throughout to indicate the force acting on particles in the center of the fluid mass. Maximum g values, often specified for rotors, are less meaningful in differential centrifugation, as most suspended particles never experience the force at maximum radius (the bottom of the tube). The yield of mitochondria is increased about 30% by rehomogenizing the 1000 g pellet (3 liters final volume, for each 2-kg batch of liver), and combining the supernatant fluids from the 1000 g centrifugation steps. Mitochondria are then recovered from the supernatant fluid by centrifugation at 11,000 g_{av} for 10 min (10,000 rpm in the JA-10 or GS-3 rotors). Mitochondrial pellets are resuspended using glass stirring rods and rubber spatulas. Traces of red blood cells evident at the bottom of the mitochondrial pellet are left behind when suspending the mitochondria. Alternatively, this debris may be removed by repeating the 1000 g centrifugation step after suspending the entire mitochondrial pellet. The mitochondrial pellets from both 2-kg batches of liver are combined and suspended in 2700 ml isolation medium, using the Tekmar unit operating at half-speed for 30 sec, to ensure dispersal of clumped mitochondria. Isolation medium is added to bring the volume of the mitochondrial suspension (from 4 kg of liver) to 9 liters, prior to pelleting of the mitochondria by centrifugation again at 11,000 g for 10 min.

For digitonin treatment, the mitochondrial pellets are suspended, as above, to a concentration of 25 g (protein)/ml. The mitochondria from 4 kg liver may be resuspended to approximately 2 liters, before sampling to determine their concentration. For this purpose, an aliquot of the mitochondrial suspension is mixed with isolation medium at room temperature and the absorbance is measured within 1 min. The concentration of bovine mitochondria can be estimated by absorption measurement (light scattering) at 550 nm, using the relationship 1 A_{550} unit corresponds to a mitochondrial concentration (in isolation medium) of 250 μg/ml. After adjusting the mitochondrial suspension to a concentration of 25 mg/ml, 0.25 volume digitonin solution is added (final digitonin concentration is 100 μg/ml), and the mixture is stirred for 15 min. The microsomal membrane (rough endoplasmic reticulum) vesicles that cosediment with mitochondria are a major source of contaminating cytoplasmic ribosomes. Digitonin treatment disrupts microsomal membranes, converting them to more slowly sedimenting forms that are easily removed by differential centrifugation. This treatment also ruptures lysosomes (releasing nucleases which must be removed by washing the mitochondria) and tends to strip the mitochondrial outer membrane, as well. The suspension is then brought to 6 liters with isolation medium, and mitochondria are washed by centrifugation again at 11,000 g for 10 min.

The mitochondrial pellets are suspended to a concentration of approximately 20 mg/ml, using the Tekmar unit as above, and sampled to determine the approximate yield at this point, by measuring the absorbance at 550 nm. The volume of the suspension is then brought to 6 liters, for a final wash by centrifugation at 11,000 g for 10 min. The average yield of mitochondria from 4 kg liver ranges around 50 g (protein). If ribosomes are to be isolated directly, the mitochondria are suspended in monosome buffer (below) to a concentration of 20 mg(protein)/ml. Alternatively, for freezing and storage of the mitochondria, they are suspended in a minimal volume (recorded) of 4× mitochondrial freezing buffer, and aliquots (record volume and amount of mitochondria) are frozen by immersion of the sample storage containers in dry ice/2-propanol. Frozen mitochondria are stored at −70°.

Isolation of Mitochondrial Ribosomes

Solutions

Monosome buffer: 100 mM KCl, 20 mM MgCl$_2$, 5 mM 2-mercaptoethanol, 20 mM Tris, pH 7.5

Dissociation buffer: 300 mM KCl, 5 mM MgCl$_2$, 5 mM 2-mercaptoetha-
nol, 10 mM Tris, pH 7.5

Monosome freeze buffer: 25 mM KCl, 5 mM MgCl$_2$, 5 mM 2-mercapto-
ethanol, 10 mM Tris, pH 7.5

Subunit freeze buffer: 25 mM KCl, 2.5 mM MgCl$_2$, 5 mM 2-mercapto-
ethanol, 10 mM Tris, pH 7.5

Load buffer: 40 mM KCl, 15 mM MgCl$_2$, 5 mM 2-mercaptoethanol,
1.6% v/v Triton X-100, 10 mM Tris, pH 7.5

Elute buffer: 300 mM KCl, 40 mM MgCl$_2$, 5 mM 2-mercaptoethanol,
10 mM Tris, pH 7.5

Triton X-100, 16% (v/v)

Puromycin solution: monosome buffer or load buffer, containing 1 mM
puromycin, 5 μg/ml heparin, pH 7.5

For sucrose density gradients and solutions containing sucrose, use
RNase-free sucrose (Sigma, St. Louis, MO); if ribosomes are to be fixed with
glutaraldehyde or formaldehyde, triethanolamine should be substituted for
Tris as the buffering agent.

Procedure. If ribosomes are to be concentrated and recovered by high-
speed centrifugation alone, the mitochondrial samples are adjusted to a
concentration of 22 mg (protein)/ml with monosome buffer. Detergent
treatment of mitochondria at higher concentrations leads to reduced recoveries of ribosomes. If the mitochondria are in 4× mitochondrial freezing
buffer, dilute to 1× before adjusting the protein concentration with monosome buffer. The scale of the operation is limited by the available rotor
volume; one Beckman 45Ti rotor can accommodate the lysate from 6 g
mitochondria (300 ml) over 20 ml cushions of 34% sucrose in monosome
buffer. If, for large-scale preparations (below), ribosomes are to be adsorbed
onto DEAE-cellulose, the mitochondrial samples are diluted to 1× mitochondrial freeze buffer (MFB) and then adjusted to a concentration of 20
mg/ml by the addition of 1× MFB.

The mitochondria (final concentration 20 mg/ml) are lysed by addition
of one-ninth volume of 16% Triton X-100 while stirring for 2 min, and the
lysate is clarified by centrifugation for 10 min at 60,000 g_{av}. Ribosomes are
recovered from the clarified lysate by centrifugation over 20 ml cushions
of 34% sucrose in monosome buffer, at 150,000 g_{av} (45,000 rpm in the 45Ti
rotor) for 15 hr.

For puromycin treatment (optional) the ribosome pellets are resuspended in 10 ml puromycin solution and incubated at 25° for 5 min to
discharge nascent peptides. This treatment usually results in a 5 to 10%
increased yield of ribosomes, presumably by removal of hydrophobic nascent peptides from active ribosomes that may otherwise cause them to

partition with membrane proteins in detergent micelles. Incubation of ribosomes with [^{3}H]puromycin under these conditions reveals that up to 10% of ribosomes carry nascent peptides that are detected as [^{3}H]peptidylpuromycin. If the puromycin treatment is omitted, ribosome pellets are suspended in 10 ml monosome buffer.

After clarification of the sample by centrifugation at 26,000 g_{av} for 10 min (20,000 rpm in the Beckman Ty65 rotor), the ribosomes (1–2 ml per tube) are layered onto 10–30% sucrose gradients in monosome buffer and centrifuged 5 hr at 96,500 g_{av} (27,000 rpm in the Beckman SW28 rotor). The sucrose gradients are monitored at 254 or 260 nm during fractionation using the ISCO (Lincoln, NE) Model UA-5 sucrose gradient monitoring system. Fractions containing native 28 S and 39 S subunits, as well as the 55 S monoribosomes, are pooled separately and diluted by the addition of an equal volume of monosome buffer, prior to recovery of the ribosomal particles by centrifugation at 150,000 g for 8–15 hr. Ribosomal subunits and monoribosomes are suspended in minimal volumes of subunit or monoribosome freeze buffer, respectively, and aliquots are frozen in dry ice/2-propanol for storage indefinitely at −70°. Typical yields (per gram mitochondrial protein) are about 2 A_{260} units of 55 S ribosomes and about 1.5 A_{260} units of 28 S and 39 S subunits. One A_{260} unit corresponds to 32 pmol of 55 S ribosomes and 77 or 55 pmol of 28 S or 39 S subribosomal particles, respectively.

For the preparation of derived 28 S and 39 S subribosomal particles, 55 S monoribosomes are suspended in dissociation buffer, and the subunits are separated by centrifugation for 15 hr at 55,000 g_{av} (20,000 rpm in the SW28 rotor) over 10–30% sucrose gradients containing this buffer. Fractions containing 28 S or 39 S subunits are pooled separately and diluted with an equal volume of monoribosome buffer prior to recovery by centrifugation, as above.

Large-Scale Preparation of Ribosomes

Equipment and Materials

Fibrous DEAE-cellulose (Whatman, Clifton, NJ, DE32)
Amicon (Danvers, MA) DC-10 hollow fiber concentration system
Amicon H1P100 hollow fiber membrane cartridge
Beckman Ti14 zonal rotor
Double-chamber sucrose gradient mixing device, 1 liter capacity

Treatment with DEAE-Cellulose. Mitochondrial samples totaling 120 g (protein) are thawed quickly in a cool water bath, with constant, gentle agitation to hasten thawing and ensure that the mitochondrial suspension remains at 0–2°. The mitochondrial suspension is diluted to 1× mitochondrial freeze buffer (MFB) and then adjusted to a concentration of 22.2 mg/

ml by the addition of 1× MFB, to a volume of 5.4 liters. Mitochondria are lysed by the addition of 600 ml of 16% Triton X-100 (final concentration 1.6%), with stirring for 2 min. After centrifugation of the lysate for 30 min at 11,000 g_{av} (10,000 rpm in two Beckman JA-10 rotors), the supernatant fluid is stirred (gently) with 120 g fibrous DEAE-cellulose (Whatman DE32, equilibrated in load buffer) for 15 min. Mitochondrial proteins that do not bind to DEAE under these conditions are removed by filtration (Whatman No. 541 or 1 filter paper), followed by washing of the DEAE with 1.2 liters load buffer containing 1.6% Triton X-100 and filtration to incipient dryness. Do not allow the DEAE to dry. Immediately transfer the DEAE into a beaker containing 1.2 liters of elute buffer and stir gently for 30 min, before filtering the desorbed ribosomes through the Whatman filter paper. Rinse the DEAE by filtration of an additional 600 ml elute buffer, combining the filtrates (1.8 liters final volume) for concentration by ultrafiltration (below). Besides the intended concentration of the ribosome sample and elimination of proteins not binding to DEAE under these conditions, preparation of mitochondrial ribosomes using DEAE provides an additional benefit. Any residual 80 S cytoplasmic ribosomes or subribosomal particles are efficiently removed by this procedure, because they remain bound to the DEAE under these elution conditions, by virtue of their higher RNA/protein ratio.

Concentration of Ribosomes by Ultrafiltration. The H1P100 hollow fiber membrane cartridge (Amicon) is used with the Model DC-10 hollow fiber concentration system (Amicon) to concentrate solutions of mitochondrial ribosomes, according to the manufacturer's instructions. This cartridge has a nominal molecular weight cutoff of 100,000, so ribosomes and subunits are retained while smaller particles and proteins in the filtrate are voided during concentration. During the operation of this system the DEAE eluate is recirculated through the cartridge, maintaining a pressure gradient (10–15 psi) across the cartridge and sufficient flow rate to prevent clogging of the hollow fibers by buildup of deposited ribosomes. After concentration to a final volume of 200–300 ml, ribosomes are recovered by centrifugation over 20 ml cushions of 34% (w/v) sucrose in monosome buffer, at 150,000 g_{av} (45,000 rpm in the Beckman 45Ti rotor) for 15 hr.

Because a hollow fiber cartridge system is used to concentrate the ribosome eluate, the preparation of ribosomes can be scaled up readily, within the limits of centrifuge and rotor capacities. Additional 120-g batches of mitochondria can be processed, each resulting in 6 liters of lysate requiring centrifugation in two JA-10 or GS-3 rotors. The ribosome concentrate from each 120 g batch of mitochondria can be accommodated in a single 45Ti rotor.

Isolation of Ribosomes and Subribosomal Particles by Zonal Centrifugation. The zonal rotor is loaded with the sucrose gradient (below) before the ribosomes are resuspended. Ribosome pellets are suspended in 10 ml puromycin solution, for puromycin treatment (optional; see above procedure), or in 10–20 ml monosome buffer by homogenization with Teflon pestles in glass homogenizers. After puromycin treatment, the sample is diluted to 20 ml with monosome buffer, and the suspension is clarified by centrifugation at 26,000 g_{av} for 10 min (20,000 rpm in the Beckman Ty65 rotor). The clarified ribosome sample is held on ice in a conical tube until ready for loading into the zonal rotor.

The Ti14 rotor, buffers, and sucrose solutions are cooled to 2°–4° before starting. The Ti14 rotor contains a total volume of 650 ml, of which 540 ml will consist of a hyperbolic 10–30% (w/v) sucrose gradient over a 30–20 ml cushion of 41% (w/v) sucrose, and a ribosome sample of 20–30 ml, overlaid with monosome buffer. The rotor is filled with 60 ml monosome buffer before placement in the centrifuge and operated according to the instructions in the rotor manual. The rotating seal assembly is installed after bringing the rotor to a speed of 3000–4000 rpm.

The sucrose gradient can be formed using two Erlenmeyer flasks in tandem. One of these, containing 500 ml of 10% (w/v) sucrose in monosome buffer and a stirring bar, rests on a stirring motor in a cold box. This flask is sealed with a rubber stopper that has an outlet line originating just above the stir bar and leading to a Masterflex pump (Cole Parmer Instr. Co., Miles, IL), by which the sucrose gradient will be pumped into the spinning rotor. The inlet line drains the second flask, containing 500 ml of 41% sucrose in monosome buffer, as the gradient is pumped into the edge line of the rotor at a rate of 20 ml/min. As the gradient is introduced at the outer edge of the rotor, it displaces monosome buffer in the rotor through the center line, into a 1-liter graduated cylinder, to allow monitoring of the volume of sucrose gradient being pumped into the rotor. The gradient is pumped into the rotor until 540 ml buffer has been displaced through the center line. The 41% sucrose cushion is then introduced by pumping 120 ml (or 110 ml, if the volume of the ribosome sample is 30 ml) into the edge line at a rate of 20 ml/min.

At this point, all of the buffer and 10 ml of the sucrose gradient will have been displaced through the center line, and the rotor is ready for loading the sample, through the center line, at a rate of 10 ml/min. Serious losses will occur if the sample is introduced too rapidly, by leakage across the rotating rulon seal. Such losses can be avoided by drawing the sample into the center line, by running the pump (still connected to the 41% sucrose reservoir) in reverse, at 10 ml/min. After all of the ribosome sample has entered the center inlet tube, the tube is placed into a beaker of monosome buffer, and pumping (still in reverse, out of the edge line) is continued until another 60 ml of the 41% sucrose cushion is voided (monitored by collection into a 100-ml graduated cylinder). By introducing this overlay of 60 ml monosome Buffer, the ribosome sample is brought farther from the center of the rotor, where it occupies less radial distance, and lies in a region of higher centrifugal force. The rotating seal assembly is now removed, and the rotor speed is increased to 5000 rpm. The capping mechanism is installed, the centrifuge lid is closed, and the chamber is evacuated. The rotor is left at this speed until the chamber pressure drops to 500 μm Hg, at which time the diffusion pump is turned on, and the rotor is brought to the desired speed. For the isolation of monoribosomes and native subunits, the rotor is run at 25,000 rpm for 16.5 hr.

At the end of the run, the rotor speed is reduced to 2000 rpm, the diffusion and vacuum pumps are turned off, and the capping mechanism is removed. The rotor is maintained at this speed while the seal assembly is replaced. The gradient is pumped out through the center line, by pumping 45% sucrose into the edge line at a rate of 20 ml/min. The absorbance of the gradient effluent is monitored at 254 nm while pumping through an ISCO flow cell in the ISCO UA-5 detector, and fractions are collected using a suitable fraction collector.

Fractions containing native 28 S subunits, 39 S subunits, or 55 S monoribosomes are pooled separately and recovered by centrifugation as described above. For the preparation of derived 28 S and 39 S subribosomal particles, the 55 S monoribosomes are suspended in 20 ml dissociation buffer, and the subunits are separated by zonal sucrose gradient centrifugation (as above), using a 10–30% sucrose gradient in dissociation buffer. Centrifugation in the Ti14 rotor is at 27,000 rpm for 19 hr.

Characterization of Mitochondrial Ribosomes

Buoyant Density in Cesium Chloride

The purity of mitochondrial ribosomes and subribosomal particles can be assessed by determining the buoyant density in CsCl gradients. The most common contaminants in mitochondrial ribosome fractions are likely to be cytoplasmic ribosomes and subribosomal particles, as well as multimeric enzyme particles and aggregates of membrane protein. The protein-rich mitochondrial ribosomes have buoyant densities intermediate between those of cytoplasmic ribosomes and membrane proteins, so they are easily discriminated by CsCl density gradient centrifugation. Ribosomes must be fixed with glutaraldehyde or formaldehyde to prevent their disassembly during exposure to the high concentrations of CsCl. For this reason, the buffers used should contain triethanolamine, instead of Tris, which reacts with the fixative.[3]

Procedure. All operations are carried out at 0°–5°. Ribosomes in monosome buffer (containing triethanolamine instead of Tris) and subribosomal particles in subunit buffer are fixed by addition of formaldehyde to 0.3% and dialysis for 1 to 2 hr against 100 to 200 ml monosome buffer (or subunit buffer) containing 0.3% formaldehyde. The dialyzate is saved for use in forming the CsCl gradients, as well as for refractive index measurements of the gradient fractions after centrifugation.

The formaldehyde-fixed particles are analyzed in isopycnic CsCl gradients using the step gradient technique of Brunk and Leick.[4] Because this procedure uses CsCl gradients with a preformed step in CsCl concentration, samples reach their equilibrium density positions sooner, allowing for shorter centrifugation times. Swinging-bucket rotors accommodating tubes of 5 ml volume are usually used for isopycnic centrifugation. Depending on the samples to be analyzed, CsCl gradients are designed such that the particles will form a band at their equilibrium density near the center of the tube. The amoung (g) of CsCl required to bring the density (d) of 5 ml (final volume) of dialyzate to the anticipated equilibrium density of the particles is determined using Eq. (1):

$$g_{CsCl} = 1.337V_f(d - 1) \tag{1}$$

where V_f is the final volume (5 ml for the centrifuge tube in this example). Two-thirds of this amount of CsCl is dissolved in sufficient dialyzate to form the lower layer (2.5 ml) of the step gradient. The remaining third is dissolved in the sample solution, using sufficient dialyzate to bring the volume to 2.5 ml, before layering onto the denser layer in the centrifuge tube. The volume of solvent required (V_s) for each layer of volume V_f is given by Eq. (2):

$$V_s = V_f - 0.252(g_{CsCl}) \tag{2}$$

The following example, using the Beckman SW 50.1 rotor, is designed for analysis of mixtures of mitochondrial and cytoplasmic ribosomes and/ or subribosomal particles. To resolve mitochondrial and cytoplasmic ribosomes in the same isopycnic gradient, an average density of 1.5 g/cm^3 is chosen, intermediate between the buoyant density of mitochondrial ribosomes (1.43 g/cm^3) and cytoplasmic ribosomes (1.58 g/cm^3) in CsCl solutions.[3] From Eq. (1) we determine that 3.34 g CsCl is required in 5 ml for an average density of 1.5 g/cm^3. Two-thirds of this amount (2.23 g) added to 1.94 ml [V_s, Eq. (2)] will create the lower layer of 2.5 ml final volume.

[3] M. G. Hamilton and T. W. O'Brien, *Biochemistry* **13,** 5400 (1974).
[4] C. F. Brunk and V. Leick, *Biochim. Biophys. Acta* **179,** 136 (1969).

Similarly, 1.11 g CsCl added to a sample volume of 2.22 ml will comprise the upper layer of 2.5 ml volume.

Centrifugation of the step gradients at 45,000 rpm (near the maximum allowed for this density, at 5°) in the SW50.1 rotor for 16 hr at 5° gives well-resolved, sharp peaks of ribosomes at their equilibrium density positions. Under these conditions, the nearly linear gradient ranges from 1.26 to 1.76 g/cm^3. Fractions of 0.25 ml are collected while monitoring the gradients at 254 nm, using the ISCO density gradient fractionater. A refractometer (Abbe; Fisher Scientific, Pittsburgh, PA) is used to measure the refractive index of each fraction at 25°. These values, corrected by subtracting the refractive increment (over water) of the dialyzate, are used to determine the corresponding CsCl concentration (and density), relying on information in the *International Critical Tables*[5] and *Tables of Properties of Aqueous Solutions Related to Index of Refraction*.[6] As a shortcut, Eq. (3) can be used to determine the density of CsCl at 5° (d) from the refractive index at 25° (n):

$$d = 11.25n - 14.034 \tag{3}$$

Functional Assays

Mitochondrial ribosomes isolated by the above procedures can translate artificial templates when supplemented with homologous elongation factors.[7] The small subribosomal particles bind mRNA with high affinity in the absence of initiation factors, conveniently measured by a Millipore filter assay.[8] The activity of large subribosomal particles can be assessed using the peptidyltransferase assay. The peptidyltransferase assay[9] provides a convenient means for assessing the susceptibility of mammalian mitoribosomes to antibacterial antibiotics affecting this reaction. Some of the ribosomes will contain variable amounts of peptidyl-tRNA, as isolated, if the incubation with puromycin is omitted, and this may be quantified using a peptidyl-[^{3}H]puromycin assay.

Peptidyltransferase Assay

Preparation of Ac[^{3}H]Leu-tRNA. *Escherichia coli* tRNA$_{Leu}$ is charged with [^{3}H]Leu (55 Ci/mmol) using a preparation of *E. coli* aminoacyl-

---

[5] "International Critical Tables," Vol. 3. (E. W. Washburn, ed.) McGraw-Hill, New York, 1928.

[6] M. G. Brown and A. V. Wolf, "Tables of Properties of Aqueous Solutions Related to Index of Refraction," Catalog No. 10403. American Optical Co. Buffalo, New York, 1965.

[7] N. D. Denslow and T. W. O'Brien, *Biochem. Biophys. Res. Commun.* **90,** 1257 (1979).

[8] N. D. Denslow, G. S. Michaels, J. Montoya, G. Attardi, and T. W. O'Brien, *J. Biol. Chem.* **264,** 8328 (1989).

[9] N. D. Denslow and T. W. O'Brien, *Eur. J. Biochem.* **91,** 444 (1978).

tRNA synthetase,[10] and the [^{3}H]Leu-tRNA is acetylated with acetic anhydride.[11]

Procedure. The peptidyltransferase assay is carried out in a volume of 150 μl under the following final conditions: 33 mM Tris-HCl, pH 7.5, 267 mM KCl, 13.3 mM magnesium acetate, 0.66 mM puromycin, 15–25 pmol ribosomes, 9.3 nM Ac[^{3}H]Leu-tRNA [6000 to 10,000 counts/min (cpm)], and 33% (v/v) ethanol. The assay is set up in an ice bath. Ribosomes (in 50 μl of 25 mM Tris-HCl, pH 7.5, 0.2 M KCl, and 10 mM (magnesium acetate) are added to reaction tubes containing 20 μl buffered salts (0.25 M Tris-HCl, pH 7.5, 2 M KCl, 0.25 M magnesium acetate). Water (10 μl) or antibiotics (in 10 μl water for final concentrations of 10^{-4} to 10^{-7} M) is then added, followed by 10 μl of 10 mM puromycin and 50 μl methanol. The reaction is started immediately on the addition of 10 μl Ac[^{3}H]Leu-tRNA (6000 to 10,000 cpm). Incubation is allowed to proceed at room temperature (25°) for 1–40 min. The reaction is terminated by the addition of 10 μl of 10 N KOH, followed by incubation for 3 min at 40° to remove any side products. The reaction is then neutralized by the addition of 0.1 ml of 1 M sodium phosphate, pH 7. The Ac[^{3}H]Leu-puromycin product is extracted with 1.5 ml ethyl acetate by shaking for a few minutes. After the phases separate, remove 1 ml of the ethyl acetate fraction into a counting vial, mix with Aquasol, and count. Correct counts using a 1.5 multiplication factor.

[10] Y. Nishizuka, F. Lipman, and J. Lucas-Lenard, this series, Vol. 12B, p. 708.
[11] A. L. Haenni and F. Chapeville, *Biochim. Biophys. Acta* **114,** 135 (1966).

[23] Bovine Mitochondrial Initiation and Elongation Factors

By Caryl J. Schwartzbach, Mary Farwell, Hua-Xin Liao, and Linda L. Spremulli

Introduction

The mechanism of animal mitochondrial protein synthesis is gradually being elucidated through biochemical characterization of the auxiliary factors involved in the translation process. From the information available at present, there will be a number of steps in the process of chain initiation in the animal mitochondrial system that are distinct from other protein-synthesizing systems. DNA sequence information on the animal mitochondrial genome indicates that the messenger RNAs in this system will have the start codon at or very close to the 5' end of the message. There is little, if any, 5' upstream sequence information, and initiation cannot use a Shine–Dalgarno interaction to facilitate the selection of the start codon. In addition, these messages are not capped, and a cap recognition and scanning mechanism is, therefore, not operative in this system. The process by which the initiation of animal mitochondrial protein synthesis occurs with mitochondrial mRNAs remains to be determined. To date, only a single initiation factor has been identified. This factor, initiation factor 2 (IF-2_{mt}), promotes the binding of the initiator tRNA to mitochondrial ribosomes in a template- and GTP-dependent reaction. It can be routinely assayed using a synthetic mRNA such as poly(A,U,G) as a message.

The elongation steps of protein synthesis in animal mitochondria are catalyzed by three factors. Aminoacyl-tRNA binding to the A-site of the ribosome is directed by elongation factor Tu (EF-Tu_{mt}). This factor is present in a complex with its nucleotide exchange factor EF-Ts_{mt}. The third elongation factor (EF-G_{mt}) promotes the translocation step of protein synthesis.

Materials

Common Buffers

Buffer A: 4 mM HEPES–KOH (pH 7.6), 0.44 M mannitol, 70 mM sucrose, 2 mM EDTA

Buffer B: 15 mM Tris-HCl (pH 7.6), 40 mM KCl, 15 mM MgCl$_2$, 0.8 mM EDTA, 0.05 mM spermidine, 0.05 mM spermine, 0.26 M sucrose, 6 mM 2-mercaptoethanol, 1.6% (v/v) Triton X-100

Buffer C: 20 mM Tris-HCl (pH 7.6), 100 mM KCl, 20 mM MgCl$_2$, 6 mM 2-mercaptoethanol, 1% (v/v) Triton X-100

Buffer D: 20 mM Tris-HCl (pH 7.6), 40 mM KCl, 1 mM MgCl$_2$, 0.1 mM EDTA, 6 mM 2-mercaptoethanol, 10% (v/v) glycerol, 1.0 μM GDP

Buffer for IF-2_{mt}

Buffer I1: 20 mM HEPES–KOH (pH 7.6), 50 mM KCl, 0.1 mM EDTA, 6 mM 2-mercaptoethanol, 10% glycerol

Buffers for EF-Tu · Ts_{mt}

Buffer T1: 20 mM HEPES–KOH (pH 7.0), 40 mM KCl, 1 mM MgCl$_2$, 6 mM 2-mercaptoethanol, 0.1 mM EDTA, 10% glycerol, 1.0 μM GDP (the GDP can be omitted from this buffer if desired)

Buffer T2: 20 mM HEPES–KOH (pH 7.0), 40 mM KCl, 1 mM MgCl$_2$, 6 mM 2-mercaptoethanol, 10% glycerol, 1.0 μM GDP

Buffer for EF-G$_{mt}$

Buffer G1: 20 mM HEPES–KOH (pH 7.0), 10 mM KCl, 0.1 mM EDTA, 6 mM 2-mercaptoethanol, 10% glycerol

Reagents. The GTP, GDP, yeast tRNA, and poly(A,U,G) are from P-L Biochemicals. Poly(U) and *Escherichia coli* tRNA are from Boehringer Mannheim (Indianapolis, IN). Phenylmethylsulfonyl fluoride (PMSF), digitonin, DEAE-Sepharose, S-Sepharose, Sephacryl S-200, phospho(enol)-pyruvate, glycerol, and pyruvate kinase are from Sigma (St. Louis, MO). The HCA high-performance liquid chromatography (HPLC) column (Mitsui Toatsu) is from Rainin (Woburn, MA). The TSKgel-5PW columns (DEAE and SP) are from Beckman. Nitrocellulose membrane filter paper (type HA, 0.45-μm pore size) is from Millipore (Bedford, MA). Pure nitrocellulose membrane filter paper is from Schleicher and Schuell (Keane, NH). Centriprep-10, Centricon-10, YM10, and YM30 membranes are from Amicon (Danvers, MA). [^{14}C]Phenylalanine, [^{35}S]methionine, and [^{3}H]GDP are from New England Nuclear (Boston, MA).

Preparation of Reagents. The [^{14}C]Phe-tRNA is prepared as described by Ravel and Shorey.[1] Yeast tRNA$_i^{Met}$ and tRNA$_m^{Met}$ are separated by BD-cellulose chromatography as described.[2] The [^{35}S]fMet-tRNA is prepared as described.[3] Mitochondrial ribosomes are prepared as described by Matthews *et al.*[4] Ribosomes can be stored at $-70°$ indefinitely. *Escherichia coli* ribosomes are prepared as described previously.[5] Elongation factors EF-G, EF-Tu, EF-Ts, and an EF-Tu · Ts complex are isolated from *E. coli* and further purified as described.[6,7]

High-Performance Liquid Chromatography

All high-performance liquid chromatography (HPLC) columns are attached to a two-pump Rainin HPLC system with a connecting mixer component (1.2 ml). The HPLC columns are run with a program designed on and controlled by a MacIntosh computer using the MacRabbit program from Rainin. Prior to application onto HPLC columns, all samples are filtered through a Nalgene 0.45-μm syringe filter (Schleicher and Schuell, Keene, NH).

[1] J. M. Ravel and R. A. L. Shorey, this series, Vol. 20C, p. 306.
[2] R. T. Walker and U. L. RajBhandary, *J. Biol. Chem.* **247,** 4879 (1972).
[3] M. C. Graves and L. L. Spremulli, *Arch. Biochem. Biophys.* **222,** 192 (1983).
[4] D. E. Matthews, R. A. Hessler, N. D. Denslow, J. S. Edwards, and T. W. O'Brien, *J. Biol. Chem.* **257,** 8788 (1982).
[5] N. Denslow and T. W. O'Brien, *Biochem. Biophys. Res. Commun.* **90,** 1257 (1979).
[6] J. Ravel, R. Shorey, S. Froehner, and R. Shive, *Arch. Biochem. Biophys.* **125,** 514 (1968).
[7] S. L. Eberly, V. Locklear, and L. L. Spremulli, *J. Biol. Chem.* **260,** 8721 (1985).

Initiation and Elongation Factor Assays

Determination of Mitochondrial Initiation Factor 2 Activity

Principle. The IF-2_{mt} activity is defined as the activity that stimulates the binding of fMet-tRNA$_i^{Met}$ to mitochondrial ribosomes or 28 S subunits with poly(A,U,G) as a template. The ribosomal complex formed is detected by nitrocellulose filter binding.[8]

Procedure: Ribosomal Initiation Complex Formation. The standard assay (0.1 ml) contains 50 mM Tris-HCl (pH 7.8), 35 mM KCl, 7.5 mM MgCl$_2$, 0.2 mM GTP, 1 mM phospho(enol)pyruvate, 0.1 unit pyruvate kinase, 0.1 mM spermine, 1 mM dithiothreitol (DTT), 10 μg poly(A,U,G), 5–7 pmol [^{35}S]fMet-tRNA [60,000–70,000 counts/min (cpm)/pmol], 0.25 A_{260} units of mitochondrial ribosomes, and the desired amount of IF-2_{mt}. Samples are incubated at 27° for 15 min. After incubation, each sample is applied directly onto a nitrocellulose filter that is prewet with cold wash buffer [50 mM Tris-HCl (pH 7.6), 35 mM KCl, and 7.5 mM MgCl$_2$]. Vacuum is applied, and the sample is washed with approximately 10 ml cold wash buffer with the vacuum on. Filters are dried at 80° for 7 min and counted in 5 ml βmax scintillation cocktail (ICN Biomedicals, Costa Mesa, CA). Background binding (<0.05 pmol) obtained in the absence of IF-2_{mt} is subtracted from each value. One unit is defined as the binding of 1 pmol of [^{35}S]fMet-tRNA to mitochondrial ribosomes.

Determination of Mitochondrial Elongation Factor Tu Activity

Principle. The EF-Tu$_{mt}$ activity is detected by its ability to replace *E. coli* EF-Tu is poly(U)-directed polymerization of phenylalanine on *E. coli* ribosomes.

Procedure: Poly(U)-Dependent Phenylalanine Incorporation. Assay mixtures (0.1 ml) contain 50 mM Tris-HCl (pH 7.8), 60 mM KCl, 6.5 mM MgCl$_2$, 0.1 mM spermine, 1 mM dithiothreitol (DTT), 2.5 mM phospho (enol)pyruvate, 0.17 units pyruvate kinase, 0.5 mM GTP, 12.5 μg poly(U), 35 pmol [^{14}C]Phe-tRNA, 1.9 μg partially purified *E. coli* EF-G, and 0.48 A_{260} of *E. coli* ribosomes. Dilutions of EF-Tu · Ts$_{mt}$ samples from HPLC columns must be done in buffer T1 containing at least 2 mg/ml bovine serum albumin (BSA) in order to prevent the inactivation of the EF-Tu$_{mt}$ activity. These samples are extremely fragile mechanically and should be handled gently. Mixing of the preparations should be done by gently pipetting the samples with an Eppendorf pipettor. Samples should not be vortexed or mixed by tapping. Incubation is carried out for 30 min at 37°,

---

[8] H.-X. Liao and L. L. Spremulli, *J. Biol. Chem.* **265,** 13618 (1990).

and the reaction is terminated by the addition of approximately 5 ml of 5% w/v trichloroacetic acid. Incorporation of [^{14}C]phenylalanine into polypeptide is determined essentially as described.[1] Blanks (0.3–0.4 pmol) representing the amount of label retained on the filter in the absence of EF-Tu · Ts$_{mt}$ are subtracted from each value. One unit is defined as the incorporation of 1 pmol of [^{14}C]phenylalanine into polypeptide.

Determination of Mitochondrial Elongation Factor Ts Activity

Principle. The EF-Ts$_{mt}$ activity is detected by its ability to stimulate the exchange of guanine nucleotides bound to *E. coli* EF-Tu as described by Fox *et al.*[9] For this assay, EF-Tu · Ts$_{mt}$ samples are diluted with buffer T1 minus BSA and mixed by tapping. This method inactivates the EF-Tu$_{mt}$ component of the complex and frees EF-Ts$_{mt}$ to interact with *E. coli* EF-Tu.

Procedure: Guanine Nucleotide Exchange. The EF-Tu · Ts$_{mt}$ complex is incubated 10 min at 0° in a final volume of 50 μl containing 0.3 μM *E. coli* EF-Tu · GDP, 25 mM Tris-HCl (pH 7.8), 50 mM NH$_4$Cl, 10 mM MgCl$_2$, and 5 μM [^{3}H]GDP. The reaction is terminated by placing the samples on ice and diluting them with 3.0 ml buffer containing 10 mM Tris-HCl (pH 7.4), 10 mM MgCl$_2$, and 10 mM NH$_4$Cl. The samples are then filtered through pure nitrocellulose membranes that are washed and counted.[1]

Determination of Mitochondrial Elongation Factor G Activity

Principle. The EF-G$_{mt}$ activity is detected by its ability to replace *E. coli* EF-G in poly(U)-directed polymerization of phenylalanine on *E. coli* ribosomes.

Procedure: Poly(U)-Dependent Phenylalanine Incorporation. Assay mixtures are prepared as described above for the EF-Tu · Ts$_{mt}$ assay except that they contain 21.5 μg partially purified *E. coli* EF-Tu · Ts. All dilutions of EF-G$_{mt}$ are done with buffer G1 containing 1 mg/ml BSA. Incubation is carried out for 15 min at 37°, and the reaction is terminated by the addition of 5% w/v trichloroacetic acid. Incorporation of [^{14}C]phenylalanine into polypeptide is determined as described.[1] Blanks (~0.5 pmol) representing the amount of label retained on the filter in the absence of EF-G$_{mt}$ are subtracted from each value. One unit is defined as the incorporation of 1 pmol [^{14}C]phenylalanine into polypeptide.

[9] L. Fox, J. Erion, J. Tarnowski, L. Spremulli, N. Brot, and H. Weissbach, *J. Biol. Chem.* **255,** 6018 (1980).

Isolation of Mitochondria

The preparation of digitonin-treated bovine liver mitochondria is based on the procedure described by Matthews *et al.*[4] Bovine livers (4 kg) are obtained from a local slaughterhouse, cut into large pieces, put into a 0.25 M sucrose solution on ice, and then transported back to the laboratory. All subsequent steps are carried out at 4°. Four 1-kg batches of liver are cut into smaller pieces (~2 cm), washed three times with 0.25 M sucrose, and ground in a food processor. The ground liver should have the consistency of minced meat, with pieces of approximately 0.75 cm in a liquid suspension. Following this step, the minced liver (from a 1-kg batch) is diluted with 1.5 liters buffer A and run through a continuous flow tissue homogenizer driven by a benchtop drill press at 1570 rpm.[10] The container in which the homogenate is collected should be kept on ice even when the homogenization process is carried out in a cold room.

The homogenate from each 1-kg batch is filtered through two layers of cheesecloth and distributed into three 1-liter bottles using buffer A to fill the bottles completely. After completion of the first two 1-kg batches, and while continuing the homogenization of the remaining two 1-kg batches, the first 6 liters of homogenate in six 1-liter bottles is subjected to centrifugation at 1400 rpm (600 g) for 15 min at 4°, in a Sorvall RC-3B (Dupont-Sorvall Co., Newton, CT) using an H6000A rotor. This step results in pellets containing extracellular and cellular membranous debris and nuclei. The supernatant is removed by aspiration into a filtration flask packed in ice, transferred into six 1-liter bottles, and spun at 4500 rpm (7300 g) for 24 min. This procedure pellets the remaining nuclei, the mitochondria, and other membrane-enclosed organelles. The supernatant is discarded and the pellets are stored on ice until completion of the remaining two 1-kg batches.

All pellets are suspended in buffer A at a convenient volume using a Waring blender and a Variac at a setting of 20 using five 5-sec bursts and then diluted to a total volume of 6 liters with buffer A. The crude mitochondrial suspension is subjected to centrifugation at 1400 rpm for 15 min in the H6000A rotor. The supernatant is aspirated into flasks packed in ice, avoiding the red pellet, and diluted to fill six 1-liter bottles. The supernatant is subjected to centrifugation at 4500 rpm for 24 min in the H6000A rotor. Following this procedure, the mitochondrial pellets are suspended in 3400 ml of buffer A and then diluted to 4 liters using 615 ml buffer A containing 1.3 g/liter digitonin. The suspension is stirred gently at 4° for 15 min, diluted to 6 liters with buffer A, and subjected to centrifuga-

---

[10] D. Ziegler and F. Pettit, *Biochemistry* **5**, 2932 (1966).

tion at 4500 rpm for 24 min in the H6000A rotor to pellet the mitochondria. Treatment with digitonin results in the release of the nuclear and lysosomal components as well as the removal of the outer mitochondrial membrane. The crude mitoplast pellet is suspended with a rubber policeman in 4 liters buffer A and centrifuged at 4500 rpm for 24 min in the H6000A rotor using four 1-liter bottles. These pellets are suspended in 960 ml buffer A, divided into twenty-four 40-ml centrifuge tubes, and subjected to centrifugation in an SS-34 rotor in a Sorvall RC-5B centrifuge for 15 min at 15,000 rpm (27,000 g). The supernatant is removed, and the pellets are then frozen in a dry ice/2-propanol bath and stored at $-70°$ until use. The yield from 4 kg liver tissue is typically around 120 g.

Special Precautions Required for Preparation Mitochondrial Elongation Factors

Bovine EF-Tu · Ts$_{mt}$ is difficult to prepare unless special precautions are taken as outlined below. Livers are obtained from male Angus cattle (red or black). Livers should be packed in ice within 15–30 min postmortem. The iced liver is immediately cut into large pieces and put into iced 0.25 M sucrose solution as quickly as possible. All subsequent procedures are carried out at 4°, and all samples are kept on ice as much as possible. When the pieces of liver are ground in the food processor, extra care is taken to ensure that the liver is ground with intermittent bursts and continual mixing of the contents of the food processor, so that the liver is not overly ground while permitting the preparation of pieces that are approximately 0.1–0.2 cm in size. The continuous flow homogenization procedure is carried out in a 4° cold room with the body of the unit packed in ice. In the following centrifugation procedures and digitonin treatment, extreme caution is taken to keep all samples as cold as possible. In addition when using the Waring blender, care is taken to use short bursts of mixing.

Preparation of Mitochondrial Extracts

Principle. Following isolation of the mitochondria, two methods are employed for the preparation of mitochondrial extracts. The first method is a standard technique utilizing Triton X-100 to disrupt the mitochondrial membranes. This process involves an overnight ultracentrifugation step that pellets mitochondrial ribosomes and provides a supernatant (S-100) used for preparation of IF-2$_{mt}$. The second method uses sonication followed by a centrifugation step at 30,000g. The resultant supernatant (S-30) is used for the preparation of EF-Tu · EF-Ts$_{mt}$ and EF-G$_{mt}$.

Procedure A: Mitochondrial Extract Preparation with Triton X-100. Frozen mitochondrial pellets (20–25 g) are thawed in a large mortar on ice in a small amount of buffer B. A pestle is used to grind the pellets to a uniform consistency, and buffer B containing 10^{-4} *M* PMSF is added. In total, 10 ml buffer B is added per gram original mitochondrial pellet. The suspension is stirred at 4° for 15 min and then subjected to centrifugation in a Sorvall RC-5 centrifuge with a SS-34 rotor for 45 min at 15,000 rpm (27,000 *g*). The supernatant (S-27) is brought to 0.3 *M* KCl by the addition of KCl dissolved in a small amount of buffer B, and mixed gently but quickly. Twelve 36-ml ultracentrifuge tubes are prepared containing 11 ml of a 34% sucrose cushion (34 g sucrose in 66 ml buffer C) and the S-27 preparation is layered onto the cushions. The tubes are subjected to centrifugation in an L8 ultracentrifuge with a Beckman 50.2Ti rotor at 35,000 rpm (103,000 *g*) at 4° for 14–16 h. The pellets are used for the preparation of mitochondrial ribosomes. The upper three-fourths of the supernatant (S-100) is removed and dialyzed against 10 volumes buffer I1 for 6 hr with two changes, then fast frozen in dry ice/2-propanol. The pellets, containing mitochondrial ribosomes, and the supernatant (S-100) are stored at −70°.

Procedure B: Preparation of Mitochondrial Extracts by Sonication. Mitochondrial extracts containing EF-Tu · Ts$_{mt}$ and EF-G$_{mt}$ activities are prepared from mitochondria that are used within 5 days following preparation. The most active preparations of EF-Tu · Ts$_{mt}$ are obtained from mitochondrial pellets that have been stored for a minimal length of time. All steps described below are carried out at 4°. About 30 g mitochondria is suspended in 120 ml buffer D containing 0.1 m*M* PMSF, and sonicated on ice with a Branson (Danbury, CT) Model 185 sonicator using five 20-sec bursts at 5–10 W with 20-sec cooling periods. It is extremely important that the solution remain cold throughout the sonication process. The homogenate is subjected to centrifugation at 30,000 *g* for 45 min. The supernatant fraction (S-30) is either used immediately or frozen in a dry ice/2-propanol bath and stored at −70° for a minimal amount of time. If larger amounts of EF-Tu · Ts$_{mt}$ are needed, multiple 30-g preparations are made and applied to the first column described below. In this case, care must be taken to thaw the S-30 samples sequentially just before loading onto the column.

Purification of Mitochondrial Initiation Factor 2

Step 1: DEAE-Cellulose Chromatography. In this procedure[11] the S-100 (~1 liter containing 2 g protein) is applied to two separate 90-ml DEAE-cellulose columns (2.5 × 18 cm) equilibrated with buffer I1 at a flow rate

[11] H.-X. Liao and L. Spremulli, *J. Biol. Chem.* **266,** 20714 (1991).

of 1–2 ml/min. It is also feasible to scale up the loading on this column to 20–25 mg protein/ml resin. Each column is washed with buffer I1 until the A_{280} is less than 0.1. The IF-2_{mt} activity is eluted with buffer I1 containing 0.5 M KCl. During this step, 5-ml fractions are collected at a flow rate of 2 ml/min. Active fractions are pooled, dialyzed against a 20-fold excess of buffer I1 with two changes for a total of 5–6 hr, and then frozen in dry ice/2-propanol. The samples from both columns are pooled for further purification. The yield from the DEAE-cellulose column cannot be determined since IF-2_{mt} activity cannot be detected in the S-100. About 80% of the protein in the starting sample is removed in this step.

Step 2: TSKgel DEAE-5PW Chromatography. A TSKgel preparative size DEAE-5PW HPLC column (2.15 × 15 cm) is equilibrated in buffer I1, and IF-2_{mt} (60–80 ml, 200–400 mg) is applied at a flow rate of 3 ml/min. The column is washed with buffer I1 containing 0.1 M KCl until the A_{280} returns to baseline. A linear gradient (450 ml) from 0.1 to 0.25 M KCl in buffer I1 is applied to the column at a flow rate of 3 ml/min. Fractions (4 ml) are collected in siliconized glass tubes. The IF-2_{mt} activity generally elutes at about 0.175 M KCl. Active fractions are pooled, dialyzed against a 20-fold excess of buffer I1, and fast-frozen. This crude preparation of IF-2_{mt} can be concentrated with an Amicon ultrafiltration cell or Centriprep concentrator with a YM30 membrane to 1–2 mg/ml if desired. The resultant sample is quite stable. The TSKgel DEAE-5PW step generally provides about a 60–80% recovery of activity with a 30-fold increase in specific activity.

Step 3: TSKgel SP-5PW Chromatography. The TSKgel DEAE-5PW HPLC preparation of IF-2_{mt} (30–60 ml, 5–20 mg protein) is applied to a TSKgel SP-5PW HPLC column (7.5 × 0.75 cm) equilibrated with buffer I1 at a flow rate of 0.5 ml/min. The column is washed with buffer I1 containing 0.15 M KCl until the A_{280} returns to baseline. The column is developed with a 60 ml linear gradient from 0.15 to 0.33 M KCl in buffer I1 at a flow rate of 0.5 ml/min. Fractions (0.5 ml) are collected in 1-ml plastic microcentrifuge tubes. Aliquots of each fraction are taken for assay and for protein determination. In most cases, BSA is added to the remainder of each fraction (to a final concentration of 0.5 mg/ml), and the fractions are frozen immediately. Once active fractions are identified, they are pooled, fast-frozen, and stored at $-70°$. The IF-2_{mt} activity is generally present in two peaks eluting at about 0.28 M KCl. No differences between the IF-2_{mt} present in these two peaks have been detected. Generally, about 33% of the input IF-2_{mt} activity is recovered in this step, with about a 150-fold increase in specific activity. This material is over 90% pure as determined by sodium dodecyl sulfate (SDS)–polyacrylamide gel electrophoresis.

Separation of Mitochondrial Elongation Factors Tu · Ts and G

DEAE-Sepharose Column Chromatography. In this procedure,[12,13] the supernatants (S-30) from two freshly sonicated mitochondrial homogenates (5 to 10 g total protein) are applied at a flow rate of about 2.5–3 ml/min to a 100-cm^3 DEAE-Sepharose fast flow column (17.5 × 2.7 cm) equilibrated with buffer D. A pump is used to achieve the desired flow rate while the sample is loading and the column is being developed. The column is washed in buffer D until the A_{280} is less than 0.1 and then developed with a 1 liter linear gradient from 0.04 to 0.40 M KCl in buffer D. Fractions (10 ml) are collected in siliconized glass tubes at a flow rate of 2.5 ml/min. The EF-Tu · Ts$_{mt}$ elutes at approximately 0.15 M KCl. The EF-Tu · Ts$_{mt}$ is recovered in about 90% yield with a 7-fold increase in specific activity. Fractions containing EF-Tu · Ts$_{mt}$ activity are dialyzed for 6 hr against two changes of buffer T1, fast-frozen, and stored at −70° until use. This material is stable for months. The EF-G$_{mt}$ activity elutes at approximately 0.22 M KCl. This factor is recovered in nearly a 100% yield with a 27-fold increase in specific activity. Fractions containing EF-G$_{mt}$ activity are pooled, placed in dialysis tubing, and concentrated 6- to 7-fold against solid polyethylene glycol (PEG) 8000. This sample is fast-frozen and stored at −70° until use.

Purification of Mitochondrial Elongation Factor Tu · Ts

The EF-Tu · Ts$_{mt}$ complex is extremely labile when present at low protein concentrations. Concentrations of at least 2.5 mg/ml are required to maintain the activity of this factor during handling and storage. All of the concentration steps indicated in the method below[12] are carried out as quickly as possible following the chromatographic procedure. All dilutions of this factor for assays are done in buffer T1 containing 50 mg/ml BSA. Finally, it should be noted that this factor is very sensitive to mechanical agitation. Samples should not be mixed by vortexing or by tapping the sample tubes. Rather, samples should be mixed gently using an Eppendorf pipette set at about one-third of the sample volume.

Step 1: S-Sepharose Chromatography. The sample (1–2 g protein, ~80 mg/cm^3 resin) is applied at a flow rate of 1.0 ml/min to a S-Sepharose Fast Flow column equilibrated with buffer T1. The column is washed with buffer T1 until the A_{280} is less than 0.1. The column is developed with a linear gradient (200 ml) from 0.04 to 0.40 M KCl in buffer T1. Fractions (2.0 ml) are collected in siliconized glass tubes at a flow rate of 1.0 ml/min. Active

[12] C. Schwartzbach and L. Spremulli, *J. Biol. Chem.* **264,** 19125 (1989).
[13] H. K. Chung and L. Spremulli, *J. Biol. Chem.* **265,** 21000 (1990).

fractions are pooled, dialyzed against two changes of buffer T1, fast-frozen, and stored as described above. This step generally gives a yield of about 35% although recoveries of up to 50% have been obtained. Generally, about 18-fold purification is achieved in this step.

Step 2: Mitsui Toatsu HCA (Hydroxyapatite) Chromatography. Following chromatography on S-Sepharose the sample is further purified by HPLC on hydroxyapatite (HCA), $Ca_{10}(PO_4)_6(OH)_2$. The HCA column (10 cm $\times$ 7.5 mm) is equilibrated with buffer T2. The sample ($\sim$60–70 mg) is applied to the column at a flow rate of 1.0 ml/min. The column is washed with buffer T2 until the A_{280} returns to baseline. A linear gradient (0–100 mM potassium phosphate, pH 7.0, in buffer T2) is run at 0.3 ml/min for 90 min. Fractions are collected at 1.2-min intervals (0.36 ml) in 1.5-ml plastic microcentrifuge tubes. Active fractions are pooled, concentrated as quickly as possible by Centriprep-10, and stored as described above. The sample is dialyzed against buffer T1 prior to the next chromatographic step. Chromatography on HCA generally results in a 6-fold increase in specific activity with about 30% recovery.

Step 3: TSKgel DEAE-5PW Chromatography. The HCA-purified sample is further purified by anion-exchange HPLC on TSKgel DEAE-5PW. The column (7.5 $\times$ 0.75 cm) is equilibrated with buffer T1 and the sample ($\sim$6.0 mg) is applied to the column at a flow rate of 0.75 ml/min. The column is washed with buffer T1 until the A_{280} returns to baseline. The column is developed in three steps. In the first step, a gradient from 40 to 51.5 mM KCl in buffer T1 is applied to the column over a 30-min period. In the second step, the column is washed with 51.5 mM KCl in buffer T1 for about 20 min. In the third step a linear gradient is run for 60 min from 51.5 to 90 mM KCl in buffer T1. Fractions are collected at 1.2-min intervals (0.36 ml) in 1.5-ml plastic microcentrifuge tubes. Fractions are assayed as soon as possible. Active fractions are pooled and concentrated using a Centricon-10. During this step, EF-Tu $\cdot$ Ts$_{mt}$ elutes at about 70 mM KCl as the leading shoulder of a large protein peak. Generally, the EF-Tu $\cdot$ Ts$_{mt}$ from this column is pooled conservatively in order to reduce the amount of the material in the large protein peak eluting just behind, and overlapping, the active fractions. This step provides about a 40–50% recovery of activity with about a 10-fold increase in specific activity.

Step 4: TSKgel DEAE-5PW Chromatography. The partially purified sample is then further separated from contaminating proteins by repeating the HPLC anion-exchange chromatographic step on the TSKgel DEAE-5PW column. The column (7.5 $\times$ 0.75 cm) is equilibrated with buffer T1. To reduce the concentration of KCl in the sample to 40 mM, it is diluted with buffer T1 without KCl. The sample (0.26 mg) is applied to the column at a flow rate of 0.3 ml/min. The column is washed with buffer T1 until the A_{280} returns to baseline. The column is developed using the three-step procedure described above. The fractions are collected at 1.0-min intervals (0.30 ml) in 1.5-ml plastic microcentrifuge tubes, and a Centricon-10 is used to concentrate the pooled sample, which is stored as described above. In this step, EF-Tu · Ts$_{mt}$ elutes as a clear protein peak in front of most of the residual contaminants present. Recovery from this step is close to 70% with a 2-fold increase in specific activity.

Alternative Purification of EF-Tu · Ts$_{mt}$. An alternative purification scheme for EF-Tu · Ts$_{mt}$ was developed that eliminates the HCA HPLC column. In this procedure the sample from the S-Sepharose column (Step 1) is concentrated and dialyzed against buffer T1. The sample is then subjected to two sequential passes through the TSKgel DEAE-5PW column equilibrated in buffer T1. The two columns are developed with the same stepwise gradient; first, a linear gradient of 40 to 54 mM KCl in buffer T1 using a flow rate of 0.3 ml/min for 30 min; second, a wash with 54 mM KCl in buffer T1 for 15 min; finally, a linear gradient from 54 to 85 mM KCl in buffer T1 at a flow rate of 0.25 ml/min for 120 min. The fractions (0.25 ml) are collected in 1.5-ml plastic microcentrifuge tubes and assayed for activity in polymerization. The active fractions are pooled and concentrated immediately using Centricon-10 concentrators. This material is pooled in two portions. The early fractions containing EF-Tu · Ts$_{mt}$ activity are about 75% pure, whereas the back fractions are about 50% pure.

Purification of Mitochondrial Elongation Factor G

The following steps are carried out to obtain purified EF-G$_{mt}$.[13]

Step 1: DEAE-Sepharose and Concentration. Chromatography of crude S-30 extracts on DEAE-Sepharose, performed as described above, allows the separation of EF-Tu · Ts$_{mt}$ and EF-G$_{mt}$.

Step 2: Gel-Filtration Chromatography. Following concentration, the sample is applied directly to a Sepharcryl S-200 column (800 cm^3, 4.5 × 51 cm) equilibrated in buffer G1. The column is developed with the same buffer at a flow rate of 1.5 ml/min, and 7.5-ml fractions are collected in siliconized tubes. The fractions containing EF-G$_{mt}$ activity are pooled, concentrated to a final volume of approximately 10 to 15 ml against solid PEG, and dialyzed for 5 hr against two changes of buffer G1. The sample is fast-frozen in a dry ice/2-propanol bath and stored at $-70°$. This step results in an extremely high recovery of EF-G$_{mt}$ activity with a 4-fold increase in specific activity.

Step 3: TSKgel DEAE-5PW Chromatography. The dialyzed sample (16 ml, 38 mg) is applied at a flow rate of 1 ml/min to a TSKgel DEAE-5PW column (2.15 × 15 cm) equilibrated in buffer G1. The column is developed with a three-step programmed gradient at a constant flow rate of 3 ml/min, and 5-ml fractions are collected in siliconized tubes. The first part of the gradient extends from 0.01 to 0.12 M KCl in buffer G1 over a 20-min period. This step is followed by a shallow rise for 160 min from 0.12 to 0.3 M KCl in buffer G1. Finally, the gradient ends with a 20-min rise to 0.4 M KCl in buffer G1. The material in the active fractions (eluting at about 0.17 M KCl) is pooled, concentrated against solid PEG to a volume of approximately 10 ml, and dialyzed against two changes of a 20-fold excess of buffer G1 for a total of 5 hr. The sample is fast-frozen in a dry ice/2-propanol bath and stored at $-70°$. Essentially all of the protein in the sample is retained by this resin as expected since the first purification step involved chromatography on a gravity DEAE-Sepharose resin. However, when the gradient is applied, EF-G$_{mt}$ elutes toward the back of the main protein peak. Approximately 75% of the EF-G$_{mt}$ activity is recovered in fractions containing only 10% of the applied protein. This step allows a 6-fold increase in specific activity from the previous step.

Step 4: Progel-TSK Heparin-5PW Chromatography. The TSKgel DEAE-5PW purified sample (10 ml, 4 mg) is applied at a flow rate of 1 ml/min to a Progel-TSK Heparin-5PW column (7.5 × 0.75 cm) equilibrated with buffer G1. The column is washed with buffer G1 until the A_{280} returns to baseline. The column is then developed with a programmed three-part gradient run at a constant flow rate of 0.25 ml/min. In the first step, a linear gradient from 10 to 53 mM KCl in buffer G1 is run for 50 min. This step is followed by a wash with buffer G1 containing 53 mM KCl until the A_{280} returns to baseline. Finally, a linear gradient from 53 to 150 mM KCl in buffer G1 is run for 140 min. Fractions (0.5 ml) are collected in 1.5-ml plastic microcentrifuge tubes. Active fractions (eluting at about 70 mM KCl) are pooled and dialyzed against two changes of a 20-fold excess of buffer G1 for a total of 5 hr. The sample is fast frozen in a dry ice/2-propanol bath and stored at $-70°$. The majority of the EF-G$_{mt}$ activity elutes with the front shoulder of the second protein peak. This step yields 55% of the EF-G$_{mt}$ activity with a 10-fold increase in specific activity. The EF-G$_{mt}$ activity at this stage appears to be labile on prolonged storage at $-70°$ due to low amounts of protein present (<0.5 mg).

Step 5: TSKgel SP-5PW Chromatography. In the final step, the Progel-TSK Heparin-5PW sample (10 ml, 0.2 mg) is applied at a flow rate of 1 ml/min to a TSKgel SP-5PW column (7.5 × 0.75 cm) equilibrated with buffer G1. The column is developed with a three-part gradient run at a flow rate of 0.25 ml/min. In the first part, the column is washed for 20 min with buffer G1 containing 30 mM KCl. Second, a linear gradient is run for 160 min from 30 to 150 mM KCl in buffer G1. Finally, the salt concentration is raised from 150 to 200 mM KCl in buffer G1 over a 40-min period.

Fractions (0.5 ml) are collected in 1.5-ml plastic microcentrifuge tubes, and active fractions (eluting at about 70 mM KCl) are fast-frozen in a dry ice/ 2-propanol bath and stored at $-70°$ until use. To preserve the maximum EF-G$_{mt}$ activity, this procedure is generally performed within 3 days of the previous step. The EF-G$_{mt}$ activity elutes with a symmetrical protein peak, well-separated from contaminating protein peaks. The recovery of EF-G$_{mt}$ activity from the TSKgel SP-5PW column is dependent on the amount of protein loaded. Samples containing higher amounts of protein generally give higher recoveries of activity. For this reason, we recommend that two to three samples purified up to the TSKgel DEAE-5PW stage be combined and processed together through the remainder of the purification scheme.

Summary

The procedures summarized above provide nearly homogeneous preparations of IF-2$_{mt}$, EF-Tu · Ts$_{mt}$, and EF-G$_{mt}$. The scheme developed for IF-2$_{mt}$ leads to a 24,000-fold purification of this factor with a 26% recovery of activity. Analysis by SDS–polyacrylamide gel electrophoresis and gel filtration chromatography indicates that this factor functions as a monomer with a molecular weight of about 85,000. The scheme developed EF-Tu · Ts$_{mt}$ provides a 10,000-fold purification with an overall yield of about 10%. The EF-Tu$_{mt}$ component in this complex has a molecular weight of about 46,000, whereas EF-Ts$_{mt}$ has a molecular weight of about 32,000 on SDS–polyacrylamide gel electrophoresis. The EF-Tu · Ts$_{mt}$ complex is tightly associated and appears to have a native molecular weight of about 70,000. The five-step purification procedure outlined above for EF-G$_{mt}$ results in a 14,000-fold purification of EF-G$_{mt}$ with a 2–5% recovery of activity. Analysis by SDS–polyacrylamide gel electrophoresis and gel filtration chromatography indicates that EF-G$_{mt}$ functions as a monomeric protein with an apparent molecular weight of about 80,000.

Section II

Mitochondrial Genetics and Gene Manipulation

[24] Transformation of *Saccharomyces cerevisiae* Mitochondria Using the Biolistic Gun

By RONALD A. BUTOW, R. MICHAEL HENKE, JOHN V. MORAN, SCOTT M. BELCHER, and PHILIP S. PERLMAN

Introduction

The yeast *Saccharomyces cerevisiae* has been the premier organism for the combined forward and reverse genetic analysis of nuclear gene organization, function, and expression. The ease with which defined DNA sequences can be introduced into yeast cells and maintained episomally or directed into defined sites on yeast chromosomes provides an extraordinarily tractable system for such studies.[1] Yeast has also been the premier organism for studies of the genetics and biogenesis of mitochondria.[2-4] Those studies benefit from the dispensability of a functional organelle when cells are grown on fermentable carbon sources so that mutations of nuclear or mitochondrial genes that interfere with mitochondrial function can be isolated and studied using the full range of yeast genetics methods. Finally, because mitochondrial DNA (mtDNA) is inherited biparentally in yeast crosses, unlike the uniparental or maternal inheritance observed in most other eukaryotes, direct genetic studies of mitochondrial genes can be carried out.

The reverse genetic analysis of mitochondrial genomes has only recently been added to the array of tools available for studying mitochondrial genes. In 1988, Johnston *et al.* described the transformation of yeast mitochondria using a particle gun device,[5] the so-called biolistic delivery system, developed by Sanford and colleagues.[6,7] In those experiments, defined DNA sequences bound to tungsten microprojectiles were introduced into yeast mitochondria using a particle gun. The presence of those sequences within mitochondria corrected a defective mitochondrial gene required for a functional electron transport chain. Although the transformation frequency was

---

[1] C. Guthrie and G. R. Fink (eds.), this series, Vol. 194.

[2] B. Dujon and L. Belcour, *in* "Mobile DNA," (D. E. Berg and M. M. Howe, eds.), p. 861. ASM Press, Washington, D.C.

[3] G. Attardi and G. Schatz, *Annu. Rev. Cell Biol.* **4**, 289 (1988).

[4] M. C. Costanzo and T. D. Fox, *Annu. Rev. Genet.* **24**, 91 (1990).

[5] S. A. Johnston, P. Q. Anziano, K. Shark, J. C. Sanford, and R. A. Butow, *Science* **240**, 1538 (1988).

[6] T. M. Klein, E. D. Wolf, R. Wu, and J. C. Sanford, *Nature (London)* **327**, 70 (1987).

[7] J. C. Sanford, T. M. Klein, E. D. Wolf, and N. Allen, *J. Particle Sci. Technol.* **5**, 27 (1987).

low, and is still low relative to the frequency of nuclear gene transformation, the biolistic technique nevertheless has proved to be a reliable method for mitochondrial transformation (see Refs. 8 and 9 for reviews). In this chapter, we summarize the basic techniques of biolistic transformation used in our laboratory and describe some applications of the procedure not only for the analysis of mitochondrial genes, but also as a means of exploring other features of mitochondrial biology.

Transformation Apparatus

Biolistic Gun

Early models of the "biolistic gun" used a gunpowder discharge to accelerate microprojectiles containing bound DNA into yeast cells spread on solid medium. This device has been largely replaced by a commercially available helium gun device (PDS-1000/He from Bio-Rad, Richmond, CA), shown diagramatically in Fig. 1. With this device, particle acceleration is achieved by the rapid expansion of a high-pressure helium wavefront, initiated by breaking a Kapton membrane (rupture disk); these disks are commercially available with nominal rupture pressures ranging from 450 to 2200 psi. Readers are referred to manufacturer's bulletins (Bio-Rad 1687 and 1700); features of the helium apparatus and some of the physical and chemical principles of the Biolistic system have also been described by Sanford et al.[10] and Birch and Bower.[11] Briefly, the helium wavefront rapidly displaces a second carrier membrane onto a stopping screen; that second membrane contains the microprojectiles coated with DNA. As a result of the rapid deceleration of the carrier membrane, the microprojectiles are dispersed onto a petri dish of target cells placed in a vacuum chamber. The final velocity of the microprojectiles is an important factor influencing transformation efficiency; it is determined by the helium pressure needed to burst the rupture disk used to generate the helium wavefront, by the extent of evacuation of the chamber, and by the distance between the stopping screen and the plate. We have found the procedures described in the following to be suitable for mitochondrial transformation. It would be

---

[8] R. A. Butow and T. D. Fox, *Trends Biochem. Sci.* **15,** 465 (1990).

[9] S. M. Belcher, P. S. Perlman, and R. A. Butow, *in* "Principles of Gene Transfer Using Particle Bombardment" (N.-S. Yang and P. Christou, eds.), p. 101. Oxford Univ. Press, New York.

[10] J. C. Sanford, M. J. DeVit, J. A. Russell, F. D. Smith, P. R. Harpending, M. K. Roy, and S. A. Johnston, *Technique* **3,** 3 (1991).

[11] R. G. Birch and R. Bower, *in* "Particle Bombardment Technology in Gene Transfer" (N.-S. Yang and P. Christou, eds.), p. 3. Oxford Univ. Press, New York.

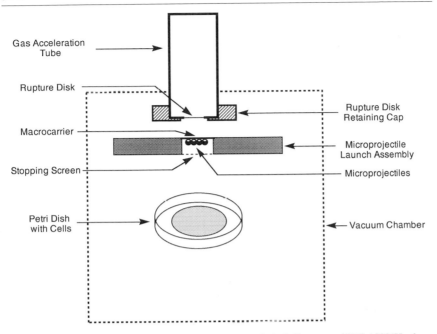

Fɪɢ. 1. Schematic of a commercially available biolistic helium gun (PDS-1000/He from Bio-Rad). See text for details.

prudent, however, for each investigator to check a number of the variables affecting transformation efficiency as there may be gun-to-gun as well as microparticle differences that could affect the desired outcome.

Cell Bombardment

With the commercially available PDS-1000/Hc apparatus, the vacuum chamber containing the petri dish of target cells should be evacuated to about 28 mm Hg in order to achieve sufficient particle velocity. The distance between the rupture disk (1100 psi rating) and the carrier disk should be set at 6 mm, and the distance between the carrier disk and the stopping screen set at 9 mm. We position the petri dish 5 cm from the carrier disk launch assembly, which results in a relatively even microprojectile bombardment of the target cells spread on the surface of the agar. With this configuration, the helium pressure regulator is then set at 1300 psi.

Microprojectiles

The initial report of yeast mitochondrial transformation described the use of 1-μm tungsten particles, and these and smaller tungsten particles are still commonly used with reasonable success. We have not made exhaustive comparisons between tungsten and other particles, nor have we systematically compared relative transformation efficiencies as a function of nominal particle size. However, we routinely use 1-μm gold particles (purchased from Bio-Rad), since, in our experience, the overall efficiency of mitochondrial transformants is greater and more dependable than is obtained with tungsten particles. We have noticed large differences in efficiency of transformation between different batches of gold particles obtained commercially; we believe that the more effective samples were actually smaller than described by the vendor. Particle size is known to be important, and this has been examined both empirically for yeast nuclear transformation by Armaleo et al.[12] and from a more general standpoint of the mechanics of particle bombardment by Klein et al.[6] Other variables that could be considered to increase the efficiency of mitochondrial transformation include the size (ploidy) of the yeast cells and mitochondrial targets by making use, in the former, of larger, polyploid yeast cells or, in the latter, of growth conditions[13] or mutant cells that affect the structure and size of the mitochondria.[14,15]

DNA Attachment to Microprojectiles

Procedures for precipitating DNA to the microprojectiles have been described by Armaleo et al.[12] We have used a variety of procedures for the preparation of DNA and have not noticed any reproducible differences in the efficiency of obtaining either nuclear or mitochondrial transformants. The following describes our standard procedure.

1. Dry 1-μm gold microprojectiles (60-mg aliquots, which are sufficient for ~200 transformations) are washed two times in 1 ml of 100% ethanol, vortexing each time for 1–2 min, and then collected by centrifugation. The particles are washed once in distilled water, suspended in 1 ml distilled water, and stored at $-20°$. These washed particles can be stored indefinitely.

2. For a transformation experiment, a 50-μl aliquot of well-suspended particles is transferred from the stock into a 0.5-ml microtube. Ten microliters DNA (1–2 μg/μl) in TE buffer (10 mM Tris-Cl, 1 mM EDTA, pH

[12] D. Armaleo, G. Ye, T. M. Klein, K. B. Shark, J. L. Sanford, and S. A. Johnston, *Curr. Genet.* **17,** 97 (1990).

[13] B. Stevens, in "Molecular Biology of the Yeast Saccharomyces," (J. N. Strathern, E. W. Jones, and J. R. Broach, eds.), p. 471. Cold Spring Harbor, Laboratory, Cold Spring Harbor, New York, 1981.

[14] L. F. Sogo and M. P. Yaffe, *J. Cell Biol.* **126,** 1361 (1994).

[15] S. M. Burgess, M. Delannoy, and R. E. Jensen, *J. Cell Biol.* **126,** 1375 (1994).

7.4) is added and the suspension vortexed for 30 sec. To the suspension is added 50 μl of 2.5 M CaCl$_2$ and 10 μl of 1 M spermidine, free base. The mixture is vortexed and then incubated at room temperature 3–5 min, and the particles are collected by centrifugation. After removal of the supernatant, the particles are washed two to three times in 200 μl of 100% ethanol and suspended in 55 μl ethanol.

3. After suspending the particles by sonication for 1–2 sec at full power, 5–10 μl is further dispersed by repeated pipetting and ejection, then removed and pipetted onto the center of a Kapton carrier membrane (ethanol-washed) that has been seated into the carrier membrane holder, and finally allowed to dry at room temperature in a desiccator. Up to five membranes at a time are prepared in this way and used as soon as possible after they are completely dry.

Transformation Procedure

General Principles

The procedure underlying the mitochondrial transformation protocol is to bombard cells with microprojectiles coated with DNA and screen survivors of the bombardment for acquisition of the exogenous DNA sequences. Because the frequency of mitochondrial transformation is very low, most experiments involve cotransformation where a plasmid is used that contains an insert of interest and, at the same time, a nuclear DNA marker (e.g., *URA3*), either on the same plasmid or on a separate vector, to allow an initial selection for nuclear transformants. This strategy effectively limits the number of cells that must be rescreened for mitochondrial transformants, as colonies of nuclear transformants must have been derived from cells that were hit by one or more microprojectiles. Although mitochondrial transformants are typically about 0.05% (or less) of the nuclear transformants, they can be easily detected as shown in Fig. 2 and described below.

In the initial description of mitochondrial transformation,[5] respiratory-deficient cells containing a deletion of a portion of the mitochondrial *COXI* gene were bombarded with DNA carrying a *COXI* fragment that, through recombination, would correct the defect. Cells that had carried out that recombination event regained the ability to respire and, thus, were selected for on glycerol-containing medium. Although this approach allows for direct selection of the transformed cells, its major drawback is that successful mitochondrial transformation depends on restoration of respiratory competence of the target cells.

The most versatile strategy to date was developed by Fox and col-

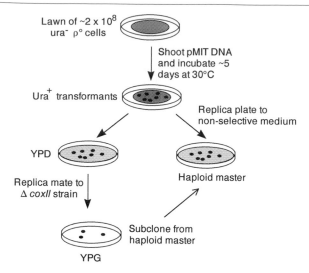

Fig. 2. Procedure for transformation of mitochondria in ρ^0 haploid petite cells of strain DBY947. A lawn of DBY947 cells (*MATα ura3-52 ade2-101*) on selective medium lacking uracil is bombarded with nominally 1-μm gold microprojectiles coated with pMIT DNA. The Ura$^+$ transformants are replica-plated to nonselective medium to obtain a master copy and then replica-mated on a YEPD plate containing a lawn of JC3-TF145 cells (*MATa kar1 ade2 lys1*) that harbor a deletion of the mitochondrial *COXII* gene. Mitochondrial transformants yield foci of respiratory-competent cells, which are detected as zones of growth on replica plates of YPG medium. The original haploid mitochondrial transformants are recovered from the master plate, subcloned, and retested for complementation of JC3-TF145 cells.

leagues,[16] in which they created "synthetic petites" by transforming the mitochondria of ρ^0 cells, which lack mtDNA, with plasmids containing DNA inserts of interest. Such mitochondrial transformants behave for all practical purposes like ρ^- petites. The synthetic petite genomes replicate even if they lack putative origins of mtDNA replication and are faithfully transmitted to progeny cells during vegetative growth. Importantly, the mitochondrial genomes of synthetic petites can be transmitted to other mitochondria in crosses, allowing for transplacements by homologous recombination, as well as for the introduction of foreign DNA sequences into the recipient mitochondria and mitochondrial genomes.

Isolation of Mitochondrial Transformants

Figure 2 illustrates the basic mitochondrial transformation procedure most commonly employed in our laboratory. A recombinant mitochondrial

[16] T. D. Fox, J. C. Sanford, and T. W. McMullin, *Proc. Natl. Acad. Sci. U.S.A.* **85,** 7288 (1988).

transformation vector, generically termed a pMIT vector,[9,17] containing the DNA of interest is used to transform the mitochondria of a ρ^0 recipient strain. The pMIT vector which we often use, pPA100, is derived from the standard yeast shuttle vector, Yep352. It contains a yeast 2-μm plasmid origin of replication, the nuclear *URA3* gene for selection of nuclear transformants, and portions of pUC18 needed for selecting and propagating *Escherichia coli* transformants. pPA100 differs from Yep352 by the inclusion of a wild-type copy of the mitochondrial *COXII* gene contained within a 2.5-kb *Pst*I fragment (from pMT36[18]). This vector permits selection for relatively frequent nuclear transformants of the target cells from Ura$^-$ to Ura$^+$. Then, relatively rare Ura$^+$ cells that are also transformed in their mitochondria are identified in a test cross to a respiratory-deficient strain carrying mtDNA deleted for a portion of the *COXII* gene[16] by scoring for diploid progeny that can grow on glycerol medium (see below).

In initial transformation experiments, a wide variation was found among laboratory strains in the efficiency of both nuclear and mitochondrial transformants. We routinely use a ρ^0 derivative of strain DBY947 (*MATα ura3-52 ade2-101*)[17] or MCC109 (*MATα kar1 ura3-52 ade2-101*),[19] each of which consistently yields greater than 1000 Ura$^+$ transformants per plate. About 2×10^8 ρ^0 cells from an overnight, early stationary phase culture of a recipient strain grown on YPD medium (2% w/v dextrose, 2% w/v Bactopeptone, and 0.02% w/v adenine) are spread on solid (3% w/v agar) minimal medium containing 5% w/v dextrose, 1% w/v casamino acids, 20 μg/ml adenine, and, for osmotic stabilization of the bombarded cells, 0.75 M sorbitol and 0.75 M mannitol. The cells are spread as a lawn in the center two-thirds of the plates, and the plates are stored briefly at 4° until used, usually within 2 hr. This number of cells has been found to be optimal in terms of yielding an acceptable number of Ura$^+$ transformants that grow into relatively large sized colonies after 3–5 days of incubation at 30°. If the lawn contains more cells, similar numbers of transformants are obtained, but the colonies are substantially smaller after 3–5 days of incubation. For a good probability of obtaining at least one mitochondrial transformant carrying a given pMIT construction, a set of five such plates is bombarded with the gold microprojectiles coated with plasmid DNA.

The Ura$^+$ transformants are first replica-plated to solid YPD medium to provide a master plate and then replica-mated to the *COXII* deletion mit$^-$ tester strain, JC3-TF145 (*MATa kar1 ade2 lys1*), spread as a lawn on

[17] P. Q. Anziano and R. A. Butow, *Proc. Natl. Acad. Sci. U.S.A.* **88**, 5592 (1991).
[18] C. Poutre and T. D. Fox, *Genetics* **115**, 637 (1987).
[19] J. C. Kennell, J. V. Moran, P. S. Perlman, R. A. Butow, and A. M. Lambowitz, *Cell (Cambridge, Mass.)* **73**, 133 (1993).

YPD medium. The mtDNA of this strain carries the cox2-17 deletion of the mitochondrial *COXII* gene from strain TF145,[16] so that the strain is a mit⁻ mutant and is unable to grow on glycerol medium. The mitochondrial transformants among the population of transformed Ura⁺ cells are identified either by the appearance of rapidly growing foci on the mating plate or as Gly⁺ regions on replicas of the mating plate on YPG medium. The haploid mitochondrial transformants are retrieved from the master plate and subcloned as required. Along with the *COXII* marker, passenger sequences (another fragment of mtDNA or some foreign DNA fragment), are conveniently placed in the pMIT vector, and these can be analyzed in various ways as outlined below.

Matings between rare cells carrying the *COXII* gene as part of a pMIT genome in the mitochondria and the tester strain grow vigorously on both the YPD mating plate and on YPG replicas of the mating plate, thus providing a sensitive test for mitochondrial transformants. In experiments where the pMIT contains both *COXII* and some other functional fragment of mtDNA, it is also possible to screen for the passenger fragment directly by carrying out a similar test cross to a strain that is gly⁻ due to a defect in the mitochondrial gene covered by the passenger fragment of mtDNA. However, we have noticed that most other segments tested yield progeny that grow less vigorously, so that screening for the *COXII* marker is a more sensitive way of finding the desired transformants. In many dozens of transformation experiments we have never found a pMIT mitochondrial transformant that did not contain both the *COXII* gene and the passenger insert. We believe that this difference in sensitivity is related to the fact that the entire *COXII* gene is present on the pMIT genome and that the petite strain has a high concentration of *COXII* mRNA; thus, complementation of the respiration defect of the tester strain occurs soon after mating and in most mated cells. For passenger sequences that are fragments of genes, the rescue of the respiration defect is delayed until recombination has reconstructed functional genes from the two deficient genomes; even then, sufficient respiration to support vigorous growth is present only in segregants that have mostly copies of the recombinant mitochondrial genomes.

Mitochondrial transformation is not limited to the use of pPA100 or other shuttle vectors resembling it. Instead, we frequently use a mixture of two plasmids. One plasmid is Yep352 containing no mtDNA sequences and the other is a standard *E. coli* vector, such as pGEM (Promega, Madison, WI) or pBS +/− (Stratagene, La Jolla, CA) that carries the *COXII* gene plus the mtDNA fragment relevant to the experiment at hand. Typically, we mix the two plasmids in a 1 to 5 mass ratio, respectively, so that every cell transformed to Ura⁺ has also been exposed to a large number of plasmids containing the mtDNA sequences. The two-plasmid version is generally preferable for several reasons. First, once a segment of mtDNA is placed in the *E. coli* vector, a family of derivatives of it can be made by site-directed mutagenesis or other *in vitro* manipulations, each of which can be transformed without having to transfer the insert to another vector. Second, the plasmid aimed at the mitochondrion is quite a bit smaller than a derivative of pPA100 containing the same insert of mtDNA. For example, in our studies of the protein encoded by intron 4α of the *COXI* gene we cloned a complete *COXI* gene containing upstream and downstream flanking sequences, the natural promoter, all of the exons, plus two introns (for a total insert of 9 kb) and have used that module to transform various mutants.[20] In those experiments, the pMIT vector lacked *COXII* so that mitochondrial transformants were scored by mating the Ura$^+$ nuclear transformants for their ability to restore glycerol growth to a strain deleted for the *COXI* gene.

Functional Analyses of Transformed Mitochondrial Sequences

In principle, mitochondrial transformation can be used for the same types of reverse genetic experiments now commonplace for nuclear genes. In this section, we describe briefly some examples where this approach has been used to study mitochondrial gene expression, including examples of the introduction of foreign DNA sequences into mitochondria.

Direct Analysis of Mitochondrial Genes Expressed in Synthetic Petite Strains

It is well documented that petite mutants of yeast transcribe their mtDNA very actively.[21] When a petite genome contains an entire gene plus its usual promoter, high levels of the transcript usually accumulate. Whenever mitochondrial posttranscriptional events needed for formation of a mature RNA do not require products of mitochondrial protein synthesis, such as processing of some tRNAs, both rRNAs, and most mRNAs that lack introns, petites carrying those genes accumulate the mature RNAs. Even some group I and group II introns splice efficiently in petite mutants,[22,23] so that RNAs spliced for those introns will accumulate. Conse-

[20] R. M. Henke, R. A. Butow, and P. S. Perlman, *EMBO J.* **14,** 5094 (1995).
[21] G. Faye, C. Kujaiwa, and H. Fukuhara, *J. Mol. Biol.* **88,** 185 (1974).
[22] H. F. Tabak, J. van der Laan, K. A. Osinga, J. P. Schouten, J. H. van Boom, and G. H. Veeneman, *Nucleic Acids Res.* **9,** 4475 (1981).
[23] L. A. M. Hensgens, A. C. Arnberg, E. Rosendall, G. van der Horst, R. van der Veen, G. B. van Ommen, and L. A. Grivell, *J. Mol. Biol.* **164,** 35 (1983).

quently, these genes are transcribed well from relevant petite mitochondrial genomes and, so, are amenable to study by mitochondrial transformation.

For example, Martin and colleagues have determined that 5' processing of mitochondrial tRNAs requires an RNA component of a mitochondrial RNase P activity that is encoded by the mitochondrial genome.[24–26] As described in detail elsewhere in this volume, they have used biolistic transformation of ρ^0 mitochondria to demonstrate that a functional RNase P RNA can be synthesized from a synthetic petite. These experiments open the way for a systematic analysis of the sequence and structural requirements of that RNase P RNA.

We have studied a self-splicing group II intron, aI5γ, of the *COXI* gene in transformed pMIT strains.[27,28] We transformed a small fragment of the *COXI* gene containing only the intron plus several hundred base pairs of flanking exon sequence (lacking the 5' and 3' untranslated regions of the mature mRNA and lacking the natural promoter for the gene). In wild-type mitochondria, the excised group II intron RNAs are quite abundant, presumably because they have a lariat structure; therefore, we tested pMIT transformants containing that portion of the *COXI* gene for splicing of the intron using Northern blots to detect the excised intron RNA. Because the *COXII* mRNA is abundant in those pMITs[17] (and, of course, in natural ρ^- strains carrying that gene), it was used to balance the amounts of RNA from each strain analyzed. Probing balanced RNA blots for aI5γ RNA, we found that petite strains carrying the wild-type intron (and mutant alleles of the intron that support splicing) contained readily detectable levels of excised intron RNA and that strains carrying splicing defective alleles lacked that signal. Probing for *COXI* spliced exons or precursor RNA (in splicing defective mutants) did not reveal any discrete signal. We interpret that finding to be a result of several factors: first, the gene fragment lacks processed 5' and 3' ends that are known to contribute to transcript stability in this system; second, the gene fragment has no strong promoter so that transcripts initiate at diverse sites in the vector.

Other potentially useful applications of synthetic petites include phenomena such as hypersuppressiveness, in which certain petite genomes are

[24] D. L. Miller and N. C. Martin, *Cell* (*Cambridge, Mass.*) **34,** 911 (1983).

[25] N. C. Martin and K. Underbrink-Lyon, *Proc. Natl. Acad. Sci. U.S.A.* **78,** 4743 (1981).

[26] M. J. Hollingsworth and N. C. Martin, *Mol. Cell. Biol.* **6,** 1058 (1986).

[27] S. M. Belcher, Ph.D. Dissertation, University of Texas Southwestern Medical Center, Dallas (1993).

[28] C. L. Peebles, S. M. Belcher, M. Zhang, R. C. Dietrich, and P. S. Perlman, *J. Biol. Chem.* **268,** 11929 (1993).

transmitted essentially uniparentally in crosses to ρ^+ strains.[29,30] So far, mitochondrial transformation has not been used to dissect hypersuppressive genomes, although this is an interesting problem that should benefit from this approach. Such studies probably cannot make use of the *COXII* "reporter" gene because its insertion into the roughly 1-kb repeating unit of a hypersuppressive petite would probably alter the phenomenon. However, it should be possible to transform ρ^0 recipient strains with cloned mtDNA from a hypersuppressive petite and detect transformants by either colony blots or test crosses scored for suppressiveness.

Transplacements

The most common strategy for studying transformed mutant alleles of mitochondrial genes is to transfer the gene into an otherwise intact mtDNA so that the effect of the mutation can be assessed in a "wild-type" context. In the yeast nucleus this operation is often called "transplacement" (for gene replacement by transformation) and we use that term here as well. Transformation with appropriately linearized DNA molecules serves as an efficient means to target transforming DNA for such gene replacements in some nuclear genomes.[31] So far, no such means has been developed for targeting sequences for gene replacement in mitochondria. However, the level of homologous recombination is quite high in the organelle so that the desired events occur at frequencies that can be detected readily by screening progeny of crosses.

In mitochondria, gene replacement events depend on two processes: first, homologous recombination forms the desired "hybrid" gene in which the incoming sequence is situated as a part of an otherwise intact mtDNA provided by a tester strain; and, second, vegetative segregation yields some progeny of the mated cells in which only mtDNAs containing the hybrid gene are present. In cases where the hybrid gene is functional, the desired transplacements are easily obtained from test crosses by selecting directly on glycerol medium for cells containing those functional recombinant genomes. In this strategy we typically include the *kar1* (karyogamy-deficient) mutation[32,33] in the nuclear genome of one or both of the strains used in the cross so that haploid strains with the desired mtDNAs are obtained

[29] W. L. Fangman and B. Dujon, *Proc. Natl. Acad. Sci. U.S.A.* **81,** 7156 (1984).
[30] H. Blanc and B. Dujon, *Proc. Natl. Acad. Sci. U.S.A.* **77,** 3942 (1980).
[31] R. Rothstein, this series, Vol. 194, p. 281.
[32] J. Conde and G. Fink, *Proc. Natl. Acad. Sci. U.S.A.* **73,** 3651 (1976).
[33] P. Nagley and A. W. Linnane, *Biochem. Biophys. Res. Commun.* **85,** 585 (1978).

(see also Ref. 34). However, in the more common cases where the incoming allele is defective, the transplacement is achieved by mating the synthetic petite or pMIT to a ρ^+ tester strain and screening for relatively rare respiration-deficient (gly⁻) recombinants. Here, it is necessary to distinguish unmated pMIT cells, which are quite common in the mating cultures, from the desired (but much rarer) recombinants that have a complete mtDNA but are defective for one gene due to an allele provided by the pMIT.

We have employed a number of manipulations to increase the yield of the desired transformants. First, we use about a 10-fold excess of the petite parent in the cross so that most of the ρ^+ cells actually mate. Second, we grow the mated cells for 10–15 hr before plating to permit substantial segregation of mtDNAs so that cells which are screened are mostly pure for one type of mtDNA or another. Third, we select among the progeny of the cross for cells that have the nuclear markers of the ρ^+ parental strain; this step largely eliminates the unmated pMIT cells from further consideration. And, fourth, we include either the nuclear *ade1* or *ade2* gene in the ρ^+ strain because colonies of ρ^+ cells are red on glucose medium while colonies of the desired respiration-deficient recombinants are tan; petite mutants of such a strain yield white colonies that are substantially smaller than the tan colonies of mit⁻ strains.

Mitochondrial transformation to form synthetic petites followed by transfer of the mutant allele to an otherwise intact mtDNA has proved to be very effective for the analysis of mitochondrial genes. For example, Fox and colleagues have systematically studied how specific nuclear-encoded factors exert translation control of the mitochondrial *COXIII* gene through interactions with the 5' untranslated leader sequence of its mRNA.[35] Using biolistic transformation, they introduced various site-directed mutations of that region of the *COXIII* gene by crossing synthetic petites containing pMIT DNA carrying the mutations into a recipient strain deleted for the *COXIII* 5' leader and a portion of the coding sequence. A similar transplacement strategy has been used by Mittelmeier and Dieckmann[36] to define upstream regions of the *COB* gene that are required for stability of its mRNA through the action of the nuclear *CBP1* gene, and for correct 5' processing of the *COB* primary transcript that includes an upstream tRNA^{Glu}.

We have used this basic approach to study the cis-acting sequences involved in the splicing of group II introns. Those studies have been a valuable addition to self-splicing experiments using the same mutant alleles

[34] P. S. Perlman, this series, Vol. 181, p. 539.
[35] M. C. Costanzo and T. D. Fox, *Mol. Cell. Biol.* **13**, 4806 (1993).
[36] T. M. Mittelmeier and C. L. Dieckmann, *Mol. Cell. Biol.* **13**, 4203 (1993).

(e.g., Refs. 28 and 37). Perhaps, more importantly, transformed point mutants have been used to isolate suppressor mutations; some of them are nearby intron mutations that compensate for the initial defect, whereas others are nuclear suppressors.[37] Genetic methods for distinguishing these types of revertants have already been reviewed in this series.[34] These methods were also used to study proteins encoded by both group I and group II introns and exon sequences that serve as target sites for mobile introns.[20,38] Finally, we have used transformation to construct derivatives of a standard strain that contain different combinations of a core set of introns.[19,38]

Foreign DNA Sequences in Mitochondria

The introduction of foreign DNA sequences into yeast mtDNA has opened several new areas of mitochondrial research. It has been known for some time that mtDNA can apparently escape from mitochondria and integrate into the nuclear genome.[39] An experimental system to study the escape process was developed by Thorsness and Fox.[40] Those investigators transformed mitochondria with DNA that included a nuclear gene such as URA3 or TRP1 along with an ARS sequence required for replication in the yeast nucleus. In strains harboring these genes in their mitochondria, they observed complementation of a nuclear gene mutation (ura3 or trp1) at frequencies of approximately 10^{-5} per cell per generation that were the result of escape of the nuclear markers, along with some mtDNA sequences, into the nucleus.[41] It is generally believed that organelle genomes are derived from prokaryotic endosymbionts and that many genes were subsequently transferred to the nucleus. Considerable support for that hypothesis has been obtained by the discovery of organisms in which a given function is expressed from a nuclear gene but a copy of the gene is still present in the organelle.[42] These experiments not only validate the earlier observations suggesting that mtDNA could move from the organelle to the nucleus,[39] but also provide a means of studying the escape process. Such studies have led to the identification of previously unknown genes that function to maintain mitochondrial integrity and morphology.[43]

A potentially powerful system to study and manipulate mitochondrial

[37] S. C. Boulanger, S. M. Belcher, U. Schmidt, S. D. Dib-Hajj, T. Schmidt, and P. S. Perlman, *Mol. Cell. Biol.* **15,** 4479 (1995).
[38] J. M. Moran, S. Zimmerly, R. Eskes, J. C. Kennell, A. M. Lambowitz, R. A. Butow, and P. S. Perlman, *Mol. Cell. Biol.* **15,** 2828 (1995).
[39] F. Farelly and R. A. Butow, *Nature (London)* **301,** 296 (1983).
[40] P. E. Thorsness and T. D. Fox, *Nature (London)* **346,** 376 (1990).
[41] P. E. Thorsness and T. D. Fox, *Genetics* **134,** 21 (1993).
[42] J. M. Nugent and J. D. Palmer, *Cell (Cambridge, Mass.)* **66,** 473 (1991).
[43] C. L. Campbell, N. Tanaka, K. H. White, and P. E. Thorsness, *Mol. Biol. Cell.* **5,** 899 (1994).

gene expression has been reported by Pinkham et al.[44] that also involves the introduction of foreign DNA sequences into mitochondria. These investigators transformed a derivative of the *COXII* gene into ρ^+ mtDNA, in which its normal (unregulated) promoter was replaced by the promoter for bacteriophage T7 RNA polymerase. Using standard nuclear transformation techniques, they also introduced the T7 RNA polymerase coding sequences engineered with a *COXIV* mitochondrial targeting peptide and under regulation of the *GAL1* or *ADH1* promoters. Importantly, expression (and subsequent import into mitochondria) of the T7 RNA polymerase had no effect on wild-type ρ^+ cells, and respiratory competency in the T7-*COXII* transformants was completely dependent on expression of the T7 RNA polymerase. These experiments represent the first example of designed regulation of expression of a mitochondrial gene and should be applicable to a broad range of issues regarding mitochondrial gene expression and the coordination between the mitochondrial and nuclear genomes.

A number of other approaches that are common features of reverse genetics studies of nuclear genomes have not yet been applied to yeast mitochondria, and we imagine that some of these approaches may be developed in the near future. For example, no foreign reporter genes have been expressed in yeast mitochondria. Similarly, gene rearrangements or expression of a given gene from an atypical site on mtDNA have not been achieved. Finally, strains carrying extra copies of a given mitochondrial gene have not yet been used to study transcriptional and, especially, post-transcriptional modes of regulation of gene expression.

Acknowledgments

This work was supported by Grants GM35510 and GM31480 from the National Institutes of Health and Grants I-0642 and I-1211 from the Robert A. Welch Foundation.

---

[44] J. L. Pinkham, A. M. Dudley, and T. L. Mason, *Mol. Cell. Biol.* **14,** 4643 (1994).

[25] Genetics and Transformation of Mitochondria in the Green Alga *Chlamydomonas*

By John E. Boynton *and* Nicholas W. Gillham

Introduction

Molecular evidence strongly supports the endosymbiotic origin of mito chondria from alpha purple bacteria.[1,2] Recent data are consistent with their monophyletic origin in spite of the extreme diversity in the size, organization, and gene content of mitochondrial genomes found in protistans, plant, animal, and fungal cells. This diversity raises intriguing questions regarding the processes by which genes from the original endosymbiont were transferred to the nuclear genome of the eukaryotic host cell during evolution. Photosynthetic eukaryotes show the greatest diversity in mitochondrial genome size and structure, ranging from the 15-kb linear genome of the green alga *Chlamydomonas reinhardtii* to genomes of certain angiosperms that exceed 1 Mb.[1-3] Models involving a large master circle and two to many subgenomic circles arising from recombination between direct repeats have been proposed for certain angiosperm species including *Brassica campestris*, *Zea mays*, *Petunia hybridia*, and *Phaseolus vulgaris*.[4,5] A second possibility is that these circular molecules might arise by recombination between repeats in large linear DNA molecules that are the *in vivo* forms of plant mitochondrial genomes.[6] Tenfold differences in mitochondrial genome size have even been observed within individual angiosperm families.[5,7] Mitochondrial genome size within the genus *Chlamydomonas* is more conserved, ranging from 22–24 kb for circular molecules to 15.8 kb linear molecules in the three species examined.[8,9]

Although the mitochondrial genomes of numerous invertebrates and

[1] M. W. Gray, *Annu. Rev. Cytol.* **141,** 233 (1992).
[2] M. W. Gray, *Curr. Opin. Gen. Dev.* **3,** 884 (1993).
[3] W. Schuster and A. Brennicke, *Annu. Rev. Plant Physiol. Plant Mol. Biol.* **45,** 61 (1994).
[4] M. R. Hanson and O. F. Folkerts, *Int. Rev. Cytol.* **141,** 129 (1992).
[5] S. Mackenzie, S. He, and A. Lyznik, *Plant Physiol.* **105,** 775 (1994).
[6] A. J. Bendich, *Curr. Genet.* **24,** 279 (1993).
[7] B. L. Ward, R. S. Anderson, and A. J. Bendich, *Cell (Cambridge, Mass.)* **25,** 793 (1981).
[8] E. M. Denovan-Wright and R. W. Lee, *Curr. Genet.* **21,** 197 (1992).
[9] M. W. Gray and P. H. Boer, *Philos. Trans. R. Soc. London* **B319,** 135 (1988).

METHODS IN ENZYMOLOGY, VOL. 264

vertebrates[10] and the liverwort *Marchantia*[11] have been completely sequenced, and the genes they encode are well characterized, the genetic systems in these organelles are poorly developed. Correlations between hereditary diseases and specific base pair substitutions and deletions in human mitochondrial genomes are the notable exceptions.[12] Chimeric mitochondrial genes associated with cytoplasmic male sterility have been particularly well documented in maize and petunia, and mutations resulting in the nonchromosomal stripe phenotype in maize are also thought to reside in the mitochondrial genome.[4,5] Mitochondrial genetics has been best characterized in baker's yeast *Saccharomyces cerevisiae* and to a lesser extent in the fission yeast *Schizosaccharomyces pombe*.[13] Cyanide-sensitive respiration is dispensable in mitochondria of baker's yeast and in the green alga *Chlamydomonas reinhardtii* because these organisms can generate energy from fermentation and photosynthesis, respectively.[13] Mitochondria of animal cells lacking DNA can be obtained by prolonged culture of cells in medium containing ethidium bromide, as well as pyruvate and uridine which are essential for growth of the cells.[14,15] Genetic dissection of mitochondrial structure and function in baker's yeast is possible because mitochondrial function is dispensable. Although certain respiratory-deficient mutants are viable in *C. reinhardtii*,[16–18] their characterization is far less extensive than in yeast. In land plants, several lethal mitochondrial mutations are maintained in the heteroplasmic state.[5]

Mitochondrial transplantation or repopulation has been successful in *Neurospora* heterokaryons,[19] *Paramecium*,[20,21] mammalian cells,[14,22–24] and

[10] D. R. Wolstenholme, *Int. Rev. Cytol.* **141**, 173 (1992).
[11] K. Oda, K. Yamoto, E. Ohta, Y. Nakamura, M. Takemura, N. Nozato, K. Akashi, T. Kanegae, Y. Ogura, T. Kohchi, and K. Ohyama, *J. Mol. Biol.* **223**, 1 (1992).
[12] D. C. Wallace, *Annu. Rev. Biochem.* **61**, 1175 (1992).
[13] N. W. Gillham, "Organelle Genes and Genomes." Oxford Univ. Press, London, 1994.
[14] M. P. King and G. Attardi, *Science* **246**, 500 (1989).
[15] P. Desjardins, E. Frost, and R. Morais, *Mol. Cell Biol.* **5**, 1163 (1985).
[16] A. Wiseman, N. W. Gillham, and J. E. Boynton, *J. Cell Biol.* **73**, 56 (1977).
[17] M.-P. Dorthu, S. Remey, M.-R. Michel-Wolwertz, L. Colleaux, D. Breyer, M.-C. Beckers, S. Englebert, C. Duyckaerts, F. E. Sluse, and R. F. Matagne, *Plant. Mol. Biol.* **18**, 759 (1992).
[18] R. F. Matagne, M.-R. Michel-Wolwertz, C. Munaut, C. Duyckaerts, and F. Sluse, *J. Cell Biol.* **108**, 1221 (1989).
[19] E. G. Diacumakos, L. Garnjobst, and E. L. Tatum, *J. Cell Biol.* **26**, 427 (1965).
[20] G. H. Beale and J. K. C. Knowles, *Mol. Gen. Genet.* **143**, 197 (1976).
[21] J. K. C. Knowles, *Exp. Cell Res.* **70**, 223 (1972).
[22] A. Chomyn, A. Martinuzzi, M. Yoneda, A. Daga, O. Hurko, D. Johns, S. T. Lai, I. Nonaka, C. Angelini, and G. Attardi, *Proc. Natl. Acad. Sci. U.S.A.* **89**, 4221 (1992).
[23] M. P. King and G. Attardi, *Cell (Cambridge, Mass.)* **52**, 811 (1988).
[24] M. P. King and G. Attardi, this volume [28].

Drosophila embryos.[25] As direct evidence for recombination in mammalian mitochondria is lacking,[10,26] the prognosis may be better for genome replacement than for substitution of specific regions of the genome during transformation. Stable incorporation of donor DNA into resident mitochondrial genomes has been demonstrated only in yeast[27-30] and *Chlamydomonas*[31] thus far. In both systems, integration of donor DNA during transformation occurs by homologous recombination.[29-31] Mitochondrial transformation in yeast is just beginning to be exploited to study expression of specific organelle genes,[32,33] and little has yet been done to take advantage of this technique in *Chlamydomonas*.

Organization of Mitochondrial Genome in *Chlamydomonas* Species

Physical Size and Conformation

Mitochondrial DNA of *C. reinhardtii* was first isolated and characterized in terms of base composition (47.5% G+C), buoyant density (1.706 g/ml in CsCl), kinetic complexity (9.78×10^6 Da), and physical conformation (4.0 to 5.4 μm, 99% linear, <1% open and closed circular molecules).[34] In cells growing synchronously on a 12 hr light–dark cycle, both mitochondrial and chloroplast DNA were observed to replicate throughout the cycle, whereas nuclear DNA replication was confined to the first 6 hr of the dark period.[35] Restriction analysis revealed that the *C. reinhardtii* mitochondrial genome consisted of a homogeneous population of linear molecules (10^7 Da) with unique ends.[36]

The entire 15.8-kb mitochondrial genome from *C. reinhardtii* (Fig. 1)

[25] E. T. Matsuura, and Y. Niki, this volume [32].

[26] D. A. Clayton, *Int. Rev. Cytol.* **141**, 217 (1992).

[27] R. A. Butow, R. M. Henke, J. V. Moran, S. M. Belcher, and P. S. Perlman, this volume [24].

[28] R. A. Butow and T. D. Fox, *Trends Biochem. Sci.* **15**, 465 (1990).

[29] T. D. Fox, J. C. Sanford, and T. W. McMullin, *Proc. Natl. Acad. Sci. U.S.A.* **85**, 7288 (1988).

[30] S. A. Johnston, P. Q. Anziano, K. Shark, J. C. Sanford, and R. A. Butow, *Science* **240**, 1538 (1988).

[31] B. L. Randolph-Anderson, J. E. Boynton, N. W. Gillham, E. H. Harris, A. M. Johnson, M.-P. Dorthu, and R. F. Matagne, *Mol. Gen. Genet.* **236**, 235 (1993).

[32] J. J. Mulero and T. D. Fox, *Mol. Biol. Cell* **4**, 1327 (1993).

[33] M. C. Constanzo and T. D. Fox, *Mol. Cell. Biol.* **13**, 4806 (1993).

[34] R. Ryan, D. Grant, K.-S. Chiang, and H. Swift, *Proc. Natl. Acad. Sci. U.S.A.* **75**, 3268 (1978).

[35] D. Grant, D. C. Swinton, and K.-S. Chiang, *Planta* **141**, 259 (1978).

[36] D. Grant and K.-S. Chiang, *Plasmid* **4**, 82 (1980).

has now been sequenced,[9,37–54] and an annotated composite sequence prepared by Dr. Michael Gray is available from GenBank (Accession No. U03843). Unique physical features of the genome organization include the presence of short dispersed repeats located in the intergenic spacer regions[44] and terminal inverted repeats of 532 nucleotides including identical single-stranded 3′ extensions of 39 to 41 nucleotides at the left and right ends.[50,54] The outermost terminal 86 nucleotides are identical to an internal 86-bp sequence located at the 3′ end of the L2b rRNA coding sequence.[54] These noncomplementary ends are consistent with the linear conformation of the genome discussed above and, together with the internal repeat, might function in replication of the linear genome.[54] Genes for both the large and small subunit mitochondrial rRNAs are divided into a number of separate subgenic coding segments that are interspersed with protein and tRNA genes (Fig. 1).[9,41] The mitochondrial genome of the interfertile *C. smithii* strain is colinear with that from *C. reinhardtii*, but it contains a 1.1-kb mobile intron (*cobII*) within the apocytochrome *b* (*cob*) gene as well as several restriction fragment length polymorphism (RFLP) markers elsewhere in the genome.[55,56]

In contrast, the distantly related sibling species pair *C. moewusii* and *C. eugametos* have circular mitochondrial genomes of 22 and 24 kb, respec-

[37] P. H. Boer, L. Bonen, R. W. Lee, and M. W. Gray, *Proc. Natl. Acad. Sci. U.S.A.* **82**, 3340 (1985).

[38] P. H. Boer and M. W. Gray, *Nucleic Acids Res.* **14**, 7506 (1986).

[39] P. H. Boer and M. W. Gray, *EMBO J.* **5**, 21 (1986).

[40] P. H. Boer and M. W. Gray, *EMBO J.* **7**, 3501 (1988).

[41] P. H. Boer and M. W. Gray, *Cell (Cambridge, Mass.)* **55**, 399 (1988).

[42] P. H. Boer and M. W. Gray, *Curr. Genet.* **14**, 583 (1988).

[43] P. H. Boer and M. W. Gray, *Nucleic Acids Res.* **17**, 3993 (1989).

[44] P. H. Boer and M. W. Gray, *Curr. Genet.* **19**, 309 (1991).

[45] U. Kuck and H. Neuhaus, *Appl. Microbiol. Biotechnol.* **23**, 462 (1986).

[46] G. Michaelis, C. Vahrenholz, and E. Pratje, *Mol. Gen. Genet.* **223**, 211 (1990).

[47] D.-P. Ma, Y.-W. Yang, and S. E. Hasnain, *Nucleic Acids Res.* **16**, 11373 (1988).

[48] D.-P. Ma, Y.-W. Yang, and S. E. Hasnain, *Nucleic Acids Res.* **17**, 1256 (1989).

[49] D.-P. Ma, Y.-W. Yang, T. Y. King, and S. E. Hasnain, *Plant Mol. Biol.* **15**, 357 (1990).

[50] D.-P. Ma, Y.-T. King, Y. Kim, W. S. Luckett, Jr., J. A. Boyle, and Y.-F. Chang, *Gene* **119**, 253 (1992).

[51] E. Pratje, S. Schnierer, and B. Dujon, *Curr. Genet.* **9**, 75 (1984).

[52] E. Pratje, C. Vahrenholz, S. Buhler, and G. Michaelis, *Curr. Genet.* **16**, 61 (1989).

[53] C. Vahrenholz, E. Pratje, G. Michaelis, and B. Dujon, *Mol. Gen. Genet.* **201**, 213 (1985).

[54] C. Vahrenholz, G. Riemen, E. Pratje, B. Dujon, and G. Michaelis, *Curr. Genet.* **24**, 241 (1993).

[55] J. E. Boynton, E. H. Harris, B. D. Burkhart, P. M. Lamerson, and N. W. Gillham, *Proc. Natl. Acad. Sci. U.S.A.* **84**, 2391 (1987).

[56] L. Colleaux, M.-R. Michel-Wolwertz, R. F. Matagne, and B. Dujon, *Mol. Gen. Genet.* **223**, 288 (1990).

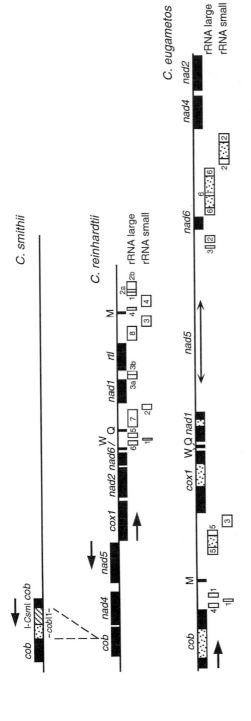

FIG. 1. Comparison of gene organization in the 15.8-kb linear mitochondrial genome of *C. reinhardtii* and the 24-kb circular mitochondrial genome of *C. eugametos*. The circular genome of *C. eugametos* was linearized between the *cob* and *nad2* genes for ease of comparison. The entire sequence of the *C. eugametos* genome is known (GenBank Accession No. UO38430), and 20 kb of the mitochondrial genome of *C. eugametos* has been sequenced [E. M. Denovan-Wright and R. W. Lee, *in* "Genetic Maps, Locus Maps of Complex Genomes" (S. J. O'Brien, ed.), 6th Ed., p. 2.170. Cold Spring Harbor Laboratory, Cold Spring Harbor, New York, 1993]. Mitochondrial genomes from both species code for apocytochrome *b* (*cob*), five NADH dehydrogenase subunits (*nad1*, *nad2*, *nad4*, *nad5*, and *nad6*), cytochrome oxidase subunit I (*cox1*), fragmented large and small subunit rRNAs, and three transfer RNAs (M, Q, and W). The mitochondrial genome of *C. reinhardtii* contains an open reading frame (*rtl*) encoding a reverse transcriptase-like protein that is absent from the mitochondrial genome of *C. eugametos*. The approximate position of the *nad5* gene in the unsequenced region of the *C. eugametos* genome is shown by ↔. The mitochondrial genome of *C. smithii*, a species fully interfertile with *C. reinhardtii*, contains a 1.1-kb type I intron (*cobI1*) (dotted/hatched box) in its *cob* gene encoding the endonuclease I-*CsmI* (hatched box) required for intron mobility. *Chlamydomonas eugametos* has type I introns (dotted box) in the *cox1*, *cob*, *nad1*, and both large and small rRNA genes. The *C. reinhardtii* map is redrawn from Harris [E. H. Harris. *in* "Genetic Maps, Locus Maps of Complex Genomes" (S. J. O'Brien, ed.), 6th Ed., p. 2.168. Cold Spring Harbor Laboratory, Cold Spring Harbor, New York, 1993] and the *C. eugametos* map was kindly provided by Drs. Denovan-Wright and Lee, Dept. of Biology, Dalhousie University.

tively (as determined by restriction mapping).[8,57] Although heterologous hybridization studies reveal that the respective gene orders are colinear, addition–deletion differences of 0.1 to 2.3 kb were observed in nine homologous fragments examined.[8] The RFLP differences in the coding regions are the result of optional group I introns.[58] The rRNA encoding genes of both mitochondrial genomes appear to be interrupted and interspersed with each other and with protein coding regions as in the linear mitochondrial genome of *C. reinhardtii*.[8,58] However, the extent of segmentation and the order of individual fragments varies between completely sequenced genes in *C. reinhardtii* and *C. eugametos*.[59] Although the order of both protein and rRNA encoding genes in the *C. moewusii/C. eugametos* mitochondrial genomes is completely rearranged compared to that of *C. reinhardtii*, with one exception (see below), the same of set proteins appears to be encoded in all three mitochondrial genomes. The presence of numerous introns within both rRNA and protein genes of the 20-kb region sequenced from *C. eugametos*[58] appears to explain the larger genome size compared to *C. reinhardtii* (Fig. 1).

Mitochondrial Gene Products

The completely sequenced mitochondrial genome of *Chlamydomonas reinhardtii* encodes only a subset of the protein subunits of the respiratory electron transport chain (Fig. 1). These include five subunits of the NADH dehydrogenase of complex I (*nad1*, *nad2*, *nad4*, *nad5*, and *nad6*), the COX1 subunit of cytochrome oxidase (*cox1*), the apocytochrome *b* (*cob*) subunit of complex III, and an open reading frame (ORF) (*rtl*) encoding a reverse transcriptase-like protein.[9] The *C. reinhardtii* mitochondrial genome also encodes the small and large subunit rRNAs in short interspersed segments, as mentioned above,[41] but only tRNA M, Q, and W genes,[42] indicating that most tRNAs must be imported. Heterologous hybridization and nearly complete sequencing of the *C. eugametos* mitochondrial genome have located all but one of the protein coding genes found in the *C. reinhardtii* mitochondrial genome on the mitochondrial genomes of *C. eugametos* and *C. moewusii*.[8,57] Although the *rtl* ORF was initially mapped on both the *C. moewusii* and *C. eugametos* genomes by heterologous hybridization using

[57] R. W. Lee, C. Dumas, C. Lemieux, and M. Turmel, *Mol. Gen. Genet.* **231**, 53 (1991).
[58] E. M. Denovan-Wright and R. W. Lee, in "Genetic Maps, Locus Maps of Complex Genomes" (S. J. O'Brien, ed.), 6th Ed., p. 2.170. Cold Spring Harbor Laboratory, Cold Spring Harbor, New York, 1993.
[59] E. M. Denovan-Wright and R. W. Lee, *J. Mol. Biol.* **241**, 298 (1994).

a probe from *C. reinhardtii*,[8] subsequent sequencing studies[58,60] failed to reveal the presence of a related ORF in these two sibling species.

Structure and organization of the fragmented rRNA genes in the *C. moewusii/C. eugametos* species pair have also been characterized in detail,[59] and, as in *C. reinhardtii*, only tRNA Q, M, and W genes have been found.[58] In both *C. reinhardtii* and *C. eugametos*, the small and large subunit rRNA segments transcribed from the fragmented genes can be assembled into the secondary structures predicted for intact molecules and held together by hydrogen bonds.[41,59] In *C. reinhardtii*, the *cob*, *nad4*, and *nad5* genes at the left end of the linear genome are transcribed from one strand, whereas the remaining genes are transcribed from the other strand (Fig. 1). In contrast, all the genes on the circular genome of *C. eugametos* sequenced to date appear to be transcribed from one strand.

Mitochondrial genes typical of other plants, fungi, and mammals such as those encoding other subunits of cytochrome oxidase (e.g., *cox2* and *cox3*) or ATP synthase (e.g., *atp6*, *atp8*, and *atp9*), or the remainder of the tRNAs,[3,11,13] are clearly absent in the completely sequenced 15.8-kb mitochondrial genome of *C. reinhardtii* and in the 20 kb sequenced of the 24-kb mitochondrial genome of *C. eugametos*. Heterologous hybridization using cloned maize probes has also failed to identify any of these genes in either the *C. eugametos* or *C. moewusii* mitochondrial genomes. As in land plants,[61] the universal code is used in the protein coding genes of the *C. reinhardtii* mitochondrial genome.[42,45]

Presence of Introns

A single mobile group I intron (*cobI1*) has been identified in the *cob* gene of *C. smithii*,[55,56] whereas in *C. eugametos*, four group I introns are found in the rRNA genes and single group I introns occur in the *cob*, *coxI*, and *nad1* genes.[58,62] The *C. smithii cobI1* intron of 1,075 bp contains a 237-codon ORF (I-*CsmI*) with 36% amino acid identity with the I-*SceI* endonuclease encoded by the mobile group I intron, ScLSU·1, from *S. cerevisiae* mitochondria.[56] This insertion is preferentially transmitted to most vegetative diploid progeny in crosses of *C. smithii* to *C. reinhardtii* lacking the *cobI1* intron[55,63] (see below). One of the rRNA introns of *C. eugametos* is found within the 530 loop region of the small subunit S_2 coding

[60] R. W. Lee, personal communication (1994).
[61] T. H. Jukes and S. Osawa, *Experientia* **46,** 1117 (1990).
[62] M. Turmel, J.-P. Mercier, and M.-J. Cote, *Nucleic Acids Res.* **21,** 5242 (1993).
[63] R. F. Matagne, D. Rongvaux, and R. Loppes, *Mol. Gen. Genet.* **214,** 257 (1988).

segment, a second is in the large subunit L$_5$ coding segment corresponding to domain II, and two are in the L$_6$ coding segment corresponding to the peptidyltransferase center.[59]

Mutations Affecting Mitochondrial Structure and Function in Chlamydomonas

Nuclear Mutations

Nine Mendelian obligate photoautotrophic mutations of *C. reinhardtii* have been isolated and characterized.[16] These dark-dier (*dk*) mutants belong to eight different complementation groups representing eight genetic loci and were grouped in three phenotypic classes based on the pleiotropic nature of the phenotypic defects. Class I mutants have gross alterations in mitochondrial inner membrane ultrastructure and are deficient in cytochrome oxidase and antimycin/rotenone-sensitive NADH–cytochrome-*c* reductase activities. Class II mutants show less severe alterations in mitochondrial ultrastructure and deficiencies in cytochrome oxidase. Class III mutants possess near normal mitochondrial ultrastructure and reduced cytochrome oxidase activity. A new Mendelian obligate photoautotrophic mutation (*dn-12*) has been reported that is deficient in cytochrome oxidase,[17] but whether *dn-12* is allelic with any of the *dk* mutants is unknown. The cytochrome oxidase-deficient *dk-97* mutant has been used to demonstrate the accumulation of tricarboxylic acid intermediates during photosynthesis and to show that glycolate and D-lactate dehydrogenase activities are linked during mitochondrial electron transport.[64] This mutant has also been used to study O$_2$ isotope discrimination resulting from the alternative oxidase.[65]

Mitochondrial Mutations

Growth of *C. reinhardtii* in acriflavine or ethidium bromide results in induction, with almost 100% efficiency, of lethal minute colony mutations when cells are plated in the light in the absence of the dyes.[66,67] These mutant cells can divide 8 to 9 times under continuous illumination before lethality occurs, during which time respiratory competence is reduced and abnormalities in mitochondrial ultrastructure develop, while chloroplast

---

[64] D. W. Husic and N. E. Tolbert, *Proc. Natl. Acad. Sci. U.S.A.* **84**, 1555 (1987).
[65] H. G. Weger, R. D. Guy, and D. H. Turpin, *Plant Physiol.* **93**, 356 (1990).
[66] N. J. Alexander, N. W. Gillham, and J. E. Boynton, *Mol. Gen. Genet.* **130**, 275 (1974).
[67] N. J. Alexander, Ph.D. Thesis, Duke University, Durham, North Carolina (1977).

structure and function remain normal. Both acriflavine and ethidium bromide were subsequently shown to cause elimination of mitochondrial DNA.[68] This suggests that the loss of mitochondrial genomes caused by these dyes results in the lethal minute phenotype. In the case of the nuclear photoautotrophic mutants described above, loss of cytochrome oxidase activity and the ability to carry out cyanide-sensitive respiration is not lethal in the light. Therefore, lethality of the minute colony mutations cannot be explained merely by the loss of the *cob* or *coxI* genes necessary for cyanide-sensitive respiration. Because elimination of mitochondrial genomes also results in loss of the mitochondrially encoded NADH subunits of Complex I necessary for both cyanide-sensitive and cyanide-insensitive respiration, their absence will obliterate ATP formation by both respiratory pathways.[69] Loss of this complex will also eliminate the shuttling of electrons necessary to generate enough NAD^+ to supply the tricarboxylic acid cycle. The essential role of the mitochondrial *nad* genes for survival is supported by analysis of the lethal minute colony mutations arising from the extension of the *cob* deletion in the obligate photoautotrophic *dum-1* mutant into the adjacent *nad4* gene.[31]

Although prolonged acriflavine exposure converts all *C. reinhardtii* cells to the lethal minute phenotype, cells plated after shorter treatments yield a mixture of normal-sized, small slowly growing, and lethal minute colonies. Among 50 slow growing colonies examined, two obligate photoautotrophic mutants (*dum-1* and *dum-2*) had deletions of approximately 1.5 kb extending from the left-hand end of the mitochondrial genome through most of the *cob* gene.[18] In contrast to wild type, where mitochondrial genomes of only monomer length are observed, the *dum-1* genomes are present as a mixture of monomer and dimer length molecules.[31] The mutant dimers appear to result from head-to-head fusions of two deleted molecules. As mentioned above, the *dum-1* mutation segregated lethal minute colony mutations at a frequency of 4% per generation. Polymerase chain reaction (PCR) analysis of these lethal minute colonies revealed that the original deletion had extended into the adjacent *nad4* gene, supporting the hypothesis presented earlier that elimination of NADH dehydrogenase is lethal. These results also emphasize the importance of the unique terminal repeat sequences[54] in the stability of the linear mitochondrial genome of *C. reinhardtii*.[31]

Several additional mitochondrial mutations and one nuclear mutation (*dn-12* discussed above) affecting mitochondrial function have been isolated following short-term growth of *C. reinhardtii* in acriflavine or ethidium

[68] N. W. Gillham, J. E. Boynton, and E. H. Harris, *Curr. Genet.* **12,** 41 (1987).
[69] A. L. Moore and J. N. Siedow, *Biochim. Biophys. Acta* **1059,** 121 (1991).

bromide.[17,70] Five mutants had varying size deletions in the *cob* genes. The *dum-3*, *dum-4*, and *dum-14* mutants had deletions of 1.6 to 1.7 kb extending from the left end of the genome to within the *cob* gene and formed dimer genomes as does *dum-1*. Shorter *cob* deletions were found in *dum-16* (1.3 to 1.4 kb) and *dum-11* (0.7 kb). Four mutants (*dum-6*, *dum-15*, *dum-18*, and *dum-19*) exhibited the characteristic pattern of inheritance of mitochondrial gene mutations (see below) but showed no obvious physical alterations in their mitochondrial genomes. Analysis of respiratory chain activities indicated that *dum-6*, *dum-18*, and *dum-19* were deficient in cytochrome oxidase and affect the *cox1* gene, whereas *dum-15* was defective in complex III and was likely a *cob* gene mutation. DNA sequencing demonstrated that *dum-6* had a GC → AT base pair substitution in *cox1*.[70]

Mutations in the *cob* gene resistant to mucidin and myxothiazol have been obtained spontaneously[71] or following treatment with manganese, which acts as a specific mitochondrial mutagen in *Chlamydomonas* as it does in yeast.[72] An eight-transmembrane helix model has been proposed for the apocytochrome *b* polypeptide of yeast, which assumes that amino acid residues involved in binding these inhibitors are all in close proximity on the same side of the mitochondrial membrane.[73] Mutations in yeast and *Rhodobacter capsulatus* resistant to these inhibitors map either within transmembrane helix III or in the loop connecting helices III and IV. The *Chlamydomonas* resistance mutations map at the same conserved positions 129 within helix III and 137 in the loop region or at position 132 within the helix where mutants have not been previously reported.

Inheritance of Mitochondrial Genes and Genomes

Uniparental Transmission of Mitochondrial Genes in Meiotic Zygotes

In experiments taking advantage of physical differences in the organelle genomes of the interfertile species *Chlamydomonas reinhardtii* and *C. smithii*, the chloroplast and mitochondrial genomes were shown to be uniparentally transmitted by opposite mating types to the progeny of meiotic zygotes.[55] In 127 tetrads scored all but one (0.08%) exhibited uniparental inheritance of a diagnostic mitochondrial DNA fragment involving the *cob*

---

[70] M. Colin, M.-P. Dorthu, F. Duby, C. Remacle, M. Dinant, M.-R. Wolwertz, C. Duyckaerts, F. Sluse, and R. F. Matagne, *Mol. Gen. Genet.* **249**, 179 (1995).

[71] P. Bennoun, M. Delosme, and U. Kuck, *Genetics* **127**, 335 (1991).

[72] P. Bennoun, M. Delosme, I. Godehardt, and U. Kuck, *Mol. Gen. Genet.* **234**, 147 (1992).

[73] A. M. Colson, B. L. Trumpower, and R. Boasseur, *in* "Molecular Description of Biological Membrane Components by Computer and Conformational Analysis" (R. Brasseur, ed.), Vol. 2, p. 147. CRC Press, Boca Raton, Florida, 1990.

gene from the mt^- parent. In contrast, of 186 tetrads from the same cross scored for transmission of chloroplast genetic markers, 172 exhibited uniparental transmission from the mt^+ parent and 14 (8.1%) transmitted chloroplast markers biparentally from both parents. These experiments demonstrate that uniparental transmission of the mitochondrial genome is even more strictly controlled in sexual crosses than is true for the chloroplast genome. When one or the other mating type was subjected to UV irradiation prior to mating, no effect was seen on transmission of the mitochondrial genome, whereas irradiation of the mt^+ parent produced the expected large increase in the frequency of zygotes transmitting chloroplast markers from the mt^- parent.[13,74] In the *C. eugametos/C. moewusii* species pair both chloroplast and mitochondrial genomes are transmitted by the mt^+ parent.[75]

Chloroplast DNA from the *C. reinhardtii* mt^- parent disappears within a matter of hours following fusion of gametes of opposite mating type.[13] In contrast, both parental mitochondrial genomes persist in the zygote, although mt^+ mitochondrial DNA declines slowly as a function of time.[76] Following transfer of these dark-incubated zygotes to the light, the remaining mitochondrial genomes from the mt^+ parent vanish within 24 hr. Hence, not only are the two organelle genomes transmitted to the zygote by opposite mating types, but chloroplast DNA destruction is triggered during gametic fusion while elimination of mitochondrial DNA occurs during zygote germination.

Reciprocal crosses of strains containing a *cob* myxothiazol resistance mutant (*MUD2*) to wild type demonstrated that this marker was also strictly uniparentally inherited from the mt^- parent in 114 tetrads analyzed.[71] Similarly, meiotic clone analysis of reciprocal crosses between two presumptive point mutations causing an obligate photoautotrophic phenotype (*dum-6* and *dum-15*, see above) showed that the mitochondrial marker carried by the mt^+ parent was only very rarely transmitted.[17] However, this strict uniparental transmission pattern appears to break down in the case of mitochondrial deletion mutations.[17,18] In crosses of wild-type mt^- to any of the *dum cob* deletion mutants in mt^+, the strict pattern of uniparental transmission of the wild-type mitochondrial phenotype is observed. In reciprocal crosses, however, the wild-type mitochondrial genome from the mt^+ parent is transmitted to 14–28% of randomly isolated meiotic products. The length of the zygote maturation period has been shown to affect the frequency of transmission of the wild-type mitochondrial phenotype from the mt^+ parent in crosses to the *dum-4 cob* deletion mutant. Zygotes germi-

[74] E. H. Harris, "The *Chlamydomonas* Sourcebook." Academic Press, San Diego, 1989.
[75] R. W. Lee, B. Langille, C. Lemieux, and P. H. Boer, *Curr. Genet.* **17,** 73 (1990).
[76] M.-C. Beckers, C. Munaut, A. Minet, and R. F. Matagne, *Curr. Genet.* **20,** 239 (1991).

nated at 3 days show strict uniparental transmission of the *dum-4* marker from the *mt⁻* parent, whereas those matured for 6 to 10 days show the much higher frequencies of transmission of the wild-type mitochondrial genome alluded to above. This suggests that mitochondrial genomes with *cob* deletions are at a strong selective disadvantage, perhaps because the deleted (*cob*) end is unstable (see above).

These findings regarding the inheritance of *dum-4* are consistent with previous observations[66] on the inheritance of the lethal minute colony phenotype induced in *mt⁻* cells by acriflavine. All of the meiotic tetrads from crosses of *mt⁻* minutes to wild-type *mt⁺* showed 4 minute : 0 wild-type progeny when zygotes were germinated between 3 and 6 days. Tetrads segregating 0 minute : 4 wild type gradually increased to about 20% after 15 days, and this was accompanied by a corresponding reduction in percent germination. Thus, when the mitochondrial genome of the *mt⁻* parent has sustained a lethal deletion (e.g., minute colonies produced by *dum-1* or *dum-4*) or is absent (e.g., minute colonies produced by acriflavine) the mechanisms ensuring uniparental transmission of mitochondrial DNA from the *mt⁻* parent fail to function in a sizable fraction of the zygotes. These zygotes appear to sense the status of their *mt⁻* mitochondrial genomes and succeed in turning off the mechanisms that cause elimination of the *mt⁺* mitochondrial genomes.

Early experiments with two *dk* mutants of *C. reinhardtii* (*dk-74* and *dk-104*) indicated that they were transmitted in a biparental but non-Mendelian fashion.[77] This non-Mendelian biparental inheritance is inconsistent with the uniparental pattern of mitochondrial DNA transmission from the *mt⁻* parent, and the genetic basis for the unusual segregation pattern observed for these two mutants remains to be determined.

Biparental Transmission and Recombination of Mitochondrial Genes in Vegetative Diploids

Vegetative diploids transmit mitochondrial genomes from both parents to their mitotic progeny,[55,78] as is true for chloroplast genetic markers.[74] Furthermore, the *cobI1* mobile intron in the *C. smithii cob* gene spreads with very high efficiency to the *C. reinhardtii* mitochondrial genome in vegetative diploids.[55,78,79] Biparental transmission of mitochondrial and chloroplast genomes in vegetative diploids is consistent with a model for uniparental inheritance of organelle genomes in *Chlamydomonas* which

[77] A. Wiseman, N. W. Gillham, and J. E. Boynton, *Mol. Gen. Genet.* **150,** 109 (1977).
[78] C. Remacle, C. Bovie, M.-R. Wichel-Wolwertz, R. Loppes, and R. F. Matagne, *Mol. Gen. Genet.* **223,** 180 (1990).
[79] C. Remacle and R. F. Matagne, *Curr. Genet.* **23,** 518 (1993).

supposes that sexual zygotic functions, including those responsible for uni-parental transmission of organelle genomes, are not activated in mitotic zygotes.[80,81] Although diploids are phenotypically mt^-, they transmit chloroplast markers from the mt^+ parent, so mt^+ is dominant to mt^- with respect to chloroplast gene transmission in diploid × haploid crosses.[80] In contrast, mt^- is dominant to mt^+ with respect to mitochondrial genome transmission in diploid × haploid crosses.[63]

Recombination between genetic and physical markers in the *C. reinhardtii* and *C. smithii* mitochondrial genomes has been demonstrated in vegetative diploids obtained from sexual crosses and artificial fusions.[79] This involved exchanges between the *MUD2* genetic marker in the *cob* gene from *C. reinhardtii* and the *C. smithii* RFLP markers *Nhe*I (located to the left of the *cobII* intron), *Nco*I (in the adjacent *nad4* gene), and *Hpa*I (in the *coxI* gene). The *cobII* intron from *C. smithii* spread frequently to intronless *cob* genes of *C. reinhardtii* in diploids derived from crosses when the *C. smithii* mitochondrial genome was donated by the mt^+ parent and virtually always in the reciprocal cross. Coconversion of the adjacent *MUD2* marker was 94%. The three *C. smithii* RFLP markers showed rates of 64–73%, and in most cases these three RFLP markers "coconverted" together. This suggests that the segregants contained the parental *C. smithii* mitochondrial genomes, as the *Hpa*I marker is 5.0 kb from the intron insertion site. However, bona fide recombinants arising from recombination between restriction sites were also observed. A substantial fraction (40 to 70%) of the diploids obtained from crosses of the *cob* point mutant *dum-15* and point mutants in the *coxI* gene segregated respiratory-competent cells capable of dark growth.[81a] A random population of these diploid cells yielded 15–20% recombinants, whereas intragenic recombinants between *coxI* point mutants arose at frequencies of 0.01 to 7%.

Cytoplasmic Exchange of Organelle Genes without Nuclear Fusion

A process of cytoduction has been described in *C. reinhardtii* whereby a large proportion of mitotic zygotes do not actually give rise to vegetative diploids, but instead undergo cytokinesis before nuclear fusion.[82] Because organelle genomes are generally biparentally transmitted in cytoduction,

[80] N. W. Gillham, J. E. Boynton, and E. H. Harris, *in* "The Molecular Biology of Plastids" (L. Bogorad and I. K. Vasil, eds.), Cell Culture and Somatic Cell Genetics of Plants, Vol. 7A, p. 55. Academic Press, San Diego, 1991.
[81] U. W. Goodenough and P. J. Ferris, *in* "Genetic Regulation of Development" (W. Loomis, ed.), p. 171. Alan R. Liss, New York, 1977.
[81a] C. Remacle, C. Colin, and R. F. Matagne, *Mol. Gen. Genet.* **249,** 185 (1995).
[82] R. F. Matagne, C. Remacle, and M. Dinant, *Proc. Natl. Acad. Sci. U.S.A.* **88,** 7447 (1991).

this process can be used to transfer genetically marked chloroplast or mitochondrial genomes from one nuclear background to another.[72,82]

Mitochondrial Transformation in *Chlamydomonas*

Comparative Features of Chloroplast and Mitochondrial Transformation

Transformation of the polyploid genomes of *Chlamydomonas* and tobacco chloroplasts and yeast mitochondria can be readily accomplished using donor DNA carrying selectable markers introduced with the biolistic particle delivery system.[28,83,84] In chloroplasts the homologous integration events yielding stable transformants are relatively rare ($\sim 10^{-5}$ to 10^{-8}), so choice of appropriate combinations of recipient strain and donor plasmid are critical to ensure strong selection for the introduced donor gene and to avoid the isolation of new nuclear mutations mimicking the phenotype of the selectable organelle marker (e.g., drug resistance). In the first approximation, use of homologous selectable markers obviates ambiguous results that might arise from problems in expressing foreign genes in chloroplasts or mitochondria and facilitates integration of the donor sequences by homologous gene replacement. Physical differences between the homologous donor and recipient genes, or closely linked to those genes (e.g., a deletion or RFLP marker), greatly facilitate identification of bona fide transformants. In addition, several foreign genes have now been expressed successfully under the control of appropriate chloroplast regulatory sequences.[84,85] A cassette containing the bacterial *aadA* coding sequence for spectinomycin resistance fused to the *atpB* promoter/leader and the *rbcL* 3' noncoding region has proved to be an extremely useful selectable marker. Where this cassette inserts in the chloroplast genome is determined by the flanking chloroplast sequences attached.

In contrast to chloroplast transformants that are routinely selected directly on the basis of their restored photosynthetic capacity or inhibitor resistance,[83] mitochondrial transformants of yeast are normally obtained in a two-step cotransformation process.[28–30] Recipient *ura⁻*, respiration-deficient cells are bombarded with a mixture of cloned nuclear DNA carrying the selectable *URA⁺* marker and cloned mitochondrial DNA with an appropriate gene to restore respiratory function. Nuclear *URA⁺* transformants are first selected on medium permissive for survival of the respiration-deficient mitochondrial genotype of the recipient strain. These trans-

---

[83] J. E. Boynton and N. W. Gillham, this series, Vol. 217, p. 510.
[84] P. Maliga, *Trends in Biotechnology* **11**, 101 (1993).
[85] M. Goldschmidt-Clermont, *Nucleic Acids Res.* **19**, 4083 (1991).

formed isolates can then be screened directly for restoration of respiratory competence or identified by crossing them to an appropriate petite tester mutant. In contrast, mitochondrial transformants of a respiratory-deficient *C. reinhardtii* mutant (*dum-1*) with a 1.5-kb deletion affecting the *cob* gene were selected directly for restored respiratory function following bombardment with total mitochondrial DNA.[31] Restoration of mitochondrial function in these transformants permitting growth on acetate-containing medium in the dark was extremely slow. Colonies appeared on the selective plates 4 to 8 weeks after transformation but grew at normal rates thereafter. By comparison, chloroplast transformants are normally detected as visible colonies on selective medium after 5 to 14 days.

Selection and Characterization of Mitochondrial Transformants in Chlamydomonas

Mitochondrial transformation in *C. reinhardtii*[31] was accomplished using biolistic methods adapted from those employed routinely for chloroplast transformation.[83] All experiments done to date used partially purified mitochondrial DNA from wild-type cells, obtained by sodium iodide equilibrium density gradient centrifugation, to complement the respiratory defect in the *cob* deletion mutant *dum-1*.[31] This DNA was precipitated onto M10 tungsten microprojectiles (Bio-Rad, Hercules, CA), and monolayers of *dum-1* cells spread on agar plates were bombarded using the PSD-1000 Particle Delivery System (Bio-Rad) under conditions standard for chloroplast transformation.[31,83] Bombarded cells were replated onto acetate-containing medium and incubated in the dark; the plates were examined weekly for growth of colonies. Use of cloned mitochondrial fragments as donor DNA instead of total mitochondrial DNA, gold microprojectiles instead of tungsten, and the PDS-1000/He particle delivery system is likely to result in an increase in the approximate 10^{-6} transformation frequency obtained in the initial experiments.[31] If suitable conditions can be established, reducing the copy number of the mitochondrial genomes in recipient cells by growing them for a short time in low concentrations of ethidium bromide might also increase the frequency of transformation. A similar approach involving use of 5-fluorodeoxyuridine to reduce chloroplast genome number increases substantially the frequency of chloroplast transformants in certain recipient strains.[83]

In the case of chloroplast transformation in *C. reinhardtii*, donor DNA likely integrates into one or a few of the 80 genome copies by recombination, and these transformed genomes segregate during division of the single organelle in each cell. Depending on growth conditions, more than one mitochondrion may exist per *Chlamydomonas* cell,[74] and very likely only one of these mitochondria should initially receive the donor DNA. Therefore, obtaining a homoplasmic mitochondrial transformant requires segregation of the transformed organelle as well as segregation of transformed genomes within that organelle. The 1-μm tungsten or gold particles coated with donor DNA are likely to lodge within or pass through the chloroplast, releasing multiple donor DNA molecules inside. These same microprojectiles are large relative to mitochondrial cross sections and are likely either to destroy the organelle or graze and rupture its outer and inner membranes, releasing a few donor molecules inside. Both the higher number of mitochondria per cell and the lower probability that a given mitochondrion receives multiple copies of donor DNA may contribute to the long quiescent period in these mitochondrial transformants. Injection of functional mitochondria into mammalian cells that are partially depleted for mitochondrial DNA results in colony formation on selective medium after 14 to 17 days and a full complement of mitochondrial genomes after 6 to 10 weeks, corresponding to 20–25 cell generations.[23] Intracellular selection of mitochondria introduced by microinjection based on drug resistance has been observed in *Paramecium*.[20,21]

In the case of transformants of the *dum-1* (*cob*) deletion mutant, individual donor mitochondrial genomes containing the wild-type *cob* gene are presumed to replicate directly, or to recombine with the resident mitochondrial genomes, restoring the deleted segments.[31] In either instance, the mitochondrial genomes with wild-type *cob* genes would continue to replicate and segregate during mitochondrial division. When the copy number of wild-type *cob* genes per mitochondrion becomes sufficiently high, functional complex III assembles on the mitochondrial inner membrane. Once a substantial fraction of the mitochondria have restored respiratory electron transport, sufficient energy would be provided for the transformant to grow in the dark.

Recombination rather than genome replacement was documented in the case of the single mitochondrial transformant of the *C. reinhardtii* *dum-1* recipient bombarded with isolated *C. smithii* mitochondrial DNA that has distinct RFLP markers.[31] This respiratory-competent transformant has the *cob* gene from *C. smithii* with the 1.1-kb intron and arose from a recombination event between the *cob* and *cox1* genes of the *C. reinhardtii* recipient. The lack of RFLP markers prevents an assessment of whether this is also true for the transformants obtained with *C. reinhardtii* donor DNA. Integration of donor DNA by homologous recombination in the *C. reinhardtii* mitochondrion is consistent with evidence for mitochondrial gene recombination in vegetative diploid progeny from crosses of *C. smithii* and *C. reinhardtii*.[70,79]

Future Prospects for Manipulating Mitochondrial Genome

The only published mitochondrial transformation experiments in *Chlamydomonas* used the respiratory-deficient *dum-1* mutation as the recipient.[31] The fact that this mutant results from a deletion that extends into the *cob* gene makes verification of bona fide transformants unequivocal. Respiratory-deficient point mutants in the *cob* and *cox1* genes of *C. reinhardtii*[17,70] might also prove to be suitable recipients. Isolation of similar mutations in *C. eugametos* and *C. moewusii* is not feasible because the wild-type strains fail to grow heterotrophically.[74] Similarly, resistance mutations like *MUD2* in the *C. reinhardtii cob* gene might be used as selectable donor markers.[71,72] Whether heterologous drug resistance markers such as the bacterial *aadA* gene encoding spectinomycin resistance[85] can be used for mitochondrial transformation remains to be determined, as virtually nothing is known about the sensitivity of *Chlamydomonas* mitochondria to antibacterial antibiotics. Furthermore, nothing is known regarding the transcriptional and translational regulatory sequences in *C. reinhardtii* mitochondria. Past experience with foreign gene expression in the *C. reinhardtii* chloroplast[85,86] indicates that selection of the proper endogenous regulatory sequences is essential.

The small size of the *C. reinhardtii* mitochondrial genome and the availability of its complete sequence may offer fertile ground for engineering new selectable markers and for examining the expression and function of both endogenous and foreign genes. The high frequency of cotransformation in chloroplasts[83] facilitates both the targeted disruption of specific genes[87] and the introduction of point mutations without direct selection.[88] Depending on the efficacy of microprojectile delivery of coprecipitated cloned fragments within the same mitochondria and the endogenous level of homologous recombination, cotransformation may also work in this organelle. Resistance mutations such as *MUD2* in the *cob* gene and the wild-type alleles of respiratory-deficient mutants such as *dum-6* or foreign drug resistance genes might serve as selectable markers. Because of the relatively small size of the mitochondrial genome in *C. reinhardtii*, one may be able to link the selected and unselected marker in the same cloned fragment, thus enhancing the probability of both markers integrating into the recipient mitochondrial genome.

[86] A. D. Blowers, G. S. Ellmore, U. Klein, and L. Bogorad, *Plant Cell* **2**, 1059 (1990).
[87] S. M. Newman, N. W. Gillham, E. H. Harris, A. M. Johnson, and J. E. Boynton, *Mol. Gen. Genet.* **230**, 65 (1991).
[88] A. Lers, P. B. Heifetz, J. E. Boynton, N. W. Gillham, and C. B. Osmond, *J. Biol. Chem.* **267**, 17494 (1992).

Whether transformation technology developed in *Chlamydomonas* can be transferred to the far larger and more complex mitochondrial genomes of land plants remains to be seen. In the case of chloroplast transformation, the same drug resistance markers that work in *Chlamydomonas*[83] have been successfully used in tobacco.[84] Clear evidence exists that land plant mitochondria have taken up and incorporated foreign DNA into their genomes multiple times during their evolution.[3,13] In addition, mitochondrial genomes actively recombine in progeny from cell fusions of different cultivars or species.[13] Both observations argue that mitochondrial transformation of land plants is likely to succeed if proper selectable markers can be identified.

Acknowledgments

This work was supported by National Institutes of Health Grant GM-19427 to J. E. B. and N. W. G. We thank Dr. Michael Gray for critical reading of the manuscript, Drs. Robert Lee and Eileen Denovan-Wright for an unpublished linear map of the mitochondrial genome of *C. eugametos*, and Dr. Heriberto Cerutti for help in preparing the comparative mitochondrial genome maps for *C. eugametos* and *C. reinhardtii*.

[26] Isolation of Avian Mitochondrial DNA-Less Cells

By Réjean Morais

Introduction

Avian mitochondrial (mt)DNA-less (rho^0, ρ^0) cells have been developed to study the contribution of the mitochondrial genome to the cell phenotype of vertebrates.[1-3] Rho^0 cell lines have been obtained following long-term exposure of primary chick embryo cells[4] and continuous chicken and quail cells[5,6] to ethidium bromide, a planar aromatic dye which binds to DNA and RNA by intercalation.[7] Ethidium bromide is a potent inhibitor of

---

[1] K. Zinkewich-Péotti, M. Parent, and R. Morais, *Cancer Res.* **50,** 6675 (1990).

[2] K. Zinkewich-Péotti, M. Parent, and R. Morais, *Cancer Lett.* **59,** 119 (1991).

[3] H. Wang, M. Parent, and R. Morais, *Gene* **140,** 155 (1993).

[4] P. Desjardins, E. Frost, and R. Morais, *Mol. Cell. Biol.* **5,** 1163 (1985).

[5] P. Desjardins, J.-M. de Muys, and R. Morais, *Somatic Cell Mol. Genet.* **12,** 133 (1986).

[6] R. Morais, P. Desjardins, C. Turmel, and K. Zinkewich-Péotti, *In Vitro Cell. Dev. Biol.* **24,** 649 (1988).

[7] M. J. Waring, *Nature (London)* **219,** 1320 (1968).

mtDNA replication and transcription in mammalian[8–10] and avian[4,11] cultured cells but does not substantially affect the expression and synthesis of nuclear DNA.[12–14] The specific inhibitory effect of the drug correlates with observations showing that, in cultured mammalian cells, ethidium bromide accumulates in mitochondria but insignificantly within the chromatin network.[15] Moreover, the activity of partially purified rat mitochondrial DNA polymerase is more strongly impaired by ethidium bromide than is that of nuclear DNA polymerase using the same template.[16]

Rho^0 avian cells proliferate in culture with mitochondria devoid of a functional respiratory chain. Cytochrome-c oxidase activity is not detectable, oxygen consumption is reduced by 95%, and analyses of reduced-minus-oxidized cytochrome spectra reveal the absence of cytochromes b and aa_3.[4,5,17] Respiration deficiency correlates with the absence of 13 essential proteins of oxidative phosphorylation and electron transport encoded in mtDNA, including three subunits of cytochrome-c oxidase. Rho^0 avian cells behave phenotypically as rho^0 yeast cells obtained following long-term exposure to ethidium bromide.[18] On short-term exposure, large nucleotide sequences are deleted within yeast mtDNA, converting cells to respiration-deficient, cytoplasmically inherited petite phenotype mutants.[19] Such mutations have never been detected in ethidium bromide-treated vertebrate cells, where, however, mtDNA rapidly attains a high negative superhelix density that is reversible on removal of ethidium bromide from the culture medium.[20]

Auxotrophy for Pyrimidines

Earlier studies had shown that populations of primary chick embryo fibroblasts (CEF) cultivated in Ham's F12 medium supplemented with

[8] M. M. K. Nass, *Proc. Natl. Acad. Sci. U.S.A.* **67,** 1926 (1970).

[9] R. D. Leibowitz, *J. Cell Biol.* **51,** 116 (1971).

[10] E. Zylber, C. Vesco, and S. Penman, *J. Mol. Biol.* **44,** 195 (1969).

[11] P. Desjardins, D. L'Abbé, B. F. Lang, and R. Morais, *J. Mol. Biol.* **207,** 625 (1989).

[12] K. Radsak, K. Kato, N. Sato, and H. Koprowski, *Exp. Cell Res.* **66,** 410 (1971).

[13] M. M. K. Nass, *Exp. Cell Res.* **72,** 211 (1972).

[14] L. Leblond-Larouche, R. Morais, and M. Zollinger, *J. Gen. Virol.* **44,** 323 (1979).

[15] J. Delic, J. Coppey, M. Ben Saada, H. Magdelénat, and M. Coppey-Moisan, *J. Cell. Pharmacol.* **3,** 126 (1992).

[16] R. R. Meyer and M. V. Simpson, *Biochem. Biophys. Res. Commun.* **34,** 238 (1969).

[17] R. Morais, *J. Cell. Physiol.* **103,** 455 (1980).

[18] E. S. Goldring, L. I. Grossman, D. Krupnick, D. R. Cryer, and J. Marmur, *J. Mol. Biol.* **52,** 323 (1970).

[19] P. P. Slonimski, G. Perrodin, and J. H. Croft, *Biochem. Biophys. Res. Commun.* **30,** 232 (1968).

[20] C. A. Smith, *Nucleic Acids Res.* **4,** 1419 (1977).

tryptose phosphate broth (TPB) were inherently resistant to the growth inhibitory effect of ethidium bromide[14,17] and chloramphenicol[14,21,22] (CAM), an antibiotic that blocks specifically the mitochondrial protein-synthesizing system in eukaryotes.[23] Growth of CEF populations exposed to ethidium bromide and CAM came to a halt after a few generations in the absence of TPB, similarly to all mammalian cell populations treated with the same drugs.[12,13,24] The active components of the broth, an enzymatic digest of animal extracts, have been found to be of pyrimidine origin in the form of ribonucleosides, ribonucleotides, and pyrimidine-containing oligonucleotides.[25] The addition of uridine or cytidine to Ham's F12 medium is sufficient to prevent the inhibitory effect of ethidium bromide and CAM on the growth of CEF and continuous quail and chicken cell lines.[5,6] Auxotrophy for pyrimidines is ascribed to a deficiency in the activity of dihydroorotate dehydrogenase, the enzyme that catalyzes the conversion of dihydroorotic acid to orotic acid within the uridylic biosynthetic pathway.[25] The dehydrogenase is located on the outer surface of the inner mitochondrial membrane[26] and requires a functional respiratory chain to transport the electrons released during the catalytic oxidation of dihydroorotic acid. In the presence of an artificial electron acceptor, the activity of the dehydrogenase is restored in mitochondrial preparations isolated from respiration-deficient CEF[25] and rho^0 cells,[6] indicating that the expression and synthesis of the dehydrogenase is maintained in those avian cells.

Cells and Culture Conditions

Quail and chicken continuous cell lines LSCC-H32 and DU249 were first described by Kaaden et al.[27] and Langlois et al.,[28] respectively. DU24 is a subclone of DU249.[6] Primary cultures of chick, quail, and duck embryo cells are prepared by the method of Dulbecco as modified by Temin and Rubin.[29] All cell lines in petri dishes are grown in Ham's F12 medium supplemented with 15% (LSCC-H32) or 10% fetal bovine serum. Penicillin

[21] R. Morais, M. Grégoire, L. Jeannotte, and D. Gravel, Biochem. Biophys. Res. Commun. 94, 71 (1980).
[22] L. Leblond-Larouche, A. Larouche, D. Guertin, and R. Morais, Biochem. Biophys. Res. Commun. 74, 977 (1977).
[23] G. Schatz and T. L. Mason, Annu. Rev. Biochem. 43, 51 (1974).
[24] B. Storrie and G. Attardi, J. Cell Biol. 56, 819 (1973).
[25] M. Grégoire, R. Morais, M. A. Quilliam, and D. Gravel, Eur. J. Biochem. 142, 49 (1984).
[26] J. J. Chen and M. E. Jones, Arch. Biochem. Biophys. 176, 82 (1976).
[27] O. R. Kaaden, S. Lange, and B. Stiburek, In Vitro 18, 827 (1982).
[28] A. J. Langlois, R. Ishizaki, G. S. Beaudreau, J. F. Kummer, J. F. Beard, and D. P. Bolognesi, Cancer Res. 36, 3894 (1976).
[29] H. M. Temin and H. Rubin, Virology 6, 669 (1958).

(100 IU/ml), streptomycin (100 μg/ml), Fungizone (0.5 μg/ml), and uridine (4 μg/ml) are routinely added to the culture medium. Cultures are maintained at 37° in a humidified incubator with an atmosphere of 95% air/5% CO_2 (v/v). Cells are passaged twice a week by trypsinization, and the medium is changed every other day. The latter operation is particularly important for rho^0 cells, which produce large amounts of lactic acid[30] that acidify the culture medium and interfere with cell growth and viability.

Development of Avian Rho^0 Cells

Avian mtDNA-less cells have been prepared from primary embryo[4] and continuous avian cells.[5,6] We describe below the procedure to obtain both finite[31] and continuous rho^0 cells.

Finite Rho^0 Cell Lines

Primary chick, quail, and duck embryo cells are prepared by trypsin treatment of 8- to 10-day-old embryos. The isolated cells are dispersed in minimum essential medium (MEM) supplemented with fetal calf serum (6%, v/v) and seeded at a concentration of 2×10^5 cells/ml into petri dishes. At confluence the cells are detached by trypsinization, suspended in Ham's F12 medium supplemented with fetal calf serum, antibiotics, and uridine and seeded at concentrations ranging from 5×10^4 to 1×10^5 cells/ml. Fibroblastoid-like cell populations are obtained after the third passage and remain phenotypically similar until senescence, which occurs for chick embryo cells after more than 40 cell divisions have been completed.

Avian embryo cell populations are exposed to ethidium bromide 24 hr following the third passage. Stock solutions of ethidium bromide (40 μg/ml) are made up in water and kept in the refrigerator for a period not exceeding 2 months. Solutions and the treated cells are sensitive to light and can be protected with aluminum foil. Concentrations of ethidium bromide higher than 1.0 μg/ml are lethal to avian embryo cell populations. Cells are generally grown in either 0.2 or 0.4 μg/ml ethidium bromide. Figure 1A,B shows typical proliferative curves of populations of CEF having long-term exposure to ethidium bromide. The doubling time for the first 3 to 4 generations is similar to control cells in the absence of uridine (Fig. 1A) but rises thereafter to over 175 hr. Under the light microscope, the cells are seen to accumulate vacuoles and to enlarge. Growth arrest is eventually followed by detachment of the cells from the plastic surface. With some

---

[30] L. Jeannotte, M.Sc. Thesis, Université de Montréal, Montréal, Québec (1982).
[31] W. I. Schaeffer, *In Vitro Cell. Dev. Biol.* **26,** 97 (1990).

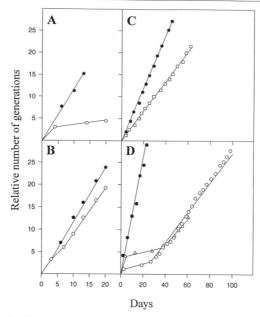

Days

FIG. 1. Effect of ethidium bromide on long-term growth of finite and continuous avian cell lines in Ham's F12 medium. Chick embryo fibroblasts were cultivated in the absence (A) and presence (B) of uridine (2 μg/ml). Cells were grown by successive transfers at inoculation densities of 0.5–1 × 10^5 cells/ml. LSCC-H32 (C) and DU24 (D) cells were cultivated in the presence of uridine (4 μg/ml). Successive transfers were at inoculation densities of 1–1.5 × 10^5 cells/ml for control cells and at 2.5 × 10^5 cells/ml for ethidium bromide-treated cells. Control cells (●); cells treated with ethidium bromide at 0.1 (△), 0.2 (○), and 0.4 μg/ml (□). Fig. 1A and B reproduced from Ref. 21 with permission from Academic Press. Fig. 1C and D reproduced from Morais et al., "Development and Characterization of Continuous Avian Cell Lines Depleted of Mitochondrial DNA," In Vitro Cellular & Developmental Biology, Vol. 24, No. 7, pp. 649–658, 1988. Copyright © 1988 by the Tissue Culture Association. Reproduced with permission of the copyright owner.

serum lots, however, avian embryo cells do show a little propagative growth, with a generation time of greater than 150 hr. The cells die in dialyzed sera, suggesting that the nondialyzed sera contain sufficient pyrimidine nucleotide precursors to sustain some cell growth. Indeed, uridine at 2 μg/ml restores the growth capacity of ethidium bromide-treated CEF cultivated in dialyzed and nondialyzed sera (Fig. 1B). The increase in cell population at each passage derives from the proliferation of most of the attached cells at seeding. No lag or adaptation period is required, indicating that the cell population as a whole is inherently resistant to the growth inhibitory effect of the drug.

The rate of growth of ethidium bromide-treated cell populations is a function of the uridine concentration in the medium. Maximal growth is attained at concentrations of uridine equal to or higher than 2.0 μg/ml.[21] Cytidine at similar concentrations supports equally well cell growth. Thymidine, uracil, and cytosine, and none of the immediate uridine precursors tested, namely, dihydroorotic acid, orotic acid, and oritidine, are able to support active cell growth.

Continuous Rho⁰ Cell Lines

As illustrated in Fig. 1C,D, the growth behavior of continuous avian cells exposed to ethidium bromide may differ from one cell line to the other. Proliferation of the fibroblastoid-like cell line LSCC-H32 is essentially similar to that of primary avian embryo cell populations (Fig. 1C). At ethidium bromide concentrations ranging from 0.1 to 0.8 μg/ml, the cell population doubling time is moderately increased as compared to control cells, and the increase in cell population derives from the proliferation of most of the attached cells at seeding: in the absence of uridine, cell division ceases after about four generations. In contrast, ethidium bromide severely affects the viability of epithelioid-like DU24 cells at concentrations as low as 0.1–0.2 μg/ml (Fig. 1D): concentrations of ethidium bromide equal to or greater than 0.4 μg/ml are toxic to DU24. After about two to four generations in the presence of 0.1–0.2 μg/ml ethidium bromide, most seeded cells are found to slough off the plastic surface. Foci sparsely distributed on the plastic surface are then seen which develop into colonies. The mixed colonies are trypsinized and passaged. Following repeated passages, the cell population acquires a growth rate similar to that of CEF and LSCC-H32 cells.

Storage of Continuous Avian Rho⁰ Cell Lines

LSCC-H32 and DU24 cell populations exposed to ethidium bromide for at least 90 days are transferred to drug-free medium and propagated. Seeded at low cell density (500 cells/60-mm petri dish), dispersed colonies are ring cloned, propagated, and subcloned. The cells are stored frozen in Ham's F12 medium to which 20% serum and 10% (v/v) dimethyl sulfoxide (DMSO) are added. Cells (4–6×10^6) are suspended in 1-ml freezing medium, kept at 4° for 2 hr, and then transferred at −80° overnight. Cells are stored in liquid nitrogen. Under these conditions, continuous avian mtDNA-less cells have remained viable for more than 8 years.

Monitoring Mitochondrial DNA Depletion

Assay of cytochrome-c oxidase (EC 1.9.3.1) activity is an appropriate routine monitor of the inhibitory effect of ethidium bromide on mitochon-

TABLE I
CYTOCHROME-*c* OXIDASE ACTIVITY OF RESPIRATION-DEFICIENT CHICK EMBRYO FIBROBLASTS
CULTIVATED IN ETHIDIUM BROMIDE-FREE MEDIUM[a]

Ethidium bromide treatment (days)	Number of experiments	Number of cell generations at end of treatment	Cytochrome-*c* oxidase activity (% of control)			
			At end of treatment	Number of cell generations after treatment		
				10 to 20	21 to 30	31 to 35
3	3	2.40 ± 0.3	9.1 ± 3	29.2 ± 16	40.9 ± 6	52.3 ± 10
6	4	4.65 ± 0.6	1.1 ± 1	5.1 ± 5	15.9 ± 13	28.4 ± 15
13	5	8.67 ± 1.7	<0.5	<0.5	<0.5	<0.5
20	3	13.81 ± 2.4	<0.5	<0.5	<0.5	ND

[a] Values represent the mean plus or minus standard deviation of individual experiments. ND, Not done. The mean cytochrome-*c* oxidase specific activity value of control cell populations was 57 ± 24 nmol cytochrome *c* oxidized per minute milligram protein after 1 to 5 cell doublings (nine individual experiments) and 49 ± 12 nmol/min/mg protein after 31 to 55 cell doublings (10 individual experiments). Reproduced from Desjardins *et al.*[4] by permission.

drial DNA replication and transcription. In the presence of ethidium bromide, the activity decreases as a function of the number of cell generations. As shown in Table I, cytochrome-*c* oxidase activity is no longer detectable after approximately nine cell doublings. The progressive decrease in activity is consistent with the absence of any new synthesis of mitochondrially encoded cytochrome-*c* oxidase subunits. Primary CEF treated with ethidium bromide for nearly 14 cell generations and then transferred to drug-free medium remained cytochrome-*c* oxidase-deficient until senescence.

The following assay is used for measuring cytochrome-*c* oxidase activity in avian cells. Cells ($1-5 \times 10^7$) are collected by trypsinization, washed with phosphate-buffered saline (PBS), and homogenized in cold distilled water using a Dounce homogenizer (for additional information, see Ref. 32). To a 3-ml plastic or quartz cuvette the following are added:

Sodium phosphate buffer, 2 ml of 0.15 M, pH 7.5
Cell homogenate 0.1–0.5 ml
Reduced cytochrome *c*, 30 μl of 3 mM (cytochrome *c* is reduced by adding 0.1 ml of 60 mM NaBH$_4$ to 1 ml oxidized cytochrome *c*; after 30 min at room temperature, 0.1 ml of 0.1 M HCl is added to stop the reaction)
Water to complete to 3 ml

[32] L. Leblond-Larouche and R. Morais, *Eur. J. Biochem.* **65**, 423 (1976).

TABLE II

EFFECT OF ETHIDIUM BROMIDE ON MITOCHONDRIAL DNA COPY NUMBER OF CHICK
EMBRYO FIBROBLASTS

Ethidium bromide treatment (days)	Mitochondrial DNA copy number/cell[a]			
	End of treatment	Days in ethidium bromide-free medium		
		14	28	42
0	604 ± 134 (11)	209 ± 132 (6)	242 ± 68 (5)	412 ± 92 (2)
3	79 ± 29 (6)	25 ± 24 (4)	37 ± 30 (2)	116 (1)
6	9 ± 2 (4)	1.6 ± 1.2 (3)	1.2 ± 0.7 (2)	4 (1)
13	0.5 ± 0.2 (2)	0.1 ± 0.1 (2)	0.9 ± 1.0 (4)	0.1 (1)
20	0.4 ± 0.3 (7)	ND	ND	ND
43[b]	0.1 ± 0.1 (5)			

[a] Values represent the mean plus or minus standard deviation of individual experiments. The number of experiments is given in parentheses. ND, Not done.

[b] Mean value from cell populations cultivated in ethidium bromide-containing medium for 35 to 52 days. Reproduced from Desjardins et al.[4] by permission.

The oxidation of cytochrome c is assayed spectrophotometrically at 550 nm, at room temperature. Cytochrome-c oxidase specific activity is calculated using an extinction coefficient of 19.6×10^6 cm^2 mol^{-1}.

The inhibitory effect of ethidium bromide on mtDNA replication can be followed as a function of time by measuring the mtDNA content of avian cells. As shown in Table II, the number of mtDNA copies per cell is progressively diluted during succeeding cell generations by a factor of $1/2^n$, where n is the number of cell doublings. In CEF, where a mean value of 600 mtDNA copies/cell has been calculated,[4] populations exposed to ethidium bromide for 10 generations should contain about one copy of mtDNA/cell, and one copy/1×10^6 cells after 30 cell doublings.

Mitochondrial DNA in total cellular DNA extracted from avian cells[4] can be quantitated as a function of cell generations using either DNA–DNA reassociation kinetics[33] or Southern blot hybridization analyses.[34] We have used both methods and have reported for each of them a detailed description of the experimental procedure followed.[4,6]

Concluding Remarks

Under the experimental conditions described here, all cell colonies picked at random following long-term exposure of continuous avian cell

[33] P. A. Sharp, U. Pettersson, and J. Sambrook, *J. Mol. Biol.* **86,** 709 (1974).
[34] E. M. Southern, *J. Mol. Biol.* **98,** 503 (1975).

lines to ethidium bromide have been found to be devoid of characteristic mtDNA molecules. Two of them, cell lines DUS3 and HCF7, have been maintained in culture and frozen at intervals over the last 8–10 years. More recently, human mtDNA-less cell lines have been developed following long-term exposure to ethidium bromide.[35] In addition to uridine, pyruvate is also required for cell growth. However, the need of pyruvate may be relatively nonspecific, as suggested.[36–39] Ham's F12 medium does contain pyruvate at 110 μg/ml, a concentration found in most other chemically defined growth media and in mammalian sera. We have observed that avian mtDNA-less cells also require pyruvate when cultivated in media devoid of this glycolytic intermediate.

Acknowledgments

This research was supported by grants from the Medical Research Council of Canada and the Cancer Research Society, Inc., Montréal. I thank former students, L. Leblond-Larouche, M. Grégoire, and P. Desjardins, for help in this work. I am grateful to R. Cedergren for suggestions concerning the manuscript.

[35] M. P. King and G. Attardi, *Cell (Cambridge, Mass.)* **52,** 811 (1988).
[36] R. Morais, K. Zinkewich-Péotti, M. Parent, H. Wang, F. Babai, and M. Zollinger, *Cancer Res.* **54,** 3889 (1994).
[37] C. D. Whitfield, *in* "Molecular Cell Genetics" (M. M. Gottesman, ed.), p. 545. Wiley, New York, 1985.
[38] C. Van den Bogert, J. N. Spelbrink, and H. L. Dekker, *J. Cell. Physiol.* **152,** 632 (1992).
[39] R. D. Martinus, A. W. Linnane, and P. Nagley, *Biochem. Mol. Biol. Int.* **31,** 997 (1993).

[27] Isolation of Human Cell Lines Lacking Mitochondrial DNA

By Michael P. King and Giuseppe Attardi

Introduction

Very little is known about the regulation of mitochondrial gene expression and the contribution of the nuclear genome to this process in mammalian cells. Furthermore, almost nothing is known about the reverse process: the influence of the mitochondrial genome and mitochondrial function on nuclear gene expression. Elucidation of the coordinate regulation of these two genomes is essential for an understanding of mitochondrial biogenesis. Additionally, a broad spectrum of human diseases associated with defects in mitochondrial function have been identified. In many cases, these deficiencies have been shown to be caused by mutations encoded by the mitochondrial DNA (mtDNA).[1] An understanding of the molecular genetic mechanisms by which these mutations act will provide insights into the etiology, pathogenesis, and ultimately the treatment of these diseases, and will enhance our knowledge of mitochondrial biogenesis.

To investigate nucleomitochondrial interactions and the molecular pathogenetic mechanisms of mtDNA mutations, it would be useful to manipulate the mtDNA complement of a cell, move mitochondria from one cellular environment to another, or introduce new genes into mitochondria. As an initial step towards these goals, we have isolated human cell lines that completely lack mtDNA (ρ^0 cell lines).[2] We describe here the theory and the methods utilized to isolate such cell lines.

Rationale

The rationale for the isolation of human ρ^0 cell lines is based on the use of an inhibitor of mtDNA replication, such as the DNA intercalating dye ethidium bromide (3,8-diamino-5-ethyl-6-phenylphenanthridinium bromide), to deplete the cells of their mtDNA. Ethidium bromide was initially shown to induce in *Saccharomyces cerevisiae* the formation of the respiratory-deficient *petite* phenotype,[3] associated with deletions of the mtDNA,[4] and was subsequently shown to cause the complete loss of mtDNA in treated yeast cells.[5,6] Ethidium bromide has also been shown to have effects on mammalian mtDNA replication. Relatively low concentrations of ethidium bromide (0.1 to 2 μg/ml) result in either partial or complete inhibition of mtDNA replication, but have no effect on the replication of nuclear DNA.[7-10] Wiseman and Attardi[11] demonstrated that even very low levels of ethidium bromide (20 ng/ml) result in nearly complete inhibition of mtDNA synthesis and reductions in the steady-state levels of

[1] D. C. Wallace, *Proc. Natl. Acad. Sci. U.S.A.* **91,** 8739 (1994).

[2] M. P. King and G. Attardi, *Science* **246,** 500 (1989).

[3] P. P. Slonimski, G. Perrodin, and J. H. Croft, *Biochem. Biophys. Res. Commun.* **30,** 232 (1968).

[4] P. Borst and L. A. Grivell, *Cell (Cambridge, Mass.)* **15,** 705 (1978).

[5] E. S. Goldring, L. I. Grossman, D. Krupnick, D. R. Cryer, and J. Marmur, *J. Mol. Biol.* **52,** 323 (1970).

[6] P. Nagley and A. W. Linnane, *Biochem. Biophys. Res. Commun.* **39,** 989 (1970).

[7] M. M. K. Nass, *Proc. Natl. Acad. Sci. U.S.A.* **67,** 1926 (1970).

[8] M. M. K. Nass, *Exp. Cell Res.* **72,** 211 (1972).

[9] R. D. Leibowitz, *J. Cell Biol.* **51,** 116 (1971).

[10] C. A. Smith, J. M. Jordan, and J. Vinograd, *J. Mol. Biol.* **59,** 255 (1971).

[11] A. Wiseman and G. Attardi, *Mol. Gen. Genet.* **167,** 51 (1978).

mtDNA in human cell lines. It was observed that the mtDNA content becomes progressively diluted as the cells continue to divide. Thus, after n cell doublings, the mtDNA content was reduced by the factor $1/2^n$.

In yeast, mitochondrial respiratory chain function is dispensable if the cells are provided with a fermentable energy source, which allows the cells to use glycolysis to generate sufficient ATP for cellular functions. In contrast, mammalian cells are unable to sustain continuous growth in the presence of ethidium bromide or other inhibitors of mitochondrial gene expression, even when provided with high levels of glucose.[8,10–14] An explanation for this difference in behavior was provided in the late 1970s by Morais and colleagues, who demonstrated that avian cell lines can become adapted to grow indefinitely in the presence of ethidium bromide or chloramphenicol if the medium is supplemented with a source of pyrimidines.[15–17] The requirement for pyrimidines is explained by the fact that dihydroorotate dehydrogenase, an enzyme in the pyrimidine biosynthetic pathway, is located in the mitochondrial inner membrane and requires mitochondrial electron transport for normal function.[18] The pyrimidine auxotrophy was later observed in human ρ^0 cell lines, as shown in Fig. 1A for the cell line 143B206. The growth rate of the wild-type parental cell line 143B.TK$^-$ is not affected by the absence of uridine. In contrast, the 143B206 cell line undergoes less than one population doubling in the absence of uridine, whereas its growth rate approaches that of the parental cell line in the presence of 50 μg/ml uridine.

In addition to their growth dependence on pyrimidines, the human ρ^0 cell lines were shown to have a growth requirement for pyruvate[2] (Fig. 1B). As in the absence of pyrimidines, the ρ^0 cells undergo less than one population doubling in the absence of pyruvate. In the presence of pyruvate at 100 μg/ml, the growth rate of this cell line is similar to that of the parental line, whose growth is not affected by the absence of pyruvate. The explanation for the growth dependence on pyruvate is not obvious since these cells should be able to generate a large amount of pyruvate from the glycolytic breakdown of glucose. It was proposed that the absence of a

[12] F. C. Firkin and A. W. Linnane, *Biochem. Biophys. Res. Commun.* **32,** 398 (1968).
[13] G. Soslau and M. M. K. Nass, *J. Cell Biol.* **51,** 514 (1971).
[14] M. E. King, G. C. Godman, and D. W. King, *J. Cell Biol.* **53,** 127 (1972).
[15] L. Leblond-Larouche, A. Larouche, D. Guertin, and R. Morais, *Biochem. Biophys. Res. Commun.* **74,** 977 (1977).
[16] R. Morais and L. Giguère, *J. Cell. Physiol.* **101,** 77 (1979).
[17] R. Morais, M. Grégoire, L. Jeannotte, and D. Gravel, *Biochem. Biophys. Res. Commun.* **94,** 7 (1980).
[18] M. Grégoire, R. Morais, M. A. Quilliam, and D. Gravel, *Eur. J. Biochem.* **142,** 49 (1984).

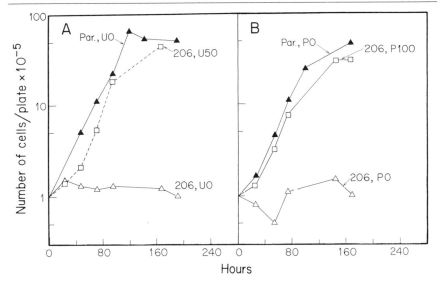

FIG. 1. Growth of 143B (Par) ρ^+ cells or 143B206 (206) ρ^0 cells in the presence or absence of uridine (U) or pyruvate (P). Multiple series of 100-mm cell culture dishes were seeded with 10^5 cells in 10 ml DMEM supplemented with 5% dialyzed FBS, 100 $\mu g/ml$ bromodeoxyuridine/ml, and 0 or 50 $\mu g/ml$ uridine (A), or in 10 ml pyruvate-deficient DMEM supplemented with 5% dialyzed FBS, 100 $\mu g/ml$ bromodeoxyuridine, 50 $\mu g/ml$ uridine, and 0 or 100 $\mu g/ml$ pyruvate (B). At various time intervals, cells from individual plates were trypsinized and counted. Adapted with permission from M. P. King and G. Attardi, "Human Cells Lacking mtDNA: Repopulation with Exogenous Mitochondria by Complementation," *Science*, Vol. 246, pp. 500–503. Copyright 1989 American Association for the Advancement of Science.

functional respiratory chain in these cells prevents the normal oxidation (via the cytosol–mitochondrial shuttle systems) of the cytoplasmically produced NADH.[2] The reduction of pyruvate to lactate through the activity of the NAD-linked lactate dehydrogenase could provide a means for oxidizing the excess cytoplasmic NADH. This reduces the amount of pyruvate that is available for entering into the tricarboxylic acid (TCA) cycle. A requirement for intermediates of the TCA cycle in cells with a respiratory-deficient phenotype arising from mutation or treatment with metabolic inhibitors has been reported previously.[19–22]

[19] I. E. Scheffler, *J. Cell. Physiol.* **83**, 219 (1974).
[20] L. De Francesco, I. E. Scheffler, and M. J. Bissell, *J. Biol. Chem.* **251**, 4588 (1976).
[21] M. Donnelly and I. E. Scheffler, *J. Cell. Physiol.* **89**, 39 (1976).
[22] N. Howell and R. Sager, *Somatic Cell Genet.* **5**, 833 (1979).

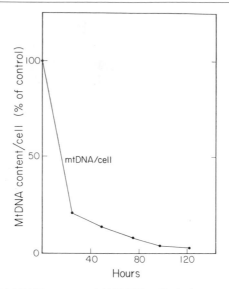

F<small>IG</small>. 2. Mitochondrial DNA content of 143B.TK⁻ cells during exposure to ethidium bromide. A culture of 143B cells was grown in DMEM supplemented with 50 ng/ml ethidium bromide and 50 μg/ml uridine. At the indicated times, total cellular DNA was isolated from samples of the culture, and mtDNA content was quantified as described.[23] Modified from King and Attardi,[23] copyright held by Cell Press.

Theory

Our general strategy for the isolation of human cell lines lacking mtDNA is to grow cells in the presence of a low concentration of ethidium bromide in medium containing a high glucose concentration and supplemented with pyrimidines and pyruvate. During cell growth, because mtDNA replication is inhibited, the initial number of mtDNA molecules present in these cells is decreased by a factor of two each time the cells divide. The decrease in mtDNA content that occurs during the initial ethidium bromide treatment is illustrated in Fig. 2.[23] Mass cultures of the cell line 143B.TK⁻ were exposed to ethidium bromide at a concentration of 50 ng/ml for different lengths of time. As a result of this treatment, the level of mtDNA decreased progressively as compared to the level found in untreated cells, reaching 3% of the initial value after 5 days of exposure of the cells to the drug. During the course of this experiment, the cells maintained a normal population doubling time of 17 hr. Thus, one would expect a level of mtDNA of $1/2^7$, or less than 1%, of the normal level of mtDNA, quite similar to the level actually found in these cells.

[23] M. P. King and G. Attardi, *Cell (Cambridge, Mass.)* **52,** 811 (1988).

The cells are maintained in ethidium bromide until the vast majority of the cells would be expected to have no mtDNA remaining. For cells possessing on average 10,000 mtDNA molecules, each cell in the population would be expected to possess less than one mtDNA molecule after 14 cell generations growth in the presence of ethidium bromide; in other words, after 14 cell generations each cell would be expected to have $10,000/2^{14}$, or 0.6, mtDNA molecules remaining. Of course, one continues cell growth under these conditions for additional cell generations in order to eliminate any residual mtDNA that may be present in these cells. At this point, the cells can be removed from the presence of ethidium bromide. However, because the cells lack a functional respiratory chain, they must be maintained in the presence of glucose, pyrimidines, and pyruvate. At this point, the cells should be subcloned to minimize any genetic or phenotypic variation that is present in the cell population.

Reagents

To create human cell lines lacking mtDNA, the following stock solutions are required. A 10 mg/ml solution of ethidium bromide (Sigma, St. Louis, MO) is made up in distilled water and should be filter sterilized. After filter sterilization, this solution should be serially diluted to a working concentration of 5 μg/ml with sterile distilled water. If solutions are filter sterilized at concentrations lower than 10 mg/ml, a significant fraction of the ethidium bromide in solution binds to the filter, reducing the final concentration of ethidium bromide. The stock and diluted solution should be stored at 4°, protected from light. Fresh solutions should be made every six months. Uridine (Sigma) should be made as a 10 mg/ml stock in distilled water and should be filter sterilized. If required, and 100 mM solution of pyruvate should be prepared in distilled water and sterilized by filtration. Both the uridine and pyruvate solutions should be stored at $-20°$, although, for routine usage, they are stable for at least several weeks at 4°.

For some experiments, dialyzed fetal bovine serum (FBS) is required. This is prepared by dialysis of the FBS in dialysis tubing (3500 molecular weight cutoff) in 0.137 M NaCl, 5 mM KCl, 25 mM Tris-HCl, pH 7.5 at 4°. Typically, a ratio of 10 : 1, buffer to serum, is utilized, and the buffer is changed once per hour for 10–12 hr.

Isolation of ρ^0 Cell Lines

We were first successful in isolating human ρ^0 cells from the 143B.TK$^-$ cell line.[2] The 143B.TK$^-$ cell line is available from the American Type Culture Collection (Rockville, MD, ATCC CRL 8303). It is a thymidine kinase-deficient derivative of an osteosarcoma cell line and, thus, is resistant to the growth inhibitory effects of bromodeoxyuridine. It is cultured in Dulbecco's modified Eagle's medium (DMEM) containing 4500 mg/liter glucose and 1 mM pyruvate and supplemented with 5% FBS (v/v) and 100 μg bromodeoxyuridine/ml. To isolate ρ^0 variants of the 143B.TK$^-$ cell line, cells were grown for an extended period of time in DMEM supplemented with 5% FBS, 100 μg/ml bromodeoxyuridine, 50 ng/ml ethidium bromide, and 50 μg/ml uridine. The 143B cell line has approximately 9100 copies of mtDNA/cell.[2] Therefore, this cell line should be grown for a minimum of 14 cell generations in the presence of the mtDNA replication inhibitor ethidium bromide. In practice, it is desirable to grow this cell line for 3–4 weeks in the presence of ethidium bromide in order to isolate cells that completely lack mtDNA. After ethidium bromide treatment, the cells are maintained in DMEM supplemented with 5% FBS, 100 μg/ml bromodeoxyuridine, and 50 μg/ml uridine. At this stage the cells should be plated at low density, and individual clones should be isolated to minimize any genetic variation that may exist in the cell population.

Although we have not observed any mutagenic effects of ethidium bromide in any isolation of ρ^0 cells that we have conducted, it is reported to be a mutagen.[24,25] Therefore, it may be advisable to repopulate the isolated, clonal ρ^0 derivatives with mtDNA from a known source[26] in order to verify that no mutations have occurred in genes involved in mitochondrial biogenesis and function. The cells, repopulated with exogenous mtDNA, should be examined for various parameters of mitochondrial function such as the rate of oxygen consumption of intact cells,[2] activity of respiratory complexes by polarographic measurements,[27] and mitochondrial protein synthesis.[28] If a mutation is present in any of the genes responsible for mitochondrial biogenesis and respiratory chain function in the chosen derivative, the levels of mitochondrial function may fall outside normal levels. The presence of such mutations would not be advantageous for most uses of the ρ^0 cells.

Verification of ρ^0 State

A number of methods are available to verify that the ethidium bromide-treated cells lack mtDNA. DNA transfer hybridization analysis may be

[24] J. T. MacGregor and I. J. Johnson, *Mutat. Res.* **48**, 103 (1977).
[25] G. S. Probst, R. E. McMahon, L. E. Hill, C. Z. Thompson, J. K. Epp, and S. B. Neal, *Environ. Mutagen.* **3**, 11 (1981).
[26] M. P. King and G. Attardi, this volume [28].
[27] G. Hofhaus, R. M. Shakeley, and G. Attardi, this volume [41].
[28] A. Chomyn, this volume [18].

utilized to examine total DNA isolated from presumptive ρ^0 cells for the presence of intact mtDNA. However, it should be noted that human cells have mtDNA-related sequences present in their nuclear genomes.[29] Therefore, any result derived from this method should be interpreted appropriately. Similarly, polymerase chain reaction (PCR)-based methods may be utilized to detect residual mtDNA sequences in the ρ^0 cells. As with hybridization-based analyses, results should be interpreted with caution.

A more reliable method to determine if any cells in the population possess mtDNA is to utilize the unique growth requirements of the ρ^0 cells. In particular, selecting for growth in the absence of a pyrimidine source has been found to be a sensitive indicator of the respiratory status, and thus the mtDNA content of cells. Cells lacking mtDNA and thus deficient in respiratory chain function are unable to grow in the absence of pyrimidines. With this method, it is possible to select for even a single cell possessing mtDNA which may be present in the cell population. It had previously been shown that microinjection of a single mitochondrion into ρ^0 cells leads to a complete repopulation of the recipient cell with the exogenous mtDNA.[2] The method utilized for selection of these mitochondrial transformants was growth in the absence of uridine. Thus, this method should have the sensitivity to detect even a few mtDNA molecules that may be present in the initial cell population.

In an experiment of this kind carried out with the 143B206 cell line, 6 $\times 10^8$ cells were grown in DMEM supplemented with 5% dialyzed FBS and 100 μg/ml bromodeoxyuridine, and the medium was changed at regular intervals (usually 3–7 days). Dialyzed FBS was utilized because even the low levels of pyrimidines present in sera permit limited growth of ρ^0 cells. As illustrated in Fig. 2, the ρ^0 cells undergo less than one population doubling in the complete absence of pyrimidines. After 5–7 days, the cells began to die and detach from the growth surface. After 1 month, almost all cells had died and detached from the growth surface. This procedure has been used with the 143B206 cell line on numerous occasions, with the same result. Furthermore, in a large number of experiments carried out in our laboratories and elsewhere, involving the repopulation of the 143B206 ρ^0 cell line with exogenous mitochondria[26] and utilizing greater than 3 $\times$ 10^8 recipient cells, no single clone able to grow in the absence of uridine and containing the mtDNA of the 143B.TK$^-$ parent of the ρ^0 cells has been identified. However, it should be noted that, in experiments involving long-term cultures of the 143B206 cells in medium not supplemented with uridine and using dialyzed FBS, the occasional appearance of extremely slowly growing clones has been observed. Analysis by PCR has failed to detect

---

[29] G. Hu and W. G. Thilly, *Gene* **147**, 197 (1994), and references cited therein.

any human mtDNA in these clones, pointing to selection of rare variants exhibiting a minimal capacity to produce UTP from endogenous sources or scavenging uridine from the culture medium.

Isolation of ρ^0 Derivatives of Other Human Cell Lines

Most work so far has been carried out with the 143B.TK$^-$ cell line; however, ρ^0 derivatives of other human cell lines are being made successfully. These cell lines include 143B12-3A2, an adenine phosphoribosyltransferase-deficient derivative of the 143B.TK$^-$ cell line[30]; GM701, an SV40 (simian virus 40)-transformed derivative of a fibroblast cell line[31]; HeLa[32]; A549, a human lung carcinoma cell line[33]; and Namalwa, a nonadherent lymphoblastoid cell line.[34] When working with cell lines that utilize a medium other than DMEM, it should be remembered that the composition of the growth medium should be examined to ensure that it contains sufficient glucose (4500 mg/ml) and pyruvate (110 μg/ml) to support optimal growth of the ρ^0 cells.

Despite the numerous ρ^0 cell lines that have been isolated, it has not been possible to isolate ρ^0 derivatives of all human cell lines. In particular, we have failed to isolate ρ^0 derivatives of HT1080, a fibrosarcoma cell line; HeLa and its derivatives, including HeLa F315, HeLa BU25, and HeLa S3; VA$_2$B and several of its derivatives; Hep 3B, a hepatoma cell line; and various fibroblast cell lines. For the majority of cell lines in which we failed to isolate ρ^0 derivatives, the cell lines initially developed reduced mtDNA levels but subsequently regained normal levels of mtDNA. Typically, no alterations in the rates of growth were observed in these cells. Thus, these cell lines are able to quickly develop resistance to the inhibitory effects of ethidium bromide on mtDNA replication. It should be noted that, in most cases, our methods of analysis would not have detected a subpopulation of cells that had lost their mtDNA in a population with normal mtDNA levels. For two groups of cell lines, HeLa and its various derivatives and VA$_2$B and several of its derivatives, the cells would cease to grow after

---

[30] A. Chomyn, G. Meola, N. Bresolin, S. T. Lai, G. Scarlato, and G. Attardi, *Mol. Cell. Biol.* **11**, 2236 (1991).

[31] G. Attardi, M. P. King, A. Chomyn, and P. Loguercio Polosa, *in* "Progress in Neuropathology" (T. Sato and S. DiMauro, eds.), p. 75. Raven, New York, 1991.

[32] J.-I. Hayashi, S. Ohta, A. Kikuchi, M. Takemitsu, Y.-i. Goto, and I. Nonaka, *Proc. Natl. Acad. Sci. U.S.A.* **88**, 10614 (1991).

[33] A. G. Bodnar, J. M. Cooper, I. J. Holt, J. V. Leonard, and A. H. V. Schapira, *Am. J. Hum. Genet.* **53**, 663 (1993).

[34] R. D. Martinus, A. W. Linnane, and P. Nagley, *Biochem. Mol. Biol. Int.* **31**, 997 (1993).

several days of exposure to ethidium bromide and then would die over the course of the following 2 weeks.

Because of the relative rarity of cell lines that have the ability to lose their mtDNA and maintain cell growth, we have developed a rapid screening method to identify such cell lines. Parallel cultures of the cell line to be tested are grown in DMEM supplemented with dialyzed FBS, 50 ng/ml ethidium bromide, and with or without 50 μg/ml uridine. If the cell line stops growth and dies in the absence of uridine, while the culture in the presence of uridine continues growth, it is a strong candidate to be able to lose its mtDNA and become ρ^0. If, however, both cultures continue cell growth, it is likely that the cells have developed resistance to ethidium bromide and have maintained or regained normal levels of mtDNA. Of course, if both cultures stop growth and die, it is not possible to isolate ρ^0 derivatives. Again, it should be noted that this method will not detect a subpopulation of cells that have lost their mtDNA among a population with normal mtDNA levels. Any cell line which, according to the above test, is a good candidate for becoming mtDNA-less should be analyzed for some parameters of mitochondrial function, like rate of oxygen consumption, to verify the absence of any nuclear mutation affecting the oxidative phosphorylation apparatus or the protein synthesizing machinery. This test is particularly important if the cell line has been previously mutagenized for any purpose, in particular to isolate variants carrying a recessive drug-resistant nuclear mutation to be used for selection of transformants in ρ^0 cell repopulation experiments.[26]

Acknowledgments

This work was supported by Grants GM11726 (G. A.) and EY10085 (M. P. K.) from the National Institutes of Health and from the Muscular Dystrophy Association (M. P. K.). M. P. K. is an Investigator of the American Heart Association, New York City Affiliate.

[28] Mitochondria-Mediated Transformation of Human ρ^0 Cells

By Michael P. King and Giuseppe Attardi

Introduction

The absence of traditional genetic tools has considerably hindered the study of mammalian mitochondrial biogenesis. Mammalian mitochondrial

DNA (mtDNA) is maternally inherited, and there is no evidence for the occurrence of recombination of this DNA. The development of a variety of methods for DNA-mediated transformation of nuclei has greatly facilitated investigations on the mechanisms and regulation of nuclear gene expression in eukaryotic cells. The absence of analogous methods in mitochondria has been the major stumbling block to the *in vivo* dissection of the mechanisms of replication and expression of the mitochondrial genome and to the development of mitochondrial genetics of mammalian cells. Because of this problem, we have developed new approaches for mitochondria-mediated transformation of human cells, providing a foundation for mammalian mitochondrial genetics. In this chapter, we describe methods for the transfer of mitochondria from one cellular environment to another, with particular emphasis on methods for the transfer of mitochondria into human cells lacking mtDNA (ρ^0 cells).

The transfer of mitochondria from one cellular environment to another for genetic studies was pioneered by Tatum and colleagues[1] in *Neurospora crassa*. Similar techniques were subsequently applied to genetic studies of *Paramecium*.[2,3] The relatively large sizes of these organisms facilitated the introduction into them of subcellular fractions containing mitochondria by microinjection. These investigations utilized mtDNA-encoded mutations specifying morphological or growth characteristics, or associated with specific drug resistance markers, for the purpose of isolating mitochondrial transformants among the injected cells.

The development of efficient, large-scale methods to obtain enucleated mammalian cells,[4–6] termed cytoplasts, provided a basis for the transfer of mitochondria into cells by the fusion of cytoplasts with the cells of interest to form cytoplasmic hybrids, termed cybrids.[7–10] In the first use of this procedure for mitochondrial genetic studies, Eisenstadt and colleagues demonstrated cytoplasmic transfer of chloramphenicol resistance to a recipient sensitive cell line.[10] The microinjection procedures initially developed for *Neurospora* and *Paramecium* have also been adapted for mammalian

[1] E. G. Diacumakos, L. Garnjobst, and E. L. Tatum, *J. Cell Biol.* **26**, 427 (1965).
[2] G. H. Beale, J. K. C. Knowles, and A. Tait, *Nature (London)* **235**, 396 (1972).
[3] J. K. C. Knowles and A. Tait, *Mol. Gen. Genet.* **117**, 53 (1972).
[4] D. M. Prescott, D. Myerson, and J. Wallace, *Exp. Cell Res.* **71**, 480 (1972).
[5] G. Poste, *Exp. Cell Res.* **73**, 273 (1972).
[6] W. E. Wright and L. Hayflick, *Exp. Cell Res.* **74**, 187 (1972).
[7] G. Poste and P. Reeve, *Nature (London) New Biol.* **229**, 123 (1971).
[8] G. Poste and P. Reeve, *Exp. Cell Res.* **73**, 287 (1972).
[9] G. Veomett, D. M. Prescott, J. Shay, and K. R. Porter, *Proc. Natl. Acad. Sci. U.S.A.* **71**, 1999 (1974).
[10] C. L. Bunn, D. C. Wallace, and J. M. Eisenstadt, *Proc. Natl. Acad. Sci. U.S.A.* **71**, 1681 (1974).

cells.[11] In that study, mitochondria containing a mtDNA-encoded chloramphenicol resistance marker were introduced into sensitive cells by microinjection.

The methods described above required donor mitochondria containing a selectable marker, and replacement of the endogenous mtDNA by the donor mtDNA was usually incomplete. The requirement for a selectable marker in donor mitochondria limited the utility of these systems for mitochondrial transfer since there are only a small number of selectable markers associated with resistance to specific drugs available. Cells carrying these markers are also very difficult to isolate, sometimes requiring months of culture to obtain drug-resistant cell lines. Residual levels of recipient cell mtDNA in the cybrids also limits the usefulness of these methods. For these reasons, and to provide a more general approach to mitochondrial genetic studies, we have developed the use of human cell lines lacking mtDNA (ρ^0 cell lines) as recipient cells in mitochondrial transfer studies.[12] We define "transmitochondrial cell lines" as the cell constructs in which the total mtDNA complement of a cell has been substituted with exogenous mtDNA.

Rationale

The procedures previously described for the introduction of mitochondria into cells possessing mtDNA by cytoplast fusion or by microinjection of single mitochondria can also be utilized to introduce mitochondria, and thus reintroduce mtDNA, into ρ^0 cells. A procedure utilizing blood platelets, naturally occurring anucleate cells, as mitochondrial donors for repopulating ρ^0 cells[13] is described elsewhere in this volume.[14] The growth dependence of ρ^0 cell lines on pyruvate and pyrimidines[12,15] provides two valuable selectable markers to isolate cells repopulated with exogenous mitochondria containing at least partially functional mtDNA. The mitochondrial transformants are selected by growth either in the absence of a pyrimidine source or in the absence of pyruvate.

As for all cytoplast fusions, the selection for mitochondrial transformants should be combined with a suitable nuclear marker to exclude nucleated mitochondrial donor cells or cell hybrids from the selected cell

[11] M. P. King and G. Attardi, *Cell* (*Cambridge, Mass.*) **52,** 811 (1988).
[12] M. P. King and G. Attardi, *Science* **246,** 500 (1989).
[13] A. Chomyn, S. T. Lai, R. Shakeley, N. Bresolin, G. Scarlato, and G. Attardi, *Am. J. Hum. Genet.* **54,** 966 (1994).
[14] A. Chomyn, this volume [29].
[15] M. P. King and G. Attardi, this volume [27].

population. In fact, most enucleation procedures leave a residual number of nucleated cells. These nucleated cells, or cell hybrids formed from the fusion of these nucleated cells with the recipient ρ^0 cells, would be expected to be pyrimidine and pyruvate independent and would be selected along with the cybrid cells. To exclude these cells and cell hybrids, recessive nuclear drug resistance markers in the recipient cells are utilized in the selection. A dominant drug resistance marker in the recipient cells would not exclude cell hybrids from the selected cell population. The most commonly used recessive mutations include resistance to bromodeoxyuridine, due in general to the absence of a functional thymidine kinase, or resistance to guanine analogs (6-thioguanine or 8-azaguanine), due to the absence of a functional guanine–hypoxanthine phosphoribosyltransferase. Fusion of a thymidine kinase- or guanine–hypoxanthine phosphoribosyltransferase-deficient cell with a cell that has thymidine kinase or hypoxanthine phosphoribosyltransferase activity would result in a hybrid that would be sensitive to treatment with the appropriate selective agent.

It should be emphasized that the use of selective mitochondrial and nuclear markers is not an absolute requirement for mitochondrial transfer studies, but such approaches do simplify the isolation of true mitochondrial transformants. As an alternative to selective markers, presumptive transmitochondrial cell lines can be screened for polymorphic mtDNA and nuclear DNA markers to verify their identity. Human mtDNA is very polymorphic,[16] and restriction fragment length differences that often exist between the donor and recipient mtDNA can be utilized for the identification of the mtDNA genotype. Similarly, the nuclei of donor cells can be distinguished from those of recipient cells by screening presumptive transformants for variable number tandem repeat markers[17] or dinucleotide repeat markers.[18] We have successfully utilized all three marker types to verify that presumptive transmitochondrial cell lines have the expected mtDNA and nuclear DNA markers.[11,12,19]

Formation of Transmitochondrial Cell Lines

We detail the methods that we use for transfer of mitochondria into ρ^0 cells by fusion of cytoplasts with ρ^0 cells. In particular, we describe proce-

[16] R. L. Cann, M. Stoneking, and A. C. Wilson, *Nature* (*London*) **325,** 31 (1987).
[17] Y. Nakamura, M. Leppert, P. O'Connell, R. Wolff, T. Holm, M. Culver, C. Martin, E. Fujimoto, M. Hoff, E. Kumlin, and R. White, *Science* **235,** 1616 (1987).
[18] J. L. Weber and P. E. May, *Am. J. Hum. Genet.* **44,** 388 (1989).
[19] J. P. Masucci, M. Davidson, Y. Koga, E. A. Schon, and M. P. King, *Mol. Cell. Biol.* **15,** 2872 (1995).

dures that we use for the enucleation of the mitochondrial donor cells, then describe the methods for fusion of the resulting cytoplasts with the ρ^0 cells. A generalized strategy for the isolation of transmitochondrial cell lines by cytoplast fusion is illustrated in Fig. 1.

Reagents. Cytochalasin B (Sigma, St. Louis, MO) is prepared as a 1 mg/ml solution in 95 or 100% ethanol. Stored at $-20°$, this solution is stable for more than 1 year. Ficoll 400 (molecular weight 400,000; Pharmacia Biotech, Piscataway, NJ) is prepared as a 50% (w/w) solution in distilled water. The Ficoll is dissolved by stirring overnight at $37°$ and then sterilized by autoclaving. Some slight discoloration may occur during this process, but this does not affect the performance of the Ficoll in the described procedures.

We have had the best results in cell fusion with polyethylene glycol 1500 (PEG) from BDH Limited (Poole, England). Precisely, 10 g PEG is weighed in a 25-ml Erlenmeyer flask, and this material is sterilized by autoclaving. This material may be stored at room temperature until required. A 40 to 50% (w/v) solution of PEG (see below) is prepared by first liquefying the PEG by briefly heating the Erlenmeyer flask in a Bunsen burner flame. The liquefied PEG is allowed to cool, and, before it solidifies, 8 ml Dulbecco's modified Eagle's medium (DMEM, sterile) and 2 ml dimethyl sulfoxide (DMSO, sterilized by autoclaving) are added. The pH of the final solution is adjusted to approximately pH 7.4 with 10 N NaOH using the phenol red pH indicator dye (present in the DMEM) to visually estimate the pH of the solution. The exact pH of the solution has been found not to be critical for cell fusions, although if the pH is excessively alkaline or acidic changes in the fusion efficiencies have been observed. This PEG solution may be stored at $4°$ up to 1 month. As the concentration of PEG is critical for successful cell fusions, any evaporation that occurs during storage will adversely affect the results. Therefore, the PEG solution is kept in a tightly sealed container if it is not used immediately. If stored at $4°$, the PEG should be redissolved at $37°$ prior to use. We have observed variation from batch to batch in the optimum concentration of PEG for cell fusion. Therefore, it is advisable to test the individual batches with a standard ρ^0 cell line and mitochondrial donor (e.g., the 143B206 ρ^0 cells and a known human fibroblast strain).

Dialyzed serum [either bovine serum or fetal bovine serum (FBS)] is prepared by dialysis of the serum in dialysis tubing (3500 molecular weight cutoff) against 0.137 M NaCl, 5 mM KCl, 25 mM Trish-HCl, pH 7.5 at $4°$. We routinely perform dialysis with buffer to serum ratios of 10 to 1. The buffer is changed hourly, and at least 10 changes of buffer are performed. Following dialysis, the serum is sterilized by filtration through a 0.2-μm

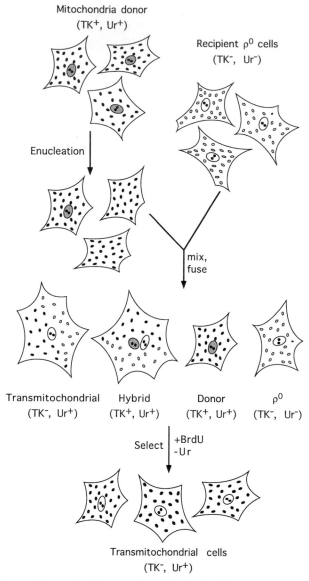

FIG. 1. Diagram illustrating the procedures for the isolation of transmitochondrial cell lines. The mitochondrial donor cells, which are thymidine kinase positive (TK^+) and are able to synthesize uridylic acid (Ur^+), are enucleated and mixed with the recipient 143B206 ρ^0 cells, which are TK^- and Ur^-. The cytoplasts are fused with the ρ^0 cells in the presence of polyethylene glycol (PEG). The transmitochondrial cells are selected from the resulting cell population by their ability to grow in the absence of uridine (to select for mitochondrial function) and in the presence of bromodeoxyuridine (BrdU, to select against contaminating cell hybrids and donor cells).

cellulose nitrate filter (Nalge Co., Rochester, NY). The sterile serum is stored at $-20°$ until use.

Enucleation Procedures

Methods for the large-scale enucleation of cells with cytochalasin B have been reviewed previously.[20,21] Readers are referred to those articles for theoretical descriptions of enucleation procedures, full descriptions of a variety of methods to enucleate cells, and solutions to specific problems that may be encountered. We describe in detail two procedures that we routinely use to enucleate cells. One method, the monolayer technique, utilizes small numbers of cells attached to the growth surface of a culture dish and is ideal if limited numbers of donor cells are available. The other procedure, the gradient technique, requires centrifugation of cells through Ficoll gradients and is best suited for enucleation of large numbers ($>10^7$) of cells.

Monolayer Technique. The monolayer technique is ideal for virtually any cell which grows attached to the growth surface. Our variation is based on a method initially described by Follett[22] in which cells, grown in culture dishes, are exposed to cytochalasin B. These cultures are subsequently centrifuged to facilitate enucleation of the cells. It is arguably the simplest, most versatile procedure, requiring the least amount of material. The main disadvantage of this method is that it does not work for cells which grow in suspension or which attach poorly to the growth surface. In addition, the number of cells that can be enucleated with this procedure is relatively small.

Polycarbonate or polypropylene 250-ml wide-mouth centrifuge bottles with screw-top caps are sterilized by autoclaving. The caps should be auto-claved separately from the bottle to prevent damage to the centrifuge bottle. The bottles are prepared for the enucleation procedure by the sterile addition of 30 ml DMEM, 2 ml bovine serum, and 0.32 ml cytochalasin B (1 mg/ml) to each. The caps are placed on the bottles, and the bottles are maintained at $37°$ prior to use.

The cells to be enucleated (from a few hundred to $\sim10^5$ cells) are seeded on a culture dish (35×15 mm; Nunc Inc., Naperville, IL). Typically, the cells are grown for at least 24 hr on the dishes to promote maximal attachment to the growth surface. However, one should prevent the cells from becoming confluent. The culture dish is prepared for centrifugation by wiping the outside of the bottom half of the dish (containing the cells) with 70%

---

[20] G. E. Veomett, *in* "Techniques in Somatic Cell Genetics" (J. W. Shay, ed.), p. 67. Plenum, New York, 1982.

[21] J. W. Shay, this series, Vol. 151, p. 221.

[22] E. A. C. Follett, *Exp. Cell Res.* **84,** 72 (1974).

(v/v) ethanol for the purpose of sterilization. Alternatively, the dish can be kept sterile during cell culturing by maintaining it within a larger, sterile culture dish. The medium is removed from the dish and the dish (without top) is placed upside down within the centrifuge bottle. We have found it convenient to manipulate the dish with 300-mm forceps.

With 32 ml medium in the centrifuge bottle, it is not possible to completely expose the bottom of the culture dish to the medium; a small pocket of air is trapped within the dish. It is not necessary to eliminate this air pocket prior to centrifugation. If larger volumes of medium are used to minimize or eliminate this air pocket, the culture dish is usually found right side up at the conclusion of the centrifugation. Thirty-two milliliters of medium has been found to be the optimal amount to maximize exposure of the cells to the medium without resulting in inversion of the culture dish during centrifugation.

The rotor (GSA, DuPont, Wilmington, DE) and centrifuge are prewarmed to 37° by centrifugation for 30–45 min at 8000 rpm. In some instances, the HS-4 swinging-bucket rotor (DuPont) has been used successfully. The optimal time and speed of centrifugation must be established for each cell type. For myoblasts and fibroblasts, the centrifuge bottle with the culture dish is placed in the prewarmed rotor and centrifuged approximately 20 min (interval between the time when the rotor reaches the desired speed and the time when the centrifuge is turned off). Typically, speeds of 6500 to 7200 rpm are used, with the brake on during deceleration.

After centrifugation, the centrifuge bottle is removed from the rotor, and the culture plate is removed from the bottles with forceps. A small amount of medium is maintained in the plate to keep the cells moist in order to maintain cell viability. The outside of the dish, including the top edge, is wiped with a sterile wiper, then moistened with 95% (v/v) ethanol, to remove any medium and to dry it. A sterile top is placed onto the dish. If the enucleated cells are not going to be used immediately, complete culture medium (medium supplemented with the appropriate concentration of serum) should be added to the dish, and the cells placed in a CO_2 incubator.

Gradient Technique. The Ficoll gradient technique that we utilize is that described by Wigler and Weinstein.[23] A discontinuous Ficoll gradient is prepared in a 14×89 mm polypropylene thin-wall ultracentrifuge tube, which is suitable for centrifugation in the SW 41 swinging-bucket ultracentrifuge rotor (Beckman Instruments Inc., Fullerton, CA). Ficoll solutions are prepared in DMEM with cytochalasin B as described by Veomett[20] (Table I). The centrifuge tube, sterilized by autoclaving, is filled with 2 ml

---

[23] M. H. Wigler and I. B. Weinstein, *Biochem. Biophys. Res. Commun.* **63,** 669 (1975).

TABLE I

FICOLL STOCK SOLUTIONS FOR PREPARATION OF GRADIENTS FOR CELL ENUCLEATION[a]

Final Ficoll concentration (%)	Volume added (ml)			
	Ficoll stock (50%, w/w)	4× DMEM	Distilled water	Cytochalasin B (1 mg/ml)
25	12.5	6.25	6.0	0.25
17	8.5	6.25	10.0	0.25
16	8.0	6.25	10.5	0.25
15	7.5	6.25	11.0	0.25
12.5	6.25	6.25	12.25	0.25
10	5.0	6.25	13.5	0.25

[a] Modified from Ref. 20.

of 25% Ficoll, 2 ml of 17% Ficoll, 0.5 ml of 16% Ficoll, 0.5 ml of 15% Ficoll, and 2 ml of 12.5% Ficoll. It should be noted that the volumes may need to be adjusted depending on the manufacturer of the centrifuge tube. It is recommended that the interfaces of the Ficoll layers be marked on the outside of the centrifuge tube to facilitate their identification after centrifugation. The gradient is prepared at least 4 hr but less than 24 hr prior to use. The gradient is kept at 37° in a CO_2 incubator to maintain the temperature and the pH of the gradient solution prior to use.

Cells are prepared for enucleation by suspending a washed cell pellet as a single cell suspension in the 12.5% Ficoll solution. Typically, 3 to 5 × 10^7 cells are suspended in 3.0 ml, and this suspension is layered on the gradient. Then DMEM is layered over the cells to bring the contents within 5 mm of the top of the centrifuge tube. The rotor and centrifuge are prewarmed by prior centrifugation at 25,000 rpm for at least 1 hr. Centrifugation of the gradients is performed at 28,000 rpm in the SW 41 rotor for 1 hr at 37°.

The distribution of the cellular components in the Ficoll gradient following centrifugation is highly dependent on the particular cell type being enucleated. Therefore, each layer of cellular material in the gradient should be examined by phase-contrast microscopy to determine the location of the cytoplasts. When 143B cells and derivatives are centrifuged through this gradient system, the cytoplasts and fragments of cytoplasts are found throughout the 12.5, 15, and 16% Ficoll solutions. Smaller cytoplast fragments are found near the top of the gradient, with larger cytoplasts being found in the layers with higher Ficoll concentrations. The cytoplasts and fragments of cytoplasts are recovered from the gradient with a transfer pipette, diluted with at least 10 volumes of DMEM, and centrifuged at

1100 g for 5 min to pellet the cytoplasts. Because of the smaller size and density of many of the cytoplasts, centrifugal forces greater than those typically used to pellet intact cells should be utilized.

Evaluation of Enucleation. If it is necessary to determine the fraction of cells that are enucleated, the cells can be directly examined by phase-contrast microscopy. Alternatively, cells can be fixed and stained with Giemsa stain to visualize the nuclei and to determine the efficiency of enucleation. If a low efficiency of enucleation is observed, several factors affecting this efficiency can be investigated. For the monolayer technique, these include the concentration of cytochalasin B, the centrifugal force to which the cells are subjected, the length of time of centrifugation, and the temperature. A frequently encountered problem with the monolayer technique is loss of cells from the growth surface during centrifugation. An extensive discussion of these problems and suggestions to overcome them have been reported by Veomett[20] and will not be repeated here.

Cell Fusion

Cytoplasts are fused with the recipient ρ^0 cells while the cells are either attached to the culture dish or in suspension. There is no distinct advantage or disadvantage to either procedure. For both methods, cytoplasts are mixed with an excess of the recipient cell line, and membrane fusion is promoted by the addition of PEG. We first describe the procedures we use for fusion of cytoplasts created by the monolayer technique with the ρ^0 cells. Subsequently, we describe the procedures for the fusion of cells in suspension. These methods have been adapted with modifications from those previously described.[24]

Recipient ρ^0 Cell Line. We have extensively utilized the 143B206 ρ^0 cell line for many mitochondrial transfer studies.[12,15] It has several characteristics which make it an excellent recipient cell line for such studies. In particular, this cell line lacks a functional thymidine kinase, providing a recessive nuclear marker to eliminate mitochondrial donor cells and cell hybrids from the selected cell population. Furthermore, this cell line grows on a solid substrate and forms cohesive, closely packed colonies, a feature which facilitates the isolation of subclones by the cloning ring method. Both the 143B206 ρ^0 cell line and its transformants display excellent growth properties, with population doubling times averaging less than 20 hr. This permits large quantities of cells to be easily grown for biochemical and molecular genetic studies. In addition, this cell line has been repopulated

[24] T. H. Norwood and C. J. Zeigler, *in* "Techniques in Somatic Cell Genetics" (J. W. Shay, ed.), p. 35. Plenum, New York, 1982.

with mtDNA from a number of different cell sources. Based on examination of many parameters of mitochondrial function in these transformants, no mutations have occurred in nuclear genes involved in mitochondrial biogenesis and function. Thus, this cell line is ideal for genetic studies of the mitochondrial genome.

Monolayer Fusions. The medium is removed from the dish containing the enucleated cells. We have found it unnecessary to wash the cytoplasts to remove any residual cytochalasin B that may be present. Furthermore, it is not necessary to allow for a recovery period following cytochalasin B treatment. Sufficient ρ^0 cells to fully cover the growth surface are added to the culture dish to which the enucleated cells are attached. Usually, we prepare the 143B206 ρ^0 cells during the centrifugation to enucleate the donor cells. Exponentially growing cells are trypsinized, washed, and suspended as a single cell suspension in DMEM supplemented with 5% FBS, 50 μg/ml uridine, the normal growth medium of the 143B206 cell line. The cell concentration is determined, and for this cell line 1.0 to 1.6 $\times$ 10^6 cells are added to the 35-mm diameter culture dish in 2 to 2.5 ml growth medium. The large excess of ρ^0 cells is chosen to ensure contact of each cytoplast with a ρ^0 cell, and thus to maximize the chances for fusion of a cytoplast with a ρ^0 cell. The cells are permitted to attach for at least 1 hr, but usually not more than 3 hr.

Several media required for cybrid fusions should be prepared prior to starting the procedure. In addition to the 40 to 50% PEG solution described above (the optimum concentration for a given PEG lot being chosen in preliminary tests), DMEM (without any additions) and DMEM with 10% (v/v) DMSO are required. These media should be kept at room temperature. We usually utilize a large-bore transfer pipette for the gentle addition of solutions to the dish in order to avoid dislodging the cells from the growth surface. Solutions are removed from the dishes by aspiration with a Pasteur pipette attached to a vacuum source. This facilitates the complete and rapid removal of solution that is required for certain steps of this procedure.

The mixture of cytoplasts and 143B206 ρ^0 cells on the 35-mm culture dish is washed three times with DMEM. Following removal of the last DMEM wash, the dish is placed at an angle with the lid on, to prevent the cells from dehydrating and thus to maintain their viability. After 30 sec, any residual medium that has accumulated at the bottom of the dish is removed. As the PEG concentration is critical for cell fusion, removal of nearly all traces of medium is necessary for efficient fusion to occur. Approximately 2.5 ml of the 40 to 50% PEG solution is added to the culture dish for exactly 60 sec. Immediately after PEG addition, the dish is gently swirled for approximately 20 sec to facilitate complete coverage of the bottom of the dish and to ensure mixing of the PEG solution with any residual medium at the cell surface. Polyethylene glycol is highly toxic to cells, and therefore the time of exposure to PEG is very critical. At the end of 60 sec, the PEG is rapidly removed, and 2.0–2.5 ml DMEM with 10% DMSO is quickly added to the plate. It would take considerable time to remove the PEG solution completely because of its viscosity. Therefore, only that which can be removed quickly is aspirated, and the DMEM with 10% DMSO is added to the plate and swirled as quickly as possible to dilute the remaining PEG. To ensure rapid addition of the DMEM with 10% DMSO, the transfer pipette is prepared in advance with this medium already inside the pipette. The DMEM with 10% DMSO is removed, and the plate is washed two additional times with the same medium. Then, 2.5 ml of the growth medium for the 143B206 cells is added to the culture dish. The cells are cultured overnight prior to trypsinization and replating in medium selective for mitochondrial transformants.

As mentioned above, the 143B206 ρ^0 cell line is thymidine kinase-deficient. Mitochondrial transformants are usually selected by their ability to grow in the absence of uridine and in the presence of bromodeoxyuridine. Therefore, the selective culture medium is DMEM supplemented with 5% dialyzed FBS and 100 μg/ml bromodeoxyuridine. Dialyzed FBS is used to prevent the normal low levels of pyrimidines in the serum from supporting limited cell growth of the ρ^0 cells. Typically, the cells from one 35-mm culture dish are plated onto ten 100-mm culture dishes or in five to 10 96-microwell plates (see below). However, it is always advisable to freeze in freezing medium (see below) a portion of the trypsinized and washed fusion mixture for possible further plating. The medium is changed every 3 days, and colonies are isolated with glass cloning cylinders from the culture dishes or directly from the microwell plates 10 to 14 days after plating in selective medium (see below). When 10^5 cells are used as mitochondria donors, with a good efficiency of enucleation, recovery of enucleated cells, and cell survival after PEG treatment, up to 1000 independent mitochondrial transformants may be expected from a single fusion.

Suspension Fusions. The cytoplasts isolated from the Ficoll gradients are washed once with DMEM as described above. Exponentially growing 143B206 ρ^0 cells are trypsinized, washed in DMEM, and suspended in the same medium. It is difficult to quantitate the cytoplasts because of the large number of small cytoplast fragments present. For this reason, it is difficult to determine an optimal ratio of cytoplasts to ρ^0 cells. Arbitrarily, we usually add to the cytoplast fraction a number of ρ^0 cells equal to the number of cells used to obtain the cytoplasts. Because of incomplete enucleation and variable recovery of cytoplasts, the ρ^0 cells will always be in excess. These ρ^0 cells are added to the cytoplasts, mixed, and centrifuged at 1100 g for 5 min to form a firm pellet. The DMEM is aspirated, and the pellet is centrifuged to collect the residual medium attached to the centrifuge tube. Any residual medium above the cell pellet is aspirated. As mentioned above, the PEG concentration is critical to cell fusion, so all traces of medium must be removed from the cells. The cell pellet is gently suspended in 0.2 ml of the 40 to 50% PEG solution described above. Suspension is done to increase exposure of the cells to the PEG, but it is not vigorous enough to disrupt intercellular contact and subsequent fusion. After precisely 60 sec, 5 ml DMEM is added to the cells, which are gently mixed. The cell suspension is plated in DMEM supplemented with 5% FBS and 50 μg/ml uridine at an appropriate cell density. Twenty-four hours later the medium is changed to DMEM supplemented with 5% dialyzed FBS and 100 μg/ml bromodeoxyuridine. The selective medium is changed after 2 days, and then at 3-day intervals. Colonies are isolated with glass cloning cylinders or in microwell plates after 10 to 20 days in selective medium (see below).

Isolation and Preservation of Colonies

Isolation from Culture Dishes. Colonies grown on 100-mm culture dishes are isolated by trypsinization of cells within glass cloning cylinders. Usually, after 10–14 days in selective medium, the colonies of mitochondrial transformants have reached sufficient size to be visible to the naked eye. When viewed in the light microscope, they usually appear as cohesive, closely packed colonies. In the same time frame, the vast majority of residual ρ^0 cells, contaminating mitochondrial donor cells, and cell hybrids have died and detached from the growth surface. Thus, mitochondrial transformants can be selected with little risk of cross-contamination from other cell types.

To ensure that single, independent colonies are selected, colonies detected with the naked eye are initially identified. These colonies are then viewed with an inverted microscope, and the colonies are circled with a marking pen. Any other colonies in close proximity are also marked so that they are excluded from within the cloning cylinder. We use glass cloning cylinders (6 $\times$ 8 or 8 $\times$ 8 mm; Bellco Glass Inc., Vineland, NJ) that are placed in glass petri dishes prior to use. High vacuum grease (Dow Corning, Corning, NY) is prepared by placing the grease in a glass petri dish and smoothing the surface with a forefinger. These items are sterilized by autoclaving.

Once all colonies to be selected have been identified and circled, all but 0.5 ml of the medium is removed from the culture dish. A ring is picked up with a sterile 140-mm curved hemostat and placed into the vacuum grease so that the bottom edge is in complete contact with the grease. Too little vacuum grease results in leakage of the trypsin solution from the cloning cylinder, and too much will result in cells being covered with the grease. The cylinder is then placed over the chosen colony, making sure to exclude any other colonies from within the cylinder. After all the cloning cylinders have been positioned on the dish, three drops of trypsin solution are added to each cylinder. After several minutes, the trypsin solution is pipetted up and down with a transfer pipette to dislodge the cells, and the suspension is transferred to a 60×15-mm culture dish containing 5 ml selective medium. A small amount of this medium is then used to rinse the inside of the cloning cylinder, and this is added back to the culture dish. The dish is swirled to distribute the cells evenly and is placed in a CO_2 incubator.

Using this method, more than 20 independent colonies can be isolated from a single 100-mm culture dish. Care must be taken to avoid evaporation of the residual culture medium during these manipulations. Evaporation of the medium results in cell death and is the major factor limiting the number of colonies that can be selected from a single plate. Covering the culture plate with the lid in the intervals between placement of the individual cloning cylinders increases the number of colonies that can be selected.

The selected colonies are grown in selective medium in 60-mm culture dishes until they reach confluence. This usually takes 10 to 14 days. At this stage, the cells are trypsinized, pelleted, and suspended in 2.1 ml of medium for cryopreservation. For the 143B206 ρ^0 cell line and derivatives, we use DMEM supplemented with 20% (v/v) bovine serum and 10% (v/v) glycerol or DMSO. Samples are taken for cryopreservation (1.5 ml) and for DNA analyses (0.6 ml). The cells for DNA analysis are placed in microcentrifuge tubes and pelleted, and the cell samples are stored at $-20°$.

Isolation from Microwell Plates. An alternative method for identifying and isolating transformant clones is to distribute the fusion mixture in 96-microwell plates, after appropriately diluting it so that the majority of the wells carrying clones have single clones. From the Poisson distribution, it can be calculated that, if 80% of the wells in a plate have no clones, approximately 17 wells will have single clones, and only approximately 3 will have more than one clone. Similarly, if 70% of the wells in a plate have no clones, approximately 25 wells will have single clones, and only approximately 5 will have more than one clone. The wells with more than one clone can be easily identified by microscope screening and excluded from further handling. For practical purposes, it can be estimated that, under the experimental conditions described above, 0.5 to 2% of the cytoplasts will produce transformants. Therefore, it can be estimated what proportion of the fusion mixture would be plated on different plates so as to have at least two plates with 70 to 80% negative wells. For example, if the 35-mm plate contains approximately 10^5 cytoplasts, expected to produce

500 to 2000 transformants, we usually plate 1% of the fusion mixture, appropriately diluted, in each of four 96-well plates (0.1 to 0.2 ml suspension per well), 0.5% in each of two microwell plates, and 2% in each of two other microwell plates. Plates with different portions of the mixture than those mentioned above will have to be prepared, if experience or uncertainty concerning the yield of cytoplasts and fusion efficiency with different mitochondrial donor and recipient cells so dictates. The nonutilized portion of the fusion mixture is frozen in freezing medium for possible further use. At the appropriate time (usually 2 to 3 weeks after fusion), the selected single clones are trypsinized and transferred to 60×15-mm culture plates as described for the cloning cylinder method.

Injection of Mitochondria into Cells

We have utilized microinjection to introduce mitochondria possessing a mtDNA-encoded drug resistance marker into sensitive cells[11] and to introduce mitochondria possessing mtDNA into ρ^0 cells.[12] This method should not be considered as a general method for mitochondrial transformation because of the relative difficulties of introducing mitochondria into cells by microinjection and because of the relatively low efficiency of this approach. This methodology should be utilized primarily for specific research questions concerning mtDNA heterogeneity, mitochondrial interactions within cells, mitochondrial segregation, or the nature of the regulatory factors controlling mtDNA copy number.

Preparation of Mitochondrial Suspension

Mitochondria for injection are prepared from cell homogenates by differential centrifugation. To avoid blockage of the injection pipette, it is necessary to obtain a suspension of mitochondria nearly free of nuclei, nuclear fragments, and other large cell fragments. The isolation of a suitable suspension requires conditions of centrifugation that would remove large particles from the suspension without causing significant losses of mitochondria. Furthermore, the mitochondria must be prepared in a buffer that maintains their integrity and viability, but which is not toxic to the cell when injected. Although we have been successful with the injection of mitochondria resuspended in 0.25 M sucrose, 10 mM Tris-HCl, pH 7.0 (at 25°), cells injected with this solution have very low survival, significantly decreasing the frequency of mitochondrial transformation. For this reason, we have developed KCl–sodium phosphate buffers for the suspension of mitochondria (see below).

We have successfully used mitochondria isolated from three cell lines for injection into cells. Two of these lines are derivatives of the HeLa cell line (HeLa S3[25] and HeLaBU25 10B3R[12]); the other is CAP23, a chloramphenicol-resistant derivative of the VA$_2$B cell line.[26] All three lines have been adapted for growth in spinner culture, and, therefore, significant quantities of cells are readily available. Typically, isolations are conducted starting with 2 to 3 g cells. The yield of mitochondria from this amount of starting material is far more than is utilized for injections, but these quantities facilitate the isolation procedures and filling of the injection pipette.

All procedures are performed at 4°. Cells are washed twice in 130 mM NaCl, 5 mM KCl, 1 mM MgSO$_4$. The cells are suspended in 18 ml of 10 mM Tris-HCl, pH 7.0 (at 25°), 10 mM KCl, 0.15 mM MgSO$_4$, allowed to swell for 2 min, and then homogenized with a Teflon pestle rotating at approximately 1600 rpm in a glass homogenizer (size C; Thomas Scientific, Swedesboro, NJ) until approximately 60 to 70% of the cells are broken. Cell breakage, monitored by phase-contrast microscopy, is performed to minimize shearing or fragmentation of the nucleus and subsequent release of chromatin. Typically, 5 to 12 sharp strokes are required to achieve this level of breakage without fragmentation of the nuclei. Proper fit of the Teflon pestle to the glass homogenizer is important for correct breakage of the cells. It may be necessary to try several combinations of pestles and homogenizers to obtain the desired fit and to achieve cell breakage meeting the required parameters.

The cell homogenate is brought to a final concentration of 0.25 M sucrose with 2 M sucrose, 10 mM Tris-HCl, pH 7.0 (at 25°). After thorough mixing, it is centrifuged at 1500 g for 3 min (Model 269 swinging-bucket rotor, International Equipment Corporation, Needham Heights, MA) to remove unbroken cells, nuclei, and large pieces of cytoplasm. The supernatant from this centrifugation is removed, added to a new centrifuge tube, and centrifuged again at 1500 g for 3 min. The supernatant is then centrifuged at 5900 g for 10 min in 50-ml polypropylene centrifuge tubes in a Sorvall SS-34 fixed-angle rotor. The resulting pellet is suspended well in 5 to 7.5 ml (~2.5 cell volumes) of 150 mM KCl, 10 mM sodium phosphate, pH 7.0. A resuspension buffer consisting of 75 mM KCl, 10 mM sodium phosphate, pH 7.0, is also utilized with no significant difference in the success rate of mitochondrial transfers. The suspension is centrifuged at 3800 g for 3.5 min to remove any residual nuclei or large cell fragments. The supernatant is transferred to a new centrifuge tube and centrifuged again at 3800 g for 3.5 min. The supernatant is used for injection. Although the latter two centrifugations resulted in significant losses of mitochondria, this is neces-

[25] T. T. Puck, P. I. Marcus, and S. J. Cieciura, *J. Exp. Med.* **103,** 273 (1956).
[26] C. H. Mitchell, J. M. England, and G. Attardi, *Somatic Cell Genet.* **1,** 215 (1975).

sary to minimize particulate matter that would clog the injection pipette. Viewed with a phase-contrast microscope, the supernatant appears to consist of membranous material and refractive vesicles, including mitochondria, with nuclear fragments also being occasionally visible.

Preparation of Pipette

The injection pipette is perhaps the single most critical variable for successful introduction of mitochondria into cells. The classic fine, extended tapered pipette used for the introduction of liquids into cells[27] is entirely unsuitable for the injection of mitochondria. The long, narrow extended region of the pipette traps large particles present in the mitochondrial preparation, clogging the pipette and preventing the flow of mitochondria from the pipette. The ideal pipette for injection of mitochondria has a relatively short region of taper minimizing the opportunity for trapping of large cell fragments. A large opening is created by breaking of the tip. This permits one to maintain a slow, but constant flow of the mitochondrial suspension through the pipette.

We routinely use 1 mm (560 μm inside diameter) by 4 inch standard wall borosilicate capillaries without a filament (IB100-4, World Precision Instruments Inc., Sarasota, FL) for preparing injection pipettes. A Brown-Flaming type micropipette puller (Model P-77, Sutter Instrument Company, Novato, CA) with a 1.5 mm wide trough filament or a 2 mm square box, 3 mm wide filament has been successfully utilized for the fabrication of pipettes. Pulling of the pipettes is effected with application of enough heat to permit melting of the borosilicate and initiation of the pulling cycle and a relatively short trip point before the hard pull. Too little heat results in insufficient melting to permit pulling of the pipette, whereas too much heat results in the production of a pipette with a long tapered region. A strong air flow, sufficient to effect rapid cooling, is also used to create a short region of taper. Usually, the two pipettes are not identical, and only one pipette meeting the required specifications is created with each pull. Even when a suitable program for pulling of injection pipettes is established, the conditions are usually such that results from one pull to the next are not consistent. Pipettes can be stored for up to several weeks before use.

Although we have not utilized a micropipette beveler, significant advantages would be expected from the beveling of the pipette tip.[28] Tips of relatively uniform size and shape could be created with significantly less difficulty than the procedures described here. This would facilitate penetra-

[27] M. Miranda and M. L. DePamphilis, this series, Vol. 225, p. 407.
[28] K. T. Brown and D. G. Flaming, *Brain Res.* **86,** 172 (1975).

tion of the cell and delivery of the mitochondrial suspension into the cell, perhaps minimizing damage to the cell from these procedures.

The injection pipette is filled from the rear with a 2-ml glass syringe equipped with a 3.5-inch N730 stainless steel needle (Part No. 91030, Hamilton Co., Reno, NV). The syringe is lowered into the mitochondrial suspension and the plunger pulled back to permit filling. The needle is placed on the syringe, and several drops of the solution are passed through the needle. The needle is placed into the rear of the injection pipette up to the point of the taper, and the pipette is completely filled with the mitochondrial suspension. Every attempt is made to minimize the size of the air bubble in the tapered region of the injection pipette during the filling process. After filling, the pipette is held at its base between the thumb and forefinger with the tip down, and the pipette is "flicked" with a fingernail to dislodge the air bubble in the taper. Usually, a moderate to strong force is required to displace this bubble.

A significant amount of pressure is required to force flow of the mitochondrial suspension through the injection pipette. Force is applied with a 2.5-ml threaded plunger syringe (Model 1002TPLT, Hamilton) equipped with a Luer-lock adapter connected to the injection pipette by Dow Corning medical grade Silastic tubing (0.03 inch inside diameter by 0.065 inch outside diameter). Both the syringe and Silastic tubing are filled with mineral oil. Hydraulic pressure is required to create sufficient force to initiate flow through the injection pipette. At high pressures, air compresses, making it very difficult to obtain sufficient pressure with an air-filled system.

Preparation of Cells for Injection

Cells for microinjection are plated at an appropriate concentration on gridded 60×15-mm culture dishes (Nunc). A gridded plate is found to be beneficial for maintaining orientation during injection of the cells. This permits the systematic injection of all cells within a defined area of the dish. Because a limited number of cells are present in each square of the grid, the chances of injecting cells more than one time (and thus increasing the chance of cell death) are negligible. Cell numbers are limited by plating the cells only in a region of the plate where injection could be easily accomplished. One or more glass cloning cylinders (5 to 8 mm inside diameter) are placed in the appropriate region of the plate, and approximately 1000 143B206 ρ^0 cells in approximately 0.1 ml DMEM supplemented with 5% FBS, 5 μg/ml uridine are placed inside each cylinder. The surface tension of the solution keeps it within the confines of the cylinder without use of a sealant. The following day, the cylinders are removed, and an additional 5 ml of the same medium is added to the culture dish. Injection of these cells is performed either the same or the following day.

Equipment for Microinjection

A Leitz Diavert inverted microscope with 25× oculars and phase-contrast objectives (10×, 20×, and 32×) is utilized for visually controlling microinjection. A Narishige micromanipulator with a joystick (Model MO-102, Narishige U.S.A., Sea Cliff, NY) is used to control the movement of the micropipette. The micropipette is maintained at an angle of 30° to 40° to the plane of the culture dish with a Narishige needle holder.

Injection of Cells

As mentioned previously, it is crucial to maintain an appropriate rate of flow of the mitochondrial suspension through the injection pipette. Therefore, one must apply sufficient pressure through the pipette to maintain flow, but at the same time avoid such a high rate of flow that too much solution is introduced into the cell, resulting in cell death. It is estimated that the maximum tolerance of introduced volume by injected mammalian cells is 10 to 15% of the cell volume.[29] Thus, one is quite limited in the amounts of material that can be introduced into cells.

The culture dish is taken from the incubator, and a deep scratch is made in the dish with the 143B206 ρ^0 cells with a sterile scalpel or razor blade. The dish is then placed on the microscope stage, and the cells are brought into focus at a low magnification (250×). The injection pipette, filled with the mitochondrial suspension, is lowered to just above the surface of the plate, in the region visible in the microscope. The dish and pipette are maneuvered so that the pipette is in the vicinity of the scratch. At this stage, the magnification is increased to 500× or 800×. The scratch (perpendicular to the pipette axis) is used to facilitate breakage of the tip of the injection pipette, which is required to initiate flow of the mitochondrial suspension. The tip is forced against the near or far edge of the scratch until the tip breaks. If the tapered region is too long, the pipette tip bends when pushed forward against the scratch, and the pipette usually breaks far up the taper. This creates an opening that is too large for injection of cells. Therefore, much care must be taken at this step to break enough of the tip that flow can be initiated, but not so much that too large an opening is created.

[29] J. A. Cooper, J. Bryan, B. Schwab III, C. Frieden, D. J. Loftus, and E. L. Elson, *J. Cell Biol.* **104,** 491 (1987).

After the pipette tip is broken, flow is initiated by advancing the screw plunger of the oil-filled syringe to increase the pressure through the injection pipette. This process is visually monitored in the microscope. Usually a small amount of air is present at the very tip of the pipette, and a significant amount of pressure is required to force this air through the tip. If flow does not initiate, the pressure is released by retracting the screw plunger, and the opening of the pipette is increased by breaking off an additional portion of the tip. This entire process is repeated until the air bubble passes through the tip, and flow of the mitochondrial suspension through the pipette is initiated. As soon as flow starts, the pressure is quickly reduced by retraction of the screw plunger to decrease the flow to the minimum possible rate.

The plate is moved so that the first chosen group of cells is in position for injection. The relatively large size of the pipette opening does not always facilitate what is considered to be the normal motion for injection of cells, namely, the advance of the pipette into the cell along the axis of the pipette. Instead, we find that lowering of the pipette into the cell along the z axis, with an occasional slight forward motion to promote entry of the pipette into the cell, is often more suitable for pipettes with large openings. It is not always necessary to have full penetration of the pipette tip into the cell to achieve introduction of mitochondrial suspension into the cell. Typically, cells are injected in a region of the cytoplasm adjacent to the nucleus, as this is usually the thickest portion of the cell and facilitates entry of the large pipette tip into the cell. When viewed with a phase-contrast microscope, the injection process is marked by a slight swelling of the cell in the vicinity of the pipette penetration point.

The injection motion is typically made as rapidly as possible to minimize the amount of material introduced into the cell. Because the minimum attainable flow rate is relatively high, one has to minimize the time of injection to have small injection volumes. Introduction of too much material into the cell is usually associated with a marked swelling of the cell in the vicinity of the pipette penetration point, which then propagates across the entire cell. Within several minutes, such a cell would exhibit large blebs forming on the plasma membrane, and would not usually survive.

The injection pipette nearly always clogs during the course of injections. Typically, nuclear fragments make up the material which lodges in the pipette tip and blocks the flow. The material is dislodged by increasing the pressure through the injection pipette with the oil-filled syringe in order to force the material through the pipette opening. As soon as flow reinitiates, the pressure is decreased to reduce the flow to the minimum attainable rate. If it is not possible to force the material through the tip, the pressure is released. Then, the size of the tip is increased by breaking off an additional portion. The pressure is then gradually increased to force the blocking material through the newly enlarged tip. This entire process is repeated until flow of the mitochondrial suspension through the pipette is reinitiated. Obviously, if the pipette tip is already very large, increasing the size of the opening prevents successful injections even when flow of the mitochondrial suspension is reinitiated. If the pipette is too large, a fresh pipette is filled with the mitochondrial suspension and is prepared for injections.

Injections are usually performed for up to 3 to 4 hr. As many as 2000 cells can be injected in this time. Because of the significant amount of time required for pipette preparation and maintaining flow of the mitochondrial suspension, this frequency of injection is significantly below those reported for injection of other materials into cells. A single culture plate is usually injected for a maximum of 30 to 45 min. Because we do not control the temperature or the pH of the medium during microinjections, cells gradually lose their adherence to the culture dish during the course of the injections, making it difficult to continue. No significant losses of mitochondrial viability are noted during the injection process. Mitochondrial transformants are obtained with similar frequencies among the first cells injected as among the cells injected several hours later.

Twenty-four hours after injection the medium is changed to DMEM supplemented with 10% dialyzed FBS, 100 μg/ml bromodeoxyuridine to select for mitochondrial transformants. If the injection pipette is broken during microinjections, a significant amount of the mitochondrial suspension is usually released onto the culture plate. Although viable cells have never been observed in the mitochondrial suspensions utilized for injections, bromodeoxyuridine is included to exclude the remote possibility of contamination by the mitochondrial donor cells. The selective medium is subsequently changed at 3- to 5-day intervals. Cell colonies are usually observed 10 to 15 days after injection, but some are also observed as late as 30 days after injection.

Troubleshooting Low Transformation Efficiencies

Low efficiencies of transformation usually result from low rates of survival of cells after injection. Cell survival is entirely dependent on introduction of small volumes of mitochondrial suspension, and, in turn, this can only be accomplished with well-shaped injection pipette. Therefore, great care should be taken to pull pipettes with the correct shape that can be easily broken to form a well-shaped tip which is wide enough to allow flow of the mitochondrial suspension. Formation of pipettes meeting these specifications can only be accomplished through trial and error. One should remember that successful injections are usually performed under less than optimal conditions since the optimal case is seldom achieved. The successful injection of mitochondria into cells is a combination of artistry and brute force.

Acknowledgments

This work was supported by Grants GM11726 (G. A.) and EY10085 (M. P. K.) from the National Institutes of Health and a grant from the Muscular Dystrophy Association (M. P. K.). M. P. K. is an Investigator of the American Heart Association, New York City Affiliate.

[29] Platelet-Mediated Transformation of Human Mitochondrial DNA-Less Cells

By ANNE CHOMYN

Introduction

The isolation of human cell lines completely devoid of mitochondrial DNA (mtDNA)[1] and the demonstration that these cell lines can be repopulated with exogenous mtDNA[1,2] by mitochondria-mediated transformation have represented a major advance for human mitochondrial genetics studies. In particular, mtDNA-less (ρ^0) cells have proved to be very useful for the study of mtDNA-linked diseases.[3-6]

Until relatively recently, the method routinely used for introducing mitochondria derived from cells of patients or normal human individuals into ρ^0 cells involved fusion of enucleated cells (cytoplasts) with the ρ^0 cells.[1-6] A limitation in this method is the availability of cultures (myoblasts, fibroblasts, or transformed lymphoblasts) derived from cells removed from an individual. An alternative method for introducing mitochondria into human ρ^0 cells has been developed,[7] which involves the fusion with these cells of platelets derived from the individual's blood. The advantage of this

---

[1] M. P. King and G. Attardi, *Science* **246,** 500 (1989).

[2] M. P. King and G. Attardi, this volume [28].

[3] A. Chomyn, G. Meola, N. Bresolin, S. T. Lai, G. Scarlato, and G. Attardi, *Mol. Cell. Biol.* **11,** 2235 (1991).

[4] J.-I. Hayashi, S. Ohta, A. Kikuchi, M. Takemitsu, Y.-I. Goto, and I. Nonaka, *Proc. Natl. Acad. Sci. U.S.A.* **88,** 10614 (1991).

[5] M. P. King, Y. Koga, M. Davidson, and E. A. Schon, *Mol. Cell. Biol.* **12,** 480 (1992).

[6] I. A. Trounce, S. Neill, and D. C. Wallace, *Proc. Natl. Acad. Sci. U.S.A.* **91,** 8334 (1994).

[7] A. Chomyn, S. T. Lai, R. Shakeley, N. Bresolin, G. Scarlato, and G. Attardi, *Am. J. Hum. Genet.* **54,** 966 (1994).

method is that there is no need to establish cultures derived from the cells of the patient or the normal individual. Furthermore, platelets have no nuclei, thus eliminating the need for an enucleation step. Blood is collected from the individual into a tube that contains an inhibitor of coagulation, namely, heparin. Thereafter, the platelets are isolated by differential centrifugation. We have successfully fused with ρ^0 cells platelets that were derived from blood samples up to 40 hr after the blood was drawn, and also with platelets that had been frozen and thawed.

The noninvasive nature of the collection of mitochondria donor material from the patient and the facility of the mitochondria transfer procedure have extended considerably the applicability of the ρ^0 cell transformation approach for the genetic and biochemical analysis of mtDNA-linked diseases. Furthermore, the prolonged viability of the platelet mitochondria and the possibility of using frozen platelets allow one to use blood samples collected at remote locations, thus increasing the access to desired cases of such diseases.

The following protocol uses ρ^0 cells derived from a thymidine kinase-deficient cell line (143B.TK$^-$).[1,2] The medium used to maintain the ρ^0 cell line and to select transmitochondrial cell lines contains 5-bromodeoxyuridine (BrdU). Any cells contaminating the platelet preparation, such as white blood cells, and any hybrids formed with such cells would synthesize thymidine kinase, incorporate BrdU into their DNA, and be killed. If one plans to use a human ρ^0 cell line with a different recessive drug resistance marker, one should use the appropriate drug in the selection medium.

Platelet–ρ^0 Cell Fusion and Cybrid Selection

Solutions and Media

Physiological saline: 0.15 M NaCl, 15 mM Tris-HCl, pH 7.4 (25°)

10× citrate in physiological saline: 0.15 M NaCl, 0.10 M sodium citrate (Na$_3$C$_6$H$_5$O$_7 \cdot$ 2H$_2$O; trisodium salt, dihydrate)

TD: 0.137 M NaCl, 10 mM KCl, 0.7 mM Na$_2$HPO$_4$, 25 mM Tris-HCl, pH 7.4–7.5

TD plus EDTA: TD that has been made 1 mM in ethylenediaminetetraacetic acid (EDTA; sodium salt)

Trypsin stock solution (10×): 0.5% (w/v) trypsin (1–300 porcine pancreas, ICN Biomedicals, Costa Mesa, CA) in water, stored at 4°. For use, the trypsin stock solution is diluted 10-fold into TD + EDTA.

DMEM minus Ca^{2+}: Dulbecco's modified Eagle's medium (DMEM) lacking calcium

Freezing medium: 17.5 ml DMEM, 5 ml fetal bovine serum (FBS), and 2.5 ml dimethyl sulfoxide (DMSO; sterilized by autoclaving)

42% PEG solution: Prepare this solution fresh, within a few hours of use. First, autoclave 8.4 g polyethylene glycol (PEG) 1500 (BDH Laboratory Supplies, Poole, England) in a 50-ml Erlenmeyer flask for 15 min in "slow exhaust" mode. Before the PEG cools enough to solidify, add 2 ml DMSO and 9.6 ml DMEM minus Ca^{2+}. (If the PEG has solidified before the DMSO could be added, remelt it in a 65° water bath.) Keep the PEG solution at 37° until needed.

ρ^0 cell medium for plating fusion products: DMEM supplemented with FBS to 10% of the final volume, and containing 100 μg/ml BrdU and 50 μg/ml uridine

Selective medium: DMEM supplemented with dialyzed FBS[8] to 10% of the final volume, and containing 100 μg/ml BrdU

Procedure. Have a phlebotomist draw 15–20 ml venous blood from the individual into heparin-containing tubes (green caps). We usually have an additional 15–20 ml blood collected for testing for HIV (human immune deficiency virus) antibodies and hepatitis B antigen. The blood may be kept at room temperature.

The method of isolating platelets is adapted from that of Mann *et al.*[9] The following procedures are performed at room temperature, unless otherwise specified. Manipulations are carried out in a sterile hood, using sterile reagents and pipettes. Determine the volume of blood, transfer it to a sterile 50-ml screw-capped polypropylene tube (Corning 25330-50), and add one-ninth volume of 0.1 M sodium citrate, 0.15 M NaCl. Mix well (by inversion) and centrifuge the blood at 200 g (at r_{max}) for 20 min at 12°. Collect three-fourths of the supernatant (plasma), leaving behind at least 3 mm of plasma above the red blood cell layer, to avoid drawing up any red blood cells. The platelet concentration in the plasma may be determined at this point by counting in a hemacytometer in a phase-contrast microscope. Centrifuge the plasma in a sterile 15-ml polypropylene tube (Corning 25319-15) at 1500 g (at r_{max}) for 20 min at 15°. Suspend the pellet containing the platelets in about 2 ml physiological saline. Pipette up and down gently, using a small bore pipette, to break up the clumps. Add physiological saline to a final volume of 11 ml. If desired, freeze all or a portion of the suspension by adding, per 1.05 ml platelet suspension, 0.15 ml sterile DMSO and 0.30 ml FBS. Distribute into 1.8-ml freezing vials and freeze, allowing the temperature to drop slowly (at a rate of ~1°/min), as is done when freezing cell cultures.

For the fusion with ρ^0 cells, transfer 7 ml of the platelet suspension (in

---

[8] A. Chomyn, this volume [18].
[9] V. M. Mann, J. M. Cooper, D. Krige, S. E. Daniel, A. H. V. Schapira, and C. D. Marsden, *Brain* **115,** 333 (1992).

physiological saline) to a fresh sterile 15-ml tube and centrifuge it at 1500 g for 15 min at 15°. Meanwhile, collect at least 10^6 ρ^0 cells by trypsinization. We detach the cells from the petri dish by washing them first with TD and then incubating them with trypsin solution for a few minutes at room temperature. The detached cells are then transferred with a pipette into a tube containing DMEM supplemented with 5 or 10% bovine calf serum, to inactivate the trypsin. The cells are then pelleted by centrifugation at 180 g (at r_{max}) in a clinical centrifuge for 5 min. Suspend the ρ^0 cells in DMEM minus Ca^{2+} at 5×10^5 cells/ml. At the end of the 1500 g centrifugation of the platelet suspension, aspirate carefully almost all the supernatant, leaving behind 50–100 μl over the undisturbed pellet. Very gently add 2 ml of the ρ^0 cell suspension (10^6 cells). Do not disturb the platelet pellet during the addition of cells. Centrifuge the cells onto the platelet pellet at 180 g in a clinical centrifuge for 10 min at room temperature. Aspirate carefully all the supernatant.

Add 0.1 ml of the 42% PEG solution to the tube containing the pellet of cells and platelets, and suspend these carefully. We use a 0.1- or 0.2-ml glass tissue culture pipette and a rubber bulb to make the addition of the 42% PEG solution and to suspend the pellet. When the pellet is nearly completely suspended, begin timing 1 min. Continue to suspend the pellet for 20 to 30 sec more. At the end of the timed minute, add 10 ml ρ^0 cell medium. Mix well and make a 1:10 dilution of the cell suspension in the same medium. Plate the cells in this 1:10 dilution in 96-well microwell plates, making at least three further dilutions, as follows:

To 10 ml of the 1:10 dilution add another 10 ml ρ^0 cell medium and seed a 96-well plate with this suspension, at 0.2 ml per well.

To 2.5 ml of the 1:10 dilution add 7.5 ml ρ^0 cell medium and seed a 96-well plate with this suspension, at 0.1 ml per well.

To 1.0 ml of the 1:10 dilution add 9.0 ml ρ^0 cell medium and seed a 96-well plate with this suspension, at 0.1 ml per well.

Additional plates at any of these dilutions may be seeded, if so desired. Incubate the microwell plates in a 37° incubator that maintains an atmosphere of 5% CO_2/95% air (v/v).

Any remaining cells that have not been plated may be collected by centrifugation, suspended in 3.0 ml freezing medium, divided into two freezing vials, frozen slowly, as is done when freezing cell cultures, and then stored in liquid nitrogen. These frozen cells may serve as a reserve supply of fused cells, should the plated cells be lost.

Two or three days after plating the cells, replace the medium in the microwell plates with selective (uridine-free) medium. Replace the medium every three to five days. Ten to twelve days after starting selection, colonies should be visible and may be ready to be transferred to petri dishes. For this purpose, the microwell plates are carefully screened, and the wells containing single colonies are noted. Microwell plates that have a high proportion of "negative" wells (preferably >70–80%) should be preferentially used for clone isolation, to reduce the probability of having more than one clone per well. For each colony to be transferred, wash the well with about 0.2 ml TD. Remove the TD and add 0.1 ml trypsin solution. After the cells have started to detach, transfer the cells directly with a Pasteur pipette to a 35-mm petri dish containing 2 ml selective medium or to a 60-mm petri dish containing 4–5 ml selective medium. Continue to grow and transfer the clone until a sufficient number of cells for freezing or for desired experiments has been reached.

In a typical experiment, we start with 16–18 ml blood, recover 33–45% of that volume as plasma/citrate solution, and use 60% of those platelets for the fusion to ρ^0 cells. The yield of cybrids is usually 600 to 1000.

Additional Remarks

1. If frozen platelets are to be used for fusion, thaw the vial of platelets (in 10% (v/v) DMSO and 20% (v/v) serum) rapidly in a 37° water bath, dilute with 10 ml physiological saline, and centrifuge at 1500 g for 15 min at 15°. Aspirate most of the supernatant from the platelet pellet, and then centrifuge 10^6 ρ^0 cells onto the platelet pellet as described above. Aspirate all the supernatant, and continue as for a fresh platelet preparation.

2. The PEG concentration may need to be adjusted, depending on the toxicity of the particular source or batch used. In the past we have used 50% PEG for fusion; however, newer batches of PEG were found to be too toxic at 50%, and thus the concentration used for fusion has been reduced (to 42% in the experiments described above). A compromise in the concentration of PEG must, therefore, be made between one that will produce fusions at a high frequency and one that is not too toxic to the cells.

3. As an alternative to plating the cells in 96-well plates immediately after fusion, the cells may be plated at various dilutions in 6- or 10-cm petri dishes. Cybrid colonies may then be picked using the glass ring method[10] or by scraping and withdrawing the cells with a Pipetman and yellow tip.[11]

4. Because the experiments described above involve the use of human blood, the usual precautions should be taken. At the minimum, wear a

---

[10] L. C. M. Reid, this series, Vol. 58 [12].
[11] L. H. Thompson, this series, Vol. 58 [26].

laboratory coat and gloves, and autoclave all the waste. Follow established institutional procedures for work with biohazardous materials.

Acknowledgment

This work was supported by National Institutes of Health Grant GM-11726 to Giuseppe Attardi.

[30] Use of Ethidium Bromide to Manipulate Ratio of Mutated and Wild-Type Mitochondrial DNA in Cultured Cells

By MICHAEL P. KING

Introduction

Many mutations of mammalian mitochondrial DNA (mtDNA) have been described. In many cases, they exist in heteroplasmic form, that is, both mutated and wild-type genomes are present within cells. To investigate the effects of these mutations on mitochondrial function, it is desirable to isolate clonal cell lines that contain exclusively mutated or exclusively wild-type mtDNA, both derived from the cells being studied. Furthermore, it is often desirable to isolate a range of mutated mtDNA levels. A correlation of levels of mutation with changes in mitochondrial function will frequently reveal the nature of the molecular mechanism of the mutation.

A system has been described for the characterization of mutations of the mtDNA.[1-4] This system is based on the isolation of human cell lines that completely lack mtDNA (ρ^0 cells) and the methods developed to introduce mitochondria, and thus reintroduce mtDNA, into these cells. This system has been used by a number of groups to investigate the pathogenetic mechanisms of mutations of mtDNA. In some cases, after transfer of mitochondria to ρ^0 cells, it is not possible to isolate a full range of mutation levels in the resulting cybrids, or cytoplasmic hybrids. Presumably, the levels of mutation in the transmitochondrial cell lines reflect the range of mutation that is present in individual cells of the mitochondrial donor cell

[1] M. P. King and G. Attardi, Science **246,** 500 (1989).
[2] M. P. King and G. Attardi, this volume [27].
[3] M. P. King and G. Attardi, this volume [28].
[4] A. Chomyn, this volume [29].

population. If the growth of primary cell cultures or cell lines derived from patients is severely affected by impaired mitochondrial function, such donor cultures would not be expected to possess high levels of mutation. Similarly, one usually selects for mitochondrial transformants of the ρ^0 cells by selecting for mitochondrial respiratory chain function (i.e., cell growth in the absence of uridine or pyruvate). Therefore, it may not always be possible to isolate cells that possess high levels of mutated genomes if the mutation results in a severe deficiency of mitochondrial respiratory chain function. If one is working with mutations isolated from *in vitro* cell cultures, the donor cell line is often clonal in origin, and thus may possess a limited range of mutation levels. Therefore, to compare the phenotype of cells possessing 100% wild-type and 100% mutated mtDNA, or to investigate the threshold level of mutation which results in phenotypic changes, it may be necessary to manipulate the mtDNA complement of cells in order to isolate cell lines that possess varying levels of mutation.

Effect of Ethidium Bromide on Cells

The DNA intercalating dye ethidium bromide is a well-known inhibitor of mtDNA replication. Previous work in mammalian cells has demonstrated that high concentrations of ethidium bromide (0.1 to 2 μg/ml) result in either partial or complete inhibition of mtDNA replication but have no effect on the replication of the nuclear DNA.[5-8] The partially purified mitochondrial DNA polymerase has been shown to be inhibited by ethidium bromide to a much greater extent than the nuclear DNA polymerase.[9] Possibly, inhibition of mtDNA replication occurs because of inhibition of the mitochondrial DNA polymerase by ethidium bromide. This inhibition of mtDNA synthesis, along with resulting decreases in mtDNA levels, is a reversible phenomenon.[6,10] Even low levels of ethidium bromide (20 ng/ml) result in near complete inhibition of mtDNA synthesis and reductions in the steady-state levels of mtDNA in human cell lines.[10] It was observed that the mtDNA content becomes progressively diluted as the cells continue to divide. Thus, after n cell doublings, the mtDNA content was reduced by the factor $1/2^n$. Nass[6] has reported that the preexisting mtDNA population in cells treated for four cell generations with ethidium bromide is largely retained by the cells. Furthermore, this DNA, while altered in

[5] M. M. K. Nass, *Proc. Natl. Acad. Sci. U.S.A.* **67,** 1926 (1970).
[6] M. M. K. Nass, *Exp. Cell Res.* **72,** 211 (1972).
[7] R. D. Leibowitz, *J. Cell Biol.* **51,** 116 (1971).
[8] C. A. Smith, J. M. Jordan, and J. Vinograd, *J. Mol. Biol.* **59,** 255 (1971).
[9] R. R. Meyer and M. V. Simpson, *Biochem. Biophys. Res. Commun.* **34,** 238 (1969).
[10] A. Wiseman and G. Attardi, *Mol. Gen. Genet.* **167,** 51 (1978).

topological structure, was largely intact. This result is significantly different from that in yeast, where ethidium bromide treatment results in degradation of the mtDNA.[11,12] Thus, it seems that even though mammalian mtDNA replication is inhibited, the mtDNA is stable in the presence of ethidium bromide. Therefore, ethidium bromide is an effective agent to manipulate the copy number of mammalian mtDNA without producing the mutational effects observed in yeast.

Rationale

It is possible to maintain human cells that completely lack mtDNA, and thus lack any mitochondrial respiratory chain function. Because of the absence of a respiratory chain, these cells develop growth requirements for pyrimidines and pyruvate. If cells are provided with these substances, they are able to maintain normal cell growth. To manipulate the mtDNA population of a heteroplasmic cell population, the cell is treated with ethidium bromide to reduce its mtDNA content. This cell population is provided with uridine and pyruvate to maintain cell viability and growth. The cells are maintained in the presence of ethidium bromide until the mtDNA content is reduced to the level where there is on average one mtDNA molecule per cell. At this point, the ethidium bromide is removed from the culture medium, and the remaining mtDNA is permitted to repopulate the cell. It had been shown previously that cells partially depleted of mtDNA rapidly regain normal levels of mtDNA.[1,10] Thus, if one starts with a cell population that possesses 50% wild-type and 50% mutated mtDNA, one would expect approximately half of the cells to be homoplasmic with mutated genomes and half to be homoplasmic for wild-type genomes after such treatment.

One can estimate the length of time to treat with ethidium bromide, based on the number of mtDNA molecules present in the cells and the cell generation time. The ethidium bromide blocks mtDNA replication, and the starting mtDNA population is decreased by a factor of two each time the cells divide. Thus, after one cell generation each cell has on average one-half the number of mtDNA molecules; after two cell generations, only one-quarter the mtDNA content; etc. For the 143B cell line and its derivatives, which possess approximately 9000 mtDNA molecules per cell and have a population doubling time averaging 19 hr,[1] one would treat the

[11] E. S. Goldring, L. I. Grossman, D. Krupnick, D. R. Cryer, and J. Marmur, *J. Mol. Biol.* **52,** 323 (1970).
[12] P. S. Perlman and H. R. Mahler, *Nature (London) New Biol.* **231,** 12 (1971).

cells for approximately 14 cell generations, or about 11 days, for the cells to have on average less than one mtDNA molecule per cell.

A shorter period may also be chosen for ethidium bromide treatment so that more than one mtDNA molecule remains in each cell. After removal of the ethidium bromide, the remaining molecules repopulate the cells. The proportion of mutated to wild-type mtDNA present in the cells at the time of release from ethidium bromide varies, resulting in varying levels of mutated mtDNA being present in the cells after mtDNA repopulation occurs. Thus, it is possible to obtain a broad range of mutated mtDNA levels in the cell lines derived from such treatments.

Experimental

To conduct experiments to manipulate the mtDNA content of cells, the following stock solutions are required. A 10 mg/ml solution of ethidium bromide (Sigma, St. Louis, MO) made up in distilled water should be filter sterilized. After filter sterilization, this solution should be serially diluted with sterile distilled water to a working concentration of 5 μg/ml. If the solution is filter sterilized at the latter concentration, a significant fraction of the ethidium bromide in the solution binds to the filter and, thus, the final concentration of ethidium bromide in solution is significantly reduced. The stock and diluted solutions should be stored at 4° and should be protected from light. Fresh solution should be made every 3 to 6 months. Uridine (Sigma) should be made as a 10 mg/ml stock in distilled water and should be filter sterilized. If required, a 100 mM stock solution of pyruvate should be prepared in distilled water and sterilized by filtration. Both the uridine and pyruvate stock solutions should be stored at $-20°$, although for routine usage they are stable for at least several weeks at 4°.

To illustrate the use of ethidium bromide to manipulate the mtDNA content of cells, we describe the treatment of two cell lines that possess wild-type and deleted mitochondrial genomes. Two clonal cell lines that possess intermediate levels of deleted mtDNA are chosen for these experiments. These cell lines had previously been obtained by fusion of enucleated fibroblasts derived from patients possessing mtDNA deletions with the ρ^0 cell line 143B206. One cell line, 206/F4 3.13, possesses 51% of a 7.5-kb mtDNA deletion, whereas the other, 206/Flp 39, possesses 48% of a 1.9-kb mtDNA deletion. These cell lines are grown in Dulbecco's modified Eagle's medium (DMEM) supplemented with 5% fetal bovine serum (FBS), 50 ng/ml ethidium bromide, and 50 μg/ml uridine. The cells are grown in the presence of ethidium bromide for 9 days for 206/Flp 39 and for 10 days for 206/F4 3.13. A shorter time is chosen for 206/Flp 39 in order to produce derivatives with a broad range of deleted mtDNA levels, and a longer time is chosen for 206/F4 3.13 in order to isolate clones with exclusively deleted or wild-type mitochondrial genomes. After ethidium bromide treatment, the cells are replated in DMEM supplemented with 5% FBS and 50 μg/ ml uridine and grown for several days, during which time the residual mtDNA would repopulate the cells. The cells are trypsinized and replated at low density (100 cells per 100-mm plate) in the same medium. Ten to fourteen days later, individual colonies are selected with glass cloning cylinders and grown for genetic analysis.

Total DNA is extracted from cell samples of each clone and subjected to DNA transfer hybridization analysis. The shorter, 9 day, treatment with ethidium bromide resulted in a range of different levels of mtDNA deletion being present in clones derived from 206/Flp 39. As shown for a representative analysis of these clonal cell lines in Fig. 1A, they possess either 100% wild-type mtDNA (No. 35), 100% deleted mtDNA (No. 39), or varying intermediate levels of deleted mtDNA (Nos. 33, 36, 37, 38, and 40). One

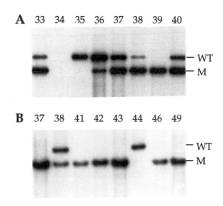

FIG. 1. Mitochondrial DNA deletion analysis of selected clones. Shown are autoradiograms resulting from DNA transfer hybridization analysis of representative clones of 206/Flp 39 (A) and 206/F4 3.13 (B) after treatment with ethidium bromide as described in the text. Total DNA samples were digested with *Eco*RI (A) or *Pvu*II (B) and electrophoresed through 1% agarose gels. DNA was transferred to Zeta-Probe GT membranes (Bio-Rad, Hercules, CA) and hybridized with ^{32}P-labeled specific probes derived from cloned fragments of mtDNA [mp18.HE2 (A) or mp8.M4 (B); M. P. King and G. Attardi, *J. Biol. Chem.* **268,** 10228 (1993)] according to the manufacturer's suggested protocol. The numbers at the top of the autoradiograms indicate the identification number of the derived clonal cell lines. The 7.4-kb wild-type mtDNA (WT) and 5.5-kb deleted mtDNA (M) restriction digestion products are indicated for the clones derived from 206/Flp 39, and the 16.6-kb wild-type mtDNA (WT) and 9.0-kb deleted mtDNA (M) restriction digestion products are noted for the clones derived from 206/F4 3.13. The mtDNA probes were obtained by extension of the universal M13 primer according to the method of Sucov *et al.* [H. Sucov, M. S. Benson, J. J. Robinson, R. J. Britten, F. Wilt, and E. H. Davidson, *Dev. Biol.* **120,** 507 (1987)].

clone, No. 34, had apparently lost all its mtDNA at the time of release from ethidium bromide and possessed no mtDNA at the time of analysis. The clonal lines derived from 206/F4 3.13 were initially subjected to dot-blot analyses to select for clones possessing mtDNA. Approximately one-half of these clones were found to possess mtDNA. Those cell lines that have mtDNA are subjected to DNA transfer hybridization analysis to determine the levels of mtDNA deletions present, and a representative analysis is shown in Fig. 1B. The majority of the clonal lines (Nos. 37, 41, 42, 43, 46, and 49) possessed 100% deleted genomes, whereas one possessed 100% wild-type mtDNA (No. 44) and one possessed an intermediate level of deleted mtDNA (No. 38).

These results demonstrate that it is possible to manipulate the relative proportions of deleted and wild-type mtDNA levels in cells. We have also utilized these methods to manipulate the levels of point mutations of mtDNA. Cells possessing intermediate levels of various mtDNA point mutations have been subjected to ethidium bromide treatments as described above. Following these treatments, the cells were subcloned and subjected to genetic analysis. Results similar to those described above for deletions of mtDNA were observed for mtDNA point mutations. With longer periods of ethidium bromide treatment, we were able to isolate cell lines that were homoplasmic for either wild-type or mutated mtDNA, and, with shorter treatments, we were able to isolate a range of mutated mtDNA levels.

Acknowledgments

The author thanks Edgar Davidson for helpful comments on the manuscript and Yasutoshi Koga for performing the DNA transfer hybridization analyses. This work was supported by grants from the National Institutes of Health and the Muscular Dystrophy Association. M. P. K. is an Investigator of the American Heart Association, New York City Affiliate.

[31] Construction of Heteroplasmic Mice Containing Two Mitochondrial DNA Genotypes by Micromanipulation of Single-Cell Embryos

By PHILIP J. LAIPIS

Introduction

The presence of two or more distinct mitochondrial (mt) DNA genomes within an organism is defined as heteroplasmy. Heteroplasmy was first observed in mammals as restriction fragment length heterogeneity within dairy cow pedigrees[1,2]; we proposed that heteroplasmy was obligatory in generating sequence variation in mammalian mtDNA.[1-3] This observation was rapidly extended to other higher eukaryotes, including insects, reptiles, and fish.[4-6] Sequence differences among heteroplasmic mtDNA genomes range from one or more nucleotide substitutions to deletions, insertions, and duplications of one to several thousand nucleotides; multiple classes of mtDNA genomes within an individual organism have also been reported.

Heteroplasmy appears important in the creation of normal variation within the mtDNA of a mammalian species, as well as for the phenotypic variation seen in individuals with mixtures of normal and defective mtDNA genotypes.[7,8] Despite the importance of heteroplasmy, the mechanisms responsible for creation and loss of the heteroplasmic state are poorly understood. Naturally occurring heteroplasmic mammals are rare. In addition, it is usually difficult to separate the contributions of mutation events creating a new (heteroplasmic) mtDNA molecule from back mutations and the subsequent segregation of that new genotype in a pedigree. This problem is particularly difficult for the most commonly observed heteroplasmic event, the rapid addition and loss of short repetitive sequence elements,

[1] W. W. Hauswirth and P. J. Laipis, *in* "Mitochondrial Genes" (P. Slonimiski, P. Borst, and G. Attardi, eds.), p. 137. Cold Spring Harbor Laboratory, Cold Spring Harbor, New York, 1982.

[2] W. W. Hauswirth and P. J. Laipis, *Proc. Natl. Acad. Sci. U.S.A.* **79,** 4686 (1982).

[3] G. S. Michaels, W. W. Hauswirth, and P. J. Laipis, *Dev. Biol.* **94,** 246 (1982).

[4] M. Solignac, M. Monnerot, and J. C. Mounolou, *Proc. Natl. Acad. Sci. U.S.A.* **80,** 6942 (1983).

[5] L. D. Densmore, J. W. Wright, and W. M. Brown, *Genetics* **110,** 689 (1985).

[6] E. Bermingham, T. Lamb, and J. C. Avise, *J. Hered.* **77,** 249 (1986).

[7] P. Lestienne, *Biochimie* **74,** 123 (1992).

[8] D. C. Wallace, *Annu. Rev. Biochem.* **61,** 1175 (1992).

where the forward and back mutation rates are high.[9,10] To examine the segregation rate of mtDNA genomes in mammals, we have developed techniques to produce mice containing two genetically distinct, functional mitochondria within individual cells of the animals. The methods may also be useful for creating mice containing *in vitro* modified mtDNA.

Experimental Procedures

Basic Requirements

Production of heteroplasmic mice requires an animal research facility capable of producing transgenic mice by pronuclear injection plus the additional micropipette synthesis equipment required for blastocyst injection of embryonic stem cells.[11,12] The basic procedure is very similar to the method developed by McGrath and Solter and collaborators for pronuclear transfer between single-cell embryos.[13] A useful source of general information is Volume 225, "Guide to Techniques in Mouse Development," in this series,[12] in particular, the chapters by Latham and Solter[14] and Barton and Surani.[15]

Equipment

Embryo manipulation is easiest with an inverted microscope; two micromanipulators are also needed. Both Narashigi and Leitz micromanipulators have been used. Nomarski optics capable of 400× magnification (or another contrast-enhancing technique) are necessary to clearly visualize and avoid the pronuclei and to reorient the opened zona pellucida for cytoplast transfer (see below). At least one stereoscopic microscope is needed for embryo isolation and for the surgical transfer of manipulated embryos into the oviducts of pseudopregnant female recipients. The equipment for pipette manufacture includes a micropipette puller (Kopf 750 or Sutter P-87 have been used), a microforge for pipette breaking, bending, and nib pulling (DeFonbrune), and a grinding device using either a metal wheel with diamond dust or a diamond wheel to bevel pipets (Narashigi EG-4). A power

[9] W. W. Hauswirth, M. J. Van de Walle, P. J. Laipis, and P. D. Olivo, *Cell (Cambridge, Mass.)* **37,** 1001 (1984).

[10] R. G. Harrison, D. M. Rand, and W. C. Wheeler, *Mol. Biol. Evol.* **4,** 144 (1987).

[11] B. Hogan, F. Constantini, and E. Lacy, "Manipulating the Mouse Embryo." Cold Spring Harbor Laboratory, Cold Spring Harbor, New York, 1986.

[12] P. M. Wassarman and M. L. DePamphilis (eds.), this series, Vol. 225.

[13] J. McGrath and D. Solter, *Science* **220,** 1300 (1983).

[14] K. E. Latham and D. Solter, this series, Vol. 225, p. 719.

[15] S. C. Barton and M. A. Surani, this series, Vol. 225, p. 732.

source for embryo electrofusion (Fusion Pulser, Biokemiai Labor Szerviz, 1165 Budapest, Zselyi, Aladar u.31., Hungary) or inactivated Sendai virus (see below) is needed. A CO_2 incubator to hold isolated embryos and allow fusion and subsequent cell division is necessary, along with facilities for embryo medium preparation, including an osmometer. Identification of heteroplasmic founder animals requires Southern blotting or polymerase chain reaction (PCR) methods.

Animals

Two inbred strains whose mtDNA genomes have been completely sequenced are used to produce heteroplasmic founder mice. C57BL/6J mice belong to the "old inbred group,"[16] from which C3H mtDNA has been sequenced.[17] No differences have been detected between the published C3H mtDNA sequence and C57 mtDNA. The mitochondria of the NMRI/BOM mouse strain are derived from NZB mice; its mtDNA has also been sequenced in its entirety.[18,19] The two mtDNA genomes differ at 106 nucleotides, that is, 106 linked genetic loci. The strains also differ in coat color markers (C57, *a/a*, *C/C*, *B/B*; NMRI, *A/A*, *c/c*, *B/B*), litter size, and maternal behavior, important features in distinguishing experimental animals from carrier embryos and in husbandry. Animals are housed under specific pathogen-free conditions in microisolator cages, on a 12-hr (7 am to 7 pm) light cycle. Sufficient breeding pairs are maintained so that a steady-state population of 50 females (assuming use of 10 females/week) of each strain are available. Stud males (30 C57, 15 NMRI) are individually caged. In addition, 30 vasectomized males plus 50–80 C57BL/6 × DBA/2 F_1 female mice are also required, to provide a source of pseudopregnant recipient females. Recipient females weighing 24–28 g are sufficiently mature to be competent mothers; oviduct transfers are more difficult in heavier females because of large fat deposits.

Media, Chemicals, and Special Reagents

The M2 and M16 media are prepared as previously described[11]; adjustment to 285–290 mOs*m*ol is critical to prevent swelling of embryos. Cytochalasin D (Sigma, St. Louis, MO; 1 mg/ml in water) and colchicine [Sigma

[16] S. D. Ferris, R. D. Sage, and A. C. Wilson, *Nature (London)* **295,** 163 (1982).
[17] M. J. Bibb, R. A. Van Etten, C. T. Wright, M. W. Walberg, and D. A. Clayton, *Cell (Cambridge, Mass.)* **26,** 167 (1981).
[18] B. E. Loveland, C. R. Wang, E. Hermel, H. Yonekawa, and K. Fischer-Lindahl, *Cell (Cambridge, Mass.)* **60,** 971 (1990).
[19] S. D. Ferris, U. Ritte, K. Fischer Lindahl, E. M. Prager, and A. C. Wilson, *Nucleic Acids Res.* **11,** 2917 (1983).

D-6279; 100 μg/ml in 95% (v/v) ethanol] are stored in aliquots at $-20°$. Avertin stock (10 g tribromoethyl alcohol in 10 ml tertiary amyl alcohol) is stored dark at 4°; 1.2% v/v working solutions, prepared by adding 60 μl to 5.0 ml of 0.85% w/v sterile NaCl, are discarded after 2 weeks. Mice are anesthetized by intraperitoneal (i.p.) injection of 25 μl/g body weight. Hormones for superovulation (when required) and hyaluronidase are prepared as previously described.[11] Both pregnant mare serum (PMS) and human chorionic gonadotropin (hCG) are diluted to 50 IU/ml in sterile saline immediately before use; 5 IU is injected i.p. into 24- to 30-day-old female mice according to the schedule described below. Inactivated Sendai virus (2500–3000 hemagglutinating units/ml) was generously provided by Dr. Davor Solter; not all Sendai virus isolates are effective in fusing the large karyoplast/cytoplast volumes produced in these experiments.

Micropipette Manufacture

Both holding and suction-transfer pipettes are crafted from R-6 glass capillary tubing (1.0 mm O.D., 0.65 mm I.D., 150 mm length; Drummond Scientific Co., Broomall, PA). Tubing is acid washed (20% v/v concentrated HNO$_3$; 80% v/v concentrated HCl) and rinsed well before use. Conditions for pulling pipettes are determined empirically; they vary with each puller and individual electrode.

Holding pipettes are pulled to yield an outer diameter of 100–120 μm over 3–4 mm of length. The pipette is broken on a soft glass anvil in a microforge[11,14] (medium heat, air, 0.3 mm filament) to give a flat 100 μm O.D. end; the end is smoothly fire-polished in the microforge to a final inner diameter of 35–38 μm (as above, no air) and then bent to a 20° angle 1.5 mm from the end (Fig. 1A). Holding pipettes can be used repeatedly if cleaned in chromic acid, distilled water, and ethanol after use.

Suction-transfer pipettes are pulled to yield an outer diameter of 20–25 μm over 1.0 mm of length and broken (low heat, air, 0.06-mm filament) at 20–25 μm O.D. Pipettes are beveled at a 45° angle on a diamond grinder. They are inserted into a Leitz micromanipulator instrument holder attached to a 30-ml plastic syringe and rinsed briefly in chromic acid, distilled water, and ethanol to remove grinding debris. Pipettes are sharpened by dipping for 4–9 sec in 15% (v/v) hydrofluoric acid while expelling air from the tip, and immediately rinsed well (three times with three changes of distilled water, three times with ethanol). Exact HF treatment times must be determined empirically; suitable pipettes appear sharp, that is, have no discernable wall thickness at their beveled end and appear somewhat roughened on their outer surface, but retain a smoothed-edged profile at the beveled end. A sharp nib is pulled on the leading edge of the beveled,

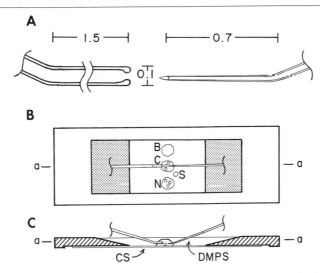

FIG. 1. Pipette dimensions and setup. (A) Magnified side view of holding (left) and suction-transfer (right) pipettes; dimensions are in millimeters. (B) Top view of micromanipulation setup, showing aluminum cutout slide, holding and suction-transfer pipettes, three 5-μl drops of dCC/M2 containing, respectively, no embryos (B), C57 embryos (C), and NMRI embryos (N), and a 0.5-μl drop of Sendai virus (S). (C) Side view of cross section of setup (at a–a of B); the holding and suction-transfer pipettes are oriented as in (A). CS, Coverslip; DMPS, dimethylpolysiloxane.

sharpened tip by *briefly* touching and then rapidly withdrawing the tip from the 0.06-mm filament (low heat, no air) in the microforge. This is most easily accomplished by hand pressure on the microforge frame while the pipette is held in a vertical position above the filament. Finally, a 20° bend is made 70 μm from the beveled, sharpened tip, at a 90° angle to the bevel (Fig. 1A). Before use, a 50% (v/v) Nonidet P-40 (NP-40)/distilled water solution is drawn into a few suction-transfer pipettes and allowed to stand for 5–10 min before extensive rinsing (four times with three changes) with distilled water. After brief air-drying in a dust-free container, pipettes are ready for use. If pipettes are not used within several weeks of manufacture, they should be washed with chromic acid, washed with concentrated HCl, rinsed well in distilled water, and recoated with NP-40 before use. The acid washes and NP-40 treatment greatly reduce the sticking of embryo membranes to the inner pipette wall; this problem seems more acute in this procedure than in pronuclear transfer or blastocyst injection, perhaps because of the large volume of the transferred cytoplast and the resulting extensive close contact with the pipette interior.

Isolation of Embryos

Embryos can be isolated from either natural-mated or superovulated female mice. Advantages of both approaches have been discussed[11]; both types of embryos have yielded heteroplasmic founder mice. Although superovulation yields vary, many embryos are obtained from a single female (over 100 eggs have been obtained from a 28-day-old C57 female). However, the quality of embryos is sometimes reduced, and many unfertilized oocytes may be present if well-rested stud males are not used, necessitating sorting during micromanipulation. For superovulation, three females (each, C57 and NMRI) between 24 and 30 days of age are injected with 5 IU PMS between 2 and 5 pm. Forty-eight hours later, 5 IU HCG is injected, and at 5 pm a single female is placed with a stud male of the same strain. For natural matings, 3–4 females are placed with a stud male; depending on colony experience and season, 40 random-cycling females should yield 4–8 mated females. In either case, females are examined for copulatory plugs before 11 am the next day. Mating takes place at approximately midnight. Naturally plugged females (and all superovulated females) are sacrificed by CO_2, and oviducts (attached to a small length of uterus) are removed to M2 medium containing 0.3 mg/ml hyaluronidase. Under a dissecting microscope at 12–50× magnification, the ampulla is torn open using No. 5 Dumont tweezers, and the single-cell embryos are freed from attached cumulus cells, transferred though several washes of M2 into M16 media, and placed in a 37° CO_2 incubator. Superovulated females should yield 15–30 embryos, although much higher numbers are common; natural-mated C57 and NMRI females yield 6–10 and 10–18 embryos, respectively. With experience, embryos can easily be ready for manipulation by 12 noon, leaving a 12-hr span for manipulation before the breakdown of the pronuclear membranes that precedes division to the two-cell embryo renders the embryos too fragile for manipulation.

Cytoplast Transfer between Embryos

Six to eight embryos of each strain are transferred from M16 to CC/M2 (M2 plus 8.3 μg/ml cytochalasin D, 0.17 μg/ml colchicine) and incubated for 25–30 min at 37° in the CO_2 incubator. While the embryos are incubating, a siliconized 25 × 60 mm No. 1 coverslip is stuck to an aluminum cutout slide (Fig. 1B,C) with sterile petroleum jelly and securely taped to the aluminum slide. The holding pipette, held in a Leitz micromanipulator instrument holder connected by air-filled Tygon tubing to a 60-ml plastic syringe, is mounted in the left micromanipulator. Similarly, the suction-transfer pipette is filled with dimethylpolysiloxane (Sigma, DMPS-5×), inserted into an instrument holder connected by air-filled Tygon tubing to a 5- or 10-ml plastic syringe, and mounted in the right micromanipulator. The holding and suction-transfer pipettes are adjusted to a level position and centered over the objective. Both pipettes are lowered into a petri dish containing dCC/M2 (CC/M2:M2, 1:4 v/v); 50 100 μl dCC/M2 is drawn into the holding pipette and 0.1–0.2 nl (i.e., 200–400 μm distance) into the suction-transfer pipette. The syringes are carefully adjusted for no movement of the medium. We and others have used oil-filled syringes or the DeFonbrune–Beaudouin syringes successfully[14,15]; air-filled tubing and syringes are equally effective if the medium–air or DMPS–air interface occurs at an internal pipette diameter of approximately 0.5 mm or greater.

After 25–30 min in CC/M2, embryos are transferred to dCC/M2. Working rapidly to minimize evaporation, three 5-μl drops of dCC/M2 are placed on the siliconized coverslip, embryos are transferred to two of the drops, a 0.5-μl drop of inactivated Sendai virus is added, and the drops are covered with 1.0 ml DMPS (Fig. 1B,C). The slide is moved to the microscope and pipettes lowered into the empty drop through the DMPS and checked for neutral pressure. The slide is then moved so that the instruments enter the drop containing embryos of the first strain (e.g., C57), and manipulation is begun.

Figure 2 shows the actual procedure. The embryos are examined under high magnification for two pronuclei, polar bodies, and any abnormalities. A suitable C57 embryo is seized by the holding pipette and oriented so the polar body is opposite the opening of the holding pipette. Suction is applied, and the zona pellucida (ZP) and polar body are pulled into the holding pipette bore, serving as a cushion. Focus is adjusted so the ZP in the holding pipette bore is sharp; the sharpened nib of the suction-transfer pipette is brought into sharp focus by vertical adjustment of the micromanipulator; this indicates that the embryo midline and nib are in the same plane. Using a moderately fast motion, the suction-transfer pipette is driven into the embryo until the nib enters the bore of the holding pipette. The ZP should be cut and the embryo plasma membrane stretched around the suction-transfer pipette opening. Gentle suction is applied. It will be necessary to adjust focus and the position of the suction-transfer pipette as embryo cytoplasm, in the form of a large cytoplast, is withdrawn, to avoid drawing in the pronuclei. When 60–80% of the cytoplasmic volume has been removed, the suction-transfer pipette can be withdrawn slowly from the embryo; the plasma membrane should seal as the membranes stretch. The "reduced" C57 embryo in the holding pipette is moved to a vacant area of the medium drop, and holding pipette suction is released. As this is the first embryo to have its cytoplasm reduced, the resulting cytoplast in the suction-transfer pipette bore can be gently expelled from

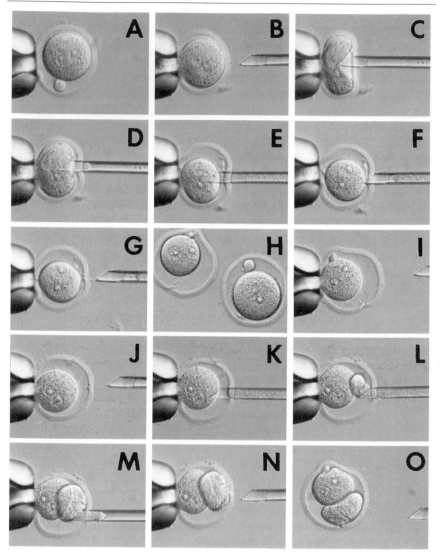

FIG. 2. Construction of heteroplasmic embryos by cytoplast transfer. (A–C) Examination, orientation, and puncture of zona pellucida of a C57 embryo. (D–G) Withdrawal of C57 cytoplast, enclosed by plasma membrane, with avoidance of male and female pronuclei. Note sealing of karyoplast and cytoplast membranes in (G). (H) A "reduced" NMRI karyoplast (left) compared to a normal NMRI embryo (right); over 60% of the cytoplasmic volume has been removed. (I–K) Orientation and insertion of the suction-transfer pipette into the zona pellucida opening of a previously "reduced" NMRI embryo. (L–N) Cytoplast injection into the "reduced" NMRI embryo. (O) An unfused heteroplasmic embryo with a large area of membrane contact.

the pipette. The two instruments are centered, and the slide is shifted to move them into the second drop, containing NMRI embryos.

In the second drop of medium, an NMRI embryo is similarly "reduced." After releasing the reduced NMRI embryo in a vacant area, the two instruments are centered, with the suction-transfer pipette still containing a "donor" NMRI cytoplast. Subsequent steps depend on the fusion method. If Sendai virus is to be used to fuse the NMRI donor cytoplast to the reduced C57 karyoplast, then the slide is moved to bring the suction-transfer pipette into the droplet containing Sendai virus. Then 50–100 pl inactivated Sendai virus is drawn into the bore (the cytoplast moves further into the pipette). If electrofusion is used, this step is omitted. In either case, the slide is moved to return the instruments to the drop containing the C57 embryos. The previously reduced embryo is picked up by the holding pipette and maneuvered so that the ZP, which is usually still deformed by previous suction, can be drawn into the holding pipette bore in the same orientation as in the initial puncture and suction operation. This should place the cut in the ZP in the plane of focus, allowing the suction-transfer pipette to be gently reinserted (once an opening is cut, internal pressure is released and the pipette cannot make a new hole). The NMRI donor cytoplast (and the Sendai virus, if present) is gently expelled into the reduced C57 embryo. Ideally, the C57 karyoplast and the NMRI cytoplast will be in close contact over a considerable surface. The potential heteroplasmic embryo is moved to another area of the medium drop and released. A second C57 embryo is selected, cytoplasm removed, and the resulting C57 donor cytoplast used to fill the first reduced NMRI embryo. This process is repeated until all embryos in the drops have been used, rejected for abnormalities, or broken during the manipulation process. With experience, cytoplasts can be transferred between a pair of embryos in 20–30 min.

When all suitable embryos are used, the slide is removed from the microscope, the DMPS removed, and the potential heteroplasmic embryos transferred into two deep well-slides containing M2, separated by nuclear (i.e., the reduced) genotype. The holding and suction-transfer pipettes can be lowered into a petri dish containing dCC/M2, flushed with fresh medium, and used for the next set of embryos. If Sendai virus was used to fuse the heteroplasmic embryos, the embryos are then transferred to M16 and incubated overnight to assess survival and division to the two-cell stage, before transfer to pseudopregnant foster mothers. If heteroplasmic embryos are to be created by electrofusion, groups of 3–4 (of the same nuclear genotype) are transferred to a deep well-slide on the stereomicroscope stage. The deep well-slide contains M2 and the previously adjusted electrodes of the electrofusion device (Fig. 3). The electrodes are adjusted flat on the well-slide bottom and 110–120 μm apart. One at a time, embryos

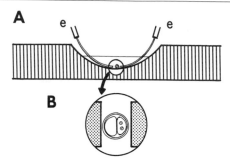

FIG. 3. Heteroplasmic embryo electrofusion. (A) Side view of well slide, showing electrodes (e) of electrofusion device with embryo between electrodes. (B) Magnified top view, showing orientation of the plane of contact between cytoplast (left) and karyoplast (right).

are pushed into the space between the electrodes and oriented so that the plane of contact between the karyoplast and the donor cytoplast is perpendicular to the field (Fig. 3B). A single pulse (100 V DC, 100 μsec) is given, the embryo is removed, and the next embryo is oriented and pulsed. Fusion generally occurs within 1 hr. If embryos do not fuse, additional pulses can be given, although repeated pulses reduce the frequency of subsequent division to the two-cell stage. Pulsed embryos are transferred to M16 medium and incubated overnight, as with Sendai virus-fused embryos. As a final step, 40–60 C57BL/6 $\times$ DBA/2 F_1 females (in groups of 3–4) are placed with vasectomized males during the late afternoon–early evening of the same day that embryo manipulation is carried out. These females will thus be retarded by 24 hr in their pregnancy, compared to the manipulated embryos.

Implantation in Pseudopregnant Foster Females

The F_1 females are checked for copulatory plugs the following morning and pseudopregnant fosters separated. The potential heteroplasmic embryos, which have been incubated at 37° overnight, are carefully examined in the microscope for fusion and subsequent division to the two-cell stage. Although 80–90% of unmanipulated embryos incubated in M2 will divide to two cells by 10 am, fused heteroplasmic embryos are often delayed in development. Some embryos will have divided to the two-cell stage, whereas others will remain as one-cell embryos; the remainder will not have fused or will have divided abnormally. Normal-appearing, fused, one- and two-cell embryos should be separated from abnormal and unfused embryos. Both types of fused embryos will potentially yield live heteroplasmic pups; abnormal appearing embryos have not given viable pups, whereas unfused embryos (or "reduced" embryos without an introduced cytoplast) can give nonheterpolasmic pups if 40–50% of the original cytoplasmic volume remains. Fused embryos of each genotype (i.e., C57 donor cytoplast into NMRI recipient karyoplast, NMRI cytoplast into C57 karyoplast) are transferred to M2 medium in groups of four to six, depending on the number of pseudopregnant foster mothers available. An additional six to eight normal two-cell embryos of the opposite genotype (C57 or NMRI, respectively) or a different genotype with distinctive coat color (e.g., DBA or 129) are added, for a total of twelve two-cell embryos per foster mother. These additional embryos are necessary for the efficient establishment and progress of pregnancy and subsequent maternal care, as well as allowing immediate visual identification of litters containing potential heteroplasmic pups.

Plugged pseudopregnant females are anesthetized with Avertin, the oviduct exposed, and the twelve embryos (fused heteroplasmic embryos plus genetically distinct carriers) are transferred into the ampulla using the methods developed for pronuclear-injected embryos.[11] Pregnancy can usually be determined by 12–14 days posttransfer; pups are delivered 20–21 days posttransfer. Fostering of pups is usually not required if F_1 host females are used. At 2–3 weeks of age, pups with the distinctive coat color of the "recipient" embryo can be numbered and a segment of tail removed for DNA preparation. The presence of the donor mtDNA genotype is confirmed and the ratio of the two mtDNAs determined, either by PCR and digestion of the amplified product with a diagnostic restriction enzyme or by restriction enzyme digestion of total DNA followed by Southern blotting and probing with an appropriate mtDNA sequence. Female mice can be bred to produce heteroplasmic offspring.

Results: Efficiency and Important Factors

A total of sixteen heteroplasmic mice have been born and analyzed; another two pups with coat color indicating potential heteroplasmy were seen but disappeared soon after birth. Nine pups were NMRI → C57 heteroplasmics (three females); seven were C57 → NMRI heteroplasmics (three females). After several initial experiments, where fusion efficiency using Sendai virus appeared low and unfused embryos may not have been completely separated from fused, two-cell, heteroplasmic embryos (an unfused embryo is difficult to distinguish from a two-cell embryo unless the nuclei can be seen), all mice born with the coat color of the recipient karyoplast proved to be heteroplasmic. The degree of heteroplasmy (in tail DNA) ranged from 5 to 80% of the introduced mtDNA genotype.

The success frequency depends on a number of factors, including embryo manipulation technique, fusion efficiency, embryo damage (i.e., poor division to two-cell stage), and oviduct transfer efficiency. In one extensive series of experiments using Sendai virus fusion, 770 embryos were isolated, carefully examined, and manipulated, resulting in 283 potential fusion embryos (37%). Only 96 (34%) of these fused and yielded normal-appearing, two-cell embryos after overnight culture. Oviductal transfer was successful in 31 of 48 pseudopregnant females (65%), resulting in a further 35% loss of the "good" two-cell fusion. In addition, not all embryos in successful transfers survive. The birth rate of carrier embryos (also cultured in M16 overnight) in the successful transfers was 54%. There was no additional loss due to cytochalasin and colcemide treatment, as the birth rate of carrier embryos, incubated in CC/M2 and dCC/M2, was between 50 and 60% in a series of control experiments. Thus, the absolute frequency of heteroplasmic mouse construction using Sendai virus is low, with nine heteroplasmic mice resulting from the 96 "good" fusions, or an overall 9% recovery.

Eleven mice resulted from 10 successful manipulation sessions, out of 87 using Sendai virus. Although Sendai virus can be an extremely effective fusing agent (in two separate experiments, two heteroplasmic mice resulted from the transfer of four potential fusion embryos, a 50% yield), many transfers yielded nothing. This frequency is much lower than that reported for pronuclear transfer.[14,15] One important difference may be the major disruption of the embryo, since 50% or more of the cytoplasm is being transferred, compared to pronuclear transfer, which involves much less volume. Significant stirring of cytoplasmic contents occurs, which may not be easily reversible. A corollary is that the increased area of surface–surface contact, covered with Sendai virus, may be disruptive to early developmental stages, either by changing plasma membrane properties or causing extensive endocytosis. Sendai virus titer is critical for efficient fusion[14,15]; significant embryo swelling and lysis occurred when excess virus was used.

Electrofusion has also been used for embryo fusion, particularly in the construction of tetraploid embryos.[20,21] In a second set of experiments involving heteroplasmic embryos created by electrofusion rather than Sendai virus, oviduct transfer of 25 fused embryos resulted in five heteroplasmic mice, a success rate of 20%. One additional pup with a coat indicating potential heteroplasmic origin was seen but disappeared soon after birth. Other factors were similar, with four of seven oviduct transfers yielding pups (57%) and 13 out of 26 carrier embryos born (50%). Although the results are more limited, electrofusion appears easier and more controllable than Sendai virus.

[20] A. Nagy, E. Gocza, E. M. Diaz, V. R. Prideaux, E. Ivanyi, M. Markkula, and J. Rossant, *Development (Cambridge, UK)* **110,** 815 (1990).
[21] M. H. Kaufman and S. Webb, *Development (Cambridge, UK)* **110,** 1121 (1990).

Conclusions

Mitochondrial DNA heteroplasmy can be artificially created in mice by using a combination of methods developed for altering the nuclear genotype of mouse embryos. Although difficult, the procedures are feasible for any laboratory experienced in transgenic mouse production. Artificial heteroplasmic mice differ from naturally occurring heteroplasmic animals in having multiple sequence differences in their mtDNA genomes, making it easy to distinguish the rate of segregation from mutation. Studies with these animals suggest several unexpected events can occur during mtDNA segregation and selection in mammals.

Acknowledgments

I am indebted to Drs. Davor Solter and especially Rudolf Jaenisch, who provided advice, reagents, and access to equipment. Dr. John Aronson, Dr. Ted Choi, and Ms. Jessica Dausman patiently instructed me in embryo manipulation. Work was supported by National Institutes of Health Grant GM 33537.

[32] Mitochondria-Mediated Transformation of *Drosophila*

By ETSUKO T. MATSUURA and YUZO NIKI

Introduction

The relationship between mitochondrial and nuclear genomes can be studied by the introduction of foreign mitochondrial genomes into a cell or an organism. Elsewhere in this volume, Butow *et al.*, Boynton and Gillham, King and Attardi, and Chomyn have described several elegant mitochondrial transformation experiments that have been carried out in unicellular systems since 1988.[1–3] In contrast, the establishment of a novel combinaton of nuclear and mitochondrial genomes in multicellular organisms was not successful until 1989.[4] Although substitution of the nuclear genome by the repeated crossing of males is theoretically possible, it is practically impossible owing to the fact that most interspecific hybrids are sterile. The absence of appropriate mitochondrial genetic markers that

[1] T. D. Fox, J. C. Sanford, and T. W. McMullin, *Proc. Natl. Acad. Sci. U.S.A.* **85,** 7288 (1988).
[2] S. A. Johnston, P. Q. Anziano, K. Shark, J. C. Sanford, and R. A. Butow, *Science* **240,** 1538 (1988).
[3] M. P. King and G. Attardi, *Cell* (*Cambridge, Mass.*) **52,** 811 (1988).
[4] Y. Niki, S. I. Chigusa, and E. T. Matsuura, *Nature* (*London*) **341,** 551 (1989).

show phenotypic expression makes it difficult to detect variant types of mitochondria. Furthermore, recombination between mitochondrial DNA (mtDNA) molecules has not been reported in animals. All of these factors account for the fact that manipulation of mitochondrial genomes has not been carried out in multicellular organisms.

Instead, heteroplasmy in which more than two types of mtDNA coexist has provided experimental systems to study the transmission of mtDNA in various animals.[5–8] Heteroplasmic individuals have been found in natural populations but usually occur at only a very low frequency. By inducing an artificially constructed heteroplasmic state,[9] replacement of mtDNA in a multicellular organism was first accomplished in *Drosophila*.[4]

Drosophila, especially *Drosophila melanogaster*, is one of the most well-studied organisms in genetics, and a variety of classic genetic and molecular biological methods have been developed to study and manipulate the *Drosophila* genome. Many species of *Drosophila*, from various taxonomic groups, are available and can be easily grown under laboratory conditions. Studies on mitochondria and mtDNA using *Drosophila* have been widely conducted since the 1970s, ranging from the molecular analysis of mtDNA itself[10–14] to evolutionary and population studies.[15–19] The entire mitochondrial genome (16,019 bp) was sequenced in *Drosophila yakuba*, which is closely related to *D. melanogaster*.[13] In various species, including *D. melanogaster*, many parts of the genome have been sequenced.[20–26]

[5] L. D. Densmore, J. W. Wright, and W. M. Brown, *Genetics* **110,** 689 (1985).

[6] P. J. Laipis, M. J. Van de Walle, and W. W. Hauswirth, *Proc. Natl. Acad. Sci. U.S.A.* **85,** 8107 (1988).

[7] D. M. Rand and R. G. Harrison, *Genetics* **114,** 955 (1986).

[8] E. Zouros, K. R. Freeman, A. O. Ball, and G. H. Pogson, *Nature (London)* **359,** 412 (1992).

[9] E. T. Matsuura, S. I. Chigusa, and Y. Niki, *Genetics* **122,** 663 (1989).

[10] M. L. Polan, S. Friedman, J. G. Gall, and W. Gehring, *J. Cell Biol.* **56,** 580 (1973).

[11] C. K. Klukas and I. B. Dawid, *Cell (Cambridge, Mass.)* **9,** 615 (1976).

[12] J. M. Goddard and D. R. Wolstenholme, *Proc. Natl. Acad. Sci. U.S.A.* **75,** 3886 (1978).

[13] D. O. Clary and D. R. Wolstenholme, *J. Mol. Evol.* **22,** 252 (1985).

[14] F. Beziat, F. Morel, A. Volz-Lingenhol, N. S. Paul, and S. Alziari, *Nucleic Acids Res.* **21,** 387 (1993).

[15] R. DeSalle, L. V. Giddings, and A. R. Templeton, *Heredity* **56,** 75 (1986).

[16] L. R. Hale and R. S. Singh, *Proc. Natl. Acad. Sci. U.S.A.* **83,** 8813 (1986).

[17] R. Kondo, Y. Satta, E. T. Matsuura, H. Ishiwa, N. Takahata, and S. I. Chigusa, *Genetics* **126,** 657 (1990).

[18] M. Solignac, M. Monnerot, and J.-C. Mounolou, *J. Mol. Evol.* **23,** 31 (1986).

[19] E. Barrio, A. Latorre, A. Moya, and F. J. Ayala, *Mol. Biol. Evol.* **9,** 621 (1992).

[20] M. H. L. de Bruijn, *Nature (London)* **304,** 234 (1983).

[21] D. O. Clary and D. R. Wolstenholme, *J. Mol. Evol.* **25,** 116 (1987).

[22] Y. Satta, H. Ishiwa, and S. I. Chigusa, *Mol. Biol. Evol.* **4,** 638 (1987).

[23] R. Garesse, *Genetics* **118,** 649 (1988).

We have developed an experimental system for the construction of heteroplasmic individuals of *Drosophila*. The technique is based on the transplantation of germ plasm into eggs to introduce foreign mtDNA into germ-line cells.[9] This system makes it possible to construct heteroplasmy by intra- and interspecific germ-plasm transplantation. The transmission of mtDNA can be easily traced over many generations because of the relatively short generation time of *Drosophila*. Individual *Drosophila* females usually produce enough progeny for the analysis of mtDNA in the next generation. In *Drosophila*, it is even possible to isolate the mtDNA from a single individual.

We analyzed the mtDNA in several heteroplasmic lines for more than 10 successive generations and found selective transmission of mtDNA under supposedly nonselective conditions.[4,27,28] Surprisingly, in some cases, the endogenous mtDNA was completely replaced with the exogenous mtDNA that had been introduced by the transplantation of germ plasm.[4,28] Replacement occurred even when the introduced mtDNA had been derived from a different species. This was the first observation of transformation of a whole multicellular organism by a mitochondrial genome.[4]

In this chapter, we describe the methods for achieving heteroplasmy and the experimental conditions for replacing mtDNA. We also describe the potential usefulness of mitochondria-mediated transformation for the further study of mitochondrial genetics.

Transplantation of Germ Plasm

In higher animals and insects, mitochondria are maternally inherited, and there is only an exceptional occurrence of paternal leakage. During *Drosophila* oogenesis, most of the mitochondria are synthesized in nurse cells and accumulate in the oocytes. Both nurse cells and oocytes are germ-line cells that are segregated from other somatic cells at a very early embryonic stage. At fertilization, a sperm enters into an egg through the micropyle at the anterior pole of the egg. The paternal leakage of mitochondria may occur only when mitochondria in a sperm are incorporated into pole cells at the posterior pole. It is thus very rare that paternal mtDNA is transmitted to the progeny.[17]

Heteroplasmy can occur because of paternal leakage of mitochondria

[24] M. Monnerot, M. Solignac, and D. R. Wolstenholme, *J. Mol. Evol.* **30,** 500 (1990).
[25] K. Tamura, *Mol. Biol. Evol.* **9,** 814 (1992).
[26] D. L. Lewis, C. L. Farr, A. L. Farquhar, and L. S. Kaguni, *Mol. Biol. Evol.* **11,** 523 (1994).
[27] E. T. Matsuura, S. I. Chigusa, and Y. Niki, *Jpn. J. Genet.* **65,** 87 (1990).
[28] E. T. Matsuura, Y. Niki, and S. I. Chigusa, *Jpn. J. Genet.* **66,** 197 (1991).

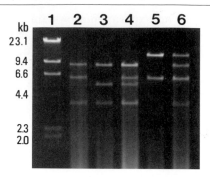

FIG. 1. Examples of mtDNA restriction patterns of the *Drosophila* strains used. The enzyme *Hae*III was used to distinguish the mtDNA of a recipient strain of *D. melanogaster* from the mtDNA of donor strains of *D. melanogaster* and *D. mauritiana*. The restriction patterns of heteroplasmy are also shown. Lane 1 contains size markers (lambda phage DNA digested with *Hind*III); lane 2, a recipient strain of *D. melanogaster* (*bw;e^{11}*); lane 3, a donor strain of *D. melanogaster* (HJ6); lane 4, heteroplasmy by the two strains of *D. melanogaster*; lane 5, a donor strain of *D. mauritiana* (g20); lane 6, heteroplasmy by *D. melanogaster* and *D. mauritiana*.

or mutation of mtDNA in the germ-line cells within an individual. Heteroplasmy is found in natural populations at very low frequencies. Therefore, to study mtDNA transmission genetics using heteroplasmy, it is more convenient to introduce exogenous mitochondria artificially than to use the heteroplasmy found in nature. We applied the germ-plasm transfer technique to introduce mitochondria into the germ-line cells of appropriate recipients. The rationale behind our system comes from the following lines of evidence. First, pole cells (the precursors of germ-line cells) are segregated from other somatic cells by the function of the germ plasm.[29] Second, the germ plasm includes many mitochondria that are located near polar granules which are the main component of the germ plasm.[30] Third, the function of the germ plasm to induce pole cells is not species-specific among *Drosophila*.[31] Therefore, it should be possible to introduce exogenous mitochondria by both intra- and interspecific transplantation.

Strains Used

The incorporation of exogenous mitochondria can be detected by examining the restriction patterns of the mtDNA. The mtDNA from the recipient and donor strains should be easily discriminated by diagnostic bands when they are separated on agarose gels. Figure 1 shows examples in which *D.*

---

[29] K. Illmensee and A. P. Mahowald, *Proc. Natl. Acad. Sci. U.S.A.* **71,** 1016 (1974).
[30] A. P. Mahowald, *J. Exp. Zool.* **167,** 237 (1968).
[31] A. P. Mahowald, K. Illmensee, and F. R. Turner, *J. Cell Biol.* **70,** 358 (1976).

melanogaster and *D. mauritiana* were used as donors for transplantation into *D. melanogaster* recipients. To confirm the absence of cotransplantation of donor nuclei, we used a recipient strain such as the *bw;e^{11}* strain of *D. melanogaster*, in which the chromosomes are marked with recessive genetic markers.

Transplantation Procedure

The procedures essential for successful germ-plasm transfer are described below.[9] The principles underlying the basic techniques of cytoplasmic injections into *Drosophila* eggs have been described by Santamaria.[32]

1. Collect donor and recipient eggs on 2% (w/v) agar plated in plastic dishes. Allow the flies to lay eggs on the plates for 30 to 40 min.
2. Dechorionate manually 20 to 30 eggs for both the donor and the recipient with a pair of watchmaker's forceps under a dissecting microscope. Take care not to injure the eggs.
3. Select healthy eggs and transfer them onto double-stick tape mounted on a glass slide. Donor eggs should be oriented perpendicular and recipient eggs oriented parallel to the axis of the needle of a micromanipulator.
4. Dry the donor and recipient eggs in a small box containing calcium chloride or silica gel. Appropriate drying of the eggs is critical for successful injection. The time necessary will vary depending on the species and strain of *Drosophila*. For example, approximately 10 min is required for *D. melanogaster* eggs, whereas less than 10 min is required for the eggs of *D. simulans* and *D. mauritiana*.
5. Drop Fluoro-silicone oil (FL1,000 c.s., Shin-Etsu, Tokyo, Japan) on the eggs and put a drop of sterilized *Drosophila* Ringer's solution on the outside of the double-stick tape.
6. Kill the embryos that have developed to the syncytial blastoderm stage by impaling them with a fine tungsten needle.
7. Mount the slide on the stage of an inverted microscope.
8. Slowly introduce the tip of a glass needle (5 to 10 μm in diameter) into the posterior pole of a recipient egg and slowly withdraw the germ plasm. Wash the needle several times with *Drosophila* Ringer's solution.

---

[32] P. Santamaria, *in* "*Drosophila*: A Practical Approach" (D. B. Roberts, ed.), p. 159. IRL Press, Oxford, 1986.

9. Introduce the same needle into the dorsal side of a donor egg and then move the tip of the needle to the posterior pole. Slowly withdraw the germ plasm.

10. Reinsert the needle at the posterior pole of the recipient egg and carefully introduce the donor germ plasm through the same site as was used in step 8.

11. Wash the needle several times with *Drosophila* Ringer's solution. Repeat steps 8 through 11 for the remaining eggs.

12. When the series of injections is complete, kill all of the noninjected eggs remaining on the slide.

13. Remove the Ringer's solution with absorbent filter paper and transfer the slides to a moist chamber to incubate at 25° until the injected recipient eggs hatch.

14. Wash the hatched larvae with *Drosophila* Ringer's solution and transfer them onto a piece of sterilized filter paper. Then transfer them onto the *Drosophila* medium.

Efficiency of Incorporation of Donor Mitochondria

Usually, less than 10% of the injected recipient eggs will develop into adult flies. The adult females (founders) are crossed to males of a recipient strain, and the cytoplasmic states of their progeny (G_0) are examined. In one set of experiments,[9] for example, 25 adults were obtained from 415 injected eggs (6%) for intraspecific transplantation in *D. melanogaster*, and 32 adults from 302 eggs (11%) were obtained for interspecific transplantation of *D. mauritiana* germ plasm into *D. melanogaster* eggs. Among them, 10 females from the intraspecific transplantation and 11 females from the interspecific transplantation were fertile. Nine females from each case produced enough progeny for subsequent analyses.

To confirm whether donor mitochondria were incorporated into the germ-line cells of recipient individuals, the mtDNA was extracted from a mass of G_0 flies produced by each founder female. Mitochondria were prepared from the homogenate by differential centrifugation, and the mtDNA was extracted through sodium dodecyl sulfate (SDS) lysis of mitochondria followed by phenol extraction. Appropriate restriction enzymes were applied to distinguish the recipient from the donor mtDNA, and the mtDNA was separated on agarose gels. In the cases described above, the donor-derived mtDNA was detected in the progeny of six of nine and five of nine founders from the intra- and interspecific transplantation, respectively. In other transplantation experiments shown in Table I, donor-derived mtDNA was detected in the progeny of 10–100% of the founder females.

TABLE I

ANALYSIS OF G_0 FEMALES AND REPLACEMENT OF ENDOGENOUS MITOCHONDRIAL DNA
IN THEIR PROGENY[a]

Species combination		Number of females examined	Number of heteroplasmic females	Replacement	Ref.
Recipient	Donor				
D. melanogaster	D. melanogaster	16	10	+	c
		43	17	−	d
		43	16	+	d
		24	20	+	e
		135	2	−	e
D. melanogaster	D. simulans(siII)[b]	10	8	−	f
		116	71	−	d
D. melanogaster	D. simulans(siIII)	91	59	+	g
D. melanogaster	D. mauritiana(maI)[b]	27	18	+	c
		37	28	+	d
		40	23	+	d
		20	7	+	h
		33	30	+	h
D. melanogaster	D. mauritiana(maII)	14	7	+	f
		29	15		f
D. melanogaster	D. sechellia	118	64	−	i
D. simulans(siII)	D. mauritiana(maI)	14	9	−	h
D. mauritiana(maI)	D. simulans(siII)	11	2	−	h

[a] The total number of isofemale lines which were initially established and those which were heteroplasmic are shown for each combination. For some species combinations, the results of several experiments using different recipient or donor strains are shown separately. For each combination of recipient and donor, four to six lines were examined for changes in the relative proportion of the two types of mtDNA for more than 10 generations. Replacement of endogenous mtDNA is indicated by a plus symbol (+), whereas the lack of replacement is indicated by a minus sign (−).

[b] *Drosophila simulans* and *D. mauritiana* are classified into cytoplasmic races based on mtDNA restriction patterns. Three races (*siI*, *siII*, and *siIII*) belong to *D. simulans,* and two races (*maI* and *maII*) belong to *D. mauritiana* [M. Solignac, M. Monnerot, and J.-C. Mounolou, *J. Mol. Evol.* **23,** 31 (1986)].

[c] E. T. Matsuura, S. I. Chigusa, and Y. Niki, *Genetics* **122,** 663 (1989).

[d] Y. Tsujimoto, Y. Niki, and E. T. Matsuura, *Jpn. J. Genet.* **66,** 609 (1991).

[e] Y. Nagata, Y. Tsujimoto, and E. T. Matsuura, unpublished data (1992).

[f] Y. Niki and E. T. Matsuura, unpublished data (1988).

[g] N. Yamamoto, Y. Niki, and E. T. Matsuura, unpublished data (1991).

[h] E. T. Matsuura, Y. T. Tanaka, and N. Yamamoto, in preparation.

[i] Y. Tsujimoto and E. T. Matsuura, unpublished data (1992).

The incidence of donor-derived mtDNA varied among strain-combinations (data not shown).

Other Procedures for Introduction of Exogenous Mitochondria

As described above, the efficiency of incorporation of donor mitochondria was not necessarily high in germ-plasm transplantation. Both recipient and donor germ plasm may contribute to the induction of pole cells. To assure that pole cells are induced only by donor-derived germ plasm, the germ plasm can be introduced at the anterior pole of the recipient egg. When ectopic pole cells are transplanted to the posterior pole of a recipient embryo in which the germ-line cells are degenerated, only the transplanted pole cells develop into functional germ cells.[33] However, this procedure is complex, and the yield of adult flies is low. Consequently, we did not use this approach.

Instead of germ-plasm transplantation, cytoplasmic transplantation has been carried out to induce heteroplasmy in *D. simulans*.[34] Although the origin of the cytoplasm was not clear, heteroplasmy was induced, and the results obtained were similar to those that we obtained.

Another possible way to introduce foreign mitochondria or mtDNA into germ-line cells would be to inject purified mitochondria directly into eggs. If this can be done, it may provide a basis for a method by which manipulated mitochondrial genomes could be introduced into the cells. However, even though mitochondria were carefully purified in several different buffers,[3] the injection of purified mitochondria into the posterior pole of eggs has not resulted in the successful introduction of donor mtDNA.[35]

Analysis of Transmission of Exogenous Mitochondrial DNA in Heteroplasmic Lines

Figure 2 provides a schematic representation of the procedure for establishing heteroplasmic lines. When two types of mtDNA were detected in mtDNA samples from the G_0 flies from a founder female, the G_0 females were individually examined for their cytoplasmic state as follows. Many isofemale lines were established from individual G_0 females and

---

[33] Y. Niki, *Dev. Biol.* **113,** 255 (1986).

[34] E. de Stordeur, M. Solignac, M. Monnerot, and J.-C. Mounolou, *Mol. Gen. Genet.* **220,** 127 (1989).

[35] M. Dejima, T. Sakai, and E. T. Matsuura, unpublished data (1993).

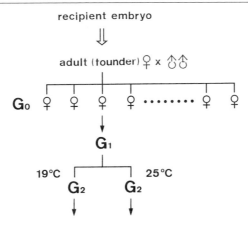

FIG. 2. Procedure for establishment and maintenance of heteroplasmic lines. [Modified from E. T. Matsuura, *Jpn. J. Genet.* **66**, 683 (1991) with permission.]

the mtDNA of their progeny (G_1) was examined. Usually 50 to 100 flies were used for the mtDNA extraction. Heteroplasmic females (G_0) were defined when both recipient- and donor-derived mtDNA were detected in mtDNA samples from the G_1 flies. The amount of donor-derived mtDNA detected in the progeny varied among strain combinations and also varied among individuals from the same combination. Table I shows the number of heteroplasmic G_0 females found following the intra- and interspecific transplantation of germ plasm. These heteroplasmic lines were maintained by brother–sister matings using more than 50 flies for every successive generation.

To compare the mode of mtDNA transmission at different temperatures, we divided the progeny of each heteroplasmic line into two groups at either the G_2 or G_3 generation (Fig. 2). One was maintained at 19° and the other at 25° in the same manner. It should be noted that the two lines established from an original heteroplasmic line were genetically identical. During the maintenance of heteroplasmic lines at the two different temperatures, the relative proportion of donor and recipient mtDNA was monitored for more than 10 generations. The mtDNA was usually extracted from the mass of flies which were used as the parents of the next generation. The relative proportion of each type of mtDNA was estimated from the intensity of the diagnostic bands for each type of mtDNA by photographing ethidium bromide-stained gels and scanning the negative with a densitometer. To confirm the heteroplasmic state in each individual, mtDNA was extracted from the progeny of individual females.

Replacement of Endogenous Mitochondrial DNA
 with Exogenous Mitochondrial DNA

In the heteroplasmic state, one of the two types of mtDNA was transmitted selectively, regardless of whether the two types of mtDNA originated from the same or different species. We observed the selective transmission of mtDNA in all cases of heteroplasmy constructed (Table I). One of the most interesting features revealed in our system was that the selection was dependent on the temperature at which the heteroplasmic lines were maintained.[28,36,37] Both the preference for the type of mtDNA and the intensity of selection were dependent on temperature. The selection was sometimes strong enough to completely replace the endogenous mtDNA in an individual with the exogenous mtDNA within 10 to 15 generations.[28,36,37] In most cases, the replacement was observed in all of the individuals within a heteroplasmic line.

The occurrence of replacement is indicated in Table I. The complete replacement of endogenous mtDNA was observed in all of the heteroplasmic lines examined in the following three combinations of recipient and donor (Table I): (i) *D. melanogaster* as both recipient and donor, (ii) *D. melanogaster* as recipient and race *si*III of *D. simulans* as donor, (iii) *D. melanogaster* as recipient and race *ma*I of *D. mauritiana* as donor. In the intraspecific combinations within *D. melanogaster*, replacement was observed at both 19° and 25°, although the intensity of selection varied with temperature. In both of the interspecific combinations described above, the donor-derived mtDNA was selectively transmitted and eventually replaced the recipient mtDNA at 25°. At 19°, however, transmission of the donor mtDNA depended on the strain of the recipient. When *D. melanogaster* was used as the recipient and either *D. melanogaster* or race *ma*II of *D. mauritiana* was the donor, only one heteroplasmic line in each exhibited replacement of recipient mtDNA by donor mtDNA.

A typical example of temperature-dependent replacement of mtDNA is shown in Fig. 3 in which the *bw;e^{11}* strain of *D. melanogaster* was used as the recipient and the g20 strain of *D. mauritiana*(*ma*I) was used as the donor. When the heteroplasmic lines were maintained at 25°, the proportion of donor mtDNA increased and recipient *D. melanogaster* mtDNA was replaced after more than 10 generations. In contrast, when the lines were maintained at 19°, donor mtDNA was lost. Individuals possessing only exogenous mtDNA were as viable and fertile as those of the *bw;e^{11}* strain under the experimental conditions. The electron-transport activity of trans-

[36] Y. Tsujimoto, Y. Niki, and E. T. Matsuura, *Jpn. J. Genet.* **66,** 609 (1991).
[37] E. T. Matsuura, Y. Niki, and S. I. Chigusa, *Jpn. J. Genet.* **68,** 127 (1993).

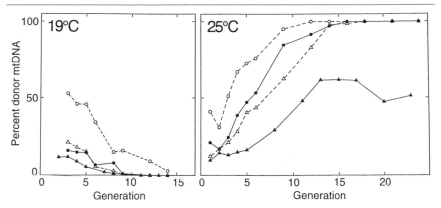

FIG. 3. Examples of changes in the proportion of donor-derived mtDNA in heteroplasmic lines. Heteroplasmy was constructed using *D. melanogaster* (*bw;e^{11}*) as the recipient and race *ma*I of *D. mauritiana* (g20) as the donor. The same four lines were maintained at both 19° and 25°. [Reprinted with permission from E. T. Matsuura, *Jpn. J. Genet.* **66**, 683 (1991).]

formed mitochondria is not significantly different from those of the *bw;e^{11}* strain.[38] In the reciprocal combination, where *D. mauritiana* is used as the recipient and *D. melanogaster* is used as the donor, heteroplasmy was not successfully obtained. It may be that the *D. melanogaster* mitochondrial genome cannot be maintained under a *D. mauritiana* nuclear genome.

Factors Affecting Replacement of Endogenous Mitochondrial DNA

In our studies, replacement of endogenous mtDNA was found to depend on two factors, namely, temperature and the combination of donor and recipient species. For mtDNA replacement to occur, it is necessary to choose the appropriate combination of nuclear and mitochondrial genomes. In addition, the direction and intensity of selection for a given combination of nuclear genome and mtDNA can be controlled by changing the temperature at which the heteroplasmic lines are maintained.

The involvement of the nuclear genome in determining selection has been clearly demonstrated in experiments employing reciprocal transplantation both within and between species[36,39] and nuclear substitution in *D. melanogaster*.[39] For one pair of coexisting types of mtDNA, one type of mtDNA was preferentially transmitted under one nuclear genome, but the other type of mtDNA was advantageous under a different nuclear

---

[38] Y. Nagata and E. T. Matsuura, *Jpn. J. Genet.* **66**, 255 (1991).
[39] E. T. Matsuura, Y. T. Tanaka, and N. Yamamoto, in preparation.

genome. These findings indicate that the genotype of the nuclear genome affects the selective transmission of mtDNA.

The mitochondrial genome itself also has an effect on selection. In experiments using *D. simulans* (*si*III) and *D. mauritiana* (*ma*I) as donors and *D. melanogaster* as recipient, the donor mtDNA was selectively transmitted under the nuclear genome of *D. melanogaster* at 25° in both cases. The intensity of selection, however, differed between the two different donor species.[40,42] As these two types of mtDNA are closely related and have almost identical coding sequences,[41] some part of the noncoding region of the mtDNA may be partly responsible for this temperature-dependent selection.

Although the molecular mechanisms of the temperature-dependent selective transmission of mtDNA are not currently known, experiments suggest that the mechanisms of mtDNA replication may be involved.[40,42] The genetic and molecular mechanisms underlying the selective transmission of mtDNA will be elucidated by further studies in *Drosophila*. When these mechanisms are understood, it should be possible to predict which mtDNA will be selectively transmitted, and whether there will be replacement of mtDNA. *Drosophila* that is transformed to possess only exogenous mtDNA will provide a novel experimental system for the study of the genetic and functional relationships between nuclear and mitochondrial genomes in multicellular animals.

Acknowledgments

The authors thank Prof. Sadao I. Chigusa, Ochanomizu University, Tokyo, for encouragement and valuable comments during the course of this study. We also thank Yuka Nagata, Yuko T. Tanaka, Noriko Yamamoto, Chieko Akiyama, Miori Dejima, Tokiko Sakai, and Eiko N. Hatano for their collaboration and contribution of unpublished data. This work was supported in part by Grants-in-Aid for Scientific Research from the Ministry of Education, Science and Culture of Japan.

[40] N. Yamamoto, Y. Niki, and E. T. Matsuura, unpublished data (1992).
[41] Y. Satta and N. Takahata, *Proc. Natl. Acad. Sci. U.S.A.* **87,** 9558 (1990).
[42] T. Sakai and E. T. Matsuura, unpublished data (1994).

[33] Allotopic Expression of Mitochondrial ATP Synthase Genes in Nucleus of *Saccharomyces cerevisiae*

By ROBYN E. GRAY, RUBY H. P. LAW, RODNEY J. DEVENISH, and PHILLIP NAGLEY

Introduction

Yeast has been the focus of many investigations of both the function of mitochondria and the assembly of mitochondrial enzyme complexes. Most genes encoding mitochondrial proteins are located in the nucleus. A small number of inner membrane proteins, however, is consistently encoded in the mitochondrial DNA (mtDNA) of many organisms. These include subunits 6 and 8 of the ATP synthase (mtATPase) complex, subunits I, II, and III of the cytochrome-*c* oxidase complex, and the cytochrome *b* apoprotein of the coenzyme QH_2–cytochrome-*c* reductase (*bc*$_1$ complex).[1] It has been proposed that the extreme hydrophobic nature of this group of membrane proteolipids limits their site of synthesis to within the mitochondrion.[2] However, one proteolipid, subunit 9 of mtATPase, has been found to be expressed in either the nucleus or the mitochondrion, depending on the organism. For example, subunit 9 of the yeast *Saccharomyces cerevisiae* is encoded by the *oli1* gene[3] in mitochondria, whereas the homologous subunit 9 of the filamentous fungus *Neurospora crassa* is encoded in the nucleus.[4] Nuclear expression of subunit 9 and subsequent delivery of the protein product to the mitochondria is associated with the presence of a 66-amino acid, positively charged, hydrophilic leader sequence (N9L) fused to the amino terminus of the mature protein. The leader sequence serves to confer an overall polar character to the precursor, maintaining its solubility within the aqueous environment of the cytosol as well as directing the protein to the mitochondria and facilitating its import into the matrix. Once the precursor has reached the matrix, the N-terminal leader sequence is cleaved before assembly of mature subunit 9 into mtATPase.[4]

Many studies of mitochondrially encoded proteins have relied on mutations within the mtDNA occurring *in vivo*.[5] Analysis was then made of the effect of such extranuclearly inherited mutations on the structure, assembly,

[1] G. Attardi and G. Schatz, *Annu. Rev. Cell Biol.* **4**, 289 (1988).
[2] G. von Heijne, *EMBO J.* **5**, 1335 (1986).
[3] W. Sebald, E. Wachter, and A. Tzagoloff, *Eur. J. Biochem.* **100**, 599 (1979).
[4] A. Viebrock, A. Perz, and W. Sebald, *EMBO J.* **1**, 565 (1982).
[5] P. Nagley, *Trends Genet.* **4**, 46 (1988).

and function of the altered mitochondrial complexes. Clearly, such studies would be profitably extended by methods which allowed the *in vitro* mutagenesis of mitochondrial genes to be followed by the efficient introduction into mitochondria of such mutated genes. Such an approach has been hampered by the lack of an efficient system for the direct transfection of mtDNA into the mitochondria of host yeast cells. Advances have been made using the biolistic technique, enabling microscopic projectiles coated with DNA to be shot directly into target cells.[6] This technique has not yet been applied to the systematic structure–function analysis of mitochondrially encoded proteins.

To circumvent the problems associated with achieving the direct transfection of DNA into mitochondria, we devised an alternative strategy modeled on the extramitochondrial synthesis and import route characteristic of nuclearly encoded mitochondrial proteins. This approach involves the expression in the nucleus of a mitochondrial gene incorporated into a yeast expression vector, and subsequent import of the protein product into the mitochondria where it is available for assembly into a fully functional enzyme complex. This strategy has been termed allotopic expression.[7] Such a system allows advanced techniques of yeast nuclear gene manipulation to be applied to the systematic molecular genetic analysis of proteins that are normally encoded in the mitochondria and whose assembly and function are to be assessed within the mitochondrial compartment.

In this chapter we describe molecular genetic methods for achieving allotopic expression and analysis of the outcomes mainly at the cellular level. We focus on subunit 8 of mtATPase of yeast. Technical aspects of the biochemical analysis of mitochondrial enzyme complexes and their assembly, particularly mtATPase,[8] as well as protein import,[9] that play an important role in allotopic expression studies are detailed elsewhere.

Molecular Genetic Rationale and Key Genetic Tests

Gene Manipulations

A primary factor to be considered in order to achieve optimal expression in the nucleus of a mitochondrial gene is the differences in the codon dictionary existing between the mitochondrial and nuclear genomes. In the

[6] R. A. Butow and T. D. Fox, *Trends Biochem. Sci.* **15**, 465 (1990).
[7] P. Nagley and R. J. Devenish, *Trends Biochem. Sci.* **14**, 31 (1989).
[8] R. H. P. Law, S. Manon, R. J. Devenish, and P. Nagley, this series, Vol. 260 [10].
[9] R. H. P. Law and P. Nagley, *Methods Mol. Biol.* **37**, 293 (1995).

case of *S. cerevisiae*,[10] these differences are as follows: CUN codes for threonine in mitochondria and for leucine in the nucleus, AUA codes for methionine in mitochondria and for isoleucine in the nucleus, and UGA codes for tryptophan in mitochondria and for stop in the nucleus. Any such differences existing between the mitochondrially encoded gene and the form of the gene required for accurate expression in the nucleocytosolic system may be corrected either by complete chemical synthesis of the gene[11] or by *in vitro* mutagenesis of cloned mtDNA.

In addition there is also a difference in the codon preference between mitochondrial and nucleocytosolic translation systems. In undertaking the allotopic expression of mtATPase subunit 8, the *aap1* gene encoding that subunit was redesigned[11] to accommodate the differences in both codon dictionary and preferred codon usage[12] between yeast mitochondria and nucleus. The resulting artificial gene encoding subunit 8 (Y8) contained changes to 30 of the total 48 codons. However, it is likely that changes to accommodate differences in preferred codon usage may be of lesser importance. Thus, in one study it was found that nuclear expression of the gene encoding mtATPase subunit 8 from the filamentous fungus *Aspergillus nidulans* in the mutant yeast strain M31 (see below) lacking endogenous subunit 8 was able to restore the ability of that strain to utilize a nonfermentable substrate.[13] In this case no sequence changes were made to the native fungal mitochondrial gene to allow for differences in preferred codon usage between the *A. nidulans* mitochondrial and *S. cerevisiae* nucleocytosolic translation systems. It is of interest to note that there is 50% amino acid sequence divergence between the two protein products; conserved features are confined to overall structural motifs.[13]

A key step in establishing allotopic expression of a mitochondrial gene is to fuse that gene (recoded if necessary) to a DNA sequence encoding a cleavable N-terminal leader sequence which will target the gene product into mitochondria. Subunit 8 has been successfully targeted into mitochondria by fusing the chemically synthesized gene[11] encoding Y8 to a DNA sequence encoding the N-terminal leader sequence of mtATPase subunit 9 (N9L) from *N. crassa*.[14] In the case of the precursor protein,[15] denoted

[10] S. G. Bonitz, R. Berlani, G. Coruzzi, M. Li, G. Macino, F. G. Nobrega, M. P. Nobrega, B. E. Thalenfeld, and A. Tzagoloff, *Proc. Natl. Acad. Sci. U.S.A.* **77,** 3167 (1980).

[11] D. P. Gearing, G. L. McMullen, and P. Nagley, *Biochem. Int.* **10,** 907 (1985).

[12] I. G. Macreadie, C. E. Novitski, R. J. Maxwell, U. John, B. G. Ooi, G. L. McMullen, H. B. Lukins, A. W. Linnane, and P. Nagley, *Nucleic Acids Res.* **11,** 4435 (1983).

[13] A. F. L. Straffon, P. Nagley, and R. J. Devenish, *Biochem. Biophys. Res. Commun.* **203,** 1567 (1994).

[14] D. P. Gearing and P. Nagley, *EMBO J.* **5,** 3651 (1986).

[15] R. H. P. Law, L. B. Farrell, D. Nero, R. J. Devenish, and P. Nagley, *FEBS Lett.* **236,** 501 (1988).

N9L/Y8-1, expression of this fusion gene results in a chimeric protein comprising the N-terminal leader of 66 amino acids, the first 5 amino acids of the mature *N. crassa* subunit 9, two serine residues arising from the DNA sequence at the point of fusion, and the 48 amino acids of yeast subunit 8 (Fig. 1B). This precursor is processed by isolated yeast mitochondria at the natural matrix cleavage site of the *N. crassa* subunit 9 precursor to generate mature subunit 8 protein bearing an additional seven residues at the N terminus.[14] A second Y8 construct, N9L/Y8-2 (Fig. 1E), which encodes a chimeric protein comprising the 66 amino acids of N9L fused directly to the 48 amino acids of Y8,[15] is processed to remove 3 amino acids from the N terminus of Y8 to yield a 45-amino acid mature protein.[16] Expression in the nucleus of the novel genes encoding N9L/Y8-1 and N9L/Y8-2 leads to restoration of function of a host yeast strain unable to synthesize subunit 8 in mitochondria.[15,17]

A second mitochondrial gene encoding a mtATPase subunit, namely, the yeast *oli1* gene encoding the 76-amino acid protein subunit 9, has been recoded for expression in the nucleus.[18] The successful *in vitro* expression and import into mitochondria of the chimeric precursor N9L/Y9-1 has been described.[15,18] The incorporation of allotopically expressed Y9 into functional mtATPase complexes, as shown by growth *in vivo* of a Y9-deficient strain on nonfermentable substrates, is yet to be demonstrated. However, the N9L leader sequence has been successfully utilized to target functional bI4 maturase protein to yeast mitochondria *in vivo*.[19]

Allotopic expression allows variant forms of the relocated genes to be generated by site-directed mutagenesis using standard methods. To simplify and facilitate the subcloning of variant gene constructs into propagation and expression vectors, we have found it convenient to place Y8 within a cassette bounded by a pair of specially selected restriction sites not present in either the gene encoding N9L/Y8-1 or its variants, or the vectors used for propagation and expression. The strategy developed[20] converts the genes encoding N9L/Y8-1 and N9L/Y8-2 into cassettes with *Bam*HI and *Not*I sites at the 5' and 3' termini respectively (Fig. 1A). The *Bam*HI/*Not*I cassette bearing the coding region for N9L/Y8-1 cloned into propagation and expression vectors can be readily replaced by *Bam*HI/*Not*I cassettes

---

[16] M. Galanis, R. H. P. Law, L. M. O'Keeffe, R. J. Devenish, and P. Nagley, *Biochem. Int.* **22**, 1059 (1990).

[17] P. Nagley, L. B. Farrell, D. P. Gearing, D. Nero, S. Meltzer, and R. J. Devenish, *Proc. Natl. Acad. Sci. U.S.A.* **85**, 2091 (1988).

[18] L. B. Farrell, D. P. Gearing, and P. Nagley, *Eur. J. Biochem.* **173**, 131 (1988).

[19] J. Banroques, A. Delahodde, and C. Jacq, *Cell (Cambridge, Mass.)* **46**, 837 (1986).

[20] T. Papakonstantinou, M. Galanis, P. Nagley, and R. J. Devenish, *Biochim. Biophys. Acta* **1144**, 22 (1993).

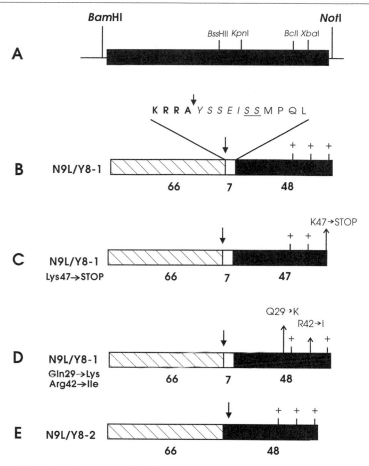

FIG. 1. Schematics of gene encoding N9L/Y8-1 and structures of chimeric import precursors of Y8 and some variants. (A) The DNA encoding the precursor construct N9L/Y8-1 is flanked by restriction sites *Bam*HI and *Not*I at the 5' and 3' ends of the gene cassette, respectively; also indicated are unique restriction enzyme recognition sites, *Kpn*I, *Bcl*I, and *Xba*I, introduced into the artificial DNA sequence encoding Y8. The *Bss*HII site lies close to the 3' end of the DNA segment encoding the N9L leader (see Fig. 2). Schematics of chimeric precursors are aligned below. (B) Construct N9L/Y8-1. Segments of the precursor are indicated as N9L leader (hatched), Y8 protein (solid), and bridging sequence at fusion point (unshaded). The amino acid sequence at the fusion point is detailed above; single-letter amino acid symbols indicate C-terminal residues of leader N9L (bold), N-terminal residues of subunit 8 (not bold), N-terminal residues of *N. crassa* subunit 9 (italics), and two residues arising from DNA at the point of fusion (italics, underlined). A downward pointing arrow (↓) indicates the site at which the matrix protease cleaves the precursor. (C) Construct N9L/Y8-1[Lys-47 → STP]. (D) Construct N9L/Y8-1[Gln-29 → Lys, Arg-42 → Ile]. (E) Construct N9L/Y8-2, a direct fusion of N9L and Y8 (there is no bridging sequence). Other symbols: +, positively charged residue; ↑, site of mutation. Numbers in bold indicate the number of amino acids in the relevant segment of each construct.

TABLE I
MEDIA FOR GROWTH OF YEAST CELLS[a]

Medium[b]	Composition
YEP	1% Bacto-yeast extract, 1% Bacto-peptone
YEPD	YEP, 2% glucose
YEPE	YEP, 2% (v/v) ethanol
YEPGal	YEP, 2% galactose
YEPEGal	YEPE, 0.5% galactose
SaccE	1% Bacto-yeast extract, 0.12% $(NH_4)_2SO_4$, 0.1% KH_2PO_4, 0.01% $CaCl_2$, 0.0005% $FeCl_3$, 0.07% $MgCl_2$, 0.05% $NaCl_2$, 2% (v/v) ethanol, supplements as required[b]
SaccEGal	SaccE, 0.5% galactose
SD	0.67% Yeast nitrogen base, 2% glucose, supplements as required[c]
SE	0.67% Yeast nitrogen base, 2% (v/v) ethanol, supplements as required[c]

[a] All concentrations shown are % (w/v) except where indicated.
[b] Solid media contain 1.5% Bacto-agar.
[c] Supplements are designated here as follows: Ade, adenine (100 μg/ml); His, histidine (50 μg/ml).

bearing variations in the coding region of Y8. In addition, a *Bcl*I site is present in the chemically synthesized Y8 gene[11]; later, *Kpn*I and *Xba*I sites were introduced by site-directed mutagenesis into the interior of the Y8 coding region[20] without alteration of encoded Y8 (Fig. 1A). The introduction of these sites allows for replacement of selected portions of the Y8 gene to generate variant forms of the gene bearing more than one mutation.[20] Some variant forms of Y8 discussed in this chapter are shown in Fig. 1.

Yeast Strains, Maintenance, and Genetic Tests

Standard methods for handling yeast cells for genetic analysis are described elsewhere.[21] The composition of media used in these studies is listed in Table I.

The *mit*⁻ yeast strains used in our laboratory as hosts for allotopic expression of a mitochondrially encoded protein component of the mtATPase complex carry individual base changes within the gene encoding that protein. These mutations either prevent the synthesis of the gene product or result in the production of a truncated or otherwise disabled protein that is unable to be assembled into a functional mtATPase complex. For example, the *mit*⁻ strain M31 used in our laboratory for the allotopic

[21] C. Guthrie and G. R. Fink (eds.), this series, Vol. 194.

expression of subunit 8 of yeast mtATPase carries mutations in the *aap1* gene encoding subunit 8. In this strain,[12] which is derived from strain J69-1B [*MATα ade1 his6 (rho+)*], the *aap1* gene carries a nonsense mutation arising from a C → T substitution at nucleotide +7 that changes the third codon from CAA (Gln) to TAA (stop). The predicted *aap1* gene product is therefore two amino acids in length. In other words, M31 cells effectively lack endogenous subunit 8. As a result M31 is unable to assemble a functional mtATPase enzyme complex and therefore cannot utilize a nonfermentable carbon source such as ethanol for cellular growth.[12]

Many *mit⁻* strains readily undergo spontaneous reversion to the wild-type phenotype. It is therefore of importance to ensure that the null phenotype is stable. Stability is tested by growing cells for 24 hr at 28° with vigorous shaking to promote good aeration in a nonselective liquid medium containing a fermentable carbon source, namely, YEPD. Portions of this culture are then spread on solid YEPE medium and incubated at 28° for 3–4 days. The generation of revertants is shown by the appearance of colonies on YEPE plates. In the case of M31 the spontaneous reversion rate to the *rho+* phenotype is exceedingly low ($<10^{-7}$).[17] This is probably due to the combination of the nonsense mutation at nucleotide +7 together with an additional frameshift mutation arising from the insertion of a T between nucleotides +9 and +10 in the *aap1* coding region.[12]

Yeast strains containing a fully functional mitochondrial genome are termed *rho+*. In a growing culture of yeast cells the nonreversible mutation from *rho+* to *rho⁻*, arising from the loss of extensive regions of functional mtDNA, occurs at a significant frequency that often generates 1 to 2% of *petite* cells. However, within a population of *mit⁻* cells the rate at which *petite* cells arise is considerably higher. *Petite* cells that have no detectable mtDNA (*rho⁰*) may also be found in populations of *mit⁻* cells. Because neither *rho⁻* nor *rho⁰* mutants are able to carry out mitochondrial protein synthesis, they are unable to assemble functional mtATPase complexes. Therefore, the maintenance of the *mit⁻* mitochondrial genome within a cellular population used for allotopic expression is of prime importance. Some *mit⁻* strains, including M31, show instability during storage at 4°. Frequent selection by restreaking of these strains onto fresh solid YEPD medium, following by testing of individual isolates (see next paragraph), ensures maintenance of *mit⁻* cells at 50–80% of the cell population. For long-term storage, freezing in 15% glycerol at −70° yields viable cells of which a high proportion retain *mit⁻* mtDNA genomes. An alternative strategy to achieve stabilization and maintenance of mtDNA in M31 cells allotopically rescued by a 2-μm based vector expressing Y8 is discussed below.

The nuclear *ade1* mutation of the J69-1B nuclear background is a useful tool for assessing the status of the mitochondrial genome of *mit⁻* M31 cells. The accumulation of a red pigment[22] provides a convenient color indicator for detection of the Ade⁻ phenotype conferred by the *ade1* mutant allele. On solid YEPD medium, colonies of respiratory-competent *ade1 rho⁺* cells of the parent strain J69-1B are large and red, whereas colonies of respiratory-deficient *petite* (*rho⁻* and *rho⁰*) cells are small and white. Importantly, *ade1 mit⁻* colonies of strain M31 grown on YEPD are intermediate in size, red-pigmented, and characteristically irregular in shape. Therefore, the proportion of *mit⁻* cells in a culture of M31 may be readily determined by plating cells on solid YEPD medium and scoring the frequency of red-pigmented colonies compared to white colonies. This characteristic has enabled the development of an alternative strategy to stabilize and maintain the mtDNA genome, thereby providing a reliable source of *mit⁻* cells. M31 cells which are growing on YEPE due to the presence of a 2-μm-based vector (see following section) that encodes a functional gene encoding N9L/Y8-1, and that also contains the selectable *ADE1* marker, are transferred to nonselective YEPD liquid medium and grown for 24 hr at 28°. Portions of this culture, diluted appropriately, are then spread on solid YEPD medium and incubated at 28° for a further 3–4 days. After this time three types of colonies are apparent[17]: large white Ade⁺ colonies that have retained the plasmid; small white *petite* colonies; and red-pigmented, irregularly shaped colonies containing *ade1 mit⁻* cells which have lost the plasmid. The loss of the plasmid from the last-named cells is confirmed by plating onto appropriately supplemented, solid glucose minimal media: SD/His and SD/His/Ade. M31 cells lacking the plasmid grow only on SD/His/Ade medium and are readily identified by this means.

The genetic procedure used to demonstrate the retention of *mit⁻* DNA in the M31 cells is described as follows. In general, testing for maintenance of the *mit⁻* genome within the cells of a population is carried out by crossing the putative *mit⁻* strain to a *rho⁻* tester strain that retains a relatively small segment of the wild-type mitochondrial genome in which lies the gene encompassing the lesion of the *mit⁻* strain. Neither the *mit⁻* nor *rho⁻* tester cells can grow on nonfermentable substrate. In such a test cross, recombination between the *mit⁻* allele and the relevant segment of *rho⁻* mtDNA generates a *rho⁺* genome, allowing diploids to grow on nonfermentable substrate. Specifically, testing for retention of the *mit⁻* mtDNA genome in M31 cells is performed by crossing M31 to the *rho⁻* tester strain G5 [*MATa ade1 lys2 trp1* (*rho⁻*)][12], which retains a wild-type *aap1* gene in a short

[22] E. W. Jones and G. R. Fink, *in* "The Molecular Biology of the Yeast *Saccharomyces: Metabolism and Gene Expression*" (J. N. Strathern, E. W. Jones, and J. R. Broach, eds.), p. 181. Cold Spring Harbor Laboratory, Cold Spring Harbor, New York, 1982.

mtDNA segment (681 bp). The colonies putatively containing M31 cells are inoculated, using a toothpick to make a small patch, in a grid array on a YEPD plate. Single drops of a culture of strain G5 freshly grown overnight at 28° in 10 ml YEPD are spotted onto each patch of M31 cells, and the YEPD plate is incubated at 28° overnight. A multipronged transfer device can be used for both M31 and G5 cells, if large numbers of putative *mit⁻* isolates warrant it. Diploid progeny are selected on the basis of auxotrophic complementation by replica-plating each patch onto solid SD/Ade medium and incubating plates at 28° for 3–4 days. Patches containing diploid cells growing on SD/Ade are then replica-plated onto solid YEPE medium and incubated at 28° for 3–4 days to test for the presence of those diploid cells in which recombination between mitochondrial genomes has occurred leading to the formation of *rho⁺* mtDNA. The growth of these diploid cells on YEPE indicates that the *mit⁻* mitochondrial genome has indeed been retained in the putative M31 cells initially picked.

Vectors

Both the copy number of a vector as well as the strength of the promoter at the expression site influence the level of expression of the inserted gene of interest. Considering initially the copy number, the yeast expression vector pPD72, currently used for routine allotopic expression of Y8, is a multicopy vector utilizing DNA replication functions from the endogenous 2-μm plasmid of yeast (Table II). Such 2-μm-based vectors maintain an average copy number of 10–40 per cell,[23] permitting relatively high levels of expression of the inserted gene. Cells transformed with a vector of this type should, in general, be grown in a medium in which selection of the vector-based genetic marker is maintained, as growth under nonselective conditions results in loss of the vector from a high proportion of cells. In some circumstances, however, where plasmid loss is desired, this can be readily achieved by growth of cells on nonselective medium such as YEPD. Where required, a reduction in the expression level of Y8 and some variants has been achieved by expression of the N9L/Y8-1 construct on a low copy number yeast expression vector (pGD2, Table II) which carries chromosomal replicator (*ARS1*) and centromere (*CEN4*) sequences. Cells transformed with pGD2 show high stability owing to enhanced retention of the vector compared to cells transformed with pPD72. For regulated allotopic expression, the high stability conferred by the presence of low copy replication control involving elements in vectors derived from chromosomal DNA (*ARS, CEN*) is strongly recommended (see relevant section below).

[23] M. A. Romanos, C. A. Scorer, and J. J. Clare, *Yeast* **8,** 423 (1992).

TABLE II
FEATURES OF YEAST EXPRESSION VECTORS USED FOR ALLOTOPIC EXPRESSION OF
MITOCHONDRIAL ATP SYNTHASE SUBUNIT 8

Vector	Yeast markers[a]	Promoter	Copy number	Replication control	Source of vector or reference
pPD72	*ADE1b LEU2d URA3*	*PGK1*	High	2 μm	b
pCI1	*ADE1b URA3*	*CUP1B*	High	2 μm	c, d
pGD2	*ADE1 URA3*	*PGK1*	Low	*CEN4 ARS1*	e, f
pED121	*ADE1 URA3*	*GAL1*	Low	*CEN4 ARS1*	g

[a] *ADE1b* designates an *ADE1* gene cassette lacking the single internal *Bam*HI site of the natural gene.

[b] T. Papakonstantinou, M. Galanis, P. Nagley, and R. J. Devenish, *Biochim. Biophys. Acta* **1144,** 22 (1993).

[c] pCI1 was derived from cloning of the *ADE1b* gene into the unique *Bst*EII site within the *LEU2d* gene of pYEULCBX.[d] The N9L/Y8-1 cassette was introduced as a *Bam*HI/ *Eco*RI fragment into the polylinker of pYEULCBX (R. E. Gray and R. J. Devenish, unpublished data, 1992).

[d] I. G. Macreadie, P. Failla, O. Horaitis, and A. A. Azad, *Biotechnol. Lett.* **14,** 639 (1992).

[e] pGD2 was constructed by cloning a *Hin*dIII fragment from pPD72 (encompassing *ADE1* gene, *PGK1* promoter, gene encoding N9L/Y8-1, and the *PGK1* termination sequence) into the unique *Hin*dIII site of YCp50).[f]

[f] M. D. Rose and J. R. Broach, this series, Vol. 194, p. 195.

[g] R. H. P. Law, R. J. Devenish, and P. Nagley, *Eur. J. Biochem.* **188,** 421 (1990).

With respect to the promoter, in vector pPD72 the allotopically expressed Y8 gene is under the control of the *PGK1* promoter (Table II). Although this promoter system provides constitutive expression, the level of expression in glucose is higher than that in ethanol. For most purposes the expression level in ethanol is adequate, although, as discussed in the following section, there are some instances in which a higher level of expression is desirable under derepressing conditions using nonfermentable carbon substrates. In such cases the gene has been placed under control of the *CUP1B* promoter in vector pCI1 (Table II).

The genetic marker utilized for selection of transformant yeast strains depends on the background of the relevant host mutant strain. Vectors referred to in this chapter utilize *ADE1* for complementation of the *ade1* mutation of strain M31.

Allotopic Expression in Yeast Cells

Yeast expression vectors encoding mitochondrial genes recoded for nuclear expression, or variant forms of such recoded genes, are introduced into the *mit⁻* yeast strain that is unable to synthesize the corresponding protein product. The restoration of the ability of that *mit⁻* strain to grow on a nonfermentable substrate such as ethanol, by allotopic expression of the recoded gene, is termed rescue of that strain. The following describes the procedures used for transformation of *mit⁻* yeast strains, as well as selection and testing of transformants for allotopic expression of the encoded mitochondrial gene. Using these techniques, the effect of an introduced mutation on the function of the protein product may be analyzed in the first instance by *in vivo* allotopic expression of the altered gene.

Transformation of mit⁻ Strains

Introduction of expression vectors into M31 is routinely performed according to the polyethylene glycol-mediated transformation procedure.[24] In our experience this method minimizes the induction of *petite* cells in cultures of M31 during the transformation procedure. The lithium acetate-mediated method[25] may also be used to good effect. Procedures that rely on cell wall digestion to generate DNA-permeable spheroplasts[26,27] have been found to result in a high rate of conversion of recipient cells to the cytoplasmic *petite* state. The suitability of these methods for the transformation of different strains varies; therefore, prior testing of procedures with each strain is recommended.

Following transformation of M31 with the multicopy expression vector pPD72 encoding N9L/Y8-1 or variant forms thereof (Fig. 1), transformants are initially selected by plating onto solid SD/His medium. The morphology of colonies that appear after 4–5 days of growth at 28° provides a distinction between the following: colonies containing transformed Ade⁺ cells, which generally appear large and white; those containing *petite* cells, which appear small and white; and a background lawn of red-pigmented microcolonies containing untransformed Ade⁻ *mit⁻* cells. Cells displaying an Ade⁺ phenotype are initially retested for the ability to grow in the absence of added adenine. It is advisable to pick cells from distinct white colonies of different sizes. Samples taken from at least 30 individual colonies are inoculated in a grid array on a fresh SD/His plate. The plate is incubated at 28° for 3–4 days, after which time growing cells (i.e., confirmed as Ade⁺) are assessed for functional allotopic expression of the relocated mitochondrial gene using the procedures described below.

[24] R. J. Klebe, J. V. Harriss, Z. D. Sharp, and M. G. Douglas, *Gene* **25,** 333 (1983).
[25] R. H. Schiestl and R. D. Gietz, *Curr. Genet.* **16,** 339 (1989).
[26] J. D. Beggs, *Nature (London)* **275,** 104 (1978).
[27] A. Hinnen, J. B. Hicks, and G. R. Fink, *Proc. Natl. Acad. Sci. U.S.A.* **75,** 1929 (1978).

Evaluation of Allotopic Expression

Following transformation and selection of Ade$^+$ transformants, evaluation of *in vivo* allotopic expression of the mtATPase subunit in terms of restoration of oxidative phosphorylation as well as the import and subsequent assembly of that subunit into functional mtATPase is made on the basis of four criteria: (1) growth of the transformants on nonfermentable substrate, such as ethanol; (2) the dependence of the respiratory-competent phenotype of transformants on the presence of the yeast expression vector containing the allotopically expressed gene; (3) comparison of the growth rate of respiratory-competent transformants to that of the reference wild-type strain; and (4) demonstration of the assembly of the allotopically expressed imported protein into mtATPase.

Primary Assessment of Growth on Nonfermentable Substrates. The first criterion is evaluated by transfer to solid YEPE medium of the transformants showing the Ade$^+$ phenotype. The ability to grow is assessed after incubation for 4–7 days at 28°. For all variants, testing at 18° and 36° is recommended because the introduced mutations may confer a temperature-conditional phenotype.

The results obtained for the growth assessment of some wild-type and variant forms of Y8 in the assay described above are shown in Table III. In the case of transformants showing strong growth on YEPE plates (e.g., Y8-1 and Y8-2) the analysis is continued as described in the following section. Where transformants fail to grow on YEPE plates (e.g., Y8-1[K47 → STP] and Y8-1[Q29 → K, R42 → I]), the Ade$^+$ cells are initially tested to determine whether the failure to grow is due to mutation of host cells to the cytoplasmic *petite* state or reflects the inability of the expressed variant gene to achieve rescue. This is readily done by crossing the respiratory-deficient Ade$^+$ transformants to the tester *rho*$^-$ strain G5 (see section above on genetic tests). If the *mit*$^-$ genome proves to be intact in the M31 host, then function of the N9L/Y8 variant is further probed by application of two approaches. An increase in the level of expression is achieved by transferring the cassette containing the variant gene to an expression vector in which it will be under transcriptional control of a stronger promoter such as *CUP1B* (Table II). The other approach, which is described in detail in a later section, is to incorporate a second copy of the N9L leader sequence into the chimeric precursor for that Y8 variant, to be expressed from the vector pPD72; this enables a considerable enhancement of the import efficiency of the precursor into mitochondria. In both approaches the variant gene in its modified vector context is then reintroduced into M31 and Ade$^+$ transformants tested for functional rescue based on allotopic expression.

TABLE III

GROWTH PROPERTIES OF M31 CELLS TRANSFORMED WITH PLASMID pPD72 ENABLING ALLOTOPIC
EXPRESSION OF Y8 AND VARIANTS AS SINGLE OR DOUBLE LEADER CONSTRUCTS[a,b]

Expressed construct[c]	Primary growth assessment[d]	Generation time (hr)[e]		Difference[f] single/double
		Single leader	Double leader	
Y8-1	+	6.31 ± 0.54	5.65 ± 0.32	$0.05 < p < 0.1$
Y8-1[K47 → STP]	−	NA	5.78 ± 0.55	NA
Y8-2	+	6.92 ± 0.35	6.77 ± 0.20	$0.2 < p < 0.5$
Y8-1[Q29 → K, R42 → I]	−	NA	NA	NA

[a] Data were compiled from M. Galanis, Ph.D. Thesis, Monash University (1992), and T. Papakonstantinou, R. J. Devenish, and P. Nagley, unpublished data (1993).

[b] Single leader construct denotes a precursor containing one copy of N9L leader upstream of Y8 or a variant; the double leader construct bears a duplicated N9L leader (N9LD) upstream of Y8 or variant.

[c] Detailed in Fig. 1.

[d] Primary assessment of growth is carried out by plating M31 cells transformed with pPD72 bearing the relevant single leader construct on solid YEPE medium, and determining the extent of growth after 4 days at 28°. A plus symbol (+) denotes substantial growth; a minus sign (−) denotes negligible growth.

[e] Growth rates were measured for cells growing in liquid SaccE/His medium at 28° with aeration, and are expressed as mean generation times, with standard deviation, for three separate experiments in each case. NA, Not applicable.

[f] The level of statistical significance (Student's t-test) is indicated for differences between growth rates of cells bearing the corresponding single and double leader constructs. NA, Not applicable.

For example, the variant gene encoding N9L/Y8-1[K47 → STP] (Fig. 1C) under the control of the *PGK1* promoter in pPD72 fails to restore growth when allotopically expressed in M31. For this variant, high level allotopic expression in the presence of ethanol could be achieved by placing the variant gene under control of the *CUP1B* promoter (Table II), leading to the rescue of M31 (R. E. Gray, unpublished observations, 1994). The same variant also rescues M31 when expressed under control of the *PGK1* promoter but endowed with a second copy of the N9L leader sequence (Table III). In contrast, the variant N9L/Y8-1[Q29 → K, R42 → I] has never been observed to rescue strain M31, either under *CUP1B* promoter control or in a double leader configuration (Table III).

Verification of Dependence of Functional Allotopic Expression on Presence of Specific Yeast Expression Vector. The second criterion is satisfied by demonstrating the dependence of the respiratory-competent growth phenotype of transformants on the presence of the yeast expression vector incorporating the allotopically expressed gene. This is achieved by promoting loss of that expression vector through vegetative segregation during growth on nonselective media. The 2-μm-based vectors are conveniently suited to such genetic analysis. In the case of M31 transformants containing a pPD72 vector construct, transformants showing strong growth on YEPE are transferred to nonselective YEPD liquid medium and grown for 24 hr at 28° with vigorous shaking to promote good aeration. Portions of the culture are then spread, after appropriate dilution, on solid YEPD medium. Incubation of plates at 28° for 3–4 days allows colonies to form. These colonies are tested simultaneously for loss of the vector and for loss of the rescued phenotype, by replica plating onto solid SD/His and YEPE medium, respectively. Useful additional control solid media for replica plating at this stage include YEPD and SD/His/Ade. Analysis of the growth of colonies after 3–4 days at 28° allows a correlation to be made between retention of the plasmid vector and the rescued phenotype.

When assessing allotopic expression of variant forms of Y8, confirmation of the identity of the variant gene construct within transformant yeast strains is obtained by the following means. Total yeast DNA is isolated.[28] The vector is amplified by transforming E. coli with this yeast DNA, selecting bacteria acquiring ampicillin resistance. Plasmid DNA that bears the variant gene construct is prepared,[29] and the sequence of the relevant segment of the plasmid bearing the predicted mutation is determined.

Assessment of Cellular Growth Rate. A more precise comparative analysis of mutant strains is made by assessing the growth rate of allotopically rescued cells in liquid medium containing nonfermentable substrate. Cells are picked from colonies on solid YEPE medium, pregrown in liquid SaccE/His medium, and actively growing cells are inoculated into further liquid SaccE/His medium and incubated with shaking at 28°. The increase in cell density is monitored using standard procedures.[20] The generation times obtained for wild-type Y8-1 and some variant forms are shown in Table III. The sensitivity of this assay is sufficient to detect a statistically significant difference between the growth rates obtained for N9L/Y8-1 and N9L/Y8-2 constructs expressed using pPD72.

Assessment of growth rates at 18° and 36° can provide further information relating to function of variant forms of Y8. In this test, the formation in cells of a functional oxidative phosphorylation system is assessed over a range of temperatures; 36° is chosen as being at the upper limit of the normal temperature range. Many temperature-conditional *mit⁻* mutants of yeast have been isolated that are unable to grow at 36° on nonfermentable

[28] R. W. Davis, M. Thomas, J. Cameron, T. P. St. John, S. Scherer, and R. A. Padgett, this series, Vol. 65, p. 404.

[29] J. Sambrook, E. F. Fritsch, and T. Maniatis, "Molecular Cloning: A Laboratory Manual." Cold Spring Harbor Laboratory, Cold Spring Harbor, New York, 1989.

substrates yet appear to grow normally on the same substrates at 28°. Growth at higher temperatures is not recommended, not only because of induction of *petite* mutants, but also because there is evidence to suggest that higher temperatures (around 44°) lead to impairment of the coupling of oxidative phosphorylation to electron transport in wild-type cells.[30] At 18°, the lower end of the temperature range, it has been suggested that the decrease in activity of mtATPase observed implies a requirement for a higher level of functional enzyme for maintenance of respiratory function.[31] Therefore, strong growth of a rescued strain at 18° provides evidence for the production of relatively high levels of functional mtATPase enzyme complex, whereas a reduction in the growth rate at 18° may suggest either that there is some impairment of mtATPase function or that there has been a reduction in the level of functional enzyme complexes produced.[31]

Assembly Assay. The assembly of imported allotopically expressed proteins into a functional enzyme complex is shown by *in vivo* labeling of transformed and wild-type cells with [^{35}S]sulfate. The labeling is carried out by growing allotopically rescued cells both in the presence and in the absence of cycloheximide, which suppresses the labeling of cytoplasmically synthesized proteins.[17] The analysis of labeled proteins involves harvesting of the cells, preparation of mitochondria, immunoprecipitation of the assembled mtATPase complex with anti-F_1 subunit monoclonal antibody, and finally analysis of immunoprecipitates by sodium dodecyl sulfate (SDS)–polyacrylamide gel electrophoresis followed by fluorography. These procedures are described elsewhere.[8]

Duplication of Leader Sequence

Duplication of the leader sequence has been shown to enhance markedly the efficiency of import both *in vitro*[32] and *in vivo*.[20] We describe here a particular application involving the comparative analysis of cellular growth rates of rescued yeast strains allotopically expressed Y8 variants fused to either a single leader N9L sequence or a tandemly duplicated N9LD leader sequence (Table III). These comparisons provide information on the potential assembly and functional defects of Y8 variants.

To generate a tandem duplication of the N9L leader, we designed an N9L cassette that could be readily inserted into an N9L/Y8-1 construct whose gene is already sited in an expression vector. For this purpose it was convenient to make use of a unique restriction site already present at the junction of the leader and mature protein sequence. The procedure used

[30] E. J. Patriarca and B. Maresca, *Exp. Cell Res.* **190,** 57 (1990).
[31] A. Mukhopadhyay, M. Uh, and D. M. Mueller, *FEBS Lett.* **343,** 160 (1994).
[32] M. Galanis, R. J. Devenish, and P. Nagley, *FEBS Lett.* **282,** 425 (1991).

is summarized as follows. As shown in Fig. 2A, in the gene encoding the natural precursor to *N. crassa* mtATPase subunit 9, a unique *Bss*HII site is present between DNA encoding the N9L leader and that encoding the subunit 9 protein (N9). An additional *Bss*HII site was introduced by *in vitro* mutagenesis just upstream of the 5' end of N9L (Fig. 2B), allowing the recovery of a DNA segment as a *Bss*HII fragment encoding the leader denoted N9L (Fig. 2C). The N9L segment was subsequently inserted into DNA encoding a target single leader construct (Fig. 2D) to generate the double-leader construct N9LD/Y8. As a result of these manipulations, a short bridge of three amino acids (Fig. 2D) was introduced between the two leaders (detailed in Fig. 2E). The unique *Pst*I site immediately 3' to the additional *Bss*HII site was introduced to enable the orientation of the insert to be determined. Among those plasmids recovered putatively encoding N9LD/Y8, the DNA segment encoding the second leader N9L has always been found to be inserted in the desired orientation into that encoding the single leader construct, most likely because of instability of plasmid DNA containing closely apposed inverted repeat sequences.

The use of double leader constructs for the delivery of passenger proteins to the mitochondria has been found to enhance the function of allotopically expressed forms of Y8 *in vivo*. For example, as shown in Table III, the difference between the generation time for the strain in which Y8-1 (Fig. 1B) is delivered to the mitochondria by the double leader and that for the single leader construct is significant ($0.05 < p < 0.1$). This result indicates that the provision of the double leader has increased the efficiency of delivery of the chimeric precursor to the mitochondria, resulting in enhanced bioenergetic function of the rescued cells. However, the difference in growth rate conferred by the double leader form of the allotopically expressed variant Y8-2, compared to that conferred by the single leader form, is not significant ($0.2 < p < 0.5$). Hence, following processing of the chimeric precursor (Fig. 2E), the Y8-2 variant confers on cells an impaired bioenergetic function that cannot be overcome by enhanced delivery of the precursor to the mitochondria. In contrast, the allotopically expressed variant Y8-1[K47 → STP], when fused to a single copy of N9L, is unable to restore growth of strain M31 on ethanol. However, the same construct fused to the double leader N9LD not only restores growth of M31 on ethanol but also confers on cells a rate of growth approaching that conferred by allotopically expressed N9LD/Y8-1. Thus, an increase in the efficiency of delivery and hence level of supply of the precursor has overcome the original growth defect. On the basis of this reasoning, supported by direct studies on import and assembly of variant Y8 proteins *in vitro*, it can be concluded that variants such as Y8-1[K47 → STP] or Y8-1[K47 → I] have defects in their rate of assembly into mtATPase.

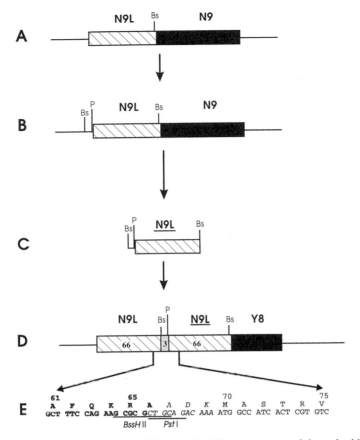

FIG. 2. Manipulations to generate DNA encoding Y8 precursor containing a double leader (N9LD). (A) The DNA sequence encoding the natural *N. crassa* mtATPase subunit 9 carries a unique *Bss*HII recognition site at the junction of the DNA sequences encoding the leader presequence (N9L, hatched) and the mature subunit 9 (N9, solid). (B) *In vitro* site-directed mutagenesis was applied to introduce both an additional *Bss*HII site and a *Pst*I site near the 5′ end of the DNA encoding N9L. (C) Digestion of the DNA segment in (B) with *Bss*HII enables recovery of a *Bss*HII cassette coding for N9L plus a short extension at the N terminus. The leader segment encoded by this cassette is denoted N9L for clarity. (D) The *Bss*HII cassette encoding N9L is inserted into the *Bss*HII site of a recipient DNA molecule encoding a single leader construct in the N9L/Y8 configuration. The resulting construct (N9LD/Y8) encodes a chimeric protein that may contain Y8 or a variant of Y8 (solid). The N terminus of the Y8 protein segment is now fused to two copies of the N9L leader that are themselves bridged by 3 amino acid residues (lightly shaded); numbers within each block indicate the number of amino acid residues. (E) Detail of sequence at the junction between N9L leaders in N9LD. Both DNA sequence and the encoded amino acid sequence are shown (amino acids are indicated by single-letter symbols). The numbers indicate the amino acid residue number relative to the initiating methionine of the upstream leader N9L, taken as residue 1. Boldface letters represent sequences derived from the upstream leader N9L, regular letters represent those of the downstream leader N9L, and italicized letters represent those of the bridging sequence introduced between the two copies of N9L. Relevant restriction enzyme recognition sites are indicated by underlines below the DNA sequence. Symbols for restriction enzyme recognition sites are as follows: Bs, *Bss*HII; P, *Pst*I.

The mechanism of enhancement of functional allotopic expression provided by the tandem duplication of the leader sequence (as in the case of N9LD/Y8-1) is apparently due to an enhanced binding capacity of the precursor to the mitochondria coupled with acceleration of import through the membrane insertion/translocation system. The binding capacity for the double leader construct N9LD/Y8-1 has been determined to be five to six times that for the corresponding single leader construct.[33] The leader duplication strategy provides a means to overcome the barrier presented by the mitochondrial membrane to the delivery of nuclear encoded proteins which are intrinsically resistant, by virtue of their hydrophobicity, to import into that organelle. The strategy has also been successfully employed in import studies *in vitro* of those Y8 variants that show little or no import into isolated mitochondria when expressed with a single leader.[32]

Regulation of Allotopic Expression Aimed at Subunit Depletion

Regulated Allotopic Expression with GAL1 Promoter

The placement of a gene under the control of a tightly regulated promoter allows the expression of the gene to be controlled by the manipulation of the cellular environment. Regulated allotopic expression of Y8 under control of the *GAL1* promoter allows production of vigorously growing allotopically rescued cells in the presence of galactose. Removal of galactose curtails the expression of the allotopically expressed protein, enabling a strategy to be developed for the systematic depletion of Y8 from respiratory-competent cells initially with fully functional mtATPase enzyme activity. Mitochondria from rescued cells initially so depleted of Y8 are very useful for studies of the function of Y8 *in vivo*, or assembly of Y8 or its variants *in vitro*.[20,34] Isolated mitochondria from wild-type yeast not depleted of Y8 do not allow assembly of imported Y8 into mtATPase.[34] Mitochondria partially depleted of Y8 contain a mixed population of assembled and incompletely assembled mtATPase complexes. It is envisaged that in an *in vitro* import/assembly system using such mitochondria, the endogenous functional mtATPase complexes would drive the import process while the incompletely assembled mtATPase complexes which lack Y8 would be available to interact with imported Y8.[34] The methods for *in vitro* studies of import[9] and assembly[8] are detailed elsewhere.

The regulated promoter elements are derived from control sequences

[33] M. Galanis, Ph.D. Thesis, Monash University, Clayton, Victoria, Australia (1992).
[34] R. H. P. Law, R. J. Devenish, and P. Nagley, *Eur. J. Biochem.* **188,** 421 (1990).

of the *GAL1* gene[35] and located on a *Bam*HI/*Eco*RI cassette in plasmid pBM150.[36] DNA encoding N9L/Y8-1 was placed under the control of the *GAL1* promoter in the yeast expression vector pED121 (Table II).[34] In our experience transformants of M31 obtained with pED121 (a low copy number vector) are stable, whereas those obtained from a high copy 2-μm-based vector bearing the gene encoding the N9L/Y8-1 construct under control of the *GAL1* promoter are very unstable. Approximately 50% of the latter transformants lose the vector even during propagation of cells on selective medium.[37]

Other inducible promoter systems[38] may be considered for specific applications. Our investigations of the *CUP1B* promoter have shown that this system does not provide the fine control obtainable with the *GAL1* promoter, in that expression of a gene under *CUP1B* control continues in the absence of added copper ions (R. E. Gray, unpublished data, 1993). Evidently there are sufficient divalent cations in most growth media to allow significant levels of expression.

For *GAL1* transcriptional control to be of effective use in allotopic expression, it is necessary for the host strain to meet the following requirements. The strain should grow on galactose only if it retains functional oxidative phosphorylation capacity. Thus *petite* or *mit⁻* cells derived from the host strain, unless rescued, should not grow on galactose. The growth characteristics on galactose of the M31 host strain and its parent *rho⁺* strain J69-1B were assessed in the following way. Cells of both strains were streaked on solid YEPGal medium. After incubation at 28° for 3–4 days it was found that J69-1B cells grew strongly, indicating that they are able to utilize galactose. However, strain M31 did not grow on YEPGal plates and thus is unable to utilize galactose. It has been observed that restoration of respiratory competence in strain M31 by expression of N9L/Y8-1 under *PGK1* control also restores the ability of that strain to utilize galactose for growth.[37]

Vector pED121 encoding N9L/Y8-1 under *GAL1* control was then introduced into strain M31 to produce strain YGL-1.[34] Transformants were selected by plating on SD/His. Cells growing on this medium, and therefore carrying the vector, were initially assessed for successful induction of Y8 in the following way. They were tested for growth on solid YEPE and YEPEGal media. After incubation at 28° for 4–5 days YGL-1 cells grew strongly on YEPEGal, indicating that functional allotopic expression of

[35] M. Johnston, *Microbiol. Rev.* **51,** 458 (1987).
[36] M. Johntson and R. W. Davis, *Mol. Cell. Biol.* **4,** 1440 (1984).
[37] R. H. P. Law, Ph.D. Thesis, Monash University, Clayton, Victoria, Australia (1990).
[38] J. C. Schneider and L. Guarente, this series, Vol. 194, p. 373.

Y8 had taken place in the presence of the inducer galactose. On the other hand, YGL-1 cells failed to grow on YEPE, indicating that in the absence of galactose allotopic expression of Y8 does not occur to any significant extent. Comparison of the growth of strains J69-1B and YGL-1 in liquid SaccEGal medium showed that the growth rate of YGL-1 does not differ significantly from that of J69-1B.[34]

Subunit Depletion from Strain YGL-1

To provide the conditions necessary for subunit depletion to occur, allotopically rescued YGL-1 cells are grown initially in liquid medium containing both galactose and ethanol (SaccEGal), then shifted to medium containing ethanol only (SaccE). Confirmation that the shift from an inducing to a noninducing medium results in a shutdown of synthesis of the inducible Y8 protein was readily obtained by measuring during the growth of YGL-1 in SaccE medium critical changes in bioenergetic parameters of mitochondria, including respiratory control, ATPase activity, oligomycin sensitivity, and ATP–P_i exchange.[34] Alternatively, the depletion of Y8 was shown by pulse-labeling of cells following a shift from SaccEGal to SaccE medium and analysis of labeled proteins.[37] A further convenient test of depletion is the so-called irreversible depletion test. Here, at different stages of growth in liquid SaccE or YEPE medium, the extent of depletion is monitored by the ability of cells to continue growth after a further shift into fresh SaccE or YEPE. Fully depleted cells are unable to continue growth after this next transfer.

The procedure developed to obtain Y8-deficient mitochondria for use in *in vitro* studies of exogenous Y8 import and assembly is summarized as follows. Cells are pregrown at 28° in 10 ml liquid SaccEGal medium until the mid-logarithmic phase of growth. Cells are harvested and then washed three times with sterile distilled water to remove galactose. Washed cells are inoculated into SaccE medium and incubated at 28° to allow further growth. Cells taken from this culture at late logarithmic phase yield mitochondria suitable for import and assembly studies.[8,9]

Concluding Remarks

The technique of allotopic expression has proved to be a powerful tool in the systematic analysis of the relationship between structure and function of the mitochondrially encoded mtATPase protein, subunit 8. In particular, functional allotopic expression of subunit 8 has permitted the implementation of a range of gene manipulation and expression strategies that focus on three distinct regions[39] of this protein: the highly conserved N-terminal functional region, the central hydrophobic region, and the positively charged C-terminal region which plays a key role in the assembly of the subunit into mtATPase. In addition, a further study has been initiated to probe the topology of subunit 8 with respect to the inner mitochondrial membrane and interactions of that subunit with other subunits of mtATPase. These studies utilize an epitope tag incorporated into the C-terminal region of subunit 8 (W. S. Nesbitt, R. H. P. Law, R. J. Devenish, and P. Nagley, unpublished data, 1994). The technique of allotopic expression provides a system to test the *in vivo* function of the tagged construct. Hence, following the primary demonstration that allotopic expression is possible, this area is now being actively exploited to address fundamental questions regarding the biogenesis of mitochondria and the assembly and function of constituent proteins.

The approach of allotopic expression developed in yeast provides a model for application to other biological systems. Specifically, it may be possible to redress deficiencies in mitochondrially encoded proteins of humans, associated with mitochondrial disease, by application of allotopic expression strategies. This could provide a route toward gene therapy for diseases arising from mtDNA mutations.

[39] R. J. Devenish, T. Papakonstaninou, M. Galanis, R. H. P. Law, A. W. Linnane, and P. Nagley, *Ann. N.Y. Acad. Sci.* **671,** 403 (1992).

[34] Allotopic Expression of Yeast Mitochondrial Maturase to Study Mitochondrial Import of Hydrophobic Proteins

By MANUEL G. CLAROS, JAVIER PEREA, and CLAUDE JACQ

Introduction

Allotopic expression of a gene implies that a new copy of this gene, localized to a new cellular compartment, for instance, from mitochondria to the nucleus, is still able to confer the corresponding wild-type phenotype. Thus, in the case of the allotopic expression of a mitochondrial gene, the corresponding cytoplasmically translated product should be targeted, in an active form, to the adequate mitochondrial subcompartment. Such a process has occurred frequently during the evolution of the eukaryotic cell, since most of the primitive mitochondrial genes have found their way into the nucleus.[1] Deliberate allotopic expression of contemporary mitochondrial genes has been successful in only a few cases in the yeast *Saccharomyces cerevisiae*.[2-4] The interest in this approach stems from two sources: (1) the desire to understand why a few genes are still sequestrated in the mitochondrial compartment, and (2) the wish to develop new tools to easily manipulate mitochondrial genes. To this end we have focused our attention on a special class of yeast mitochondrial genes that are localized in the introns of the genes coding either for the apocytochrome *b* or the subunit I of cytochrome oxidase (*COX1*). These group I intron-encoded genes can code either for DNA endonuclease or for RNA maturase. Although the DNA endonuclease activity seems to be dispensable for mitochondrial functions, the RNA maturase activities so far identified are essential.[5] RNA maturase mutations block the splicing process of the intron in which it is encoded, and thus offer the opportunity to take advantage of *in vivo* complementation assays.

This chapter is divided into two parts. We first focus on methods to transfer a mitochondrial gene to the nucleus. The bI4 RNA maturase is used as an example to show how a cytoplasmically translated form of this protein can complement RNA maturase mitochondrial mutations. In the second part we show how the RNA maturase can be used as a sensitive reporter activity, to study the mitochondrial import of hydrophobic proteins, and how this can permit the development of a genetic screen to search for genes involved in this import process.

Allotopic Expression of bI4 RNA Maturase

Principle

In vivo assays to assess the properties of a mitochondrial gene transferred to the nucleus rely on the availability of a sensitive complementation

[1] M. W. Gray, *Annu. Rev. Cell Biol.* **5,** 25 (1989).
[2] J. Banroques, A. Delahodde, and C. Jacq, *Cell (Cambridge, Mass.)* **46,** 837 (1986).
[3] P. Nagley, L. B. Farrell, D. P. Gearing, D. Nero, S. Meltzer, and R. J. Devenish, *Proc. Natl. Acad. Sci. U.S.A.* **85,** 2091 (1988).
[4] A. Delahodde, V. Goguel, A. M. Becam, F. Creusot, J. Perea, J. Banroques, and C. Jacq, *Cell (Cambridge, Mass.)* **56,** 431 (1989).
[5] Z. Kotylak, J. Lazowska, D. C. Hawthorne, and P. P. Slonimski, *in* "Achievements and Perspective of Mitochondrial Research. Volume II: Biogenesis" (E. Quagliariello, S. Papa, F. Palmieri, and C. Saccone, eds.), p. 1. Elsevier, Amsterdam, 1985.

assay. Although a full discussion of the properties of the yeast intron-encoded RNA maturases is outside of the scope of this chapter (for reviews, see Refs. 5 and 6), a summary of the main characteristics of the bI4 RNA maturase is necessary to understand the technique presented here.

Figure 1 (top) summarizes the wild-type biogenesis and activity of the bI4 RNA maturase. The bI4 RNA maturase is synthesized as a precursor corresponding to the translation of the four N-terminal apocytochrome b exons and the in-phase intron open reading frame. This precursor is proteolyzed, and its C-terminal part (corresponding to the intron-encoded protein) constitutes the RNA maturase. This RNA maturase is required for the *in vivo* RNA splicing of both bI4 and aI4 introns in, respectively, the genes coding for the apocytochrome b and for the subunit 1 of cytochrome oxidase. A protein very similar to the bI4 RNA maturase is coded in the aI4 intron. There are two known mutational contexts in which the aI4 intron-encoded protein can suppress bI4 RNA maturase deficiencies: (1) when the open reading frame of the aI4 intron is mutated (mutation *mim2*[7]) and (2) when mutated alleles of the nuclear *NAM2* gene and wild-type form of the aI4 open reading frame are present together.[8] As a result of these observations, three different genetic contexts (Fig. 1, bottom) can be used to assess the RNA maturase activity of cytoplasmically translated protein. In genetic context (1), the mitochondrial mutation V328 creates a nonsense codon in the bI4 RNA maturase coding sequence (strain CW01, Table I); in (2), a clean deletion of the bI4 intron does not prevent the formation of an active cytochrome b, but the absence of the bI4 RNA maturase blocks the RNA processing of the aI4 intron (strain CW02, Table I); in (3), the mutation G1659 in the bI4 intron leads to the synthesis of a truncated form of the bI4 RNA maturase (strain CW06, Table I), and the clean deletion of the aI4 intron prevents any putative involvement of the aI4 intron-encoded protein in the RNA splicing process of the bI4 intron (strain CW05, Table I).

Complementation Assays

Like most mutations affecting the mitochondrial genome, maturase-dependent mutations render the cell unable to derive energy from respiration. This phenotype is most easily scored as an inability to grow on nonfermentable carbon sources: such mutants grow on glucose but fail to grow on glycerol. We routinely use the following media: YPGA (1% yeast extract,

[6] A. M. Lambowitz and M. Belfort, *Annu. Rev. Biochem.* **62,** 587 (1993).
[7] G. Dujardin, C. Jacq, and P. P. Slonimski, *Nature (London)* **298,** 628 (1982).
[8] M. Labouesse, G. Dujardin, and P. P. Slonimski, *Cell (Cambridge, Mass.)* **41,** 133 (1985).

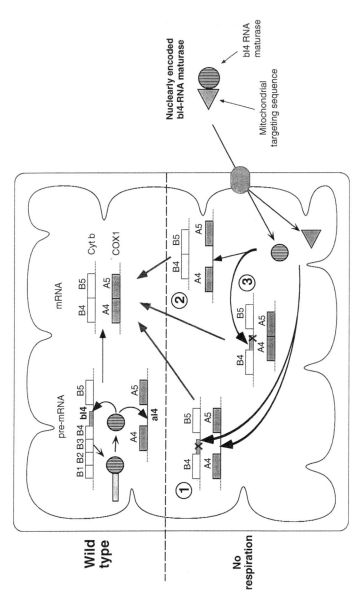

FIG. 1. General scheme used to assay the mitochondrial RNA maturase activity of the cytoplasmically translated RNA maturase. (*Top*) Wild-type context in which the bI4 RNA maturase, translated from the intronic open reading frame of the bI4 intron between exons B4 and B5 of apocytochrome *b* gene, is necessary for the RNA splicing of both the bI4 and the aI4 introns. (*Bottom*) Three mutational contexts in which the bI4 RNA maturase coding sequence is either mutated (1 and 3) or absent (2). The engineered RNA maturase coding sequence can be translated in the cytoplasm, and its translation product can be targeted to mitochondria (right). This cytoplasmically translated protein should be able to complement the three maturase deficiencies depicted in 1, 2, and 3.

TABLE I
YEAST STRAINS USED FOR RNA MATURASE COMPLEMENTATION ASSAYS

Strain	Nuclear genotype	Mitochondrial genotype	Reference
CW01	α, ade2-1, ura3-1, his3-11,15, trp1-1, leu2-3,112, can1-100	ρ^+, V328	a
CW02	α, ade2-1, ura3-1, his3-11,15, trp1-1, leu2-3,112, can1-100	ρ^+, ΔbI4	b
CW06	α, ade2-1, ura3-1, his3-11,15, trp1-1, leu2-3,112, can1-100	ρ^+, G1659	c
CW05	α, ade2-1, ura3-1, his3-11,15, trp1-1, leu2-3,112, can1-100	ρ^+, G1659 ΔaI4	c
CW04	α, ade2-1, ura3-1, his3-11,15, trp1-1, leu2-3,112, can1-100	ρ^+, mit$^+$	b

[a] M. Labouesse, C. J. Herbert, G. Dujardin, and P. P. Slonimski, *EMBO J.* **6,** 713 (1987).
[b] J. Banroques, A. Delahodde, and C. Jacq, *Cell (Cambridge, Mass.)* **46,** 837 (1986).
[c] Gift from G. Dujardin, CNRS, 91190 Gif, France.

2% peptone, 2% D-glucose, 30 μg/ml adenine), N3 (1% yeast extract, 2% peptone, 2% glycerol, 50 mM sodium potassium phosphate buffer, pH 6.3), and W0 (0.7% yeast nitrogen base without amino acids, 2% w/v D-glucose, and 30 μg/ml for requirements). For solid media, 2% agar is added. Note that the yeast extract for N3 medium should be completely free of traces of fermentable substrates in order to avoid any leaky growth of the nonrespiratory strains.

To test the maturase activity of the nuclear-encoded forms of the gene, maturase-deficient strains (Table I) are first transformed with a nuclear plasmid (Fig. 2) carrying an engineered form of the bI4 RNA maturase coding sequence. More than 20 leucine prototroph transformants are streaked on a glycerol complete medium (N3). Under these conditions, active forms of the cytoplasmically translated maturase expressed in strains like CW02 or CW05 (Table I) permit growth after 3 days at 28°. A more quantitative estimate of the maturase activity can be obtained from growth rate in liquid glycerol medium. The doubling time, in glycerol, of maturase-deficient strains transformed with a nuclear plasmid carrying the engineered RNA maturase coding sequence is about 3.5 hr.

Construction of Universal Code Equivalents of Mitochondrial Genes

The considerable differences between the rules for gene expression in nuclear and mitochondrial genes require a complete engineering of the mitochondrial gene to be expressed in the cytoplasm. The differences be-

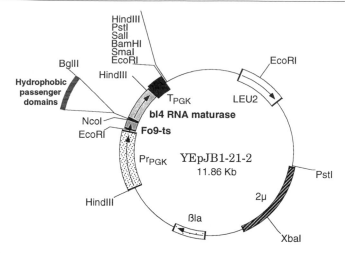

FIG. 2. Vector used to express the nuclear form of the bI4 RNA maturase and its derivatives. In vector YEpJB1-21-2 [J. Banroques, J. Perea, and C. Jacq, *EMBO J.* **6,** 1085 (1987)], the sequence coding for the $F_0$9-mts fused to the engineered maturase coding sequence is under the control of the PGK promoter and terminator. A unique *Bgl*II restriction site can be used to insert sequences coding for transmembrane hydrophobic domains of apocytochrome *b*.

tween the two genetic codes are particularly important. In *Saccharomyces cerevisiae*, the three codons UGA, AUA, and CUN correspond to different amino acids in cytoplasm and in mitochondria.[9] This implies that, for example, the correct expression of the mitochondrial apocytochrome *b* gene in the cytoplasm would require the introduction of 18 base changes,[10] that of the bI4 RNA maturase would require 12 base changes, and that of the aI3 DNA endonuclease (I-*Sce*III) would require 34 base changes.[11]

Several highly efficient methods for directed mutagenesis are now available. However, to avoid repetitive experiments, we have developed a method to introduce all the base changes in a single round of mutagenesis. The rationale is simple: the mutagenized DNA strand which contains all the mutagenic oligonucleotides is equipped at its ends with two longer oligonucleotides (A and B in Fig. 3) that introduce two new sequences for PCR (polymerase chain reaction) amplification. Thus, after a classic directed mutagenesis approach (including the two oligonucleotides A and B), addition of two primers recognizing the two new sequences of A and B will permit a selective PCR amplification of the mutated strand. For

[9] S. G. Bonitz, R. Berlani, G. Coruzzi, M. Li, G. Macino, F. G. Nobrega, M. P. Nobrega, B. E. Thalenberg, and A. Tzagoloff, *Proc. Natl. Acad. Sci. U.S.A.* **77,** 3167 (1980).

[10] M. G. Claros, J. Perea, Y. Shu, F. A. Samatey, J. L. Popot, and C. Jacq, *Eur. J. Biochem.* **228,** 762 (1995).

[11] M. Schapira, C. Desdouets, C. Jacq, and J. Perea, *Nucleic Acids Res.* **21,** 3683 (1993).

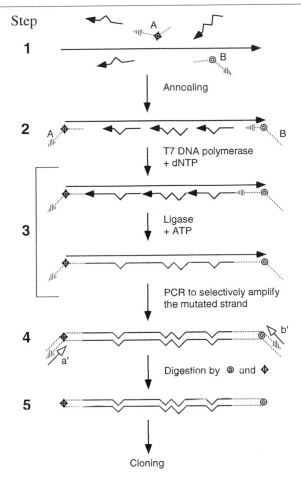

Fig. 3. Mutagenesis procedure to introduce several base changes in a single round of mutagenesis. The single-stranded phage DNA is hybridized with phosphorylated mutagenic oligonucleotides (wavy arrows) plus two oligonucleotides A and B (hatched arrows) that will hybridize partially to the ends of the mutagenized regions. After extension and ligation, the mutagenized strand can be, selectively, PCR amplified by taking advantage of the extra sequences contained in the oligonucleotides A and B.

instance, to construct a universal genetic code equivalent of the apocytochrome *b* gene, we have used the following oligonucleotides A and B (lowercase letters denote the additional sequences noncomplementary to the apocytochrome *b* sequences):

A: 5′-CATTTGATTTTCTAAATGCCATggatccggcgacggcgacggcggcccc-3′
B: 5′-cggcgccgcagagcggaacccggatccTTATTTATTAACTCTACCG-3′

These two oligonucleotides were designed to introduce two restriction sites absent from the gene coding sequence. All the other oligonucleotides were classically designed. They had 10–12 matched base pairs flanking the mutation and an approximate melting temperature of 71° to 75° for the correctly base paired oligonucleotide. We have consistently had satisfactory results using the following protocol.

Step 1: Preparation of Single-Stranded DNA Template. One hundred milliliters of 2 × YT medium[12] supplemented with ampicillin (80 μg/ml) is inoculated with 300 ml of an overnight culture of JM101 containing the recombinant phagemide (derivated from Bluescript, Strategene, La Jolla, CA). After 2 hr of shaking at 37°, 500 μl [5 × 10¹² plaque-forming units (pfu)] of a M13-MK07 stock and kanamycin (70 μg/ml) are added to the culture, and the incubation is continued overnight. Supernatant is retrieved by centrifugation at 15,000 rpm for 5 min. The phages are precipitated for 30 min at room temperature with 15 ml of a solution of 20% polyethylene glycol (PEG) 6000/2.5 M NaCl, then centrifuged 20 min at 20,000 rpm. The pellet is suspended in 15 ml of 100 mM Tris-HCl, 50 mM EDTA, and 0.2% cetyltrimethylammonium bromide (CTAB), pH 8.1 (TE buffer). After 20 min at room temperature and centrifugation at 20,000 rpm for 20 min, the pellet is dissolved in 1 ml of 1.2 M NaCl and DNA is precipitated with 2.5 ml ethanol for 3 hr at −80°. DNA is retrieved by centrifugation (30 min at 20,000 rpm), washed with 70% (v/v) ethanol, and dissolved in TE buffer at a concentration of 0.5 μg/ml.

Step 2: Annealing. Seven micrograms of single-stranded (ss) DNA template is mixed with 10 pmol of each phosphorylated oligonucleotide in 40 μl of 20 mM Tris-HCl, pH 7.5, 10 mM MgCl₂, 50 mM NaCl, and 1 mM dithiothreitol (DTT). The mixture is treated for 3 min at 93° and allowed to cool gently at room temperature for 20 min.

Step 3: Filling the Gaps. The reaction mixture is supplemented with 40 μl containing 1 mM each deoxynucleoside triphosphate, 1 mM ATP, 5 units of T7 DNA polymerase, 5 units of T4 DNA ligase, 20 mM Tris-HCl, pH 7.5, 10 mM MgCl₂, and 10 mM DTT. This mixture is incubated 10 min at 0°, 10 min at 20°, and 2 hr at 37°.

Step 4: PCR Amplification. After purification using GeneClean (BIO 101, La Jolla, CA), the DNA is subjected to a PCR to selectively amplify the mutated strand. For instance, in the case of the apocytochrome *b* gene mentioned above, the two primer oligonucleotides have the following sequence:

    a: 5′-GGGGCCGCCGTCGCCGTCGCC-3′
    b: 5′-CGGCGCCGCAGAGCGGAACCC-3′

[12] K. J. Sambrook, E. F. Fritsch, and T. Maniatis, "Molecular Cloning: A Laboratory Manual." 2nd Ed. Cold Spring Harbor Laboratory, Cold Spring Harbor, New York, 1989.

The PCR protocol is performed with 20 cycles of 1 min denaturing at 94°, 1 min annealing at 68°, and 2 min extension at 72°. After the 20 cycles, the reaction mixture is kept at 68° for 5 min.

Step 5: Analyzing the Product. The PCR product is gel purified, cleaved with the restriction enzymes corresponding to the sites introduced by the oligonucleotides A and B, cloned, and sequenced.

Comments. (1) In preliminary experiments, the mutated strand was gel purified (after labeling) before the PCR step. This turned out to be unnecessary provided the amounts of oligonucleotides used for the mutagenesis are low, and provided a GeneClean step is included to avoid any interference with the subsequent PCR step. (2) In general, 80 to 100% of the desired base changes are introduced in a single round of mutagenesis.

Choice of Mitochondrial Targeting Sequence

It is well documented that the mitochondrial targeting signal is usually contained within the amino-terminal region of the protein, and that the 20- to 80-amino acid long presequences have information to guide the proteins to different mitochondrial compartments.[13] Even if most of the N-terminal mitochondrial targeting sequences do not share any homologous block of amino acids, they have several common characteristics.[14,15] (1) The presequences are composed of two regions with a strong hydrophobic moment: the first region is folded in an amphiphilic α helix, and the second region, of variable length, could be related to the hydropathy properties of the imported protein. (2) If there is no sequence consensus, the N-terminal mitochondrial targeting sequences are enriched in Arg (13.5%), Leu (12.2%), Ser (11.2%), and Ala (14.1%); they contain at least two positively charged residues (Arg, Lys) and no acidic ones (Glu, Asp). The peptidase cleavage site seems to be surrounded by a few more frequent amino acids: Arg at positions -2 or -3, or Arg at -10 or -11 (in the last case, there is generally an apolar residue at position -8), and the first 10–15 residues of the mature protein are enriched in Pro and Ser.

Because the mitochondrial targeting of a cytoplasmically translated protein relies on the choice of an appropriate signal, we present here a brief review of the most relevant mitochondrial targeting sequences (mts) that have been used to import passenger proteins.

Cytb2-mts and Cytc₁-mts. Both precytochrome b_2 and precytochrome c_1 contain bipartite signal sequences that are processed in two sequential proteolytic steps in distinct mitochondrial compartments.[16] The most

---

[13] N. Pfanner and W. Neupert, *Annu. Rev. Biochem.* **59**, 331 (1990).

[14] G. von Heijne, *EMBO J.* **5**, 1335 (1986).

[15] G. von Heijne, J. Steppuhn, and R. G. Herrmann, *Eur. J. Biochem.* **180**, 535 (1989).

[16] E. M. Beasley, S. Muller, and G. Schatz, *EMBO J.* **12**, 2303 (1993).

amino-terminal part of the signal is responsible for targeting the proteins to mitochondria and initiating import into the matrix, whereas the second part of the signal is required for localization of the intermediate forms of the proteins to the intermembrane space (IMS). Besides being model systems to study the mechanism by which the IMS signal targets these proteins to the IMS,[17,18] these presequences were fused to the DHFR (mouse dihydrofolate reductase) to study, *in vitro*, the role of different factors of the import machinery such as chaperones, ATP, and the components of the mitochondrial receptor complex.[19-23]

CoxIV-mts and F₁β-mts. Both cytochrome oxidase subunit IV and F_1-ATPase subunit β of *S. cerevisiae* are localized in the inner membrane.[24] The 25-amino acid long mitochondrial targeting sequence of CoxIVp has been widely used for *in vitro* assays.[25-29] Artificial fusions with DHFR were useful to study the coupling between cytosolic protein synthesis and mitochondrial import,[30,31] or the import competence of mitochondria from *Schizosaccharomyces pombe*.[32] The $F_1\beta$-mts has also been fused to DHFR or CCHL (cytochrome c heme lyase) to study the relationships between the protein translocation machineries in inner and outer mitochondrial membranes.[33,34]

Mas70-mts. The import receptor Mas70p is localized in the outer mitochondrial membrane. Its targeting sequence contains two signals: the first 12 residues target the protein to the mitochondria, and the more distal

[17] E. Schwarz, T. Seytter, B. Guiard, and W. Neupert, *EMBO J.* **12,** 2295 (1993).

[18] S. Rospert, S. Müller, G. Schatz, and B. S. Glick, *J. Biol. Chem.* **269,** 17279 (1994).

[19] A. P. G. M. van Loon, A. W. Brämdli, and G. Schatz, *Cell* **44,** 801 (1986).

[20] H. Koll, B. Guiard, J. Rassow, J. Ostermann, A. L. Horwich, W. Neupert, and F. U. Hartl, *Cell (Cambridge, Mass.)* **68,** 1163 (1992).

[21] C. Wachter, G. Schatz, and B. S. Glick, *EMBO J.* **11,** 4787 (1992).

[22] M. Kiebler, P. Keil, H. Schneider, I. J. Vanderklei, N. Pfanner, and W. Neupert, *Cell (Cambridge, Mass.)* **74,** 483 (1993).

[23] W. Voos, B. D. Gambill, B. Guiard, N. Pfanner, and E. A. Craig, *J. Cell Biol.* **123,** 119 (1993).

[24] G. Schatz, *Eur. J. Biochem.* **165,** 1 (1987).

[25] S. T. Hwang, C. Wachter, and G. Schatz, *J. Biol. Chem.* **266,** 21083 (1991).

[26] M. Yang, V. Geli, W. Oppliger, K. Suda, P. James, and G. Schatz, *J. Biol. Chem.* **266,** 6416 (1991).

[27] T. Jascur, D. P. Goldenberg, D. Vestweber, and G. Schatz, *J. Biol. Chem.* **267,** 13636 (1992).

[28] M. Horst, P. Jeno, N. G. Kronidou, L. Bolliger, W. Oppliger, P. Scherer, U. Manningkrieg, T. Jascur, and G. Schatz, *EMBO J.* **12,** 3035 (1993).

[29] H. Murakami, G. Blobel, and D. Pain, *Proc. Natl. Acad. Sci. U.S.A.* **90,** 3358 (1993).

[30] M. Fujiki and K. Verner, *J. Biol. Chem.* **266,** 6841 (1991).

[31] M. Fujiki and K. Verner, *J. Biol. Chem.* **268,** 1914 (1993).

[32] A. L. Moore, A. J. Walters, J. Thorpe, A. C. Fricaud, and F. Z. Watts, *Yeast* **8,** 923 (1992).

[33] M. E. Walker, E. Valentin, and R. A. Reid, *Biochem. J.* **266,** 227 (1990).

[34] B. Segui-Real, R. A. Stuart, and W. Neupert, *FEBS Lett.* **313,** 2 (1992).

residues act as a stop-transfer signal to arrest the protein in the mitochondrial outer membrane.[35] The first 12 amino acids can be used to target, *in vivo* and *in vitro*, passenger proteins to the mitochondrial matrix.[36,37]

Sod-mts. The first hundred N-terminal amino acids of the manganese superoxide dismutase have been fused to the *URA3* gene product to misroute it to the mitochondria. This facilitated the development of an *in vivo*, genetic screen to isolate genes involved in the import process.[38]

F₀9-mts. The presequence of the F_0-ATPase subunit 9 of *N. crassa* has also been widely used, fused to DHFR, to study *in vitro* different aspects of the import apparatus.[21,22,39–42] It guides to mitochondria a hydrophobic protein composed of two transmembrane domains, and it is cleaved by a two step process.[43] The $F_0$9-mts has been successfully used to target to mitochondria cytoplasmically translated versions of *S. cerevisiae* mitochondrial ATPase subunits 8 and 9[44–46] and the bI4 RNA maturase.[2,36] When two $F_0$9-mts are tandemly duplicated, the import efficiency of passenger proteins is improved.[10,47]

To summarize, a passenger protein, for example, DHFR, can be targeted, at least *in vitro*, to the matrix by $F_0$9-mts,[48] to the intermembrane space by Cytc₁-mts,[19] to the inner membrane by internal targeting sequences of the ADP/ATP translocator,[49] or to the outer membrane by the Mas70-mts.[50]

Choice of Expression Vector

The development of plasmid vectors for expression of cloned DNA sequences in yeast commonly employed the upstream activator sequences

[35] E. C. Hurt, U. Müller, and G. Schatz, *EMBO J.* **4**, 3509 (1985).
[36] J. Banroques, J. Perea, and C. Jacq, *EMBO J.* **6**, 1085 (1987).
[37] H. M. McBride, D. G. Millar, J. M. Li, and G. C. Shore, *J. Cell Biol.* **119**, 1451 (1992).
[38] A. C. Maarse, J. Blom, L. A. Grivell, and M. Meijer, *EMBO J.* **11**, 3619 (1992).
[39] J. Ostermann, W. Voos, P. J. Kang, E. A. Craig, W. Neupert, and N. Pfanner, *FEBS Lett.* **277**, 281 (1990).
[40] J. Blom, M. Kubrich, J. Rassow, W. Voos, P. J. T. Dekker, A. C. Maarse, M. Meijer, and N. Pfanner, *Mol. Cell. Biol.* **13**, 7364 (1993).
[41] B. D. Gambill, W. Voos, P. J. Kang, B. J. Miao, T. Langer, E. A. Craig, and N. Pfanner, *J. Cell Biol.* **123**, 109 (1993).
[42] K. R. Ryan and R. E. Jensen, *J. Biol. Chem.* **268**, 23743 (1993).
[43] A. Viebrock, A. Perz, and W. Sebald, *EMBO J.* **1**, 565 (1982).
[44] L. B. Farrell, D. P. Gearing, and P. Nagley, *Eur. J. Biochem.* **173**, 131 (1988).
[45] R. H. P. Law, L. B. Farrell, D. Nero, R. J. Devenish, and P. Nagley, *FEBS Lett.* **236**, 501 (1988).
[46] R. H. P. Law, R. J. Devenish, and P. Nagley, *Eur. J. Biochem.* **188**, 421 (1990).
[47] M. Galanis, R. J. Devenish, and P. Nagley, *FEBS Lett.* **282**, 425 (1991).
[48] J. Ostermann, A. L. Horwich, W. Newpert, and F. U. Hartl, *Nature (London)* **341**, 125 (1989).
[49] C. Smagula and M. G. Douglas, *J. Biol. Chem.* **263**, 6783 (1988).
[50] J. M. Li and G. C. Shore, *Science* **256**, 1815 (1992).

(UAS) and promoters from yeast genes encoding metabolic enzymes such as 3-phosphoglycerate kinase (PGK). The PGK promoter is generally considered to be constitutive, although its expression is actually repressed as much as 30-fold on nonfermentable carbon sources.[51,52] Thus the PGK vector was not, *a priori*, the best vector to express a mitochondrial gene and estimate the respiratory ability of the transformed strain. It is nevertheless the vector (Fig. 2) that gave positive results with the RNA maturase assays. One possible explanation is that the RNA maturase is known to be present in very small amounts in the wild-type cell and that a low level of expression is more likely to mimic the wild-type context. Similarly, we have not taken into account the codon usage bias to try and increase the translation efficiency. It remains to be discovered, however, whether high levels of expression of the RNA maturase with new efficient inducible vectors could lead to different results.[53]

Comments

The experimental procedures described above have enabled us to demonstrate some new properties of the mitochondrial intron-encoded proteins. It has been shown (1) that the bI4 RNA maturase activity is contained in the 254-amino acid long protein corresponding to the C-terminal part of the precursor,[36] (2) that the protein I–SceII encoded in the aI4 intron can cleave the mitochondrial genome of an intronless allele,[4] and (3) that an *in vivo* recombinational gap-repair approach can be followed to create a large family of chimeric products of bI4 RNA maturase and I–SceII DNA endonuclease.[54]

It is interesting to note that several different mitochondrial targeting sequences can be used to import cytoplasmically forms of the RNA maturase. One of the most efficient was constituted by the first 12 residues of the Mas70-mts.[36] Surprisingly, it was also found that the nuclearly encoded RNA maturase, without any added mitochondrial targeting peptide, can be targeted to the mitochondria to lead to a weak but significant RNA maturase activity. Finally, all these experiments based on genetic assays have to be confirmed by biochemical approaches. This could be done by taking advantage of an epitope tagging approach.[55]

[51] M. F. Tuite, M. J. Dobson, N. A. Roberts, R. M. King, D. C. Burke, S. M. Kingsman, and A. J. Kingsman, *EMBO J.* **1,** 603 (1982).
[52] S. M. Kingsman, D. Cousens, C. A. Stanway, A. Chambers, M. Wilson, and A. J. Kingsman, this series, Vol. 185, p. 329.
[53] M. Schena, D. Picard, and K. R. Yamamoto, this series, Vol. 194, p. 389.
[54] V. Goguel, A. Delahodde, and C. Jacq, *Mol. Cell. Biol.* **12,** 696 (1992).
[55] P. A. Kolodziej and R. A. Young, this series, Vol. 194, p. 508.

Applications: Use of Allotopic Expression of bI4 RNA Maturase to Study Mitochondrial Import of Hydrophobic Proteins

Several features of the cytoplasmically translated RNA maturase depicted above make it an attractive reporter protein to study the mitochondrial import of passenger proteins. (1) It is a small (27-kDa), hydrophilic protein that should not affect the import process of chimeric proteins. (2) It probably contains a mitochondrial proteolytic site that triggers the maturase activity when the chimeric protein is in the mitochondrial matrix. (3) Its activity can be detected easily and sensitively by complementation assays and growth on respirable substrates. (4) The wild-type mitochondrial form of the RNA maturase is synthesized as a chimeric protein with hydrophobic transmembrane fragments of apocytochrome *b* (Fig. 1, top). Thus association of hydrophobic parts of apocytochrome *b* to the cytoplasmically translated RNA maturase should not alter its mitochondrial activity.

Principle

The principle is to construct a tripartite gene fusion (tribrid) by inserting a sequence coding for apocytochrome *b* transmembrane domains between the $F_0$9-mts and the RNA maturase coding sequence, taking advantage of the single *Bgl*II site between the two elements (Fig. 2). Mitochondrial import of the tribrid protein, made *in vivo*, can be estimated from its ability to complement the different maturase-deficient mutations (Table I and Fig. 1, bottom). Such an approach requires rigorous controls to be sure that (1) all the tribrid protein is correctly synthesized *in vivo*, (2) the RNA maturase activity is not affected by the fused peptide, and (3) positive complementations of RNA maturase mutations actually correspond to an import of the tribrid protein. This last point deserves to be studied by a completely different approach. For that purpose, the RNA maturase coding sequence can be replaced by an epitope coding sequence to trace the hydrophobic fragment. An example of the two parallel approaches is shown in Fig. 4.

Comments

When different combinations of the eight transmembrane helices of apocytochrome *b* were inserted in the tribrid protein, it appeared that insertion of more than three to four transmembrane domains completely blocks the RNA maturase complementation.[10] Interestingly, this apparent limit to the *in vivo* mitochondrial import corresponds to the characteristics of mitochondrial proteins whose gene is always localized in the mitochondrial compartment.[56]

[56] J. L. Popot and C. de Vitry, *Annu. Rev. Biophys. Biophys. Chem.* **19**, 369 (1990).

A

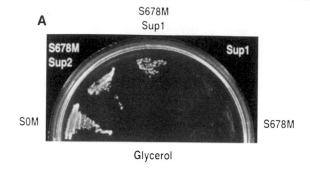

Glycerol

B

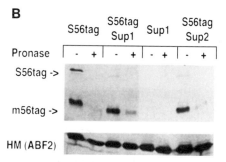

FIG. 4. The RNA maturase as a reporter activity to isolate multicopy suppressor of the blockade effect of hydrophobic peptides on the mitochondrial import. (A) The CW02 strain can grow on glycerol when it is transformed with a vector expressing S0M, which is the RNA maturase (M) fused to the $F_0$9-mts (S). This complementation effect is lost when the three transmembrane domains 6, 7, and 8 of apocytochrome b are inserted between the mitochondrial targeting sequence and the maturase (S678M). Two multicopy suppressors, Sup1 and Sup2, can restore the growth on glycerol. (B) Biochemical evidence of the effects of Sup1 and Sup2. S56M is a tribrid protein that contains the two transmembrane domains 5 and 6 of apocytochrome b. When expressed in the cytoplasm, S56M cannot complement maturase deficiencies (or very poorly), except when the genes *SUP1* or *SUP2* are overexpressed. Replacement of the maturase coding sequence by the tag coding sequence allowed analysis of the mitochondrial extracts with the specific antibodies. Purified mitochondria prepared as described (M. P. Yaffe, this series, Vol. 194, p. 627), were treated with pronase before gel analysis. Non-pronase-digested controls were also run. Clearly, in the presence of Sup1 and Sup2, a pronase-resistant form of the tagged protein can be seen. A control is carried out with the mitochondrial protein HM [F. Caron, C. Jacq, and J. Rouvière-Yaniv, *Proc. Natl. Acad. Sci. U.S.A.* **76**, 4265 (1979)], also called ABF2.

This situation offers the opportunity to develop a genetic screen to look for genes which, when mutated or when present on a multicopy plasmid, would confer the possibility to import, into mitochondria, longer hydrophobic peptides. As an example, when the yeast strain CW02 was transformed with a plasmid YEpC/3-21-2 coding for a tribrid protein S678M (containing the three transmembrane domains 6, 7, and 8^{10}), it did not grow on glycerol (Fig. 4A). Transformation of this strain with a library constructed by cloning the partially Sau3AI-digested yeast genomic DNA in the BamHI site of the multicopy vector pFL44L[57] (a gift from F. Lacroute, CNRS, 91190 Gif, France) permitted the selection of more than 100,000 uracil prototrophs. These transformants were replica-plated on glycerol (N3 broth), and 11 transformants grew after 3 days at 28° (Fig. 4A). Plasmids were isolated and used to transform yeast mutant strain CW02 (Table I). It turned out that 6 of the 11 selected plasmids could restore growth on glycerol in the absence of the plasmid YEpC/3-21-2. As expected,[8] these 6 plasmids were found to contain a copy of the $NAM2$ gene which, in the presence of the aI4 intron, can complement bI4 RNA mutations (see above). This result could also be confirmed by transforming the strain CW05 (Table I). The other 5 transformants were found to correspond to three different nuclear genes. The properties of two of these genes are presented in Fig. 4. When present on a multicopy vector, they can stimulate the complementation properties of a cytoplasmically translated RNA maturase that had been fused to the two transmembrane domains 5 and 6 of apocytochrome b. If, in this experiment, the maturase coding sequence is replaced by a DNA encoding the 9-amino acid epitope HA1 of influenza virus hemagglutinin (recognized by the monoclonal antibody 12CA5, Berkeley Antibody), it was found that a fraction of the two transmembrane domains 5 and 6 was pronase-resistant when one of the two selected genes was overexpressed (Fig. 4B). This strongly suggests that genes involved in the mitochondrial import of hydrophobic proteins can be selected by such a genetic approach. Such information could appear to be very useful if one wishes to consider alternative approaches for the treatment of mitochondrial DNA diseases.[58,59]

Acknowledgments

We thank A. Delahodde and V. Goguel for helpful discussions, G. Dujardin for yeast strains, F. Lacroute for pFL plasmids and gene library, and B. Clark for revision of the manuscript. M. G. C. was supported by a PFPI fellowship from the Spanish Ministerio de Educación y Ciencia.

[57] N. Bonneaud, O. Ozier-Kalogeropoulos, G. Li, M. Labouesse, L. Minvielle-Sebastia, and F. Lacroute, *Yeast* **7,** 609 (1991).
[58] A. Chomyn, G. Meola, N. Bresolin, S. T. Lai, G. Scarlato, and G. Attardi, *Mol. Cell. Biol.* **11,** 2236 (1991).
[59] D. C. Wallace, *Annu. Rev. Biochem.* **61,** 1175 (1992).

Section III

Mitochondrial Diseases and Aging

[35] Automated Sequencing of Mitochondrial DNA

By MASASHI TANAKA, MIKA HAYAKAWA, and TAKAYUKI OZAWA

Introduction

Analysis of the entire sequence of mitochondrial DNA (mtDNA) is essential for identification of genotypes underlying mitochondrial diseases[1] as well as for elucidation of mutational mechanisms in mtDNA.[2] This chapter describes a method for direct sequencing of mtDNA fragments amplified from genomic DNA without using cloning procedures, radioisotopes, or columns for template purification. A manual method for direct sequencing of mtDNA was reported previously.[3] The present protocol describes methods for fully automated sequencing of mtDNA with a combination of a chemical robot and a fluorescence-based automated DNA sequencer. This method permits rapid, accurate, and easy determination of entire sequence of mitochondrial genome.

Several methods for direct sequencing of PCR (polymerase chain reaction)-amplified DNA have been reported.[4-7] In the case of the primer extension-labeling method using radioisotope,[6,7] or the dye-termination method,[8] primers used in the first PCR reaction must be completely removed before sequencing. In contrast, the primer end-labeling method,[6] which does not require removal of primers, is suitable for direct sequencing of PCR-amplified DNA. However, the primer end-labeling method is not efficient in analyzing the entire sequence of mtDNA, because each of the sequencing primers must be labeled separately. To circumvent these problems, we have employed an indirect labeling method,[9] in which a

[1] T. Ozawa, *Hertz* **19,** 105 (1994).

[2] M. Tanaka and T. Ozawa, *Genomics* **22,** 327 (1994).

[3] M. Tanaka and T. Ozawa, *in* "Protocols in Molecular Neurobiology" (A. Langstaff and P. Revest, eds.), p. 25. Humana, Totawa, New Jersey, 1992.

[4] U. B. Gyllensten and H. A. Erlich, *Proc. Natl. Acad. Sci. U.S.A.* **85,** 7652 (1988).

[5] R. Higuchi, C. H. von Beroldingen, G. F. Sensabaugh, and H. A. Erlich, *Nature (London)* **332,** 543 (1988).

[6] M. A. Innis, K. B. Myambo, D. H. Gelfand, and M. A. D. Brow, *Proc. Natl. Acad. Sci. U.S.A.* **85,** 9436 (1988).

[7] M. Tanaka, W. Sato, K. Ohno, T. Yamamoto, and T. Ozawa, *Biochem. Biophys. Res. Commun.* **164,** 156 (1989).

[8] J. M. Prober, G. L. Trainor, R. J. Dam, F. W. Hobbs, C. W. Robertson, R. J. Zagursky, A. J. Cocuzza, A. A. Jensen, and K. Baumeister, *Science* **238,** 336 (1987).

[9] L. J. McBride, S. M. Koepf, R. A. Gibbs, W. Salser, P. E. Mayrand, M. W. Hunkapiller, and M. N. Kronick, *Clin. Chem.* **35,** 2196 (1989).

sequence complementary to a universal sequencing primer is introduced to templates by an asymmetric PCR method,[6] and then fluorescence-labeled universal sequencing primers are used for sequencing of these templates. These fluorescence-labeled primers are stable and are commercially available as a sequencing kit, and thus they are convenient for repeated sequencing.

The principle of fluorescence-based direct sequencing is schematically presented in Fig. 1. First, the entire mtDNA (16,569 bp) is amplified as 6 overlapping fragments of approximately 3.2–3.6 kb (Table I) by a symmetric PCR method using primers L_1 and H_1 at an equal concentration of 1 μM (Table II). Primers L_1 and H_1 are 20-mer oligonucleotides with sequences of the light (L) and the heavy (H) strands of mtDNA, respectively (Tables III and IV). Then single-stranded DNA templates (60 segments of approximately 600–1100 nucleotides covering the entire genome) are amplified from each fragment by an asymmetric PCR method using primer FL_2 at a limiting concentration (0.01 μM) and primer H_2 (the second primer for

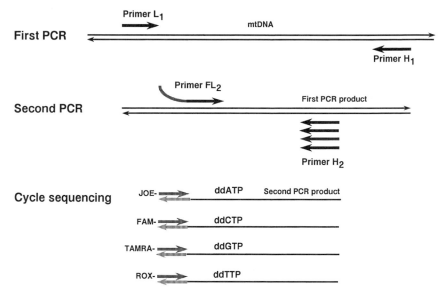

FIG. 1. Principle of fluorescence-based direct sequencing of PCR-amplified mtDNA. (*Top*) Primers L_1 and H_1 are used for the first PCR amplification of mtDNA. (*Middle*) A limiting primer FL_2 and an excess primer H_2 are used for the asymmetric second PCR amplification. The dotted part of primer FL_2 represents the forward universal sequence of −21M13. (*Bottom*) Primers labeled with four different fluorescent dyes (JOE, FAM, TAMRA, and ROX) are used for termination reactions with ddATP, ddCTP, ddGTP, and ddTTP, respectively. The striped arrow at the 3' end of the second PCR product represents the sequence complementary to the forward universal sequence.

amplification of the H strand) in an ordinary concentration (1 μM). The FL$_2$ primer is a 38-mer oligonucleotide (Table V); an 18-base sequence of a universal forward sequencing primer (−21M13, 5′-TGT AAA ACG ACG GCC AGT-3′) on the 5′ side is connected with a 20-base L-strand-specific sequence on the 3′ side. In the asymmetric PCR, the H-strand synthesis starting from the H$_2$ primer extends up to the FL$_2$ primer; thus, a complementary sequence of the universal sequencing primer can be introduced at the 3′ end of the H-strand template.

These templates are concentrated by ethanol precipitation and then can be directly used for sequencing reactions by a chemical robot equipped with a thermal cycler plate. Sequencing is based on termination reactions with dideoxynucleotides using four primers labeled with different fluorescent dyes at their 5′ end. After sequencing reactions in four different wells of the thermal cycler plate, these products are pooled by the robot into one tube containing ethanol and sodium acetate. The samples concentrated by ethanol precipitation are then applied onto a polyacrylamide gel in an automated DNA sequencer. With this system, tedious manual work for removal of previous primers is unnecessary. Twenty segments can be analyzed in a single run of the sequencer; analysis of the entire mtDNA sequence from one individual can be completed in three runs.

Experimental Methods

Materials

Oligonucleotide primers, synthesized and purified by gel filtration, are obtained from Bio-Synthesis (Lewisville, Texas). The sequences of primers L, H, and FL are listed in Tables III–V. *Taq* DNA polymerase (Takara Taq) is purchased from Takara (Ohtsu, Shiga, Japan). This enzyme is supplied at a concentration of 5 units/μl in 20 mM Tris-HCl, pH 8.0, 100 mM KCl, 0.1 mM EDTA, 1 mM dithiothreitol (DTT), 0.5% (w/v) Tween 20, 0.5% (w/v) Nonidet P-40, 50% (w/v) glycerol. A mixture of deoxynucleoside triphosphates (dNTPs, 2.5 mM each) and 10× PCR buffer (100 mM Tris-HCl, pH 8.3, 500 mM KCl, 15 mM MgCl$_2$) are also supplied by Takara. A GeneAmp PCR core reagent kit from Applied Biosystems (Foster City, CA) is also recommended.

A PRISM ready reaction dye primer cycle sequencing kit with the −21M13 sequencing primer is obtained from Applied Biosystems. The reagents in the kit include the following: A mix [375 μM ddATP (dideoxyadenosine triphosphate), 15.6 μM dATP, 62.5 μM dCTP, 93.8 μM c7dGTP (7-deaza-2′-deoxyguanosine 5′-triphosphate), 62.5 μM dTTP, 0.1 pmol/μl JOE dye primer], C mix (187.5 μM ddCTP, 62.5 μM dATP, 15.6 μM dCTP,

TABLE I
PRIMER PAIRS FOR FIRST PCR AMPLIFICATION OF SIX FRAGMENTS AND FOR SECOND
PCR AMPLIFICATION OF SIXTY SEGMENTS AND REGIONS OF MITOCHONDRIAL DNA
READABLE WITH THESE PCR PRODUCTS

Fragment segment	First PCR			Second PCR			Readable region	
	L_1 primer	H_1 primer	Length	FL_2 primer	H_2 primer	Length	From	To
Fragment A	L77	H3370	3313					
1				FL100	H742	680	121	520
2				FL398	H1014	654	419	818
3				FL700	H1696	1034	721	1120
4				FL995	H1591	634	1016	1415
5				FL1254	H1828	612	1275	1674
6				FL1485	H2060	613	1506	1905
7				FL1779	H2385	644	1800	2199
8				FL2045	H2669	662	2066	2465
9				FL2332	H3370	1076	2353	1752
10				FL2616	H3370	792	2637	3036
Fragment B	L2815	H6050	3255					
11				FL2864	H3538	712	2885	3284
12				FL3119	H3876	795	3140	3539
13				FL3455	H4222	805	3476	3875
14				FL3712	H4552	878	3733	4132
15				FL4008	H4552	582	4029	4428
16				FL4250	H4854	642	4271	4670
17				FL4529	H5528	1037	4550	4949
18				FL4827	H5528	739	4848	5247
19				FL5110	H5759	687	5131	5530
20				FL5362	H6009	685	5383	5782
Fragment C	L5582	H9133	3571					
21				FL5602	H6158	594	5623	6022
22				FL5890	H6454	602	5911	6310
23				FL6186	H6757	609	6207	6606
24				FL6452	H7005	591	6473	6872
25				FL6716	H7613	935	6737	7136
26				FL6957	H7613	694	6978	7377
27				FL7245	H7758	551	7266	7665
28				FL7497	H8144	685	7518	7917
29				FL7762	H8395	671	7783	8182
30				FL8062	H8915	891	8083	8482

(*continued*)

TABLE I (*continued*)

Fragment segment	First PCR			Second PCR			Readable region	
	L_1 primer	H_1 primer	Length	FL_2 primer	H_2 primer	Length	From	To
Fragment D	L8281	H11571	3310					
31				FL8345	H9133	826	8366	8765
32				FL8634	H9140	544	8655	9054
33				FL8913	H9483	608	8934	9333
34				FL9182	H9813	669	9203	9602
35				FL9484	H10137	691	9505	9904
36				FL9747	H10660	951	9768	10,167
37				FL10028	H10629	639	10,049	10,448
38				FL10297	H10913	654	10,318	10,717
39				FL10616	H11179	601	10,637	11,036
40				FL10910	H11519	647	10,931	11,330
Fragment E	L10796	H14378	3062					
41				FL11183	H11791	646	11,204	11,603
42				FL11467	H12000	571	11,488	11,887
43				FL11766	H12451	723	11,787	12,186
44				FL12057	H12818	799	12,078	12,477
45				FL12357	H12921	602	12,378	12,777
46				FL12602	H13203	639	12,623	13,022
47				FL12889	H13492	641	12,910	13,309
48				FL13188	H13767	617	13,209	13,608
49				FL13479	H14102	661	13,500	13,899
50				FL13721	H14321	638	13,742	14,141
Fragment F	L13901	H609	3297					
51				FL13986	H14754	806	14,007	14,406
52				FL14279	H14909	668	14,300	14,699
53				FL14559	H15162	641	14,580	14,979
54				FL14837	H15340	541	14,858	15,257
55				FL15126	H15745	667	15,147	15,546
56				FL15405	H16016	649	15,426	15,825
57				FL15696	H81	992	15,717	16,116
58				FL15948	H81	740	15,969	16,368
59				FL16200	H127	534	16,221	51
60				FL16504	H581	684	16,525	355

93.8 μM c7dGTP, 62.5 μM dTTP, 0.1 pmol/μl FAM dye primer), G mix (31.3 μM ddGTP, 62.5 μM dATP, 62.5 μM dCTP, 23.5 μM c7dGTP, 62.5 μM dTTP, 0.1 pmol/μl TAMRA dye primer), and T mix (312.5 μM ddTTP, 62.5 μM dATP, 62.5 μM dCTP, 93.8 μM c7dGTP, 15.6 μM dTTP, 0.1 pmol/μl ROX dye primer). Each of the mixes also contains 100 mM Tris-HCl (pH 8.9 at room temperature), 25 mM $(NH_4)_2SO_4$, 6.3 mM $MgCl_2$, and 0.18 units/μl of AmpliTaq DNA polymerase.

TABLE II
COMPARISON OF PCR CONDITIONS FOR FIRST AND SECOND PCR

Parameter	First PCR	Second PCR (asymmetric)
Primers	L_1, 1 μM	FL_2, 0.01 μM
	H_1, 1 μM	H_2, 1 μM
dNTPs	200 μM	20 μM
Taq DNA polymerase	1.25 units	0.5 units
Volume	50 μl	20 μl
Denaturation	94°, 15 sec	94°, 15 sec
Annealing	60°, 15 sec	60°, 15 sec
Extension	72°, 180 sec + 5 sec each	72°, 180 sec + 5 sec each
Cycles	40	40

Isolation of Genomic DNA

Genomic DNA isolated from tissues (heart, brain, platelets, whole blood cells, buccal mucosal cells) by a standard phenol–chloroform method can serve as the template for the first PCR. For tissues rich in mitochondria, such as heart and brain, the concentration of genomic DNA is adjusted to 10 ng/μl. For tissues with low mtDNA content (blood cells and buccal mucosal cells), a range of 50–100 ng/μl is recommended.

First PCR: Symmetric

1. For amplification of six mtDNA fragments from one individual, 5 μl each of primers L_1 and H_1 (10 μM) are added to 0.2-ml MicroAmp reaction tubes (Perkin-Elmer, Norwalk, CT). Then 6 μl of genomic DNA (10 ng/μl), 24 μl of 2.5 mM dNTPs, 1.5 μl of 5 units/μl *Taq* DNA polymerase, 30 μl of 10× PCR buffer, and 178.5 μl of distilled

TABLE III
L PRIMERS FOR AMPLIFICATION OF LIGHT STRAND OF
MITOCHONDRIAL DNA

Primer	Position		Sequence (5' → 3')
	From	To	
L77	77	96	ACGCGATAGCATTGCGAGAC
L2815	2815	2834	GGGCGACCTCGGAGCAGAAC
L5582	5582	5601	ACAGCTAAGGACTGCAAAAC
L8281	8281	8300	CCCCCTCTAGAGCCCACTGT
L10796	10,796	10,815	CCACTGACATGACTTTCCAA
L13901	13,901	13,920	TCTCCAACATACTCGGATTC

water are mixed in a 1.5-ml microcentrifuge tube (240 μl in total), and a 40-μl aliquot of this mixture is dispensed into each of the 0.2-ml MicroAmp tubes containing primers (final volume of 50 μl). The tubes are vortexed and centrifuged briefly. Using a GeneAmp PCR system Model 9600 (Perkin-Elmer), amplification is carried out with 40 cycles of 15 sec of denaturation at 94°, 15 sec of annealing at 60°, and 180 sec of extension at 72° with a 5-sec increment of extension time for each cycle.

2. First PCR products (5 μl) are electrophoresed on a 1% agarose gel in TAE (40 mM Tris–acetate, pH 8.0, 2 mM EDTA) containing 0.5 μg/ml ethidium bromide. The first PCR should yield a single band with the size listed in Table I.

3. The first PCR products are transferred to a 1.5-ml microcentrifuge tube, and 5 μl of 3 M sodium acetate (pH 7.4) and 100 μl ethanol are added. The mixture is kept on ice 10 min and then centrifuged at 13,000 g for 10 min at 4°. The precipitate is rinsed with 150 μl of 70% (v/v) ethanol, and then the tube is centrifuged at 13,000 g for 5 min at 4°. The precipitate is dried *in vacuo* for 10 min and then suspended in 50 μl water. For long-term storage, the first PCR prod uct should be suspended in TE (10 mM Tris-HCl, pH 8.0, 1 mM EDTA).

Second PCR: Asymmetric

1. For the amplification of 10 segments from each first PCR product, 10 μl primer mixture (0.02 μM primer FL$_2$ and 2 μM primer H$_2$) is added to each 0.2-ml MicoAmp reaction tube. Then 4 μl of the first PCR product, 1.6 μl of 2.5 mM dNTPs, 1 μl of 5 units/μl of *Taq* polymerase, 20 μl of 10× PCR buffer, and 73.4 μl of distilled water are mixed in a 1.5-ml microcentrifuge tube (100 μl in total), and 10-μl aliquots of the mixture are dispensed into each MicroAmp reaction tube. Amplification conditions are the same as in the first PCR.

2. Each product (5 μl) is analyzed by electrophoresis on a 1% agarose gel. A sharp band of a double-stranded template (about 600–1100 bp) and a single-stranded template that migrates as a faint band slower than the double-stranded template should be confirmed (Fig. 2). The second PCR product (15 μl) is transferred into a 1.5-ml microcentrifuge tube. After the addition of 1.5 μl of 3 M sodium acetate (pH 7.4) and 30 μl of ethanol, the template is kept on ice for 10 min and then centrifuged at 13,000 g for 10 min at 4°. The precipitate is rinsed with 45 μl of 70% (v/v) ethanol and then centri-fuged at 13,000 g for 5 min at 4°, dried *in vacuo* for 10 min, and suspended in 10 μl water.

TABLE IV
H Primers for Amplification of Heavy Strand
of Mitochondrial DNA

| Primer | Position | | Sequence (5′ → 3′) |
	From	To	
H81	81	100	CAGCGTCTCGCAATGCTATC
H127	127	146	AGGCAGGAATCAAAGACAGA
H581	581	600	TTTGAGGAGGTAAGCTACAT
H609	609	628	GCCCGTCTAAACATTTTCAG
H742	742	761	TTGATGCTTGTTCCTTTTGA
H1014	1014	1033	AGCCACTTTCGTAGTCTATT
H1591	1591	1610	TACACTCTGGTTCGTCCAAG
H1606	1696	1715	GGTTGTCTGGTAGTAAGGTG
H1828	1828	1847	AGGTATAGGGGTTAGTCCTT
H2060	2060	2079	GGGGATTTAGAGGGTTCTGT
H2385	2385	2404	ACTTGTTGGTTGATTGTAGA
H2669	2669	2688	GGCAGGTCAATTTCACTGGT
H3370	3370	3389	AGAATTTTTCGTTCGGTAAG
H3538	3538	3557	AGAAGAGCGATGGTGAGAGC
H3876	3876	3895	GGGTTCGGTTGGTCTCTGCT
H4222	4222	4241	GAGATTGTAATGGGTATGGA
H4552	4552	4571	TTCTAGGCCTACTCAGGTAA
H4854	4854	4873	TTTTGTCATGTGAGAAGAAG
H5528	5528	5547	TTGAAGGCTCTTGGTCTGTA
H5759	5759	5778	TCAAACCTGCCGGGGCTTCT
H6009	6009	6028	CCCAGCTCGGCTCGAATAAG
H6050	6050	6069	CGTTGTAGATGTGGTCGTTA
H6158	6158	6177	TATCGGGGGCACCGATTATT
H6454	6454	6473	GATTAGGACGGATCAGACGA
H6757	6757	6776	ATGGTGTGCTCACACGATAA
H7005	7005	7024	ACAACGTAGTACGTGTCGTG
H7613	7613	7632	ATAGGGGAAGTAGCGTCTTG

(continued)

3. To determine the concentration, 2.5 μl of the concentrated template is applied onto a 1% agarose gel. An aliquot (routinely 5 μl) of the concentrated template is diluted with distilled water to a volume of 10 μl and applied to the chemical robot.

Sequencing Reaction by Chemical Robot

In the scrooge *Taq* protocol using a chemical robot, 2 μl of either A or C mix and 0.6 μl of the template are dispensed into each well for termination reactions with the dideoxynucleoside triphosphates ddATP and ddCTP; 4 μl of either G or T mix and 1.2 μl of the template are mixed

TABLE IV (*continued*)

| Primer | Position | | Sequence (5′ → 3′) |
	From	To	
H7758	7758	7777	GACGGTTTCTATTTCCTGAG
H8144	8144	8163	TAGTATACCCCCGGTCGTGT
H8395	8395	8414	GTATGGGGGTAATTATGGTG
H8915	8915	8934	GGGTGTAGGTGTGCCTTGTG
H9133	9133	9152	ATTAAGGCGACAGCGATTTC
H9140	9140	9259	GGCTTGGATTAAGGCGACAG
H9483	9483	9502	CAGAAAAATCCTGCGAAGAA
H9813	9813	9832	ATGACGTCAACTCCGTGGAA
H10137	10,137	10,156	TTTTCTATGTAGCCGTTGAG
H10629	10,629	10,648	GGCACAATATTGGCTAAGAG
H10660	10,660	10,679	TTCGCAGGCGGCAAAGACTA
H10913	10,913	10,932	GAGGAAAAGGTTGGGGAACA
H11179	11,179	11,198	TGCCTGCGTTCAGGCGTTCT
H11519	11,519	11,538	GGGTAGGCTATGTGTTTTGT
H11571	11,571	11,590	TAGGCAGATGGAGCTTGTTA
H11791	11,791	11,810	GAGTTTGAAGTCCTTGAGAG
H12000	12,000	12,019	GGTGAGTGAGCCCCATTGTG
H12451	12,451	12,470	ATAAAGGTGGATGCGACAAT
H12818	12,818	12,837	TGCTGTGTTGGCATCTGCTC
H12921	12,921	12,940	CTATTTGTTGTGGGTCTCAT
H13203	13,203	13,222	TTTGTGTAAGGGCGCAGACT
H13492	13,492	13,511	TTGGAGTAGAAACCTGTGAG
H13767	13,767	13,786	GCCCCATTGTTGTTTGGAAG
H14102	14,102	14,121	GAGTGGGAAGAAGAAAGAGA
H14321	14,321	14,340	GTGGTGGTTGTGGTAAACTT
H14378	14,378	14,397	TTAGTGGGGTTAGCGATGGA
H14754	14,754	14,773	GGGGTTAATTTTGCGTATTG
H14909	14,909	14,928	GTTGAGGCGTCTGGTGAGTA
H15162	15,162	15,181	TACTCTCGCCCCTCAGAATG
H15340	15,340	15,359	ATCCCGTTTCGTGCAAGAAT
H15755	15,755	15,774	ACTGGTTGTCCTCCGATTCA
H16016	16,016	16,035	CCCATGAAAGAACAGAGAAT

for termination reactions with ddGTP and ddTTP. At least 8.6 μl of diluted template is necessary for the sequencing reaction using the scrooge protocol in a Molecular Biology LabStation Catalyst 800 (Applied Biosystems). The reaction conditions are according to the manufacturer's recommendation except for the following PCR conditions: a three-step profile (20 sec of denaturation at 96°, 20 sec of annealing at 60°, and 30 sec of extension at 70°) for 25 cycles followed by a two-step profile (10 sec of denaturation at 96°, 60 sec of extension at 70°) for 15 cycles.

At the end of reaction, the contents of the four wells are automatically

TABLE V

FL PRIMERS FOR ASYMMETRIC AMPLIFICATION OF MITOCHONDRIAL DNA

Primer	Position From	Position To	Sequence (5' → 3')
FL100	100	119	TGTAAAACGACGGCCAGTGGAGCCGGAGCACCCTATGT
FL398	398	417	TGTAAAACGACGGCCAGTTTTTATCTTTTGGCGGTATG
FL700	700	719	TGTAAAACGACGGCCAGTAGCATCCCCGTTCCAGTGAG
FL995	995	1014	TGTAAAACGACGGCCAGTAAAACTCCAGTTGACACAAA
FL1254	1254	1273	TGTAAAACGACGGCCAGTCTATATACCGCCATCTTCAG
FL1485	1485	1504	TGTAAAACGACGGCCAGTGCCCGTCACCCTCCTCAAGT
FL1779	1779	1798	TGTAAAACGACGGCCAGTATAGTACCGCAAGGGAAAGA
FL2045	2045	2064	TGTAAAACGACGGCCAGTACTTTAAATTTGCCCACAGA
FL2332	2332	2351	TGTAAAACGACGGCCAGTCGCATAAGCCTGCGTCAGAT
FL2616	2616	2635	TGTAAAACGACGGCCAGTAATAGGGACCTGTATGAATG
FL2864	2864	2883	TGTAAAACGACGGCCAGTTCACCAGTCAAAGCGAACTA
FL3119	3119	3138	TGTAAAACGACGGCCAGTCCCTGTACGAAAGGACAAGA
FL3455	3455	3474	TGTAAAACGACGGCCAGTCTGACGCCATAAAACTCTTC
FL3712	3712	3731	TGTAAAACGACGGCCAGTGTAGCCCAAACAATCTCATA
FL4008	4008	4027	TGTAAAACGACGGCCAGTAAACACCCTCACCACTACAA
FL4250	4250	4269	TGTAAAACGACGGCCAGTCCCCTCAAACCTAAGAAATA
FL4529	4529	4548	TGTAAAACGACGGCCAGTAGCGCTAAGCTCGCACTGAT
FL4827	4827	4846	TGTAAAACGACGGCCAGTCAAGGCACCCCTCTGACATC
FL5110	5110	5129	TGTAAAACGACGGCCAGTCCGCATTCCTACTACTCAAC
FL5362	5362	5381	TGTAAAACGACGGCCAGTACTCCACCTCAATCACACTA
FL5602	5602	5621	TGTAAAACGACGGCCAGTCCCACTCTGCATCAACTGAA
FL5890	5890	5909	TGTAAAACGACGGCCAGTCCTCACCCCCACTGATGTTC
FL6186	6186	6205	TGTAAAACGACGGCCAGTCCCCGCATAAACAACATAAG
FL6452	6452	6471	TGTAAAACGACGGCCAGTCTTCGTCTGATCCGTCCTAA
FL6716	6716	6735	TGTAAAACGACGGCCAGTAGGTATGGTCTGAGCTATGA

(continued)

pooled into a 1.5-ml tube containing a mixture of 4.5 μl of 3 M sodium acetate (pH 5.2) and 150 μl ethanol. The tubes are removed from the robot, placed on ice for 10 min, and then centrifuged at 13,000 g for 15 min at 4°. The pellet is rinsed with 150 μl ethanol and then centrifuged at 13,000 g for 5 min at 4°. The pellet is dried *in vacuo* for 10 min and stored at −20° in the dark until electrophoresis. Just prior to electrophoresis the pellet is dissolved in 5 μl of a 5:1 (v/v) mixture of deionized formamide and 50 mM EDTA (pH 8.0) containing 30 mg/ml of Blue dextran 2000 (Pharmacia, Uppsala, Sweden). The DNA samples are heated at 90° for 2 min for denaturation, immediately cooled on ice, and loaded onto a 6% polyacrylamide gel that has been preelectrophoresed for 10 min. Samples are electrophoresed at 27 W for 12 hr. Sequences are analyzed with an automated

TABLE V (*continued*)

| Primer | Position | | Sequence (5′ → 3′) |
	From	To	
FL6957	6957	6976	TGTAAAACGACGGCCAGTGGCCTGACTGGCATTGTATT
FL7245	7245	7264	TGTAAAACGACGGCCAGTACCACATGAAACATCCTATC
FL7497	7497	7516	TGTAAAACGACGGCCAGTGGCCTCCATGACTTTTTCAA
FL7762	7762	7781	TGTAAAACGACGGCCAGTGGAAATAGAAACCGTCTGAA
FL8062	8062	8081	TGTAAAACGACGGCCAGTCTTGCACTCATGAGCTGTCC
FL8345	8345	8364	TGTAAAACGACGGCCAGTCCAACACCTCTTTACAGTGA
FL8634	8634	8653	TGTAAAACGACGGCCAGTTCTCATCAACAACCGACTAA
FL8913	8913	8932	TGTAAAACGACGGCCAGTACCACAAGGCACACCTACAC
FL9182	9182	9201	TGTAAAACGACGGCCAGTGCCTCTACCTGCACGACAAC
FL9484	9484	9503	TGTAAAACGACGGCCAGTTCTTCGCAGGATTTTTCTGA
FL9747	9747	9766	TGTAAAACGACGGCCAGTTACTTCGAGTCTCCCTTCAC
FL10028	10,028	10,047	TGTAAAACGACGGCCAGTAACTAGTTTTGACAACATTC
FL10297	10,297	10,316	TGTAAAACGACGGCCAGTAAACAACTAACCTGCCACTA
FL10616	10,616	10,635	TGTAAAACGACGGCCAGTCAACACCCACTCCCTCTTAG
FL10910	10,910	10,929	TGTAAAACGACGGCCAGTAGCTGTTCCCCAACCTTTTC
FL11183	11,183	11,202	TGTAAAACGACGGCCAGTCGCCTGAACGCAGGCACATA
FL11467	11,467	11,486	TGTAAAACGACGGCCAGTAAAACTAGGCGGCTATGGTA
FL11766	11,766	11,785	TGTAAAACGACGGCCAGTGCACTCACAGTCGCATCATA
FL12057	12,057	12,076	TGTAAAACGACGGCCAGTAAAACACCCTCATGTTCATA
FL12357	12,357	12,376	TGTAAAACGACGGCCAGTAACCACCCTAACCCTGACTT
FL12602	12,602	12,621	TGTAAAACGACGGCCAGTTCATCCCTGTAGCATTGTTC
FL12889	12,889	12,908	TGTAAAACGACGGCCAGTGCCTTAGCATGATTTATCCT
FL13188	13,188	13,207	TGTAAAACGACGGCCAGTCACTCTGTTCGCAGCAGTCT
FL13479	13,479	13,498	TGTAAAACGACGGCCAGTAGGAATACCTTTCCTCACAG
FL13721	13,721	13,740	TGTAAAACGACGGCCAGTTATTCGCAGGATTTCTCATT
FL13986	13,986	14,005	TGTAAAACGACGGCCAGTACTCCTCCTAGACCTAACCT
FL14279	14,279	14,298	TGTAAAACGACGGCCAGTGACCCCTCTCCTTCATAAAT
FL14559	14,559	14,578	TGTAAAACGACGGCCAGTCGACCACACCGCTAACAATC
FL14837	14,837	14,856	TGTAAAACGACGGCCAGTTGAAACTTCGGCTCACTCCT
FL15126	15,126	15,145	TGTAAAACGACGGCCAGTCCTTCATAGGCTATGTCCTC
FL15405	15,405	15,424	TGTAAAACGACGGCCAGTTCCACCCTTACTACACAATC
FL15696	15,696	15,715	TGTAAAACGACGGCCAGTTTTCGCCCACTAAGCCAATCA
FL15948	15,948	15,967	TGTAAAACGACGGCCAGTAGGACAAATCAGAGAAAAAG
FL16200	16,200	16,219	TGTAAAACGACGGCCAGTACAAGCAAGTACAGCAATCA
FL16504	16,504	16,523	TGTAAAACGACGGCCAGTGTTCCTACTTCAGGGTCATA

DNA sequencer, Model 373A from Applied Biosystems using the analysis program version 1.2.1. The sequences up to 500 bp are recorded.

Comments

During the initial cycles of asymmetric PCR, the double-stranded product between primers FL_2 and H_2 is amplified exponentially. When the

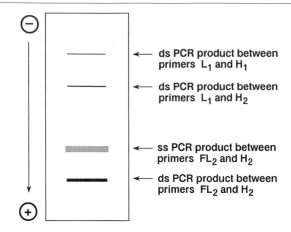

FIG. 2. Schematic diagram of products from the asymmetric second PCR in agarose gel electrophoresis. Either double-stranded (ds) PCR product between primers L_1 and H_1 or ds PCR product between primers L_1 and H_2 is not always visible. Amplification of both single-stranded (ss) PCR product between primers FL_2 and H_2 and ds PCR product between primers FL_2 and H_2 should be confirmed.

limiting primer FL_2 is depleted, the single-stranded product starting from the excess primer H_2 is then synthesized linearly with the number of cycles. The amount of FL_2 primer used for the second PCR is critical both for the amplification of templates and for the subsequent sequencing reaction with the fluorescence-labeled primer. The ratio of the limiting primer FL_2 to the excess primer H_2 is $1:100$ in the present protocol, but a ratio of $1:50$ can also be used. When the concentration of primer FL_2 is too low, the second PCR only yields PCR products between primer L_1 and primers H_1 or H_2, which do not serve as the template for the sequencing.

For the cycle sequencing method both double-stranded and single-stranded DNA can serve as templates. It should be emphasized that the main purpose of the asymmetric PCR is not to obtain single-stranded DNA template but to deplete the FL_2 primer. The unlabeled FL_2 primer competitively inhibits the binding of the fluorescence-labeled sequencing primer to the template DNA, because the template DNA has a sequence

FIG. 3. Fluorescence-based direct sequencing of PCR-amplified mtDNA. A mtDNA fragment was amplified using primers L2815 and H3538. The second PCR product was amplified using primers FL3171 and H3538. The large peak at nucleotide 370 (bottom) is due to the 3' end of template DNA. Dotted line ($\cdots\cdots$), adenine; solid line (——), cytosine; wide broken line (————), guanine; and dot–dash line (—·—·—), thymine.

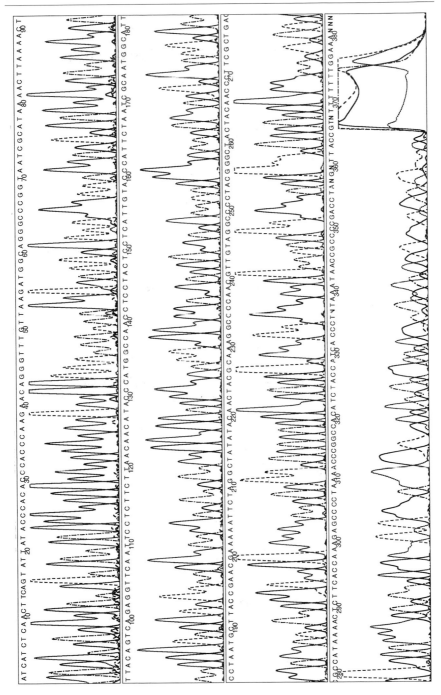

complementary to both the FL_2 primer and the sequencing primer. Because the amount of FL_2 primer in the second PCR (0.2 pmol) is comparable to the amount of the sequencing primer (0.2 pmol for A and C reactions and 0.4 pmol for G and T reactions), the depletion of the FL_2 primer is essential. When the FL_2 primer is not depleted during the second PCR and is carried over into the sequencing reaction, we observe only strong signals of free fluorescence-labeled primers and very low signals from extended strands in the sequencing gel pattern.

The ratio of template to sequencing primers is an important factor that determines the length of readable sequences. If the template contains only double-stranded DNA, the optimal concentration of the PCR product of 600 bp should be about 40–50 ng/μl. The second PCR product contains, however, both single- and double-stranded DNA, and it is difficult to determine accurately the concentration of single-stranded DNA. Therefore, the dilution ratio should be determined by comparing the sequence results from a series of template concentrations; the dilution used in the present protocol is 2-fold.

Dye primers are light-sensitive. The chemical robot should be installed in a lightproof room. After the sequencing reactions, the tubes should be kept away from sunlight by covering with aluminum foil during vacuum drying. Protection against room light, however, is not always necessary.

It is strongly recommended to install the chemical robot in an air-conditioned room, especially in laboratories in hot and humid climates, and set the room temperature at 18°. The temperature of the thermal cycler plate during the dispension of reagents and templates should be set at 6° to control the condensation of vapor onto the thermal plate.

If peaks in the first part of analyzed sequence data are low and peaks in the latter part are high, the dNTP/ddNTP ratio in the sequencing reaction is too high. In ethanol precipitation in the presence of sodium acetate, dNTPs coprecipitate with DNA. When too much dNTPs is present in the template for sequencing, chain termination with ddNTP is inhibited, resulting in low peaks in the first part of the sequencing data and high peaks in the latter part. Therefore, in the second PCR, we use dNTPs at one-tenth concentration of the first PCR (20 μM). Amplification in the second PCR is barely hindered by this low concentration of dNTPs. When the normal concentration of dNTPs (200 μM) is used, the second PCR product should be precipitated with one-fifth volume of 10 M ammonium acetate and 1 volume of 2-propanol in order to remove excess dNTPs.

When the distance between the excess primer H_2 and the limiting primer FL_2 is too small, a high peak, corresponding to the end of the template, appears in the last part of the sequence data and causes interference with the neighboring peaks (Fig. 3). Therefore, to obtain sequence data of 500 bp, the distance between primers FL_2 and H_2 should be about 600 bp.

In the sequencing results, large peaks are sometimes observed at 40–50 nucleotides from the primer peak. These are due to primer dimers formed between primer FL_2 and primer H_2 (or H_1). To reduce the precipitation of such primer dimers, it is recommended to precipitate the second PCR products with ammonium acetate and 2-propanol.

Resolution of the sequencing gel depends largely on the quality of polyacrylamide gels. The acrylamide/bisacrylamide mixture (19 : 1, w/w) from Bio-Rad (Hercules, CA) is recommended by Applied Biosystems. A premixed acrylamide solution for automated sequencing (Pageset SQC6A, from Biomate, Tokyo, Japan) is also useful. A 24-well square-tooth comb (originally designed for GeneScan, Applied Biosystems), instead of a conventional shark-tooth comb for the sequencer, is convenient both for gel preparation and for sample application.

[36] Detection and Quantification of Mitochondrial DNA Deletions

By NAY-WEI SOONG and NORMAN ARNHEIM

Introduction

In diseases such as chronic progressive external ophthalmopelgia, Kearn–Sayre's syndrome, and Pearson's syndrome, a significant proportion (20–80%) of the mitochondrial DNA (mtDNA) carry large deletions of up to 10 kb (reviewed in Ref. 1). The proportion of deleted mtDNA populations in diseased tissues can usually be measured using standard Southern blotting techniques. However, detection and measurement of low levels of deletion in tissues from nondiseased individuals present challenges not only because of the high sensitivity required but also because of the need to discriminate between the few molecules of deleted mtDNA and the overwhelming excess of wild-type mtDNA molecules. We describe in the following sections PCR (polymerase chain reaction)-based techniques used in our laboratory[2-4] for the detection and semiquantitative and quantitative

[1] D. C. Wallace, *Annu. Rev. Biochem.* **61,** 1175 (1992).
[2] G. A. Cortopassi and N. Arnheim, *Nucleic Acids Res.* **18,** 6927 (1990).
[3] N. W. Soong and N. Arnheim, in "PCR in Neuroscience" (G. Sarkar, ed.) p. 105 (1995). Academic Press, New York.
[4] N. W. Soong, D. R. Hinton, G. A. Cortopassi, and N. Arnheim, *Nat. Genet.* **2,** 318 (1992).

measurement of the levels of the mtDNA[4977] deletion, the so-called common deletion that removes 4977 bp of mtDNA sequence between two 13-bp repeats.

Polymerase Chain Reaction Assay for Detection of Mitochondrial DNA[4977] Deletion

The mtDNA[4977] deletion removes a section of mtDNA between nucleotide positions 8470 and 13,447 in the mitochondrial genomic sequence and occurs at a presumed deletion "hotspot" involving two 13-bp direct repeats beginning at these positions.[1] Two primers are designed to lie just outside the 13-bp repeats (Fig. 1). The PCR assay relies on the deletion to bring the two primers sufficiently close together to enable efficient amplification. The use of short PCR cycle times preferentially amplifies the deleted molecules over the nondeleted wild-type molecules presumably because, for the wild-type mtDNA, there is insufficient time for the extension of each primer through 5 kb of sequence. Thus, exponential amplification cannot take place. However, when the deletion removes this 5 kb of intervening sequence, the primers are brought close enough together such that each primer can be extended through the binding site for the other primer, a prerequisite for PCR.

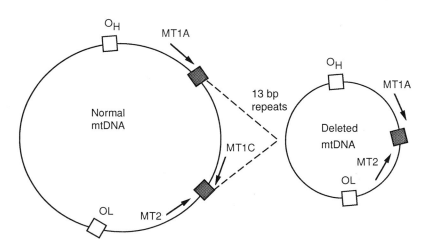

FIG. 1. A 4977-bp deletion between two 13-bp direct repeats brings primers MT1A and MT2 sufficiently close together to allow preferential amplification under short cycle times. Primers MT1C and MT2 amplify a section of undeleted, wild-type mtDNA and are used for normalization of total mtDNA. O_L and O_H denote the mitochondrial origins of replication.

PCR Analysis

The PCR is carried out in 50-μl volumes in 1× PCR buffer (50 mM KCl, 10 mM Tris-HCl, pH 8.3, 2.5 mM MgCl$_2$, 0.1 mg/ml gelatin). The primer and nucleotide concentrations are 0.5 and 187.5 μM per deoxynucleoside triphosphate (dNTP), respectively. For detection of mtDNA molecules with the 4977-bp deletion, primers MT1A and MT2 are used. Their sequences and the regions they correspond to on the mtDNA sequence map[5] are listed below.

MT1A: GAATTCCCCTAAAAATCTTTGAAAT, nucleotides
        8224–8247
MT2:   AACCTGTGAGGAAAGGTATTCCTGC, nucleotides
        13,501–13,477

An initial denaturation step of 3 min at 92° is used. Cycling is carried out with a 20-sec denaturation segment at 92° followed by a combined annealing and extension segment of 20 sec at 60°. A final extension step of 72° for 3 min is performed at the end of the appropriate number of cycles. Deletion PCR is typically carried out for 30 cycles with 2 units of *Taq* (*Thermus aquaticus*) polymerase using 1 μl of total genomic DNA (100 ng–1 μg) isolated from tissues. Some tissues may contain extremely low levels of deletion, and it may be necessary to run more cycles of amplification using two rounds of amplification (see below). In all cases, however, proper contamination controls are mandatory.

The reaction products can be resolved using 2% (w/v) agarose gels and visualized by ethidium bromide staining. With primers MT1A and MT2, a 303-bp product is amplified from mtDNA^{4977} templates. We estimate that 10^3–10^4 deleted mtDNA molecules will result in detectable levels of the 303-bp product with ethidium bromide staining at 30 cycles of PCR. Using the short cycling parameters described above, no detectable amplification of undeleted mtDNA molecules occurs even though these sequences are in vast excess.

Considerations

In practice, it should be possible to detect any large deletion using this strategy by proper design of primers located just outside the deletion break points and optimization of PCR parameters. However, the ability to amplify a small fragment should not be taken as proof of the presence of a specific

[5] S. Anderson, A. T. Bankier, B. G. Barrell, M. H. L. De Brujin, A. R. Coulson, J. Drouin, I. C. Eperon, D. P. Nierlich, B. A. Roe, F. Sanger, P. H. Schreier, A. J. H. Smith, R. Staden, and I. G. Young, *Nature* (*London*) **290**, 457 (1981).

deletion. The possibility of PCR artifacts should be explored carefully and eliminated.[2-4]

Occasionally, we encounter samples that give no detectable deletion product even with increased cycle numbers. In these cases, the levels of deletions may be extremely low, or inhibitors that interfere with PCR may be present. To eliminate the latter possibility, a tested positive control DNA sample is mixed with the questionable sample and PCR performed. If there is a significant decrease in the signal of the deletion product from that of the unmixed positive sample, then the presence of inhibitors should be suspected. Further purification of the questionable DNA sample should then be performed.

Detection of extremely low levels of deletion may still be achieved by concentrating the DNA sample, using radioactively labeled primers or nested PCR. For the mtDNA[4977] deletion, we perform a primary PCR step of 25 cycles using primers MT1A and MT3 (sequence: GCGATGAGAG-TAATAGATAGGGCTCAGGCG, nucleotides 13,580–13,551), 50 bp up-stream of MT2. One microliter of this primary reaction is then used as template for the secondary reaction using primers MT1A and MT2 for another 25 cycles. The PCR conditions for both primary and secondary amplifications are as described above. It is vital that stringent contamination controls be implemented in this procedure, as it is extremely sensitive and in principle can detect single molecules.

Semiquantitative Comparisons of Mitochondrial DNA[4977] Levels

Any differences in the intensity of the deletion product from PCR of different samples may be due to varying amounts of total mtDNA added to the reaction. Therefore, for a valid comparison of mtDNA[4977] levels to be made between different samples, the samples first have to be normalized for total mtDNA content.[2-4] This is done by using another set of primers (MT1C and MT2, Fig. 1) to amplify a region of wild-type mtDNA of similar size (324 bp) to the deleted product (303 bp). Primer MT1C has the sequence AGGCGCTATCACCACTCTTGTTCG and corresponds to nucleotides 13,176–13,198 on the Anderson mtDNA sequence.[5] The PCR conditions are identical to those for the deletion-specific reaction except that, typically, deletion-specific PCR are run for 30 cycles while control PCR for wild-type mtDNA are run for 15 cycles; the difference in cycle numbers reflect the rare occurrence of mtDNA[4977] molecules. Also, only 1 U of *Taq* polymerase is required for each control reaction. The signals of the different samples from this control reaction are visually compared on ethidium bromide-stained 2% agarose gels. Samples that give more intense signals are diluted and the control reaction performed again. Normalization is achieved through this iterative process of adjusting the DNA concentration of different samples such that they produce roughly equal intensities of wild-type PCR signals. The PCR is then performed on the normalized samples using the deletion-specific primers (MT1A and MT2). A semiquantitative comparison of the relative mtDNA^{4977} levels between different samples can then be made by a visual comparison of the deletion product signals. It is important not to run the reactions into saturation by starting with too much template or running too many cycles. A quick way to check is to run two successive dilutions of the same sample. The product signals obtained should correspond to the dilutions. If different dilutions of the same sample give similar intensities, then the reaction is saturated and higher dilutions of the sample ought to be made. In general, 2-fold dilutions should give detectable differences in signal intensities of the PCR products.

Estimating Proportion of Deleted Molecules by Limiting Dilutions

An estimate of the proportion of mtDNA^{4977} relative to wild-type mtDNA can be obtained by using the method of limiting dilutions. Control and deletion PCR are performed separately on serial dilutions of a DNA sample using the same number of cycles (we use 30 cycles). Comparison of the different dilutions at which the control and deletion signals become undetectable gives an idea of the proportion of mtDNA^{4977} present. For example, if the deletion signal disappears at 10^{-2} dilution, and the control signal disappears at 10^{-6} dilution, then the proportion of deleted to wild-type mtDNA molecules would be roughly 1 in 10,000. Using this procedure on adult heart DNA,[2] it was found that the limiting dilution for the detection of wild-type mtDNA was 1000 times less concentrated than the limiting dilution for the detection of mtDNA^{4977}. Thus, there appears to be 1 mtDNA^{4977} molecule for every 1000 wild-type mtDNA molecules in this tissue. It is assumed that the efficiencies of the deletion and control PCRs were similar as the sizes of the amplified regions were similar (324 bp for control product versus 303 bp for the deletion product).

Quantitative Analyses of Deleted Mitochondrial DNA Levels

Principle

The proportion of deleted mtDNA to wild-type mtDNA can be more precisely determined by comparing, quantitatively, the signals of deleted and control PCR products amplified from a sample to external standard curves.[4] Thus, final amount of template present in the sample can be correlated to the initial amount of starting template present. Two standard curves are used, one for deletion PCR and the other for control PCR. For each sample, deletion and control PCR are performed separately and the products quantitated relative to the respective standard curve. The use of radioactively end-labeled primers allows for sensitive quantitation of PCR products. The standard curves are prepared by amplification of a known dilution series of the corresponding template under the same conditions as the sample. It is imperative that quantitation be performed in the exponential phase of both amplifications, as this is the phase when both deletion and control products will be accumulating with similar kinetics. The accumulation of products in the exponential phase can be modeled by Eq. (1):

$$N = n(1 + E)^k \tag{1}$$

where N is the number of molecules at the end of the reaction, n is the number of template molecules, k is the number of cycles, and E is the average efficiency. The value of E has a theoretical maximum of 1 where the amount of product doubles with each cycle. This value declines during late cycles of amplification as the reaction proceeds past the exponential phase. There is a linear relationship between the logarithm of the amount of starting template n, and the logarithm of the final amount of the specific, amplified product N [Eq. (2)].

$$\log N = \log n + k \log(1 + E) \tag{2}$$

A plot of $\log N$ against $\log n$ for a fixed number of cycles should therefore generate a line with a slope of 1. For this to be true, E has to remain fairly constant over k numbers of cycles, although E does not necessarily have to have a value of 1. If the value of E is decreasing substantially with each cycle as the reaction progresses past the exponential phase, a plot will be generated with a slope significantly less than 1. Linearity is not a sufficient condition for exponential amplification. A more stringent indicator is the slope of the plot, which should be close to 1.

The standard curves are used to define the exponential range over which quantitation will be valid by ensuring that the above criterion is met. The various steps in the procedure are now described in detail for the quantitation of the mtDNA[4977] deletion. The procedure should be adaptable for the quantitation of other mtDNA deletions.

End Labeling of Primers, PCR, and Quantitation of Products

Primers are end-labeled with [γ-^{32}P]ATP using T4 polynucleotide kinase. Unincorporated nucleotides are removed by spinning through P4 columns. End-labeled primer lots are prepared to give an approximately 10× concentration for PCR (5 μM) and are diluted directly into the PCR mix. The PCR are performed with both primers end-labeled using conditions described earlier. Primers MT1C and MT2 are used for the control reaction, and primers MT1A and MT2 are used for the deletion reaction. The PCR can be performed in 96-well microtiter plates to accommodate many samples at the same time.

After PCR, 10% (5 μl) of each reaction is electrophoresed through 8% polyacrylamide gels. The gel is dried, and the counts from each specific product band are quantified with a PhosphorImager (Molecular Dynamics, Sunnyvale, CA) after 15–24 hr of exposure. In our hands, using end-labeled primers, better sensitivity and lower noise were achieved than with direct incorporation of [α-^{32}P]dATP during PCR.

Preparation of External Standards

For the construction of standard curves for deletion and control PCR, the respective PCR products are purified as a source of templates for the amplification reactions. We use genomic DNA from aged heart tissues as the initial template for these preparative PCRs. Each of the deletion and control reactions are performed and the products electrophoresed on 8% w/v polyacrylamide gels separately. The product bands are excised, electro-eluted, and concentrated by centrifuging through Centricon-10 (Amicon, Denvers, MA). Stock solutions of both purified deleted and control products are prepared in this way and stored at $-20°$. As a precaution against degradation, contamination, and loss of standards due to adherence to vessel walls during prolonged storage, the samples are diluted in solutions of *Escherichia coli* tRNA as carrier. These dilutions are then distributed into smaller volumes and stored at $-20°$. Individual aliquots are used when necessary, minimizing handling of the rest of the standard stock.

Definition of Exponential Range

The use of external standards requires that quantification be performed during the exponential phase of the reaction. The use of radiolabeled primers allows sensitive quantification of products before the exponential range has been exceeded. From our experience, when the products become clearly visible by ethidium bromide staining, the PCR is at or on the verge of surpassing the exponential range.

The range of dilutions for both deletion and control external standards over which amplification was exponential has to be determined. Operationally, this is determined as the range over which a plot of log counts versus log dilutions give a good linear fit with a slope close to 1. The incorporated counts provide a measure of the amount of accumulated product, whereas the dilution of the standard stock is proportional to the amount of starting template. Thus, this plot preserves the relationship in Eq. (2). The ranges are optimized for 15 cycles of control amplification and 30 cycles of deletion amplification. These cycle numbers are found from earlier experiments to give good, quantifiable signals for most genomic DNA samples for mtDNA[4977].

First, 10-fold serial dilutions of the external standard stocks are used as templates over a 10 log range. This allows the different phases of the amplification to be clearly distinguishable (Fig. 2 shows the plot for the deletion standards). The next step consists of amplifying 2-fold serial dilutions encompassing the suggested exponential phase. The range of dilutions that gives a slope closest to 1 (Fig. 3 shows the exponential range of deletion standards) can thus be determined with better resolution. We choose to work with known dilutions of our standards rather than quantifying the absolute amount of DNA as this may be subject to systematic errors if the initial quantification of the stock is inaccurate.

For our system the exponential dilution "window" for deletion amplification using 30 PCR cycles was determined as ranging from 10^{-6} down to $1/256 \times 10^{-6}$, all dilutions being expressed relative to the initial standard deletion stock. The window for control amplification using 15 PCR cycles ranged from 10^{-3} down to $1/128 \times 10^{-3}$ relative to the standard control stock. The 2-fold difference in the ranges (256-fold for deletion PCR versus 128-fold for control PCR) reflects small disparities in background noise and amplification efficiencies. By spectrophotometric measurements, we estimate that this corresponds to a maximum of 6.2×10^7 input molecules

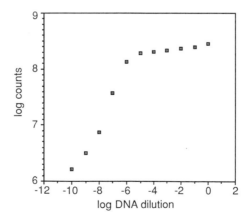

Fig. 2. Amplification of 10-fold serial dilutions of the deletion standard stock over a 10 log range. The reaction is exponential between dilutions of 10^{-6} and 10^{-10} and enters a plateau phase above a dilution of 10^{-5}. Twofold dilutions are made within this suggested exponential range and amplified as in Fig. 3.

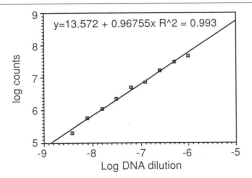

FIG. 3. Amplification of 2-fold serial dilutions of the deletion standard stock starting from 10^{-6}. A linear regression analysis is performed and shows that the amplification is exponential in this range with a slope close to 1 (0.97).

for control amplification and a maximum of 10^4 input molecules for deletion amplification. These ranges are reproducible. Thus, theoretically, we can quantitate differences in $mtDNA^{4977}$/wild-type mtDNA ratios over a 3.3×10^4 (128×256) range.

Quantitation of Samples

Preliminary deletion and control PCR with unlabeled primers are performed on dilutions of DNA samples whose levels of $mtDNA^{4977}$ are to be quantitated. The product signals on ethidium bromide-stained gels are then visually compared with those generated by amplification of the most concentrated standard dilution in the exponential window (10^{-3} for control, 10^{-6} for deletion). The samples are then diluted if necessary such that the amplification signals do not exceed that of the most concentrated standard dilution. Two different dilutions of each sample DNA that conform to this criterion are then used for quantitation to ensure that amplification remains in the exponential phase. Performing these preliminary steps eliminates handling of radioactivity until the actual quantification is to be done.

Aliquots of these sample DNA dilutions are then amplified with ^{32}P end-labeled primers in both control and deletion PCR. Both series of standards are amplified in parallel with the samples. The products are run in an 8% polyacrylamide gel, and the specific deletion or control product bands are quantified using a phosphor imager. The signal generated by each sample is then extrapolated from the appropriate standard curve to obtain the equivalent dilution of the standard stock that would have given the same signal. The ratio of the equivalent dilution for the deletion standards to that for the control standards can then be used to derive the ratio of mtDNA[4977]/wild-type mtDNA. The calculation is illustrated in the example below.

All standard curves are inspected for conformity to the conditions that the slope should be close to 1 and for good linear regression fit. This check should warn of any significant interreaction variation in amplification efficiencies. The two different dilutions of each sample should also give similar ratios. Samples for which they are significantly different are reanalyzed. Each experiment typically consists of 30 assay reactions for tissue samples (15 samples, 2 dilutions each) and 10 serial dilutions for the external standards. This is done for the control and deletion amplifications.

Relative Calibration of Standard Curves

To convert the ratio of equivalent dilutions to the ratio of mtDNA[4977]/wild-type mtDNA, the relative concentration of the control standard to the deletion standard must be known exactly. This is determined by the following experiment.

Eight 2-fold dilutions of both control and deletion standards starting from a 10^{-3} dilution of each stock are amplified separately using their respective primers for 15 cycles. To avoid differences in specific activities, only the common primer, MT2, is end-labeled. The same lot of labeled MT2 is used for both PCRs. Log plots of both PCR series are made to ensure that the amplifications remain in exponential phase using the criteria described above. The ratio of counts of the control PCR product to that of the deletion PCR product are calculated for each dilution in the range over which both PCR are exponential. For our system, the average of the eight ratios worked out to 5.4. This means that for an equivalent dilution

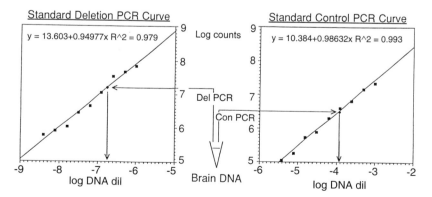

Fig. 4. Quantification of the deletion level in a brain DNA sample using the constructed deletion and control external standard curves. See text for further explanation.

of control and deletion standards, the control standard would be 5.4 times more concentrated. This value is factored into the ratio of equivalent dilutions for each sample to obtain the ratio of mtDNA^{4977}/wild-type mtDNA. This calibration exercise need only be performed once and thereafter periodically checked to ensure that degradation of the standards has not occurred. We performed the identical experiment nearly 1 year after our first calibration and obtained a similar value of 5.2.

Example: Calculation of Ratio of mtDNA^{4977} to Total Mitochondrial DNA

Two standard calibration curves that relate the dilution of the standard template to the amount of incorporated radioactivity are shown in Fig. 4. The standard curve specific for mtDNA^{4977} uses known dilutions of a 303-bp template (deletion), and similarly the calibration curve for undeleted mtDNA uses known dilutions of a 324-bp template (control). For each standard curve, the linear regression and correlation coefficient (r^2) are included. In the example shown in Fig. 4, the deletion (Del) and control (Con) PCR signals for a brain DNA sample are extrapolated from the respective standard curves to obtain the dilution of the standard templates which would have given an equivalent signal. From the deletion curve, a value of -6.75 for the Del PCR is obtained from the x axis. The antilogarithm of this value is taken to give an equivalent dilution of 1.76×10^{-7}. Similarly, for the control curve, a value of -3.93 for the Con PCR is obtained, which gives an equivalent dilution of 1.18×10^{-4}. Dividing the Del dilution by the Con dilution gives 0.149%. At identical dilutions, the control standard is 5.4 times more concentrated in amplifiable templates than the deletion standard (see section on relative calibration of standard curves). This value is therefore factored in to give a final ratio of 0.027% (0.149/5.4).

Conclusion

We describe PCR techniques to measure the proportion of mtDNA^{4977} to wild-type mtDNA. The quantitative method has been used to measure the levels of the deletion in different regions of the brains of aged individuals.[4] The method could be used for the quantitation of other rare mtDNA deletions in brain or other tissues of humans or other species. The parameters for PCR need to be individually optimized and exponential ranges of amplification determined for each PCR target. Once these conditions are determined, the technique is quite reproducible.

[37] Detection and Quantification of Point Mutations in Mitochondrial DNA by PCR

By Makoto Yoneda, Yoshinori Tanno, Shoji Tsuji, and Giuseppe Attardi

Introduction

Numerous deleterious point mutations of mitochondrial DNA (mtDNA), now amounting to more than 30, have been shown to be associated with various types of human disorders involving deficiencies in the mitochondrial oxidative phosphorylation apparatus.[1,2] Most of the mutations occur in heteroplasmic form, that is, the mutant and wild-type mtDNAs coexist in varying proportions in different tissues of an individual, causing the clinical features of the diseases. Therefore, detection and quantification of the mutant mtDNA are essential for the diagnosis of the diseases and for understanding the molecular basis of their pathogenesis.

In the present chapter, we describe three useful methods for detection and quantification of mutant mtDNA by the polymerase chain reaction (PCR). The methods described here include (a) PCR–RFLP (restriction fragment length polymorphism), (b) PCR–RSM (PCR-mediated restriction site modification), and (c) PCR–SSCP (single-strand conformation polymorphism).

Materials

The oligodeoxynucleotides used in the following experiments are synthesized with a DNA synthesizer [Shimadzu Model NS-1 (Kyoto, Japan) or Applied Biosystems Model 308B (Foster City, CA)] according to the manufacturer's instructions, and then purified through an OPC cartridge (Applied Biosystems). [γ-^{32}P]ATP (5000 Ci/mmol) and [α-^{32}P]dATP (3000 Ci/mmol) are purchased from Amersham Japan (Tokyo, Japan). *Taq* DNA polymerase, proteinase K, and T4 polynucleotide kinase are purchased from Takara (Ohtsu, Shiga, Japan), restriction enzymes from Toyobo (Tokyo, Japan), acrylamide from Bio-Rad (Richmond, CA), and deoxynucleoside triphosphates from Sigma (St. Louis, MO).

---

[1] D. C. Wallace, *Annu. Rev. Biochem.* **61,** 1175 (1992).
[2] E. Schon, M. Hirano, and S. DiMauro, *J. Bioenerg. Biomembr.* **26,** 291 (1994).

Sample Preparation for PCR

Autoptic or Bioptic Tissues

Total genomic DNA, including mtDNA, to be used as a template for PCR amplification, is extracted from small pieces (20–100 mg) of autoptic or bioptic tissues by incubation in a 1.5-ml Eppendorf tube containing 200–500 μl of 10 mM Tris-HCl, pH 7.4, 10 mM EDTA, 100 μg/ml proteinase K, and 0.5% sodium dodecyl sulfate (SDS), at 37° for 4 hr, followed by extractions with phenol–chloroform (1 : 1, v/v) and chloroform, and then precipitation with ethanol, as described previously.[3]

Monolayer Cultured Cells

Cell lysates prepared from monolayer cultured cells, without DNA extraction, can be used directly for PCR amplification. The cell lysate is prepared by a modification of the method described by Kawasaki.[4] In brief, a small number of cultured cells (10^4–10^5) are collected by trypsinization and lysed by incubation at 50° for 2 hr in 100 μl PCR buffer (10 mM Tris-HCl, pH 8.3, 50 mM KCl, 1 mM MgCl$_2$) containing 100 μg/ml proteinase K and 0.5% v/v Tween 20. After inactivation of the proteinase K by boiling for 5 min, approximately one-tenth volume (~10 μl) of the cell lysate is used in a 100-μl PCR reaction.

PCR–Restriction Fragment Length Polymorphism

To detect mutations that create or abolish a restriction enzyme site, a region spanning the mutation is amplified by PCR, using an appropriate pair of oligodeoxynucleotide primers, and then digested with the corresponding restriction enzyme. We describe here an example of detection and quantification of an A to G transition at nucleotide position 3243[5,6] (numbered according to the Cambridge standard human mtDNA sequence, referred

---

[3] T. Ozawa, M. Yoneda, M. Tanaka, K. Ohno, W. Sato, H. Suzuki, M. Nishikimi, M. Yamamoto, I. Nonaka, and S. Horai, *Biochem. Biophys. Res. Commun.* **154**, 1240 (1988).

[4] E. S. Kawasaki, *in* "PCR Protocols: A Guide to Methods and Applications" (M. A. Innis, D. H. Gelfand, J. J. Sninski, and T. J. White, eds.), p. 146. Academic Press, San Diego and New York, 1990.

[5] A. Chomyn, A. Martinuzzi, M. Yoneda, A. Daga, O. Hurko, D. Johns, S. T. Lai, I. Nonaka, C. Angelini, and G. Attardi, *Proc. Natl. Acad. Sci. U.S.A.* **89**, 4221 (1992).

[6] M. Yoneda, A. Chomyn, A. Martinuzzi, O. Hurko, and G. Attardi, *Proc. Natl. Acad. Sci. U.S.A.* **89**, 11164 (1992).

to below as Cambridge sequence),[7] in the tRNA$^{Leu(UUR)}$ gene of mtDNA, which is associated with the MELAS (mitochondrial myopathy, encephalopathy, lactic acidosis, and stroke-like episodes) syndrome.

The PCR is carried out with a DNA Thermal Cycler PCR 480 (Perkin-Elmer/Cetus, Norwalk, CT), or an equivalent instrument. The reaction mixture, oligodeoxynucleotide primers, and cycle conditions used in the PCR are described below.

Mix in a 500-μl microcentrifuge tube the following:

Target DNA: 100 ng genomic DNA or one-tenth of the reaction volume of cell lysate (10 μl)
Oligodeoxynucleotides (1 μM of each):
forward primer: nucleotides 3031–3050, 5'-GGTTCGTTTGTTCA-ACGATT-3'
reverse primer: nucleotides 3360–3341, 5'-TGCCATTGCGATTA-GAATGG-3'
Buffer: 10 μl of a solution containing 100 mM Tris-HCl, pH 8.3, 500 mM KCl, 15 mM MgCl$_2$, 0.1% gelatin
dNTPs: 200 μM of each deoxynucleoside triphosphate (dATP, dGTP, dTTP, and dCTP), adjusted to pH 7.0 with NaOH
Enzyme: 2.5 units of Taq DNA polymerase/100 μl reaction mixture
Adjust the volume to 100 μl with distilled water and overlay one drop of mineral oil

After the initial denaturation of the sample at 94° for 45 sec, the following cycle conditions are used:
Number of cycles: 30
Denaturation: 94°, 45 sec
Annealing: 47°, 90 sec
Primer extension: 72°, 45 sec
After the final incubation at 72°, the reaction mixture is incubated for an additional 5–10 min at 72°, and then kept at 4° until the sample is used.

After PCR amplification, the products are extracted with an equal volume of chloroform to remove the mineral oil, then precipitated with ethanol, and finally digested with the restriction enzyme ApaI (recognition site: 5'-GGGCCC-3') in a 50-μl volume. To check for completion of digestion of the PCR products, approximately 1 μg of an appropriate plasmid DNA fragment or marker PCR product containing at least one ApaI site is included in the reaction mixture [we use XmnI-linearized pBluescript KS(+) plasmid DNA (Stratagene, La Jolla, CA)]. After digestion, about

---

[7] S. Anderson, A. T. Bankier, B. G. Barrell, M. H. L. de Bruijn, A. R. Coulson, J. Drouin, I. C. Eperon, D. P. Nierlich, D. Roe, F. Sanger, P. H. Schreier, A. J. H. Smith, R. Staden, and I. G. Young, *Nature (London)* **290,** 457 (1981).

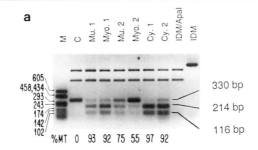

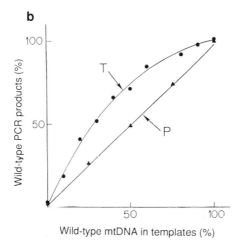

FIG. 1. (a) Detection of the MELAS 3243 mutation by PCR–RFLP in mtDNA from muscle biopsy specimens (Mu.), myoblast cultures (Myo.), and cybrids (obtained by cytoplast fusion with a human mtDNA-less cell line) derived from two different individuals carrying the MELAS mutation (Cy.). Lane C shows mtDNA from a wild-type cell line; M, *Hae*III-digested pBluescript KS(+) DNA; IDM, an internal digestion marker [*Xmn*I-linearized pBluescript KS(+) DNA] to check the completion of enzyme digestion. (b) Standard curves for quantification of the MELAS mutation in mtDNA. Curve T shows the mixed-template standard curve; P, mixed-product standard curve. See text for details. Adapted, with permission, from (a) A. Chomyn, A. Martinuzzi, M. Yoneda, A. Daga, O. Hurko, D. Johns, S. T. Lai, I. Nonaka, C. Angelini, and G. Attardi, *Proc. Natl. Acad. Sci. U.S.A.* **89,** 4221 (1992) and (b) M. Yoneda, A. Chomyn, A. Martinuzzi, O. Hurko, and G. Attardi, *Proc. Natl. Acad. Sci. U.S.A.* **89,** 11164 (1992).

10 μl of the reaction mixture is electrophoresed through a 1.5% agarose gel in TAE buffer (40 mM Tris–acetate, pH 8.0, 1 mM EDTA), and the gel is stained with ethidium bromide (5 μg/ml). If the internal marker is completely digested, one can proceed with the quantification of the experimental PCR products. Alternatively, more enzyme is added to the reaction mixture, the sample is further digested, and the test is repeated.

Figure 1a shows an ethidium bromide-stained agarose gel containing the completely digested PCR fragments from mutant and wild-type mtDNAs. The amplified product containing the MELAS 3243 mutation is cleaved into 214- and 116-bp fragments, whereas the product from the wild-type mtDNA remains uncleaved (330 bp). The proportion of the digested and undigested fragments is determined by laser densitometry of a negative photograph of the gel. The percentage of mutant mtDNA is calculated by the formula: $D_M/(D_W + D_M) \times 100$, where D_W and D_M are the densities of the bands corresponding to the wild-type and mutant mtDNAs, respectively. In the quantification of the mutation, one must consider that heteroduplexes of mutant and wild-type sequences, formed during the last annealing step of the PCR, may be resistant to the restriction enzyme digestion, and thus lead to the underestimation of the percentage of mutant mtDNA.[6,8,9]

To correct for resistance to digestion of mutant–wild-type heteroduplexes, a mixed-template standard curve should be constructed by PCR amplification of a mtDNA fragment containing about 100% mutant sequence and a mtDNA fragment containing approximately 100% wild-type sequence, and then subjecting different-ratio mixtures of the mutant and wild-type PCR products to a second cycle of PCR amplification and enzyme digestion of the final product. Figure 1b compares such a mixed-template standard curve (T), constructed for the MELAS 3243 mutation, with the mixed-product standard curve (P), obtained by mixing, in different ratios, the PCR products independently amplified from the mutant and wild-type mtDNA samples and then subjecting the mixtures to *Apa*I digestion. The discrepancy between the two curves presumably results from the formation of *Apa*I-resistant mutant–wild-type heteroduplexes during the last annealing step of the PCR. It should, however, be emphasized that the mixed-template curve, in some experiments, may approach or deviate from the mixed-product curve more than in the case illustrated in Fig. 1b. This may be influenced by several factors, like the ratio of primers to template, the PCR conditions, and the secondary structure of the single-stranded PCR products. Therefore, it is essential to construct the mixed-template standard curve under exactly the same conditions for PCR as used with the experimental sample.

An alternative approach for determining the proportion of mutant and wild-type mtDNAs by PCR–RFLP, without the influence of heteroduplex formation, is to label the PCR products after the last annealing step.[10,11]

[8] J. M. Shoffner, M. T. Lott, A. M. S. Lezza, P. Seibel, S. W. Ballinger, and D. C. Wallace, *Cell (Cambridge, Mass.)* **61,** 931 (1990).
[9] M. Yoneda, T. Miyatake, and G. Attardi, *Mol. Cell. Biol.* **14,** 2699 (1994).
[10] M. P. King, Y. Koga, M. Davidson, and E. A. Schon, *Mol. Cell. Biol.* **12,** 480 (1992).
[11] Y. Tanno, M. Yoneda, I. Nonaka, K. Tanaka, T. Miyatake, and S. Tsuji, *Biochem. Biophys. Res. Commun.* **179,** 880 (1991).

The rationale for this approach is that the only products that will be labeled in the last cycle are homoduplexes, and these products will reflect the proportion of mutant and wild-type mtDNA sequences in the overall final PCR products. For optimum results by this approach, 10 μl of the PCR reaction mixture, after the last annealing step, is mixed with fresh deoxynucleoside triphosphates (50 μM), *Taq* DNA polymerase (2.5 units), and oligodeoxynucleotide primers (20 to 50 pmol), and brought to 100 μl with PCR buffer. Labeling is carried out by adding to the mixture [α-^{32}P]dATP (10 μCi)[10] or, alternatively, one of the primers 5'-end-labeled [5 pmol, ~10^7 counts/min (cpm)],[11] and then carrying out one last cycle under the conditions described in the above-cited references, namely, 2 min at 94°, 1 min at 55°, and 12 min at 72°,[10] or 1 min at 94°, 30 sec at 50°, and 3 min at 72°,[11] respectively. The 5'-end-labeling of the primer is performed by incubating 100–200 ng primer with [γ-^{32}P]ATP (40 pmol, 200 μCi) and T4 polynucleotide kinase (20 units) in 30 μl of 50 mM Tris-HCl (pH 7.6), 10 mM MgCl$_2$, 10 mM 2-mercaptoethanol at 37° for 30 min, extracting the mixture with phenol–chloroform, and precipitating the oligodeoxynucleotide with ethanol.[12]

PCR-Mediated Restriction Site Modification

To detect mutations that do not create or abolish any restriction enzyme site, one can artificially introduce a new restriction site, which can be recognized only in either the mutant or the wild-type sequence, in the PCR step, using a mismatched primer. Fig. 2 illustrates examples of detection of mutations in the tRNA^{Lys} gene of human mtDNA, which cause the MERRF (myoclonus epilepsy associated with ragged-red fibers) syndrome, by PCR with a mismatched primer.[13,14] Most MERRF patients carry either an A to G transition at nucleotide position 8344[8,15] or a T to C transition at position 8356[14,16] in the mitochondrial tRNA^{Lys} gene. Neither of the two mutations can be recognized by any commercially available restriction enzyme.

For detecting the A to G mutation at position 8344, a segment of mtDNA is amplified by PCR, using a mismatched primer (nucleotides

[12] J. Sambrook, E. F. Fritsch, and T. Maniatis, *in* "Molecular Cloning: A Laboratory Manual." 2nd Ed. Cold Spring Harbor Laboratory, Cold Spring Harbor, New York, 1989.

[13] M. Yoneda, Y. Tanno, I. Nonaka, T. Miyatake, and S. Tsuji, *Neurology* **41**, 1838 (1991).

[14] G. Silvestri, C. T. Moraes, S. Shanske, S. J. Oh, and S. DiMauro, *Am. J. Hum. Genet.* **51**, 1213 (1992).

[15] M. Yoneda, Y. Tanno, S. Horai, T. Ozawa, T. Miyatake, and S. Tsuji, *Biochem. Int.* **21**, 789 (1990).

[16] M. Zeviani, F. Muntoni, N. Savarese, G. Serra, V. Tiranti, F. Carrara, C. Mariotti, and S. DiDonato, *Eur. J. Hum. Genet.* **1**, 80 (1992).

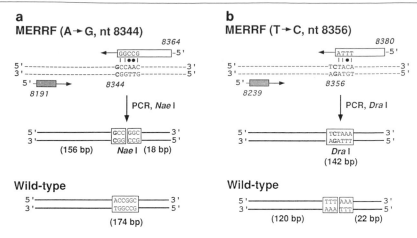

FIG. 2. Detection of two MERRF-associated tRNA^{Lys} gene mutations (an A to G transition at position 8344 and a T to C transition at position 8356) by PCR–RSM. The mismatched nucleotide pairs in the primers are indicated by filled circles. The mutations are located in residues shown by bold letters.

8364–8345, 5′-TCACTGTAAAGAGGTG<u>C</u>CGG-3′; underlined nucleotides differ from the Cambridge sequence) as a reverse primer, combined with a forward primer (nucleotides 8191–8210) (Fig. 2a).[13] The PCR is carried out with a DNA Thermal Cycler. The reaction mixture, oligodeoxynucleotide primers, and cycle conditions used in the PCR are described below.

Mix in a 500-μl microcentrifuge tube the following:

Target DNA: 100 ng genomic DNA or one-tenth of the reaction volume of cell lysate (10 μl)
Oligodeoxynucleotides (1 μM of each):
    forward primer: nucleotides 8191–8210, 5′-AAACCACAGTTT-CATGCCCA-3′
    reverse primer: nucleotides 8364–8345, 5′-TCACTGTAAA-GAGGTG<u>C</u>CGG-3′
Buffer: 10 μl of a solution containing 100 mM Tris-HCl, pH 8.3, 500 mM KCl, 15 mM MgCl$_2$, 0.1% gelatin
dNTPs: 200 μM of each deoxynucleoside triphosphate (dATP, dGTP, dTTP, and dCTP), adjusted to pH 7.0 with NaOH
Enzyme: 2.5 units of Taq DNA polymerase/100 μl reaction mixture
Adjust the volume to 100 μl with distilled water and overlay one drop of mineral oil

After the initial denaturation of the sample at 94° for 30 sec, the following cycle conditions are used:

Number of cycles: 30
Denaturation: 94°, 30 sec
Annealing: 45°, 1 min
Primer extension: 70°, 30 sec

After the final incubation at 70°, the reaction mixture is incubated for an additional 5–10 min at 70°, and then kept at 4° until the sample is used.

After PCR amplification using the mismatched primer, a new *Nae*I (or *Ngo*MI) restriction site (5'-**G**CCGGC-3'; the bold G residue in the MERRF mutation-carrying sequence corresponds to an A in the wild-type sequence) is generated in the tRNA^{Lys} mutation site, whereas the wild-type sequence is not cleaved by the enzyme. The digested PCR fragments are electrophoresed through a 2.4% NuSieve agarose/0.9% Seakem agarose (FMC Bio-Products, Rockland, ME) gel in TAE or a 5% nondenaturing polyacrylamide gel in 0.5× TBE buffer (45 mM Tris–base, 45 mM boric acid, 1 mM EDTA). The larger product derived from *Nae*I digestion of the mutant sequence is 18 base pairs smaller than that derived from the wild-type sequence. Tests for completion of digestion and quantification of the mutation are carried out according to the methods described for PCR–RFLP.

Silvestri *et al.*[14] reported a method for detecting the MERRF mutation at nucleotide position 8356 in the tRNA^{Lys} gene of mtDNA by PCR using a mismatched primer [8380–8357, 5'-ATTTAGTTGGGGCATTTCACT-TTA-3' (Fig. 2b); underlined nucleotide differs from that in the Cambridge sequence and creates a new *Dra*I site (5'-TTTAAA-3') in wild-type mtDNA: the bold C residue in the MERRF sequence corresponds to a T in the wild-type sequence] as a reverse primer, combined with a forward primer (nucleotides 8239–8263; Fig. 2b).[14] The digested PCR fragments are electrophoresed through a 12% nondenaturing polyacrylamide gel in 0.5× TBE buffer.

The optimal cycle conditions for PCR using a mismatched primer vary widely, depending on the primer and template sequences. An extension temperature one or two degrees lower than usual (70°–71°) is recommended, because the annealed products between the template DNA and the mismatched primer tend to be easily denatured during the step of primer extension.

PCR–Single-Strand Conformation Polymorphism

Although the conventional PCR–RFLP and PCR–RSM methods described above are useful techniques for identifying point mutations occurring in mtDNA, they are not applicable for the detection of all possible

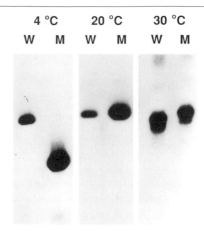

FIG. 3. Mobilities of the PCR products by PCR–SSCP at different running temperatures of the gel. The denatured PCR products were electrophoresed through a 0.5× MDE gel containing 10% glycerol. Lanes marked W contain products derived from wild-type mtDNA; M, products from mutant mtDNA. Reproduced with permission from Y. Tanno, M. Yoneda, K. Tanaka, M. Yamazaki, K. Hinokuma, K. Wakabayashi, F. Ikuta, and S. Tsuji, Quantitation of heteroplasmy of mitochondrial tRNA$^{Leu(UUR)}$ gene using PCR-SSCP, *Muscle Nerve* © 1995, John Wiley & Sons, Inc.

base substitutions. The technique of PCR–SSCP, which is based on electrophoretic mobility shifts of single-stranded DNA fragments caused by mutations in a nondenaturing gel,[17] can detect a large fraction (~50%) of point mutations. We have applied PCR–SSCP to detect the tRNA$^{Leu(UUR)}$ mutation at nucleotide position 3243 that is associated with the MELAS syndrome.[18] The reaction mixture, oligodeoxynucleotide primers, and cycle conditions used in the PCR are described below.

Mix in a 500-μl microcentrifuge tube the following:

Target DNA: 100 ng genomic DNA or one-tenth of the reaction volume of cell lysate (10 μl)

Oligodeoxynucleotides (0.05 μM of each):
forward primer: nucleotides 3210–3229, 5′-CCCACCCAAGAA-CAGGGTTT-3′
reverse primer: nucleotides 3324–3305, 5′-GAGGTTGGCCATGG-GTATGT-3′
(the 5′ end of the oligodeoxynucleotides is labeled with [γ-^{32}P]ATP and T4 polynucleotide kinase, as described above)

Buffer: 10 μl of a solution containing 100 mM Tris-HCl, pH 8.3, 500 mM KCl, 20 mM MgCl$_2$, 0.1% gelatin

[17] M. Orita, Y. Suzuki, and T. Sekiya, *Genomics* **5,** 874 (1991).
[18] Y. Tanno, M. Yoneda, K. Tanaka, M. Yamazaki, K. Hinokuma, K. Wakabayashi, F. Ikuta, and S. Tsuji, *Muscle Nerve,* in press.

dNTPs: 62.5 μM of each deoxynucleoside triphosphate (dATP, dGTP, dTTP, and dCTP), adjusted to pH 7.0 with NaOH

Enzyme: 2.5 units of *Taq* DNA polymerase/100 μl reaction mixture

Adjust the volume to 100 μl with distilled water and overlay one drop of mineral oil

After the initial denaturation of the sample at 94° for 30 sec, the following cycle conditions are used:

Number of cycles: 20

Denaturation: 94°, 30 sec

Annealing: 50° 1 min

Primer extension: 72°, 2 min

After the final incubation at 72°, the reaction mixture is incubated for an additional 5–10 min at 72°, and then kept at 4° until the sample is used.

After the PCR, 5 μl of the sample is mixed with 45 μl of gel loading dye (95% formamide, 20 mM EDTA, pH 8.0, 0.05% bromphenol blue (w/v), 0.05% xylene cyanol (w/v)), heated at 80°, applied (1 μl/lane) onto a 0.5× MDE (mutation detection enhancement) gel (AT Biochem., Malvern, PA), and then run using an SSCP-specified electrophoresis system (Atto, Tokyo, Japan). The gel temperature is controlled using a water circulation system (Coolnit Bath EL-15, Taitec, Koshigaya, Saitama, Japan).

For the accurate quantification of the PCR products separated by SSCP, it is important to avoid the trailing of molecules running faster, in order to obtain a good separation of the products derived from the wild-type and mutant mtDNAs. For this purpose, one should test several primer pairs and choose that which gives products exhibiting the least trailing. Furthermore, to obtain an adequate resolution of the PCR products on the MDE gel, one should also optimize the running temperature of the gel. Figure 3 shows the mobilities of the PCR products electrophoresed at different running temperatures (4°, 20°, and 30°) through a 0.5× MDE gel containing 10% glycerol. The mobilities of the denatured PCR products varied widely at different temperatures. The concentration of glycerol in the gel is another critical factor influencing the electrophoretic mobility of the single-stranded DNA, and it should be optimized for maximum resolution of the PCR products. The best separation of the PCR products containing the 3243 mutation from the wild-type products was obtained by electrophoresis for 18 hr at 4°, 12 W, on a 0.5× MDE gel containing 10% glycerol.

Acknowledgments

The work described in this chapter was supported by National Institutes of Health Grant GM-11726 (to G. A.), a Gosney Fellowship and a Human Frontier Science Program Fellowship (to M. Y.), and grants from the Ministry of Education, Science and Culture of Japan (to S. T. and M. Y.) and from the Uehara Memorial Foundation (to S. T.).

[38] Detection and Quantification of Oxidative Adducts of Mitochondrial DNA

By KENNETH B. BECKMAN and BRUCE N. AMES

Introduction

Analysis of oxidatively modified bases in nuclear DNA (nDNA), an effort which has focused on the adduct 8-oxo-2'-deoxyguanosine (oxo^{8}dG), has supported the role of oxidative DNA damage in cancer and other degenerative diseases.[1,2] An important source of intracellular oxygen radicals is the mitochondrial electron transport chain.[3] Because the mitochondrial genome consists of a supercoiled molecule which resides within mitochondria in close proximity to the electron transport chain, it is expected that it should be a significant target of mitochondrial oxygen radicals. Indeed, as was first reported in 1988,[4] mitochondrial DNA (mtDNA) does appear to have a higher steady-state level of oxo^{8}dG than nDNA. A wealth of research since then has uncovered the role of mtDNA mutations in a growing number of inherited and sporadic diseases.[5,6] It has been proposed that mitochondrial dysfunction, resulting in part from mutations in mtDNA, may play a central role in organismal aging.[7]

As a result of these converging lines of research, the analysis of oxidative adducts in mtDNA has been carried out in a number of different laboratories, using a variety of methods. As discussed below, the range of values reported from these studies is quite broad. Some of this variation may be due to inherent difficulties in the analysis of oxo^{8}dG. As this topic has been discussed in a number of articles,[8-11] we make only passing mention to it here. Instead, we concern ourselves mainly with the special problems posed by mtDNA, and with the variation between the reported values.

Table I summarizes the results of 11 studies of oxidative adduct levels

[1] B. N. Ames, M. K. Shigenaga, and L. S. Gold, *Environ. Health Perspect.* **5,** 35 (1993).
[2] A. P. Grollman and M. Moriya, *Trends Genet.* **9,** 246 (1993).
[3] C. Richter, *Mutat. Res.* **275,** 249 (1992).
[4] C. Richter, J. W. Park, and B. N. Ames, *Proc. Natl. Acad. Sci. U.S.A.* **85,** 6465 (1988).
[5] D. C. Wallace, *Proc. Natl. Acad. Sci. U.S.A.* **91,** 8739 (1994).
[6] R. Luft, *Proc. Natl. Acad. Sci. U.S.A.* **91,** 8731 (1994).
[7] M. K. Shigenaga, T. M. Hagen, and B. N. Ames, *Proc. Natl. Acad. Sci. U.S.A.* **91,** 10771 (1994).
[8] S. Adachi, M. Zeisig, and L. Moller, *Carcinogenesis (London)* **16,** 253 (1995).
[9] H. G. Claycamp, *Carcinogenesis (London)* **13,** 1289 (1992).
[10] G. Harris, S. Bashir, and P. G. Winyard, *Carcinogenesis (London)* **15,** 411 (1994).
[11] M. K. Shigenaga, E. N. Aboujaoude, Q. Chen, and B. N. Ames, this series, Vol. 234, p. 16.

in mtDNA, with the studies grouped by method. Values have been converted from those originally published into standard units (fmol oxo^{8}dG/ μg DNA, the number of oxo^{8}dG molecules per 10^5 dG molecules) in order to facilitate direct comparisons. Although various methodological differences exist between all of these studies, they have been broken down into four groups: (1) cesium chloride purification with high-performance liquid chromatography (HPLC) and electrochemical (EC) detection (CsCl/ HPLC/EC); (2) phenol extraction with HPLC separation and EC detection (phenol/HPLC/EC), (3) phenol extraction with HPLC separation and mass spectrometric (MS) detection (phenol/HPLC/MS), and (4) alkaline lysis with endonuclease fingerprinting (AL/EF). In Fig. 1, these four basic protocols are outlined. Within each group the values vary by less than an order of magnitude, whereas between groups they vary by more than three orders of magnitude. Here, we have analyzed differences between the methods in order to identify protocols that minimize artifacts. Although our concern is principally with avoiding artifacts that may lead to overestimates of oxidative damage, we have also indicated procedures that may lead to underestimates.

Isolation of Mitochondria

The isolation of mitochondria has been achieved with varying degrees of purification. The minimal procedure involves disruption of chilled tissues

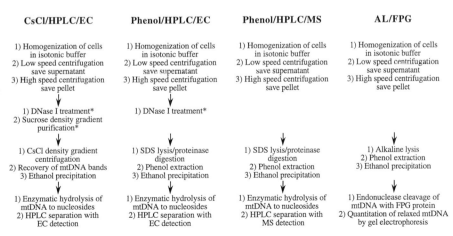

FIG. 1. Four basic methods for measuring oxidative adducts in mitochondrial DNA. An asterisk (*) indicates that the step was employed in only one of the studies which used this technique.

or cells by homogenization in ice-cold isotonic buffer, followed by low-speed centrifugation (10 min, 4°, 600–700 g) to remove nuclei and unruptured cells, and high-speed centrifugation (10 min, 4°, 10,000 g) to collect the crude mitochondrial pellet. The details of basic mitochondrial isolation have been reviewed.[12] More extensive purification of this crude pellet includes DNase I treatment to remove contaminating nDNA and sucrose gradient centrifugation to separate mitochondria from other subcellular components.

Buffer Composition

The exact buffer composition depends on the source of the experimental material studied, and is discussed elsewhere.[12] Typically, Tris (10–100 mM, pH 7.4–7.6) with EDTA (5–10 mM) is used. For osmotic balance, both sucrose (0.25 M) and mannitol/sucrose (0.21 M/0.07 M) solutions have been employed. Mannitol has been widely used as a scavenger of oxygen radicals and may therefore be preferable to sucrose.[13]

DNase I Incubation

The removal of contaminating nuclear DNA from the crude mitochondrial pellet may be achieved by incubating the suspension of intact mitochondria with DNase I.[13-15] However, it has not been determined whether DNase I treatment has a net positive or negative effect on the measured level of oxo^{8}dG, and so it is not known whether artifacts are thereby eliminated or introduced. If nDNA contains fewer oxidative adducts per microgram than mtDNA, as has been reported, then the removal of contaminating nDNA would be expected to increase the measured burden of mtDNA adducts. On the other hand, it is possible that the contaminating nDNA is unrepresentative of nDNA as a whole, and represents a more heavily oxidized fraction (e.g., from the fragmented nuclei of apoptotic cells). If this were true, then the removal of contaminating nDNA could conceivably decrease the measured level of oxo^{8}dG in mtDNA. Moreover, as the DNase I protocol involves incubating mitochondria at a physiological temperature (e.g., 37°), it is possible that artifactual oxidation could result. It has been shown that beef heart mitochondria incubated in mannitol/sucrose/phosphate buffer with added succinate will accumulate oxo^{8}dG.[16] It has not been reported, however, whether incubation of mitochondria in

[12] R. W. Estabrook and M. E. Pullman (eds.), this series, Vol. 10, pp. 74–142.
[13] Y. Higuchi and S. Linn, *J. Biol. Chem.* **270,** 7950 (1995).
[14] P. Mecocci, U. MacGarvey, and M. F. Beal, *Ann. Neurol.* **36,** 747 (1994).
[15] P. Mecocci, U. MacGarvey, A. E. Kaufman, D. Koontz, J. M. Shoffner, D. C. Wallace, and M. F. Beal, *Ann. Neurol.* **34,** 609 (1993).
[16] C. Giulivi, A. Boveris, and E. Cadenas, *Arch. Biochem. Biophys.* **316,** 909 (1995).

the absence of substrate may lead to *in vitro* oxidation of mtDNA. From the studies in Table I, it is not possible to assess the effect of DNase I incubation on the accuracy of the measurements.

Sucrose Gradient Centrifugation

In most of the studies summarized in Table I, the crude mitochondrial pellet was lysed directly after differential centrifugation or DNase I treatment. In one report, the organelles were further purified on a sucrose density gradient, which yielded values that were four to 16 times lower than those in other studies.[13] The fact that purification of the mitochondria was associated with a low value of oxo⁸dG suggests that it may have eliminated an artifact which elevated the level of oxo⁸dG in the other studies. Alternatively, purification may have resulted in the loss of a fraction of mitochondria in which the DNA was highly adducted. Unfortunately, it is impossible to conclude from Table I whether an artifact was either eliminated or introduced by sucrose density purification, as the source of mitochondria used in each of the studies is different. Sucrose density gradient purification, for example, was performed on mitochondria from cultured HeLa cells, whereas all of the other published studies have used cells from experimental animals or human subjects. As tissue culture cells contain few mitochondria, and are highly dependent on glycolysis,[17] the low level of adducts observed in HeLa cells may merely reflect a lower rate of oxygen consumption *in vivo*. A direct comparison of oxidative adducts in purified versus crude mitochondrial prepartions from a single source has yet to be performed.

Lysis of Mitochondria and Purification of Mitochondrial DNA

Following the isolation of mitochondria, mtDNA must be purified prior to analysis. As the mtDNA of vertebrates is a supercoiled molecule of roughly 16 kb, it behaves like a bacterial plasmid, and so the techniques developed for its isolation are quite similar to those used for plasmids in *E. coli*. The isolation of mtDNA for the analysis of oxidative lesions has been achieved by a number of these methods, including (1) sodium dodecyl sulfate (SDS) lysis of mitochondria and banding of the mtDNA in a CsCl density gradient, (2) SDS lysis of mitochondria and digestions of the lysate with proteinase K, followed by phenol extraction and ethanol precipitation, and (3) alkaline lysis of mitochondria, followed by phenol extraction and ethanol precipitation. The choice of method determines the forms of

[17] R. I. Freshney, "Culture of Animal Cells." Wiley–Liss, New York, 1994.

TABLE I

8-OXO-2'-DEOXYGUANOSINE IN MITOCHONDRIAL DNA[a]

| Method | Source | Mitochondrial isolation | | mtDNA purification | | Oxidative adducts (oxo⁸dG equivalents)[b] | | Ref. |
		DNase I	Sucrose gradient	Phenol	mtDNA forms	fmol/μg	per 10^5 dG	
CsCl/HPLC/EC	Rat liver	−	−	−	All	410	53	c
	HeLa cells	+	+	−	All	27	3.5	d
Phenol/HPLC/EC	Bovine heart	−	−	+	All	100	13	e
	Housefly	+	−	+	All	110	15	f
	Human brain	+	−	+	All	240	32	g
	Human brain	−	−	+	All	304	40	h
Phenol/HPLC/MS	Human muscle	−	−	+	All	3,750	500	i
	Murine liver	−	−	+	All	10,425	1,390	j
	Human heart	−	−	+	All	11,250	1,500	k
	Rat liver	−	−	+	All	3,750	500	l
	Rat heart	−	−	+	All	16,500	2,200	l
AL/EF	Porcine liver	−	−	+	Supercoiled	6.2	0.8	m
	Porcine kidney	−	−	+	Supercoiled	7.7	1.0	m
	Rat liver	+	−	+	Supercoiled	12	1.6	m

[a] The values shown are the highest control values from the papers cited.

[b] The conversion factor used was (fmol oxo⁸dG/μg DNA) × 0.13 = oxo⁸$10^5$ dG, based on 1 μg DNA ≈ 0.75 nmol dG. The results from AL/EF were originally expressed as the number of FPG-sensitive sites/10^5 bp. Assuming that all FPG-sensitive sites are oxo⁸dG, and that dG makes up 25% of nucleotides, the conversion factor used was: (FPG-sensitive sites/10^5 bp) × 2 = oxo⁸dG × 7.7 = fmol oxo⁸dG/μg DNA. As not all FPG-sensitive sites are due to oxo⁸dG, this is an upper limit.

[c] C. Richter, J. W. Park, and B. N. Ames, Proc. Natl. Acad. Sci. U.S.A. **85**, 6465 (1988).

[d] Y. Higuchi and S. Linn, J. Biol. Chem. **270**, 7950 (1995).

[e] C. Giulivi, A. Boveris, and E. Cadenas, Arch. Biochem. Biophys. **316**, 909 (1995).

[f] S. Agarwal and R. S. Sohal, Proc. Natl. Acad. Sci. U.S.A. **91**, 12332 (1994).

[g] P. Mecocci, U. MacGarvey, A. E. Kaufman, D. Koontz, J. M. Shoffner, D. C. Wallace, and M. F. Beal, Ann. Neurol. **34**, 609 (1993).

[h] P. Mecocci, U. MacGarvey, and M. F. Beal, Ann. Neurol. **36**, 747 (1994).

[i] M. Hayakawa, K. Torii, S. Sugiyama, M. Tanaka, and T. Ozawa, Biochem. Biophys. Res. Commun. **179**, 1023 (1991).

[j] M. Hayakawa, T. Ogawa, S. Sugiyama, M. Tanaka, and T. Ozawa, Biochem. Biophys. Res. Commun. **176**, 87 (1991).

[k] M. Hayakawa, K. Hattori, S. Sugiyama, and T. Ozawa, Biochem. Biophys. Res. Commun. **189**, 979 (1992).

[l] M. Takasawa, M. Hayakawa, S. Sugiyama, K. Hattori, T. Ito, and T. Ozawa, Exp. Gerontol. **28**, 269 (1993).

[m] J. Hegler, D. Bittner, S. Boiteux, and B. Epe, Carcinogenesis (London) **14**, 2309 (1993).

mtDNA purified (closed circular, nicked, open circular, or "complex") that are isolated.[13]

Cesium Chloride Density Gradient Centrifugation

The use of cesium chloride density gradients has traditionally been the method of choice for purifying high quality mtDNA in molecular cloning.[18] Although cloning-grade DNA is not required for the analysis of oxidative adducts, CsCl gradient purification has other advantages. Principally, the DNA is not subjected to organic extraction, a procedure that has been observed to cause artifactual oxidation of DNA under certain circumstances (but see below).[9] Also, if performed carefully, banding on CsCl permits the separation of closed circular DNA from other forms (nicked, open circular). Because different forms of mtDNA may be damaged to different degrees,[19] CsCl gradients may permit a detailed analysis of oxidative damage in different fractions.

A major disadvantage of the traditional CsCl gradient is that the mtDNA is exposed to both ethidium bromide and ultraviolet (UV) light during the visualization and recovery of the mtDNA, procedures that may cause oxidation of mtDNA *in vitro*.[20] Higuchi and Linn, in an extensive analysis of mtDNA extraction methods, avoided the use of ethidium bromide and UV irradiation by fractionating the CsCl gradient by drop collection and identifying fractions containing mtDNA via dot-blot Southern hybridization.[13]

Finally, it has been observed that if DNase I pretreatment of mtDNA is omitted prior to lysis of mitochondria and density centrifugation, then some forms of mtDNA are lost due to association with contaminating nDNA in the gradient. Therefore, both DNase I treatment of the mitochondrial suspension and the avoidance of ethidium bromide/UV light are advisable when using CsCl density gradients.

Sodium Dodecyl Sulfate Lysis with Proteinase Digestion and Phenol–Chloroform Extraction

A less stringent and far less strenuous method of isolating mtDNA is the use of proteinase digestion and phenol extraction.[13,15,16,21] Isolated

[18] D. P. Tapper, E. R. Van Etten, and D. A. Clayton, this series, Vol. 97, p. 426.
[19] C. Richter, personal communication, 1995.
[20] N. A. Fischer, H. E. Poulsen, and S. Loft, *Free Radicals Biol. Med.* **13**, 121 (1992).
[21] M. Hayakawa, K. Torii, S. Sugiyama, M. Tanaka, and T. Ozawa, *Biochem. Biophys. Res. Commun.* **179**, 1023 (1991).

mitochondria are lysed in buffered SDS (0.5–2.0% w/v), and the solubilized lysate is incubated with proteinase K (0.1–1.0 mg/ml, 37°–50°, 3–16 hr). The hydrolyzate is extracted twice with phenol or phenol–chloroform–isoamyl alcohol (25:24:1, v/v) and once with chloroform–isoamyl alcohol (24:1, v/v) before precipitation by standard methods. The main advantages of this method are its simplicity and the fact that all forms of mtDNA are recovered quantitatively without exposure to UV light. Moreover, assuming that the suspension of mitochondria is relatively free of contaminating nuclei, the mtDNA isolated in this fashion is generally free of contaminating nDNA. DNase I pretreatment of the mitochondrial suspension ensures that nDNA contamination is minimal.[14,15]

It is important here to comment briefly on the debate surrounding the use of phenol. It has been reported that extraction of DNA with phenol or other organic solvents, when combined with enzymatic hydrolysis of the DNA, ethanol precipitation of the hydrolyzate, and subsequent drying of the hydrolyzate under a stream of air to remove the ethanol, leads to a significant elevation of oxo^{8}dG. Other workers have therefore taken care to avoid the "phenol artifact" in their work by omitting the use of phenol entirely.[8,13] Although concentrating enzymatic hydrolyzates of DNA with air can induce significant increases in oxo^{8}dG levels, most protocols for the measurement of oxo^{8}dG do not employ this procedure. Instead, DNA is ethanol-precipitated and residual ethanol is removed by centrifugation under vacuum before enzymatic hydrolysis of the sample. Higher molecular weight components of the hydrolyzate are then removed, if necessary, by ultrafiltration. This procedure is sufficient to prevent measurable increases in oxo^{8}dG. Furthermore, studies of the effects of extraction of calf thymus DNA with fresh, high purity phenol, old phenol (stored at room temperature for 1 year), or 24:1 (v/v) chloroform–isoamyl alcohol have revealed essentially no difference in oxo^{8}dG levels. In addition, rat liver DNA that has been subjected to up to four cycles of organic extraction (where a single cycle is comprised of phenol, phenol–chloroform–isoamyl alcohol, and chloroform–isoamyl alcohol extractions) reveals no significant differences in oxo^{8}dG. Others have also reported that phenol, when used carefully, does not introduce artifactual oxidative adducts into DNA.[10] It should be noted that phenol is an antioxidant, with oxidation potentials well below that of dG. Thus, phenoxyl radicals or other contaminants that could conceivably catalyze oxidation of dG to oxo^{8}dG should be quenched by the vast molar excess of phenol used in the DNA extraction. On the basis of these observations, phenol extraction per se is judged to be associated with minimal oxidation artifacts.

Alkaline Lysis with Phenol–Chloroform Extraction

The most rapid method for isolating mtDNA is a plasmid minipreparation protocol adapted to mitochondria.[22,23] The principle is straightforward: alkaline lysis is achieved with SDS in sodium hydroxide, which causes DNA to denature. Rapid neutralization of the lysate with potassium acetate permits the selective reannealing of supercoiled mtDNA; linear DNA, including nDNA, aggregates into insoluble precipitates that are removed by centrifugation. The resulting supernatant is then extracted with phenol and precipitated. An advantage of alkaline lysis is that the mtDNA is not subject to an incubation at an elevated temperature, as is required during proteinase digestion. Therefore, any artifactual oxidation that could occur as a result of the incubation is avoided. On the other hand, a disadvantage of alkaline lysis is that it selects for the closed circular form of mtDNA at the expense of other forms. If damage is unequally distributed between different forms of mtDNA, this may lead to an underestimate of overall mtDNA oxidative adducts.

Measurement of Oxidative Lesions

Measurement of 8-Oxo-2'-deoxyguanosine with High-Performance Liquid Chromatography with Electrochemical Detection

The extraordinary sensitivity of electrochemical detection has made the analysis of oxo^{8}dG by HPLC the most frequently used technique for measuring oxidative adducts. DNA is hydrolyzed to nucleosides with nucleases (e.g., nuclease P1 from *Penicillium citrinum,* bovine pancreatic DNase I, or snake venom phosphodiesterase) and alkaline phosphatase, and individual nucleosides (including modified nucleosides) are separated and detected by HPLC with tandem UV/EC detection. The amount of oxo^{8}dG is therefore expressed relative to that of dG, providing an internal control for the efficiency of enzymatic digestion. A modification has been reported that incorporates a mathematical transformation of the chromatograph to decrease the background EC signal, thereby improving the detection limit to 1.76 fmol (0.5 pg) oxo^{8}dG. This level of sensitivity should allow for the quantification of 0.005 oxo^{8}dG/10^5 dG in 50 μg DNA.[8]

Despite the relatively good agreement between published values of oxo^{8}dG in nDNA, it must be stressed that the sensitivity of EC detection

[22] H. C. Birnboim and J. Doly, *Nucleic Acids Res.* **7,** 1513 (1979).
[23] T. K. Palva and E. T. Palva, *FEBS Lett.* **192,** 267 (1985).

demands strict attention to controls. For instance, we have observed that the analysis of very small quantities of calf thymus or salmon sperm DNA (≤ 10 μg) may be associated with 2- to 5-fold elevations in the amount of oxo[8]dG per microgram (Fig. 2, filled circles). In these experiments the efficiency of digestion (the ratio of input DNA to measured dG) was independent of the quantity of DNA, and so the results are probably not the result of inefficient release of oxo[8]dG at higher concentrations of DNA. Moreover, injection onto the HPLC column of decreasing amounts of a single hydrolyzate (50 μg DNA) did not result in an upward trend in oxo[8]dG/μg DNA (Fig. 2b, open circles), which indicated that it is not an artifact of the integration of smaller peaks. Rather, the trend may be due to a contaminant that coelutes with oxo[8]dG, and that has a relatively greater additive effect on smaller peaks. Alternatively, the trend may be due to oxidation of a relatively fixed number of dG molecules by a contaminant of the enzymes, occurring in a fashion independent of the total amount of dG.

Whatever the cause of this artifact, the point is that it is not advisable to analyze less than 20 μg of DNA under these conditions; at the very least, comparisons between samples must be performed on equivalent amounts of DNA. This caveat is particularly relevant to studies mtDNA, since it is frequently the case that the yield of mtDNA is relatively small. For instance,

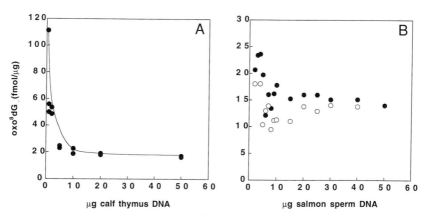

FIG. 2. Effect of the quantity of DNA analyzed on the measured level of oxo[8]dG/μg DNA. (A) Varying amounts of calf thymus DNA (0.5–50 μg) were hydrolyzed by sequential digestion with nuclease P1 and alkaline phosphatase, and the hydrolyzates were analyzed by HPLC with electrochemical detection, according to the protocol of Shigenaga *et al.*[11] (B) Varying amounts of salmon sperm DNA (2–50 μg) were hydrolyzed and analyzed by HPLC (filled circles). In addition, 50 μg of salmon sperm DNA was hydrolyzed, and varying fractions of the hydrolyzate (2–80%, equivalent to 2–40 μg) were analyzed by HPLC (open circles).

in our hands, an entire 15-g rat liver yields 20–40 μg mtDNA by SDS/
proteinase digestion/phenol extraction, whereas the brain typically yields
5–10 μg. A comparison of oxo^{8}dG in the two tissues then, if it involved
the hydrolysis of all of the DNA from both samples, might indicate a
difference in oxo^{8}dG that was an artifact of the difference in yield.

In fact, a trend toward higher oxo^{8}dG levels associated with lower yields
is evident in some published work. For example, in a study of azidothymidine (AZT)-induced mitochondrial dysfunction, it was found that orally
administered AZT at 1, 5, or 10 mg/kg increased the percentage of oxo^{8}dG
in liver mtDNA from 1.39% (control value) to 25.2, 38.1, and 28.0%, respectively.[24] However, at the same time AZT decreased the yield of oxo^{8}dG
from 105 ng/mg protein to 39, 29, and 43 ng/mg, respectively. From these
data is evident that the apparent AZT-induced oxidation of mtDNA correlates more closely with yield of mtDNA than it does with dose of AZT.
In two other studies, in which the age-associated accumulation of oxo^{8}dG
in human diaphragm and rat hearts was observed, the increased levels of
oxo^{8}dG in older animals were always associated with decreased yields.[25,26]
In rat liver, where increased age was not associated with decreased yield,
there was no increase in oxo^{8}dG with age. From these studies, a model has
been constructed in which *in vivo* oxidation of mtDNA interferes with
replication and leads to a loss of mtDNA. Although this hypothesis may
be right, it is also possible that the observed effects are merely a result of
having analyzed different amounts of DNA in different samples.

The problem of artifacts, of course, is inherent to the measurement of
trace amounts of an oxidative adduct and requires various countermeasures,
as has been discussed before. The redox active metals iron and copper are
the most important source of artifactual oxidation, as they catalyze the
generation of hydroxyl radicals and associate closely with DNA.[27,28] Therefore, an important step in avoiding artifactual oxidation is to remove traces
of iron and copper from reagents and samples by dialysis against buffers
from which metallic ions have been removed with a chelating resin such
as Chelex (Bio-Rad, Hercules, CA). Also, the inclusion of a metal chelator
throughout the enzymatic hydrolysis and chromatography steps should be
routine. Not all chelators are equally effective in this regard. Iron chelated

[24] M. Hayakawa, T. Ogawa, S. Sugiyama, M. Tanaka, and T. Ozawa, *Biochem. Biophys. Res. Commun.* **176,** 87 (1991).
[25] M. Hayakawa, K. Hattori, S. Sugiyama, and T. Ozawa, *Biochem. Biophys. Res. Commun.* **189,** 979 (1992).
[26] M. Takasawa, M. Hayakawa, S. Sugiyama, K. Hattori, T. Ito, and T. Ozawa, *Exp. Gerontol.* **28,** 269 (1993).
[27] Y. Luo, Z. Han, S. M. Chin, and S. Linn, *Proc. Natl. Acad. Sci. U.S.A.* **91,** 12438 (1994).
[28] Y. Luo, E. S. Henle, R. Chatopadhyaya, R. Jin, and S. Linn, this series, Vol. 234, p. 51.

by EDTA, for example, retains its ability to catalyze hydroxyl radical formation, and so a compound such as desferrioxamine, which chelates iron in a redox-inactive form, should be used.[11] Also, samples should not be exposed to air drying, and samples should be analyzed promptly after processing.[8–10] By standardizing procedures, such as analyzing of a constant quantity of DNA, it is possible at least to ensure precision in measurements of oxo[8]dG via HPLC/EC. Lastly, details of the sources and exact quantities of reagents used and the amounts of DNA hydrolyzed should be reported completely in all published work, in order to allow laboratories to minimize the differences between their values.

Measurements of 8-Oxo-2'-deoxyguanosine with High-Performance Liquid Chromatography/Mass Spectrometry

An alternative to EC detection of oxo[8]dG is the use of mass spectrometry. It has been combined with enzymatic hydrolysis of DNA to nucleosides and HPLC separation in a method termed ultramicro-HPLC/MS analysis, and applied to mtDNA.[24] The levels of oxo[8]dG reported with this technique are from 10- to 100-fold higher than those measured with HPLC/EC detection, despite the fact that similar tissues (mammalian organs) were used. This discrepancy, and the fact that there exists by now a fairly large database of oxo[8]dG measurements generated by a number of different laboratories with the HPLC/EC method, suggests that the HPLC/EC method is preferable to HPLC/MS.

Measurement of Oxidative Lesions with Endonuclease Fingerprinting

The cloning of repair endonucleases that are relatively specific for oxidative lesions, such as the formamidopyrimidine–DNA glycosylase (FPG protein) of *Escherichia coli,* has led to the development of a novel way to quantify oxidative adducts, termed endonuclease fingerprinting (EF). Purified DNA is incubated with saturating amounts of an endonuclease specific for the adduct of interest, resulting in a nick in the damaged DNA strand at every vulnerable site. The number of nicks are then quantified by a relaxation assay, which measures the loss of mtDNA supercoiling by electrophoresis of the DNA on an agarose gel. A Poisson formula relates the proportions of nickel to supercoiled mtDNA to the number of nicks per molecules, from which the estimate of oxidative adducts is derived. One requirement of this technique is that only supercoiled mtDNA can be used; therefore, the isolation of mtDNA for EF has been achieved by alkaline lysis. The details of EF have been described.[29]

[29] B. Epe and J. Hegler, this series, Vol. 234, p. 122.

The EF estimates of oxidative adducts in mtDNA from porcine liver and kidney and rat liver have yielded maximum values that are from 2 to 60 times lower than those derived from HPLC/EC.[30] At present, an explanation for this difference is lacking. However, it is possible that oxidative damage in mtDNA is not distributed evenly between different forms of mtDNA, and that supercoiled mtDNA represents a "cleaner" fraction of the genome. Apparently, the analysis of different forms of mtDNA (supercoiled, nicked) by HPLC/EC has revealed that supercoiled mtDNA is relatively free of oxidative adducts.[31]

Conclusions

The wide range of values in Table I clearly illustrates the need for more studies. To date, different protocols have been used in virtually every step of the analysis: the isolation of mitochondria, the purification of mtDNA from mitochondria, and the analysis of oxidative adducts in the mtDNA itself. In addition to these differences, the studies have used a number of different sources of mitochondria, including mammalian organs, tissue culture cells, and the mitochondria from entire arthropods. In the future, studies should attempt to define the effect of different protocols on the level of oxidative adducts. With the help of such studies, an optimal protocol will surely emerge. Lastly, it should be noted that although the absolute values obtained in studies of oxidative adducts may be in question, the relative differences seen between control values and those of treatment groups (e.g., older animals, animals placed under oxidative stress, animals suffering a specific pathology) are often what is of greatest interest. In a number of the studies included in Table I, such differences were noted and were statistically significant. This indicates that for comparative purposes, at least, present methods are sufficient.

Acknowledgments

This work was supported by National Institute of Environmental Health Sciences Center Grant ESO1896 and National Cancer Institute Outstanding Investigator Grant CA39910 to B. N. A. The authors thank M. K. Shigenaga, H. Helbock, and S. Linn for helpful criticism.

---

[30] J. Hegler, D. Bittner, S. Boiteux, and B. Epe, *Carcinogenesis* (*London*) **14**, 2309 (1993).
[31] C. Richter, personal communication, 1995.

[39] Use of Fibroblast and Lymphoblast Cultures for Detection of Respiratory Chain Defects

By Brian H. Robinson

Introduction

Cultured skin fibroblasts and cultured lymphoblasts have been found to be extremely useful in the diagnosis of respiratory chain defects in children with lactic acidemia.[1-3] The usage of cultured skin fibroblasts for this purpose has some disadvantages and some advantages. These revolve around the variability in expression in fibroblasts of what are now regarded as tissue-specific respiratory chain defects. Although examination of mitochondria prepared from muscle biopsy tissue is a commonly used technique to detect respiratory chain abnormalities, artifacts can arise in this system, especially in patients who have abnormal lipid storage within the muscle. We have successfully used a system of diagnosis that relies on a detailed investigation of skin fibroblast metabolism. This method of analysis looks at three parameters: the redox state of the cells, the activities of the respiratory chain enzymes themselves, and the ability of isolated mitochondria to make ATP when provided with selected substrates.

Before embarking on a description of these determinations, it is first necessary to describe briefly the important systems linking cytosolic and mitochondrial redox states and the relationship of fibroblast glycolytic rate to mitochondrial respiration in cultured cells.

Sources of Cellular ATP in Cell Culture

Choice of Medium

Glucose-Containing Medium. When cultured skin fibroblasts or cultured lymphoblasts are grown in Eagle's α-minimal essential bicarbonate-buffered medium (αMEM), plus 15% fetal calf serum, with a 5% CO_2/95% air (v/v) atmosphere, they initially derive much of their energy for growth and

---

[1] B. H. Robinson, *Adv. Hum. Genet.* **18,** 151 (1989).

[2] B. H. Robinson, *Biochim. Biophys. Acta* **1182,** 231 (1993).

[3] Y. Tatuch, J. Christodoulou, A. Feigenbaum, J. T. R. Clarke, J. Wherret, C. Smith, N. Rudd, R. Petrova-Benedict, and B. H. Robinson, *Am. J. Hum. Genet.* **50,** 852 (1992).

metabolism from converting glucose to lactic and pyruvic acids.[4,5] Within a few days of subculture in medium containing 5.5 mM glucose, the glucose disappears from the medium, with 85% of it reappearing in the medium as lactate and pyruvate. If culture is continued without further glucose supplementation, the lactate and pyruvate are then metabolized oxidatively to provide energy.[4] Cell cultures with defects in mitochondrial metabolism often go into decline or die at the point at which the glucose is exhausted.[4] We therefore recommend that cell lines suspected of harboring mitochondrial defects be cultured in a modified medium that contains 10 mM glucose rather than the usual 5.5 mM and that the medium be supplemented with 1 mM uridine.[6] The uridine supplementation is needed to overcome the block in endogenous cellular uracil synthesis which requires activity of dihydroorotate dehydrogenase, a respiratory chain-linked enzyme whose activity may be depressed by the unfavorable redox state present in respiration-deficient cells.[6]

Galactose-Containing Medium. If the glucose in the culture medium is replaced by galactose (5.5 mM), both cultured skin fibroblasts and cultured lymphoblasts will grow at almost the same rate seen with glucose-containing medium.[7] The metabolism supporting growth, however, is quite different (Fig. 1). In galactose medium very little lactate (L) and pyruvate (P) accumulate in the medium because of the restricted flow of galactose through to glucose 6-phosphate. The mode of metabolism is therefore one in which little cellular ATP synthesis derives from glycolysis, the majority being derived from mitochondrial oxidation of pyruvate. For this reason, cultured cells with defects in mitochondrial metabolism will not grow well and often will die when placed in a galactose medium.[7]

Test Procedure

The culture media required are αMEM with 10 mM glucose and 1 mM uridine and αMEM with 5 mM galactose.

1. Grow skin fibroblasts to confluence in 5 ml αMEM–glucose medium; this will give about 1 mg protein of cells in a 9-cm-diameter petri dish.
2. Trypsinize these cells ready for subculture.
3. Subculture these cells into two plates with 5 ml glucose medium each

[4] N. MacKay, B. H. Robinson, R. Brodie, and N. Rooke-Allen, *Biochim. Biophys. Acta* **762,** 198 (1983).
[5] B. H. Robinson, D. M. Glerum, W. Chow, R. Petrova-Benedict, R. Lightowlers, and R. Capaldi, *Pediatr. Res.* **28,** 549(1990).
[6] M. P. King and G. Attardi, *Science* **246,** 500 (1989).
[7] B. H. Robinson, R. Petrova-Benedict, J. R. Buncic, and D. C. Wallace, *Biochem. Med. Metab. Biol.* **48,** 122 (1992).

Glucose   Galactose

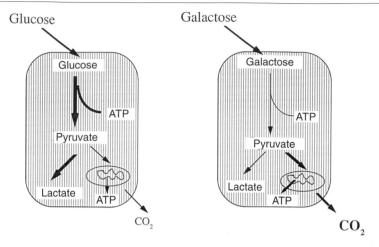

CO_2

CO_2

FIG. 1. Diagram to illustrate the direction of metabolite flow when cultured skin fibroblasts are grown in different media. When glucose is the substrate, large amounts of lactate are formed with only a small percentage of pyruvate being oxidized to CO_2. When galactose is the substrate, the bulk of the pyruvate, which is formed at a much slower rate, is converted to CO_2 and water. Thus, ATP is generated largely by glycolysis when glucose is the substrate but from mitochondria when galactose is the substrate.

and two plates with 5 ml galactose medium each. Return cells to an incubator at 37°.

4. Inspect daily for 7 days and observe cell growth and survival relative to control cell lines treated in an identical fashion.

5. Mitochondrial defects are indicated by death of cells in the galactose-containing medium.[7] These cells will become detached from the substratum and float in the medium. This may happen in the first day of culture with severe defects and after 3 or 4 days with milder defects.[7]

This test will detect the following defects in mitochondrial metabolism: (a) cytochrome oxidase deficiency (Leigh's disease phenotype), (b) complex I deficiency (L/P ratios of 70–250), (c) pyruvate dehydrogenase complex deficiency in severe cases, and (d) multiple respiratory chain defects. This test will not detect: (a) most defects due to abnormalities of mtDNA, for example, MELAS, MERRF, and NARP, (b) mild cases of pyruvate dehydrogenase (PDH) complex deficiency, (c) cytochrome oxidase deficiency: cardiomyopathy phenotype, liver-specific phenotype, and (d) complex I deficiency, if mild, with L/P ratios below 60.

Redox State of Cells with Respiratory Chain Defects

Although the reliance of culture skin fibroblasts on glycolysis does not make them an ideal system for studying oxidative metabolic defects, the ratio of accumulated lactic to pyruvic acids can give important information about the redox state of the cell and thus reveal problems in respiratory chain function. The lactate dehydrogenase (LDH) reaction in the cytosolic compartment of cultured cells is described by the equilibrium[8]

$$\frac{[\text{Lactate}]}{[\text{Pyruvate}]} \times K_{\text{LDH}} = \frac{[\text{NADH}]_c}{[\text{NAD}^+]_c}$$

where K_{LDH} is the equilibrium constant for LDH and $[\text{NADH}]_c/[\text{NAD}]_c$ describes the redox state of pyridine nucleotides in the cytosolic compartment. The ratio of accumulated lactate to pyruvate in the medium (L/P ratio) observed after incubation with glucose is approximately 25 : 1.[4] From the LDH equilibrium it follows that $[\text{NAD}]_c/[\text{NADH}]_c$ is 360 : 1 where $K_{\text{LDH}} = 1.11 \times 10^{-4}$. This value of 360 : 1 is similar to that seen in other mammalian cells and reflects the status of the $[\text{NAD}]_m/[\text{NADH}]_m$, the pyridine nucleotide redox state in the mitochondria, which is usually much more reduced. The equilibrium between the pyridine nucleotide couples in the two compartments is maintained by the glutamate/aspartate shuttle system (Fig. 2) such that $[\text{NAD}]_m/[\text{NADH}]_m$ lies between 10 : 1 and 30 : 1 in most cells.[8] The difference in the cytosolic and mitochondrial redox states is maintained by electrogenic expulsion of aspartate from the mitochondrial compartment.[9] The mitochondrial pyridine nucleotide redox state is in equilibrium with the oxidized–reduced cytochrome c couple as modulated by the ratio $[\text{ATP}]/[\text{ADP}][\text{inorganic phosphate}]$ in the cell.[10]

A typical set of measurements of L/P ratio is shown in Fig. 3. Control cells produce about 500 nmol lactate in the medium after a 1-hr incubation together with 25 nmol pyruvate, giving an L/P ratio between 20 : 1 and 30 : 1. When the PDH complex is deficient, the lactate and pyruvate production rates may be higher but the L/P ratio observed is either the same or lower than control cells. In cases where a respiratory chain defect is present, the amount of pyruvate is decreased and the amount of lactate generally increased, with the resultant L/P ratio being raised above the normal range of 18–35.[11,12]

Determination of Lactate/Pyruvate Ratio in Cultured Skin Fibroblasts

1. Grow skin fibroblasts to confluence (0.8–1.2 mg per 9-cm petri dish).
2. Aspirate culture medium.

[8] H. A. Krebs, *Symp. Soc. Exp. Biol.* **17**, 299 (1973).
[9] J. R. Williamson, B. Safer, K. F. LaNoue, C. M. Smith, and E. Walajtys, *Symp. Soc. Exp. Biol.* **17**, 241 (1973).
[10] M. Ericinska and D. F. Wilson, *J. Membr. Biol.* **70**, 1 (1982).
[11] B. H. Robinson, N. MacKay, P. Goodyer, and G. Lancaster, *Am. J. Hum. Genet.* **37**, 938 (1985).
[12] B. H. Robinson, J. Ward, P. Goodyer, and A. Beaudet, *J. Clin. Invest.* **77**, 1422 (1986).

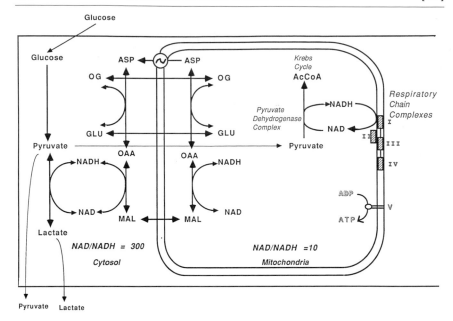

FIG. 2. Diagram to show how the mitochondrial oxidation of NADH generated by glycolysis is achieved by the use of a glutamate/asparate shuttle system and is reflected in the ratio of lactate to pyruvate. The pyridine nucleotide redox couples are maintained at different states by the operation of a shuttle in which malate (MAL), glutamate (GLU), and 2-oxoglutarate (OG) are in equilibrium between the cytosolic and mitochondrial compartments. The difference in redox states is maintained by electrogenic expulsion of aspartate from mitochondria, and the ratio of lactate to pyruvate reflects the cytosolic redox state through the lactate dehydrogenase equilibrium.

3. Wash cells with 0.5 ml phosphate-buffered saline (PBS) and aspirate.
4. Add 1 ml PBS to each plate and incubate at 37° for 1 hr.
5. Aspirate medium, add back 1 ml PBS plus 1 mM glucose, and incubate at 37° for 1 hr.
6. After 1 hr add 50 μl of 1.6 M perchloric acid (PCA) to stop metabolism.
7. Remove the PCA-acidified PBS into Eppendorf tubes (1.5 ml) and centrifuge to remove any debris.
8. Add 1 ml Biuret reagent to each plate for protein determination[13] setting up appropriate blanks and standards.

[13] A. G. Gornall, C. S. Bardawill, and M. M. David, *J. Biol. Chem.* **177,** 751 (1949).

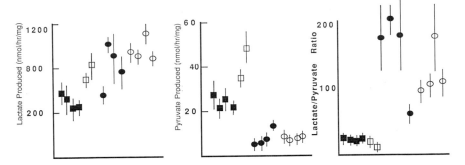

Fig. 3. Typical set of results obtained from lactate and pyruvate measurements in fibroblasts. Cultured skin fibroblasts were incubated with 1 mM glucose for 1 hr as described in the protocol, and the amounts of lactate and pyruvate produced were measured. The three graphs represent the amount of lactate, the amount of pyruvate, and the lactate/pyruvate ratio for a number of cell lines. Cell lines used: ■, controls; □, pyruvate dehydrogenase complex deficiency; ●, cytochrome oxidase deficiency; ○, NADH–cytochrome-c reductase deficiency (rotenone-sensitive). Results are depicted as means ± S.E.M. for five observations.

9. Take 25-μl aliquots of the PCA extract and determine lactate spectrophotometrically.[14]
10. Take 50 μl aliquots of PCA extract and determine pyruvate content by enzyme fluorimetric analysis.[15] If a fluorimeter is not available for this determination, the PCA extract can be neutralized with 6 N triethanolamine (TEA)/0.5 N K$_2$CO$_3$, and the pyruvate can be determined spectrophotometrically using 0.5 ml of extract in a 1-ml determination system.

Dependence of Lactate/Pyruvate Ratio on Confluence State

Strictly one should be able to measure the L/P ratio in any fibroblast culture independent of the number of milligrams of fibroblast protein, but the above protocol calls for a measurement of protein. This is simply because the L/P ratio obtained from fibroblasts does vary in control cells with the state of confluence (Fig. 4). When cells are rapidly dividing they tend to have more ATP demand and therefore a more oxidized redox state than when they are confluent. For a 9-cm plate (petri dish), cells come into confluence at about 0.8 mg protein per plate and growth begins to slow down. If the cells are allowed to remain on the plate for 1 or 2 weeks after

[14] I. Gutman and A. W. Wahlefeld, *in* "Methods of Enzymatic Analysis" (H. U. Bergemeyer, ed.), Vol. 3, p. 1464. Academic Press, New York, 1974.
[15] R. W. Von Korff, this series, Vol. 13, p. 519.

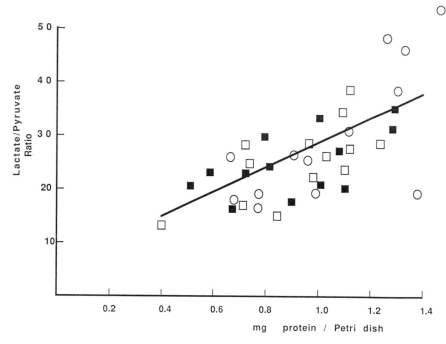

FIG. 4. Dependence of the lactate/pyruvate ratio in cultured skin fibroblasts on the state of confluence. The lactate/pyruvate ratio was determined for three different control cell lines (□, ○, and ■) as a function of state of confluence of cells grown on a 9-cm petri dish. Each point represents a separate determination. Confluence for a 9-cm petri dish starts at 0.8 mg protein. The equation for the line is $y = 22.2x + 6.48$, where x is the protein concentration.

this has been achieved the L/P ratio will be seen to go higher. For this reason it is best to choose standard conditions of confluence if two different cell lines are going to be compared. We recommend the values obtained between 0.8 and 1.1 mg of protein for a 9-cm-diameter petri dish of cells.

The passage number of the fibroblast culture should also be noted. This should not be a problem as the L/P ratio does not change much over the lifetime of fibroblast culture. As cells begin to approach sensecence, L/P ratios tend to become elevated out of the normal range for a set protein concentration. The dependence of the L/P ratio on the state of confluence for Fig. 4 can be described as:

L/P ratio = 22(protein concentration) + 6.48

It is advisable for each investigator, however, to make his/her own graphic relationship.

Lactate/Pyruvate Ratio in Cultured Lymphoblasts

A similar procedure can be used to obtain an L/P ratio for cultured lymphoblasts. However, the values obtained for respiratory chain-deficient cell lines are not nearly as elevated as they are for cultured skin fibroblasts. The reason for this is probably related to the fact that lymphoblasts are a transformed cell line and do not go into the semiquiescent state of confluence. Thus, there is a constant high demand for ATP that tends to keep the redox states in a rather oxidized condition.[16] This constraint makes it difficult to use L/P ratios in lymphoblasts for the detection of respiratory chain defects.

Lactate/Pyruvate Ratios in Amniocyte Cultures

Both our laboratory and others have investigated the use of L/P ratio measurements in amniocytes for the prenatal detection of respiratory chain defects. Our laboratory has successfully predicted a complex I-deficient case based on elevated L/P ratios in amniocytes; nine other cases have been predicted normal and were unaffected.[17]

Measurement of ATP Synthesis in Digitonin-Permeabilized Skin Fibroblasts

Mitochondrial preparations can be made by conventional methods from cultured skin fibroblasts. However, these preparations are rarely well coupled and cannot be used in the way that heart and liver mitochondria are for estimating ATP synthetic rates with the use of an oxygen electrode. Methods have therefore been developed to permeabilize the fibroblast cell membrane with digitonin so that mitochondrial oxidative phosphorylation can be assessed indirectly.[12,18]

Method

1. Grow fibroblast cells from a patient and a control in 9-cm petri dishes until confluent (0.8–1.1 mg cell protein).
2. Aspirate medium from the cultures and replace with 1.5 ml of 0.25 M sucrose, 20 mM MOPS, 0.05 mg/ml digitonin, pH 7.4, leave for 3 min at room temperature, and drain.

[16] S. J. Hyslop and B. H. Robinson, unpublished observations.
[17] T. Myint, R. Petrova-Benedict, and B. H. Robinson, unpublished observations.
[18] B. H. Robinson, *in* "Mitochondrial Disorders in Neurology" (A. H. V. Shapira and S. DiMauro, eds.), p. 168. Butterworth/Heinemann, London, 1994.

3. Replace digitonin medium with 0.25 M sucrose, 20 mM MOPS, 20 mM EDTA, pH 7.4.
4. After 5 min, drain and replace with 1.5 ml medium containing 0.25 M sucrose, 20 mM MOPS, 1 mM EDTA, 5 mM inorganic phosphate, 1 mM ADP (high purity, ATP-free).
5. Arrange the plates so that one plate has no substrate, only the medium above; one plate has 5 mM pyruvate and 1 mM L-malate; one plate has 5 mM glutamate plus 1 mM L-malate; one plate has 1 μM rotenone and 10 mM succinate; one plate has 2 mM antimycin, 2 mM ascorbate, and 0.1 mM tetramethylphenylenediamine (TMPD).
6. Incubate 1 hr at 37°, then add 50 μl of 1.6 M perchloric acid.
7. Incubation buffer is removed, added to Eppendorf tubes, and centrifuged to remove any cell debris.
8. The acidified solution is neutralized with 6 N TEA/0.5 N K$_2$CO$_3$.
9. Aliquots of the incubation buffer are assayed for ATP by enzyme spectrophotometric/fluorimetric methods.[19]
10. Protein is measured by the addition of 1 ml Biuret reagent to the acid-precipitated cells.

The results of this assay system are expressed as nanomoles ATP per hour per milligram cell protein. Typically, with no added substrates, 40–50 nmol ATP is produced as a result of oxidation of endogenous substrate remaining after the washing of the digitonin-permeabilized cells (Fig. 5). With NAD$^+$-linked substrates 140–170 nmol are produced, with succinate/rotenone a similar amount, and with ascorbate/TMPD about 100–120 nmol. Reference to Fig. 5 shows that pyruvate dehydrogenase complex deficiency can be differentiated from complex I deficiency because glutamate/malate oxidation is normal in the former but not in the latter case. In complex IV deficiency and complex V deficiency (homoplasmic for mtDNA 8993 mutation), the rate of ATP synthesis is decreaased with all substrates. A variation on this method that utilizes digitonin-permeabilized trypsinized fibroblasts also works well.[20]

ATP Synthesis in Lymphoblast Mitochondria

Isolation of Lymphoblast Mitochondria

Lymphoblast mitochondria are prepared as follows. The method is a modification of that described by Bourgeron et al.[21]

[19] J. R. Williamson and B. E. Corkey, this series, Vol. 13, p. 434.
[20] R. J. A. Wanders, J. P. N. Ruiter, and F. A. Wijburg, Biochim. Biophys. Acta 1181, 219 (1993).
[21] T. Bourgeron, D. Chretien, A. Rotig, A. Munnich, and P. Rustin, Biochem. Biophys. Res. Commun. 186, 16 (1992).

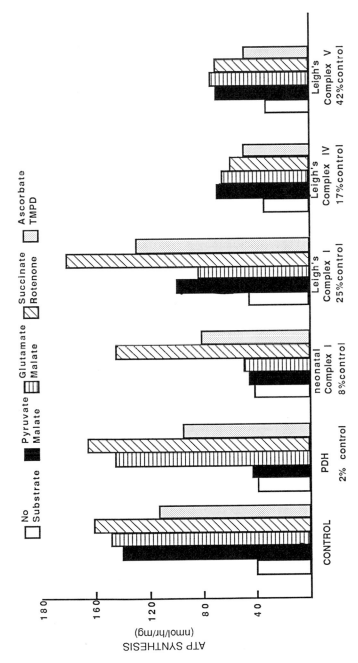

FIG. 5. ATP synthesis in digitonin-treated cultured skin fibroblasts. ATP synthesis was measured in a series of cultured skin fibroblasts using the substrates pyruvate/malate, glutamate/malate, succinate plus rotenone, and ascorbate/TMPD after treatment with digitonin to permeabilize the cells. Cell lines used were as indicated: from a control subject, from a patient with fatal neonatal pyruvate dehydrogenase complex deficiency, from a patient with fatal neonatal lactic acidosis and complex I deficiency, from a patient with Leigh's disease and complex I deficiency, from a patient with Leigh's disease and complex IV deficiency, and from a patient with Leigh's disease and complex V deficiency (mtDNA 8993 mutation-homoplasmic). Bar graphs represent the mean of at least four separate determinations for each cell line for each substrate.

1. Wash 2×10^7 lymphoblasts with 10 ml 0.15 M sucrose, 1 mM EDTA, 20 mM MOPS, 1 g/liter bovine serum albumin (fatty acid-free BSA) serially by centrifugation three times at 2000 rpm for 5 min at room temperature.
2. Suspend washed lymphoblasts in the above medium plus 0.1 mg/ml digitonin for 3 min at 4°.
3. Disrupt cells with 30 strokes of a glass/Teflon homogenizer.
4. Centrifuge twice at 2500 g for 5 min and centrifuge the resulting supernatant a further 10 min at 10,000 g.
5. Suspend the crude mitochondrial pellet in 0.5 M sucrose, 20 mM MOPS, 1 mM EDTA, pH 7.4 at 4°.

Assay of ATP Synthesis

1. Add 30 μg (protein) fresh lymphoblast mitochondria to 200 μl of a solution containing 0.25 M sucrose, 2 mM MOPS, 1 mM EDTA, 5 mM potassium phosphate, and 1 mM ADP at pH 7.4.
2. Set up the above in 1-ml microcentrifuge tubes (a) with no added substrate, (b) with 5 mM pyruvate and 0.1 mM L-malate, (c) with 5 mM glutamate and 0.1 mM L-malate, (d) with 5 mM succinate and 1 μM rotenone, and (e) with 2 mM ascorbate and 0.1 mM TMPD.
3. Incubate tubes at 37° for 1 hr, then add 10 μl of 1.6 M perchloric acid to stop the reaction.
4. Centrifuge the tubes to remove precipitated protein.
5. Use 100 μl aliquots of the resulting solution to assay for ATP content by fluorimetric assay.[19]

The results using lymphoblast mitochondria are comparable to those obtained with cultured skin fibroblasts treated with digitonin. This technique has been used successfully to demonstrate ATP synthetic defects in mtDNA-encoded defects leading to Lebers hereditary optic neuropathy and in the mtDNA defect at nucleotide 9997 leading to hypertrophic cardiomyopathy.[22,23]

[22] A. Majander, K. Kiopone, M.-L. Savontaus, E. Nikoskelainen, and M. Wikstrom, *FEBS Lett.* **292,** 289 (1991).
[23] F. Merante, I. Tein, L. Benson, and B. H. Robinson, *Am. J. Hum. Genet.* **55,** 437 (1994).

[40] Use of Myoblast Cultures to Study Mitochondrial Myopathies

By ERIC A. SHOUBRIDGE, TIM JOHNS, and LOUISE BOULET

Introduction

Mutations in mitochondrial DNA (mtDNA) that affect mitochondrial protein translation (large-scale deletions, tRNA point mutations) are associated with many of the clinical entities classified as mitochondrial encephalomyopathies.[1] Skeletal muscle fibers in patients with these diseases contain heteroplasmic mixtures of wild-type and mutant mtDNAs, inhomogeneously distributed along their length. The relative proportion of each is related to the severity of the biochemical defect and the clinical phenotype. Primary muscle cultures from these patients are valuable tools with which to study the phenotypic expression of pathogenic mtDNA mutations, particularly since methods to create specific mtDNA mutations in mammalian cells have not been established, nor have immortal human muscle cell lines been isolated.

Skeletal muscle fibers themselves are postmitotic, multinuclear syncytia that cannot be grown in tissue culture. They are, however, surrounded by mononuclear, undifferentiated myoblasts called satellite cells that will proliferate in culture. Satellite cells remain dormant in a peripheral (satellite) position between the muscle fiber plasma membrane and the basal lamina during most of postnatal life. They are stimulated to reenter the cell cycle following muscle injury, where they function to repair damaged myofibers; they also play a role in muscle growth, maintaining a relatively constant nuclear/cytoplasmic ratio in the mature myofiber. Satellite cells can be dissociated from muscle biopsy specimens with proteolytic enzymes or alternatively allowed to grow out of muscle explants in culture.

Although a number of different cell types are often available to study the expression of pathogenic mtDNA mutations (fibroblasts, lymphocytes, cybrids), two important advantages are gained by using primary myoblast cultures. First, myoblasts can be induced to fuse in culture to form multinucleate myotubes that terminally differentiate and contract, a process which mimics normal muscle development. It is thus possible to study the effects of growth versus terminal differentiation on the expression of mtDNA

---

[1] G. Karpati and E. A. Shoubridge, *Curr. Neurol.* **13**, 133 (1993).

mutations.[2] One can also determine the threshold for phenotypic expression of pathogenic mtDNA mutants, as the degree of heteroplasmy in individual myotubes can be manipulated by seeding cultures with myoblasts containing different proportions of mutant and wild-type mtDNAs. Second, because satellite cells are dormant most of the time, it is likely that the turnover of mitochondria and mtDNA is extremely slow as compared to that in differentiated muscle fibers. Thus, the relative proportions of mutant and wild-type mtDNAs in the satellite cell population likely reflects the degree of heteroplasmy that existed in the myoblast precursor population during embryological development. A comparison with adult muscle can provide information on the segregation of mtDNAs subsequent to terminal differentiation.[2]

In this chapter we detail the procedures we have used for (i) direct cloning of primary human muscle satellite cells, (ii) immunofluorescent staining of myoblasts for fluorescence-activated cell sorting (FACS), and (iii) analysis of mitochondrial translation products in myoblasts and myotubes. The culture methods we have used are based on the pioneering work carried out in the laboratories of Blau and Ham.[3,4]

Materials

Solutions and Reagents for Cell Culture

Solution A: 10 mM glucose, 30 mM HEPES, 130 mM NaCl, 3.0 mM KCl, 0.0033 mM phenol red; make up in Milli-Q water, adjust to pH 7.6 with NaOH, and sterile filter

Solution A plus 0.05% (w/v) crystalline trypsin (GIBCO, Grand Island, NY) and 0.02% (w/v) EDTA; sterile filter and store frozen in aliquots at $-20°$ for up to 1 month

Phosphate-buffered saline, Mg^{2+}- and Ca^{2+}-free (PBS): 8 g NaCl, 0.2 g KCl, 1.15 g $Na_2H PO_4$, 0.2 g KH_2PO_4; dissolve in 1 liter Milli-Q water and autoclave

PBS plus 0.05% (w/v) crystalline trypsin (GIBCO) and 0.02% (w/v) EDTA; sterile filter and store at 4° for 1 week

PBS plus 0.5% bovine serum albumin (BSA); sterile filter and store at 4°

Dexamethasone (Sigma, St. Louis, MO; 0.039 mg/ml); dissolve in Milli-Q water and freeze at $-20°$ in 10-ml aliquots

[2] L. Boulet, G. Karpati, and E. A. Shoubridge, *Am. J. Hum. Genet.* **51,** 1187 (1992).
[3] H. M. Blau and C. Webster, *Proc. Natl. Acad. Sci. U.S.A.* **78,** 5623 (1981).
[4] R. G. Ham, J. A. St. Clair, C. Webster, and H. M. Blau, *In Vitro Cell. Dev. Biol.* **24,** 833 (1988).

Epidermal growth factor, human recombinant (EGF, Collaborative Research 40052, Bedford, MA), stock of 10 μg/ml; dissolve in Milli-Q water and freeze in 1-ml aliquots

5.1H11 mouse monoclonal antibody (anti-N-CAM): this antibody is not commercially available to our knowledge (Dr. F. Walsh, Guy's Hospital, London, UK, kindly made the hybridoma cells available to us); the undiluted hybridoma supernatant is used to label myoblasts

Fluorescein-conjugated F(ab')$_2$ fragment, rabbit anti-mouse immunoglobulin G (IgG) (fluorescein isothiocyanate, FITC) (Cappel 13110082); dilute 1:100 in PBS plus 0.05% BSA

Cell Culture Media

Supplemented growth medium (SGM)[4] is Ham's F10 medium (GLUTAMAX, GIBCO) with the following supplements added to 1 liter of liquid: 500 mg bovine serum albumin (Sigma; 0.5 mg/ml, final concentration), 500 mg Pedersen fetuin (Sigma; 0.5 mg/ml, final concentration), 180 mg insulin (Sigma; 0.18 mg/ml, final concentration), 10 ml of stock solution of dexamethasone (Sigma; 0.39 μg/ml, final concentration), and 1.0 ml of stock solution of EGF (10.0 ng/ml, final concentration); stir at 37° for at least 1 hr and sterile filter, then add 15% fetal calf serum (FCS) (heat inactivated, 30 min at 56°) and gentamicin (GIBCO; 2.5 μg/ml)

Fusion medium: Dulbecco's modified Eagle's Medium (DMEM) plus 2% horse serum (heat inactivated) and 2.5 μg/ml gentamicin

Freezing medium: Ham's F10 medium plus 20% FCS and 10% dimethyl sulfoxide (DMSO)

Washing medium: DMEM or Ham's F10 plus 5% newborn calf serum and 2.5 μg/ml gentamicin

Notes on Culture Media. Ham has reported consistently better clonal growth of human muscle satellite cells in MCDB 120 medium.[4] At the time we started our work this medium was not readily obtainable. It is now available commercially from Clonetics (San Diego, CA). The active component in fetuin is likely a trace contaminant, as the biological activity can be separated from purified Pedersen fetuin.[5] It is likely that a trace contaminant also contributes to the effect of insulin, because the concentration is much higher than physiological.[4] Human satellite cells can be successfully cloned in serum-free MCDB 120 containing the five supplements listed above; however, growth is superior in medium that is also supplemented with 15% FCS.[4]

[5] Z. Nie, D. Jellinek, and R. G. Ham, *Biochem. Biophys. Res. Commun.* **178,** 959 (1991).

Procedures

Direct Cloning of Human Muscle Satellite Cells

1. Obtain skeletal muscle (~50–100 mg) aseptically from a biopsy specimen and place in a 15-ml sterile tube containing 10 ml solution A. Hanks' balanced salt solution or other sterile saline solutions can also be used. Place on ice immediately and store at 4°. The muscle can be kept for up to 5 days without significant loss of proliferative capacity of the satellite cells, although we usually proceed with the culture within 1 or 2 days.

2. Transfer the muscle sample to a preweighed 60-mm tissue culture dish and rinse with 5 ml solution A. Cut the biopsy into small pieces (1 mm^2) with a pair of small, sterile scissors in 5 ml, solution A plus trypsin and EDTA. Transfer the minced tissue to a Wheaton (Millville, NJ) trypsinizing flask containing a magnetic stirring bar. Rinse the culture dish twice more with 5 ml solution A plus trypsin and EDTA, each time transferring the tissue fragments and solution to the trypsinizing flask.

3. Stir the tissue pieces vigorously for 15–20 min on a stirring plate at 37°. This is accomplished by placing a water bath (a Pyrex dish is suitable) on top of the magnetic stirrer in a laminar flow hood. Allow the tissue debris to settle for about 1 min and decant the solution containing the dissociated cells into a 50-ml centrifuge tube containing an equal volume (15 ml) of washing medium.

4. Repeat step 3 two more times (three times in all), each time adding 15 ml solution A plus trypsin and EDTA to the remaining tissue in the trypsinizing flask.

5. Spin the cells at approximately 500 g (~1450 rpm on a Beckman GP benchtop centrifuge, Mississauga, Ont.) for 10 min and discard the supernatant. Resuspend the pellet in 2 ml SGM. Pool the cell suspensions from each dissociation.

6. The suspension of dissociated cells will contain many erythrocytes and small tissue fragments. To clone myoblasts directly it is necessary to estimate the number of satellite cells that were present in the original biopsy material because they cannot be easily counted in the suspension. We have used an estimate of 500 satellite cells/mg wet weight of skeletal muscle. Accordingly, dilute the cell suspension and plate in 0.2 ml growth medium in 96-well plates (Costar, Cambridge, MA). Because the number of satellite cells in the sample is a rough estimate, we routinely dilute the original sample to theoretically 2.5, 5, and 10 cells/ml, which would give 0.5, 1, or 2 cells per well on a 96-well plate. We set up duplicate plates at each dilution, which in practice usually allows us to isolate 40–50 clones. It is also possible to plate out the cells at this stage at very low dilution and to rescue colonies subsequently using cloning rings; however, we have

found cloning by limiting dilution to be a straightforward and simple procedure. Plate the remaining cells in a 60-mm tissue culture dish in 5 ml SGM. This produces a bulk culture that can be sorted by FACS to eliminate contaminating fibroblasts or recloned if necessary. Cells are incubated in a humidified CO_2 incubator at 37°.

7. Single cells should be visible 3–4 days after plating using an inverted phase-contrast microscope. It is usually simpler to wait a few more days to score wells containing clones, which are then observable as small colonies. Wells that were seeded with more than one cell are usually easy to score because more than one focus is present. The medium in positive wells should be changed every week. After 1–3 weeks, when clones are about two-thirds confluent, transfer the colonies to 24-well plates as follows. Aspirate the medium and add a drop of PBS plus trypsin and EDTA. Place the plate at 37° for a couple of minutes until the cells round up. Using a Pipetman, disperse the cells and transfer to a well containing 1.5 ml SGM.

8. When the clones are again two-thirds confluent in the 24-well plate, aspirate the SGM, wash with 0.5 ml PBS, then add 30 μl PBS plus trypsin and EDTA to each well. Aspirate the trypsin solution and place plates in the incubator for a couple of minutes at 37° until the cells round up, then disperse with a pipette. At this stage it is useful to test for fusion competence of the clones. We usually transfer two-thirds of the cells to a 60-mm tissue culture dish in 5 ml SGM to expand the clone and one-third of the cells to a 24-well plate in 1.5 ml SGM for the fusion test.

9. Myoblasts rarely fuse in SGM unless it becomes depleted of growth factors; therefore, it is necessary to test the fusion competence in depleted (fusion) medium. Grow the cells that were transferred to the 24-well plate to confluence and replace SGM with fusion medium. Score for the presence of multinucleate myotubes 2–3 days later. Figure 1 shows a phase-contrast micrograph of a clonal myoblast culture and extensive myoblast fusion in the same culture 48 hr after switching to fusion medium. We have observed large differences in the degree of fusion among different clones obtained from the same biopsy, and we routinely note which clones are fusing well at this stage for use in later experiments. It is also possible to fix the cells and stain the nuclei of these cells to score fusion if a quantitative index of fusion is desired. In our experience less than 10% of clones do not show any evidence of cell fusion, and these are scored as fibroblasts.

10. The final step involves freezing and recovery of myoblasts. After the cells have reached about two-thirds confluence on the 60-mm plate they can be frozen. (It is also possible to freeze them earlier if a small number of cells at early passage is required.) Trypsinize the myoblasts using about 1 ml PBS plus trypsin and EDTA. Aspirate the trypsin solution and place the cells in the incubator until the cells start to round up. Stop the

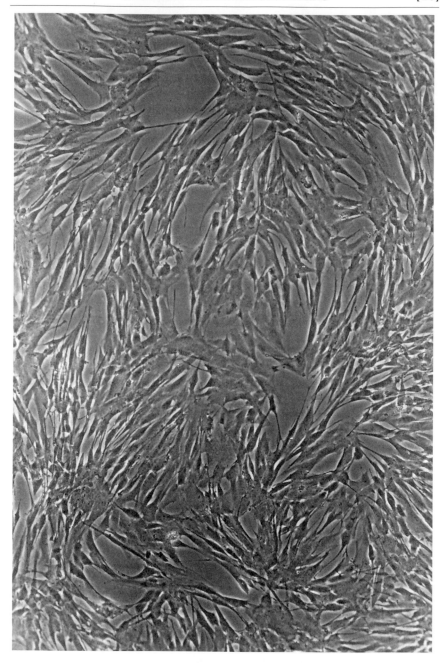

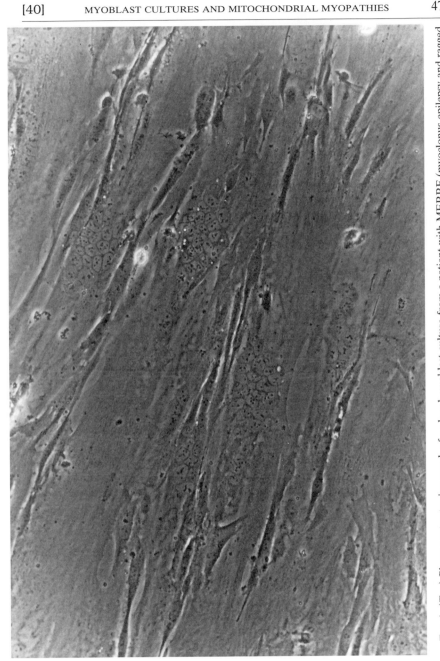

FIG. 1. (*Top*) Phase-contrast micrograph of a clonal myoblast culture from a patient with MERRF (myoclonus epilepsy and ragged red fibers) and the A to G mutation at position 8344 in tRNA^{Lys}. (*Bottom*) Same culture 48 hr after replacement of SGM with fusion medium. Most myoblasts have fused to form multinucleate myotubes. This clone was homoplasmic for the tRNA^{8344} mutation.

reaction with 5 ml GM, then rinse the plate and disperse the cells with an additional 5 ml SGM. Spin the cells at approximately 500 g and gently suspend in 1.5 ml of cold freezing medium. Transfer to freezing vials (Nunc) and place in a foam rack at −70° overnight. After 24 hr the cells can be transferred to a liquid nitrogen freezer. To thaw myoblasts, quickly warm the frozen vial in a 37° water bath until the contents have almost thawed. Immerse the vial in 70% (v/v) ethanol to sterilize and transfer the entire contents to a tissue culture plate containing SGM. After the cells have plated down (3–4 hr or overnight) change the medium to eliminate the DMSO.

Fluorescence-Activated Cell Sorting of Human Satellite Cells

It is sometimes useful to have a large population of pure skeletal myoblasts from a biopsy sample at an early passage. In this case purifying myoblasts using a FACS machine is a much more efficient alternative to cloning. This technique is based on the use of a mouse monoclonal antibody (5.1H11) raised against human myotubes and subsequently identified as anti-N-CAM (neural cell adhesion molecule).[6,7] N-CAM is present on the surface of primary human satellite cells, but not on skeletal fibroblasts that contaminate primary muscle cell cultures. The method we have used is similar to that described by Webster *et al.*[6] Below we describe our method for immunostaining myoblasts with fluorescein for sorting. All procedures are performed using sterile technique.

1. Expand the cells in the bulk myoblast culture until there are 4–5 × 10^6 cells. Although fewer cells can be used, it is better to start with several million because there are unavoidable losses during the sorting procedure. Dissociate the cells in PBS plus trypsin and EDTA and resuspend in SGM. Centrifuge at 4° for 10 min at 11,000 rpm in a microcentrifuge and discard the supernatant.

2. Transfer the cells to a sterile, 1.5-ml Eppendorf tube at 4 × 10^6 cells per tube and suspend in 1.0 ml PBS plus 0.5% BSA. Centrifuge 1 min (setting 4 on a Fisher microcentrifuge) at room temperature and discard the supernatant.

3. Suspend the cells as above and add 100 μl anti-NCAM antibody (undiluted supernatant from hybridoma cells). Mix and incubate on ice in the dark for 45 min, mixing every 15 min.

4. Wash the cells 3 times with PBS plus 0.5% BSA. Centrifuge for 1 min at setting 4 on the microcentrifuge between each wash.

[6] C. Webster, G. K. Pavlath, D. R. Parks, F. S. Walsh, and H. M. Blau, *Exp. Cell Res.* **174,** 252 (1988).
[7] F. S. Walsh, *Adv. Exp. Med. Biol.* **280,** 41 (1990).

5. After the last wash suspend the cell pellet in 1.0 ml PBS plus 0.5% BSA and add 100 μl FITC (diluted 1 : 100), mix, and incubate on ice in the dark for 45 min.

6. Wash the cells 3 times with PBS plus 0.5% BSA and suspend in SGM at a concentration of 4–5 $\times$ 10^6 cells/ml for sorting.

7. The details of the actual sorting depend on the type of FACS machine available and are best dealt with by reference to the original paper[6] and consultation with an experienced operator of a FACS machine. We have used a single laser machine (Becton-Dickinson, FACSTAR). Clonal cultures of myoblasts and skeletal muscle fibroblasts are useful to establish parameters which allow sorting of essentially pure (>99.5%) myoblast populations. The negative cells (skeletal muscle fibroblasts) can be collected at the same time and used to establish a fibroblast culture.

Analysis of Mitochondrial Translation Products

Most reported mitochondrial myopathies are associated with tRNA point mutations or large-scale deletions (usually involving several protein-coding and tRNA genes). Both types of mutations would be predicted to affect the translation of mtDNA-encoded proteins. The rate of translation of the 13 proteins encoded in mtDNA can be readily investigated by labeling myoblasts or myotubes with [^{35}S]methionine in the presence of emetine, an inhibitor of cytoplasmic translation. The specificity of the labeling can be tested by adding chloramphenicol, which inhibits all mitochondrial translation, to the labeling mixture. The method we have used is based on that developed by Attardi and Ching.[8]

Solutions for Labeling Translation Products

DMEM without methionine (GIBCO)
Emetine (Sigma), stock solution of 2 mg/ml; make up fresh and filter sterilize
Chloramphenicol (Sigma), stock solution of 2 mg/ml; make up fresh and filter sterilize
[^{35}S]methionine (specific activity >1000 Ci/mmol)

Procedure

1. Grow myoblasts to subconfluence in a 60-mm tissue culture dish. For myotubes, switch the medium to fusion medium when the cells have become confluent and label 48–72 hr later.

---

[8] G. Attardi and E. Ching, this series, Vol. 56, p. 66.

2. Wash the cells twice with PBS and incubate in 2 ml DMEM without methionine for 30 min. Add 100 μl emetine (final concentration 100 μg/ml) and incubate for a further 5 min. Add chloramphenicol (final concentration 100 μg/ml) plus emetine to a parallel, negative control plate. All incubations are done in a humidified CO_2 incubator at 37°.

3. Add 200–400 μCi [^{35}S]methionine to each plate and incubate for up to 1 hr. Aspirate the medium, replace with regular DMEM, and incubate for 10 min.

4. Wash the cells gently three times in PBS, trypsinize, and pellet the cells in excess PBS on a benchtop centrifuge at approximately 1000 g at room temperature. Note that a brief trypsinization period permits isolation of essentially pure myotubes, free of any unfused mononuclear myoblasts, which are more tightly attached to tissue culture plastic.

5. Suspend the cells in 200 μl PBS and determine the protein concentration. Pellet cells (30–50 μg protein) at top speed in a microcentrifuge and suspend in 10 μl water plus 10 μl gel loading solution [186 mM Tris-HCl, pH 6.7, 15% glycerol, 7% sodium dodecyl sulfate (SDS), 6% mercaptoethanol, 0.005% bromphenol blue]. Sonicate the suspension for 3 sec (60% on a

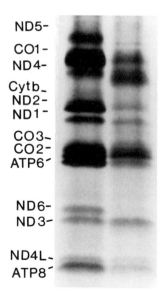

FIG. 2. Analysis of mitochondrial translation products in myotubes derived from either control myoblasts (left-hand lane) or myoblasts homoplasmic for the A to G mutation at position 8344 in tRNA^{Lys} (right-hand lane), demonstrating the translation defect associated with this mutation. Mitochondrial translation products were assigned as in Chomyn et al.[10] A discussion of the nature of the translation defect and the threshold for expression of this mutation can be found in Boulet et al.[2]

VibraCell sonicator, Sonics and Materials Inc., Danbury, CT) and spin at top speed in a microcentrifuge for around 10 min.

6. Run the samples on a 12–20% Laemmli[9] gradient gel. Gels can be processed for fluorography for analysis on X-ray film or direct autoradiography for analysis on a phosphorimager (Molecular Dynamics) or similar device.[2] Figure 2 shows an analysis of the mitochondrial translation products in myotubes from a control and from a patient with the A to G mutation at position 8344 in tRNA^{Lys}.[10]

Additional Comments

The proliferative potential of myoblasts varies with age. It is reported that one can obtain about 60 cell divisions from a fetal myoblast clone (1.1×10^{18} cells), but only 20 (1×10^6 cells) from a 50-year-old subject.[6] It is, however, important to note that fusion potential declines dramatically as the cells approach senescence, so if myoblast fusion is essential to the experiment, the cells have to be used at an early passage. It is probably wise to limit the number of cell divisions to about half the estimated lifetime. In practice this can be difficult to estimate since individual myoblast clones from the same patient can vary greatly in proliferative capacity and fusion competence. It is important to score this as clones are characterized and to perform the experiments after as few population doublings as possible. The fusion medium we have used does not permit good survival of myotubes beyond 72 hr. Long-term survival (up to 3 weeks) can be promoted by addition of insulin (10 μg/ml), EGF (10 ng/ml), and BSA (0.5 mg/ml) to the fusion medium.[11]

Acknowledgments

This research was supported by grants from the Medical Research Council of Canada and the Muscular Dystrophy Association of Canada to E. A. S. We thank K. Pshyk and K. McDonald for technical assistance with myoblast staining and FACS analysis.

[9] U. K. Laemmli, *Nature* (*London*) **227**, 680 (1970).
[10] A. Chomyn, G. Meola, N. Bresolin, S. T. Lai, G. Scarlato, and G. Attardi, *Mol. Cell. Biol.* **11**, 2236 (1991).
[11] J. A. St. Clair, S. D. Meyer-Demarest, and R. G. Ham, *Muscle Nerve* **15**, 774 (1992).

[41] Use of Polarography to Detect Respiration Defects in Cell Cultures

By Götz Hofhaus, Rebecca M. Shakeley, and Giuseppe Attardi

Introduction

The rapidly expanding identification of bioenergetic dysfunctions associated with diseases or aging and the development of patient-derived cultured cell systems have greatly increased the demand for methods suitable for testing the oxidative phosphorylation capacity of intact cells or cells permeabilized with digitonin.[1-5] The latter cells, in which the mitochondria and cytoskeleton organization of the cell are substantially preserved, and in which it is possible to test a variety of respiratory substrates, cofactors, and inhibitors, have proved to be particularly useful for identifying the step in the oxidative phosphorylation process that is affected in a particular bioenergetic disorder. In this chapter, we describe the methods used in our laboratory for testing the respiratory capacity of digitonin-permeabilized cultured human cells.

Rationale

Digitonin, by binding to cholesterol in the eukaryotic plasma membrane,[6] creates in the latter pores through which the soluble components of the cell can be released.[1-5] Because the intracellular membranes have a cholesterol content substantially lower than the plasma membrane,[7] treatment of cells with low concentrations of the detergent permeabilizes selectively the plasma membrane, leaving the mitochondria, other cell organelles, and the cytoskeleton substantially intact.[2] In particular, as the outer mito-

---

[1] W. P. Dubinsky and R. S. Cockrell, *FEBS Lett.* **59,** 39 (1975).
[2] G. Fiskum, S. W. Craig, G. L. Decker, and A. L. Lehninger, *Proc. Natl. Acad. Sci. U.S.A.* **77,** 3430 (1980).
[3] S. I. Harris, R. S. Balaban, L. Barrett, and L. J. Mandel, *J. Biol. Chem.* **256,** 10319 (1981).
[4] L. D. Granger and A. L. Lehninger, *J. Cell Biol.* **95,** 527 (1982).
[5] W. G. E. J. Schoonen, A. Handayani Wanamarta, J. M. van der Klei-van Moorsel, C. Jakobs, and H. Joenje, *J. Biol. Chem.* **265,** 11118 (1990).
[6] T. J. Scallen and A. E. Dietert, *J. Cell Biol.* **40,** 802 (1969).
[7] A. Colbeau, J. Nachbaur, and P. A. Vignais, *Biochim. Biophys. Acta* **249,** 462 (1971).

chondrial membrane has a relatively low cholesterol content and the inner membrane lacks cholesterol,[7] the mitochondria can be functionally preserved. Therfore, if the permeabilized cells are suspended in buffer containing ADP, phosphate, and a respiratory substrate, they can carry out state 3 respiration, and the O_2 consumption measured in an oxygraph with a Clark electrode will reflect the activity of the respiratory enzyme(s).[3–5,8,9]

Figure 1 shows typical polarographic patterns obtained with digitonin-permeabilized cells of the human cell lines 143B.TK$^-$ and VA$_2$B. With malate or glutamate as a substrate, the corresponding dehydrogenase generates NADH, which is oxidized in three discrete steps by the NADH–ubiquinone oxidoreductase [EC 1.6.99.3 NADH dehydrogenase] or respiratory NADH dehydrogenase (Complex I), the ubiquinol–cytochrome-c oxidoreductase (Complex III), and the cytochrome-c oxidase (Complex IV). The overall reaction is measured by the O_2 consumption in the terminal step:

$$\text{NADH} \xrightarrow{\text{Complex I}} \text{ubiquinone} \xrightarrow{\text{Complex III}} \text{cytochrome } c \xrightarrow{\text{Complex IV}} O_2$$

Rotenone, a specific inhibitor of Complex I, will inhibit completely the malate- or glutamate-dependent respiration. Using succinate or glycerol 3-phosphate (G-3-P) as a substrate, the corresponding dehydrogenase will feed electrons directly into the ubiquinone pool, bypassing Complex I. Antimycin, a specific inhibitor of Complex III, will inhibit completely the succinate- or G-3-P-dependent respiration. Using N,N,N',N'-tetramethyl-p-phenylenediamine (TMPD) and ascorbate, the latter reduces TMPD to transfer electrons to the endogenous cytochrome c, and this is then oxidized by Complex IV. This reaction is inhibited by KCN, an inhibitor of Complex IV.

It is clear from what is explained above that, in digitonin-permeabilized cells, the malate- or glutamate-dependent respiration depends on the activities of Complex I, Complex III, and Complex IV, the succinate or G-3-P-dependent respiration depends on the activities of Complex III and Complex IV, whereas the TMPD/ascorbate-dependent respiration depends only on Complex IV activity. However, one must exclude a limitation in the transport of the chosen substrate into the mitochondria or in the activity of the corresponding dehydrogenase. In wild-type cells, Complex I activity is usually rate limiting in respiration, whereas, when Complex I is bypassed, Complex III is rate limiting.

[8] G. Hofhaus and G. Attardi, *EMBO J.* **12,** 3043 (1993).
[9] G. Hofhaus and G. Attardi, *Mol. Cell. Biol.* **15,** 964 (1995).

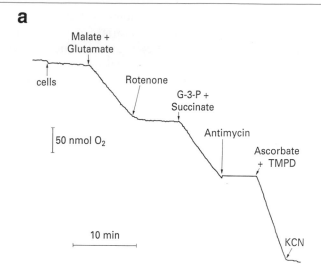

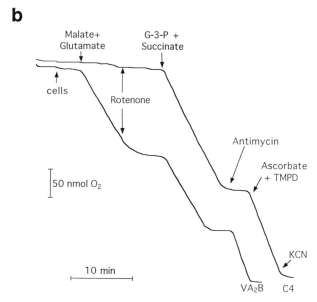

FIG. 1. Oxygen consumption by digitonin-permeabilized 143B.TK⁻ cells (a) and VA₂B cells (b), as well as NADH-dependent-respiration defective C4 cells (b). (b) Reproduced from G. Hofhaus and G. Attardi, *EMBO J.* **8,** 3043 (1993), by permission of Oxford University Press.

Materials

Equipment

Oxygen consumption is measured with a Clark electrode in a water-jacketed chamber connected to a circulating water bath. We use a Gilson 5/6 Oxygraph (Gilson Medical Electronics, Inc., Middleton, WI) or a YSI Model 5300 Biological Oxygen Monitor (Yellow Springs Instrument Co., Yellow Springs, OH), with a 1.50- or 1.85-ml chamber from Gilson Medical Electronics. The chamber is sealed with a stopper containing a capillary port for additions. To introduce substrates and inhibitors into the chamber, Hamilton syringes fitted with 2.5-inch-long needles (Hamilton Co., Reno, NV), are used. *Note:* The recommended length of the needles is especially important for dispensing near the bottom of the chamber, just above the magnetic stirrer, the ethanol solutions of rotenone and antimycin: these solutions, in fact, because of their lower specific density as compared to the respiration medium, tend to float along the needle to the surface; for the same reason, it is advisable to "shoot" the solutions into the chamber.

Media and Reagents

Control experiments failed to show any significant difference in polarographic data after adjustment of the pH of the stock solutions of substrates with NaOH or KOH.

Media

Medium A: 20 mM HEPES (adjusted to pH 7.1 with NaOH or KOH) 250 mM sucrose, 10 mM MgCl$_2$; store in aliquots at $-20°$

Respiration medium: medium A, 1 mM ADP, 2 mM potassium phosphate (from a 0.5 M stock, pH 7.1); to be prepared fresh

10% (w/v) Bovine serum albumin (BSA) solution (in distilled water); store at $-20°$

Reagents

10% Digitonin in dimethyl sulfoxide (digitonin should be recrystallized from an ethanol solution[10]); store in aliquots at $-20°$

Substrate stock solutions (adjusted to pH ~7.0 with NaOH or KOH) and stored in aliquots at $-20°$, unless otherwise indicated:

1 M Glutamate
1 M Malate
1 M Glycerol 3-phosphate

---

[10] E. Kun, E. Kirsten, and W. N. Piper, this series, Vol. 55, p. 115.

1 *M* Succinate

1 *M* Ascorbate

40 m*M* TMPD (Fluka Chem. Corp., Ronkonkoma, NY): prepare fresh
in medium A and protect from light; a slight blue color, indicative
of autoxidation, may develop over a few hours, but this change will
not affect results

Inhibitor concentrated solutions (stored at $-20°$, unless otherwise specified):

2 m*M* Rotenone (Sigma, St. Louis, MO) in absolute ethanol

4 m*M* Antimycin A (Sigma) in absolute ethanol

200 m*M* KCN in medium A (prepared fresh and adjusted to pH 8)

Inhibitor working solutions:

100 μ*M* in absolute ethanol

10 μ*M* in absolute ethanol

Polarographic Assays

Calibration of Instrument

Prior to the assay, start the circulating water bath (set at $37°$), the
chamber stirrer, the oxygraph, and the chart recorder. Remove from the
chamber the half-saturated KCl solution, which is normally kept in it under
nonoperating conditions, rinse the chamber several times with distilled
water, and then fill it with distilled water that has been prewarmed to $37°$
in a flask in a water bath under frequent shaking, in order to equilibrate
it with O_2 (the concentration of O_2 in distilled water under saturating
conditions at $37°$ is 217 nmol/ml). Avoid introducing air bubbles. Using
the oxygraph sensitivity control, set the maximum reading on the chart,
and wait for several minutes until the reading is stabilized. Then add a few
grams of sodium dithionite to set the zero O_2 point. The chamber is then
rinsed again several times with distilled water and filled with O_2-saturated
distilled water, which should bring the pen back to the original maximum reading.

Cell Permeabilization

While the above operations are being performed, one can start the
collection of the cells. With large-size cells, like HeLa, 143B.TK⁻ [parent
of mtDNA-less ($\rho°$) 206 cells], $\rho°$206, transmitochondrial cell lines derived
therefrom,[11] and fibroblasts, 5×10^6 cells are sufficient; with smaller cells,

---

[11] M. P. King and G. Attardi, this volume [28].

like lymphoblasts or mouse L cells, use $1-2 \times 10^7$ cells. It is advisable to change the medium of the cultures the day before the measurements, and to remove from the medium any metabolic inhibitor for at least 24 hr, unless its effects on respiration need to be studied.

The optimum conditions for cell permeabilization by digitonin depend on both the absolute concentration of the detergent and the ratio of detergent to membranes, and may vary for different cell lines. Accordingly, for each cell line, one must define the conditions which, on one hand, allow a complete release of the soluble components of the cell and, on the other, do not affect the respiratory function of mitochondria. In the latter connection, it is critical that the outer mitochondrial membrane not be damaged, to avoid any possible loss of cytochrome c from the intermembrane space. To establish the optimum conditions of digitonin treatment with the cell line under investigation, one should use the trypan blue exclusion test, which discriminates between permeabilized (stained) and nonpermeabilized cells, and measure the O_2 consumption of the permeabilized cells in the absence or presence of external substrates. For the trypan blue exclusion test, a 0.4% (w/v) solution of trypan blue in phosphate-buffered saline or equivalent medium is mixed in the proportion of $1:10$ with a portion of the cell suspension, and, after 5 min, the suspension is examined in a hemocytometer under the microscope. A useful test for defining the range of concentrations of digitonin per milligram of cellular protein that yields optimal state 3 respiration rates is the measurement of O_2 consumption in the presence of ADP, P_i, and either malate and glutamate, or succinate and G-3-P (in the latter cases, after addition of rotenone)[3,4] in cells permeabilized with different concentrations of digitonin. The digitonin permeabilization conditions described below have been optimized for 143B.TK$^-$ and VA$_2$B cells. For this purpose, we have adopted the criteria of absence of O_2 consumption in the absence of substrates and maximal ascorbate/TMPD-supported cytochrome-c oxidase activity after full permeabilization of the cells to trypan blue and their full depletion of substrates. The appropriate amount of cells are trypsinized (if grown on solid substrate), collected by centrifugation in a 15-ml capped Corning polypropylene tube, and thoroughly suspended in 1 ml medium A. After addition of 1 ml of the same medium containing digitonin at a concentration of 100 μg/ml, the tube is placed in a table rotator or gently hand-shaken at room temperature. After 1 min exposure to digitonin, the cell suspension is diluted with 8 ml medium A, and the cells are collected by centrifugation at $350g_{av}$ for 3 min on a tabletop International (Needham Heights, MA) clinical centrifuge. The pellet is suspended in 10 ml medium A, and the suspension, after removal of a small sample for trypan blue exclusion test is centrifuged under the same conditions. During this time, the 1.50 (1.85) ml oxygraph chamber is emptied and refilled with 1.45 (1.80) ml air-equilibrated respiration medium pre-warmed to 37°. The final cell pellet is suspended in 100 μl of the same medium, and the suspension is introduced into the chamber. After 1 min of stirring, two 10-μl aliquots are removed and introduced into the two chambers of a hemocytometer for cell counting, and the chamber is stoppered, taking care to remove any air bubbles. While the cells are being counted and checked for trypan blue staining, the temperature of the respiration buffer in the chamber becomes equilibrated. If the cells have been fully depleted of substrates, the baseline should give a constant reading or nearly so. In particular, in the absence of exogenous substrates, O_2 consumption rate should be less than 0.15 fmol/min/cell.

Measurements

The substrates and inhibitors are introduced into the chamber through the port of the stopper, using separate Hamilton syringes fitted with a needle long enough to allow the deposition of the reagent solution near the bottom of the chamber. The amounts of reagent indicated below are for a 1.5-ml chamber. First, add 7.5 μl each of the 1 M glutamate and 1 M malate stock solutions (final concentration in the chamber, ~5 mM of each substrate) (see Note 1). After adequate time for accurate slope measurement, add 1.5 μl of the 100 μM rotenone solution (final concentration ~0.1 μM), to inhibit Complex I.

After maximal inhibition of O_2 consumption has been reached, add 7.5 μl each of the 1 M G-3-P and 1 M succinate stock solutions (final concentration ~5mM of each substrate) (see Note 1), and, after appropriate time, 3 μl of 10 μM antimycin (final concentration ~20 nM), to inhibit Complex III (see Note 2). After maximal inhibition of respiration has been obtained, add 15 μl of 1 M ascorbate stock solution (final concentration ~10 mM), and immediately thereafter 7.5 μl of 40 mM TMPD (final concentration ~200 μM). Finally, after an appropriate time, add 7.5 μl of 200 mM KCN solution (final concentration ~0.1 mM), to inhibit complex IV. From the measurements described above the rates of malate plus glutamate-depen-

TABLE I

	O_2 consumption (fmol/min/cell)		
Substrate	HeLa cells	143B.TK$^-$ cells	Fibroblasts
Malate/glutamate	2.4	2.5	1.3
Succinate/G-3-P	3.0	3.4	2.0
Ascorbate/TMPD	6.7	6.5	4.2

dent, succinate plus G-3-P-dependent, and ascorbate plus TMPD-dependent respiration are calculated by subtraction of the rotenone-resistant, antimycin-resistant, and KCN-resistant O_2 consumption, respectively.

Note 1: To ensure that the transport of a given substrate into mitochondria or the activity of the corresponding dehydrogenase is not limiting, it is desirable to use simultaneously two substrates feeding electrons into Complex I (e.g., malate and glutamate) and two substrates feeding electrons into Complex III (succinate and G-3-P).

Note 2: The inhibitors used in the polarographic assays are very powerful and tend (especially rotenone and antimycin) to stick to the Hamilton syringes and to the glass and membrane of the measurement chamber; therefore, inhibitor residues can affect subsequent measurements. Accordingly, one should use a separate syringe for each inhibitor. After use, both the syringes and the chamber should be rinsed carefully with ethanol, and then water. It is also advisable to wash the chamber with a 10% BSA solution to adsorb any residual rotenone and antimycin, and then extensively with water.

Results

Table I gives representative values of O_2 consumption with different substrates by cultured human cells, obtained using the protocol described above. With the cultured cells listed, the respiration supported by malate/glutamate was approximately 95% sensitive to rotenone, that supported by succinate/G-3-P, around 100% sensitive to antimycin, and that supported by ascorbate/TMPD, about 95% sensitive to KCN.

Polarographic analysis carried out as described above has led to the identification of several VA_2B cell derivatives deficient in Complex I due to different mutations in mtDNA-encoded subunits of the enzyme.[8,9] Figure 1b shows the polarographic evidence of a defect in NADH-dependent respiration in one such mutant (C4).[8]

Acknowledgments

This work was supported by National Institutes of Health Grant GM-11726 (to G. A.) and by fellowships from the Deutsche Forschungsgemeinshaft and the Muscular Dystrophy Association (to G. H.). We are indebted to P. C. Lee and Gaetano Villani for helpful discussions.

[42] Assessment of Mitochondrial Oxidative Phosphorylation in Patient Muscle Biopsies, Lymphoblasts, and Transmitochondrial Cell Lines

By IAN A. TROUNCE, YOON L. KIM, ALBERT S. JUN, and DOUGLAS C. WALLACE

Introduction

Investigation of oxidative phosphorylation (OX-PHOS) in mitochondrial diseases has traditionally focused on the muscle biopsy. While muscle may not be as severely involved as the central nervous system (CNS) in many mitochondrial disease patients, respiratory chain abnormalities can usually be demonstrated in skeletal muscle mitochondria.[1–3] Skeletal muscle biopsy remains a valuable resource for biochemical studies since it is an easily accessed tissue, and if several grams can be obtained then milligram quantities of mitochondria can be isolated for a wide range of OX-PHOS investigations. However, in mitochondrial diseases the postmitotic muscle fibers commonly show secondary OX-PHOS defects that may hinder investigations of primary defects. An example is in Leigh disease caused by the mitochondrial DNA (mtDNA) np 8993 T → G transversion in the ATP6 gene.[4,5] Skeletal muscle OX-PHOS studies in different patients have revealed normal to globally decreased respiratory chain enzymes,[5,6] yet somatic cell genetic studies have shown that the mtDNA np 8993 mutation is linked to a specific defect in the H^+-ATP synthase with normal respiratory chain function.[7]

Transformed cell lines expressing OX-PHOS defects provide a powerful model system for further genetic and biochemical characterization of nuclear and mtDNA mutants, and the design and testing of therapeutic ap-

[1] D. C. Wallace, *Annu. Rev. Biochem.* **61,** 1175 (1992).
[2] L. Ernster and C.-P. Lee, in "Bioenergetics" (C. Kim, and T. Ozawa, eds.), p. 451. Plenum, New York, 1990.
[3] M. Birch-Machin, S. Jackson, R. S. Kler, and D. M. Turnbull, *Methods Toxicol.* **2,** 51 (1993).
[4] I. J. Holt, A. E. Harding, R. K. H. Petty, and J. A. Morgan-Hughes, *Am. J. Hum. Genet.* **46,** 428 (1990).
[5] J. M. Shoffner, P. M. Fernhoff, N. S. Krawiecki, D. B. Caplan, P. J. Holt, D. A. Koontz, Y. Takei, N. J. Newman, R. G. Ortiz, M. Polak, S. W. Ballinger, M. S. Lott, and D. C. Wallace, *Neurology* **42,** 2168 (1992).
[6] F. M. Santorelli, S. Shanske, A. Macaya, D. C. DeVivo, and S. DiMauro, *Ann. Neurol.* **34,** 827 (1993).
[7] I. Trounce, S. Neill, and D. C. Wallace, *Proc. Natl. Acad. Sci. U.S.A.* **91,** 8334 (1994).

proaches. Transformed cells, including cybrids, are readily grown in quantities sufficient for isolation of milligrams of well-coupled mitochondria for respiration studies and OX-PHOS enzyme assays.[7] This fact, combined with cytoplasmic or cybrid transfer[8] of mitochondria and mtDNAs into human mtDNA-less (ρ^0) cell lines[9] has allowed the linkage of specific OX-PHOS defects to individual mtDNA disease mutations.[7,10–12]

Our approach to OX-PHOS investigation combines the identification of patients with OX-PHOS defects using the more sensitive muscle biopsy studies followed by further biochemical characterization of primary defects in Epstein-Barr virus-transformed lymphoblasts. Cybrid transfer of patient mtDNAs into ρ^0 cells is then used to determine if the OX-PHOS defect is of nuclear or mtDNA origin (Fig. 1). In this chapter we present methods for the isolation of well-coupled mitochondria from skeletal muscle biopsies, lymphoblasts, and transmitochondrial cybrids, and provide protocols for the polarographic measurement of respiration and spectrophotometric assay of individual OX-PHOS complexes in these mitochondrial fractions. We then present novel methods for the production of transmitochondrial cybrids using lymphoblastoid and osteosarcoma ρ^0 cells as recipients. Suspension enucleation of cells combined with electrofusion allows the use of any cell type, including lymphoblasts, as mitochondrial donors in fusions with either lymphoblastoid or osteosarcoma ρ^0 cells (Fig. 1).

1. Isolation of Intact Mitochondria from Muscle Biopsies and Cultured Cells

Muscle Biopsies

The tough myofibrillar architecture that comprises the bulk of skeletal muscle demands relatively harsh homogenization to release mitochondria, and some tradeoff between yield and integrity of the mitochondrial fraction has to be made. Polarographic assay of respiratory control (see Section 3) is the best indicator of membrane integrity of the organelle fraction. Two approaches for isolation of intact mitochondria from skeletal muscle have been used. The first uses ionic isolation medium and a brief protease digestion step before mechanical homogenization and differential centrifuga-

[8] D. C. Wallace, C. L. Bunn, and J. M. Eisenstadt, *J. Cell Biol.* **67,** 174 (1975).

[9] M. P. King and G. Attardi, *Science* **246,** 500 (1989).

[10] A. Chomyn, G. Meola, N. Bresolin, S. T. Lai, G. Scarlato, and G. Attardi, *Mol. Cell. Biol.* **11,** 2236 (1991).

[11] J.-I. Hayashi, S. Ohta, A. Kikuchi, M. Takemitsu, Y.-I. Goto, and I. Nonaka, *Proc. Natl. Acad. Sci. U.S.A.* **88,** 10614 (1991).

[12] M. P. King, Y. Koga, M. Davidson, and E. A. Schon, *Mol. Cell. Biol.* **12,** 480 (1992).

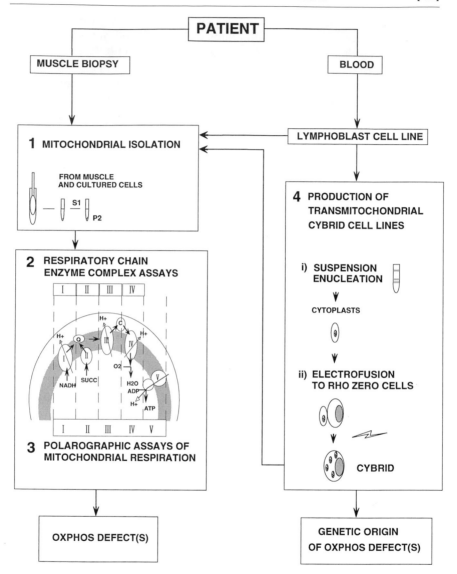

FIG. 1. Relationships among methods described in this chapter. Mitochondrial isolation and OX-PHOS studies in skeletal muscle are complemented with a somatic cell genetic approach that can define the nuclear or mtDNA origin of specific OX-PHOS defects.

tion.[13] The second, which is described here, uses a nonionic isolation medium and only mechanical tissue homogenization, followed by differential centrifugation.[14–17] All techniques require 1 to 5 g fresh muscle biopsy for consistent results.

Measured OX-PHOS capacity declines considerably with subject age in human skeletal muscle mitochondria.[18–20] Hence, age must be controlled when comparing disease and normal subject enzyme activities.

Procedure

1. Collect between 2 and 4 g of quadriceps muscle under the local anesthetic 0.5% (w/v) lidocaine hydrochloride and 1:200,000 epinephrine, and place sample directly into 50 ml ice-cold isolation buffer [210 mM mannitol, 70 mM sucrose, 1 mM EGTA, 0.5% (w/v) bovine serum albumin (BSA, fatty acid-free) and 5 mM HEPES, pH 7.2]. The fresh tissue should be processed immediately for mitochondrial isolation, with all manipulations carried out at 1°–4°. Mitochondria should be recovered within 60–90 min of biopsy.

2. The muscle specimen is placed in a preweighed tube, the isolation buffer is decanted and any remaining buffer removed by blotting, and the tube plus sample is weighed. Remove any obvious fat or connective tissue, and process in approximately 1-g batches with a Thomas (Swedesboro, NJ) tissue slicer. Collect thinly sliced muscle into 50-ml tube, suspending in 10 ml isolation buffer/g muscle.

3. Homogenize the muscle suspension in 15-ml batches using 10 passes in a Teflon–glass homogenizer with motor-driven pestle at moderate rotation speed. Alternatively, the muscle can be minced to a crude paste with a pair of fine scissors and homogenized using a blender-type homogenizer such as the Sorvall Omnimixer (Sorvall, Bloomington, DE) with microchamber.[15] Collect the 10% (w/v) homogenate and centrifuge at 1500 g for 5 min, 4°. Decant the supernatant into a fresh tube, discard the pellet, and repeat the centrifugation.

[13] C.-P. Lee, M. E. Martens, and S. H. Tsang, *Methods Toxicol.* **2**, 70 (1993).
[14] S. R. Max, J. Garbus, and H. J. Wehman, *Anal. Biochem.* **46**, 576 (1972).
[15] E. Byrne and I. Trounce, *J. Neurol. Sci.* **69**, 319 (1985).
[16] H. S. A. Sherratt, N. J. Watmough, M. A. Johnson, and D. M. Turnbull, *Methods Biochem. Anal.* **33**, 243 (1987).
[17] X. Zheng, J. M. Shoffner, A. S. Voljavec, and D. C. Wallace, *Biochim. Biophys. Acta* **1019**, 1 (1990).
[18] I. Trounce, E. Byrne, and S. Marzuki, *Lancet* **1**, 637 (1989).
[19] J. M. Cooper, V. M. Mann, and A. H. V. Schapira, *J. Neurol. Sci.* **113**, 91 (1992).
[20] D. Boffoli, S. C. Scacco, R. Vergari, G. Solarino, G. Santacroce, and S. Papa, *Biochim. Biophys. Acta* **1226**, 73 (1994).

4. Carefully decant the supernatant from the second low-speed spin, avoiding any loosely sedimented material, and centrifuge the supernatant at 8000 g for 15 min at 4°. Discard the supernatant and suspend mitochondrial pellet in 30 ml isolation buffer. Centrifuge the mitochondrial suspension again at 8000 g for 15 min.

5. Suspend the washed mitochondrial pellet in 0.1 ml isolation buffer for each gram of muscle used. Polarographic studies should be undertaken immediately using the fresh isolate. Other aliquots should be frozen for later enzyme analysis. For determination of mitochondrial protein, a 30-μl aliquot is removed and centrifuged in an Eppendorf microcentrifuge at 8000 g for 15 min, the supernatant removed, and the pellet suspended in 300 μl isolation buffer without BSA. This centrifugation is repeated, the pellet suspended in 30 μl isolation buffer without BSA, and the protein concentration determined.

Cultured Cells

Cell Lines and Culture Conditions. Lymphoblastoid cell lines are routinely established by Epstein-Barr virus (EBV) transformation of leukocytes isolated from whole blood using Ficoll–Hypaque gradients.[21] Cultures of EBV-transformed lymphoblasts and WAL-2A-ρ^+ lymphoblast cybrids (see Section 4) are grown in RPMI 1640 medium with 15% (v/v) heat-inactivated fetal calf serum (hi-FCS) and expanded for mitochondrial isolation by seeding cells into disposable plastic 2-liter roller bottles containing 250 ml medium at a density of 2×10^5 cells/ml. Depending on growth these cultures are expanded at 2- to 4-day intervals to 500 then 1500 ml using RPMI with 10% (v/v) hi-FCS, and harvested 2 to 3 days after the final passage. The 143BTK$^-$-ρ^+ osteosarcoma cybrids (see Section 4 below) are seeded into spinner bottles containing 500 ml medium at a density of 2×10^5 cells/ml. The medium is spinner minimal essential medium (SMEM) supplemented with 10% (v/v) FCS, glucose to 4 mg/ml, L-glutamine to 1 mM, and HEPES (pH 7.4) to 15 mM. Depending on growth, these cultures are expanded at 2- to 4-day intervals to 500, 1000, then 2000 ml, and harvested 2 or 3 days after the final passage. Such cultures yield $1-2 \times 10^9$ cells, around 2–4 g wet weight. We do not use antibiotics in the culture medium when expanding patient lymphoblasts, or bromodeoxyuridine (BrdU), 6-thioguanine (6-TG), or antibiotics in the culture medium when expanding TK$^-$ or HPRT$^-$ cell lines for mitochondrial isolation. This ensures that these agents do not interfere with mitochondrial function.

[21] D. C. Wallace, J. Yang, J. Ye, M. T. Lott, N. A. Oliver, and J. McCarthy, *Am. J. Hum. Genet.* **38**, 461 (1986).

The following procedure is modified from that of Moreadith and Fiskum[22] for isolation of mitochondria from ascites tumor cells. We follow the modifications of Howell et al.[23] except in using EGTA instead of EDTA and a lower concentration, and the titration of the digitonin is more controlled to achieve plasma membrane fragility without exposing organelles to the detergent. We have found the method suitable for isolating high-quality intact mitochondria from all immortal cell lines tested, including 143BTK$^-$ cells and derived cybrids; EBV-transformed lymphoblastoid cultures; WAL-2A and derived lymphoblastoid cybrids; HeLa cells; and mouse L cells. Using this procedure we obtain yields of around 2–5 mg mitochondrial protein per 10^9 cells (or 1–2.5 mg/g wet weight).

Procedure. All steps are done at $0°–4°$.

1. Harvest approximately $1–2 \times 10^9$ cells in mid log growth by centrifuging at 300 g for 3 min. Use a refrigerated centrifuge and keep cell pellets and subsequent fractions on ice. Transfer pellets to a preweighed 50-ml tube, suspending in total volume of 50 ml cold isolation buffer [210 mM mannitol, 70 mM sucrose, 1 mM EGTA, 0.5% BSA (fatty acid-free), and 5 mM HEPES, pH 7.2]. Centrifuge at 450 g for 3 min. Remove supernatant and record weight of cell pellet.

2. Suspend cells in 4 ml isolation buffer for each gram of packed cells. While swirling cell suspension in tube, add digitonin [10% (w/v) solution in dimethyl sulfoxide (DMSO)] slowly to a final concentration of 0.10 mg/ml. Mix with a pipette for 1 min and check cell permeability with trypan blue. Increase digitonin concentration in 0.05 mg/ml increments until more than 90% of the cells are permeabilized, stained blue. The optimal final digitonin concentration to achieve this varies between 0.20 and 0.40 mg/ml. Be quick, as excessive digitonin exposure can drastically reduce yield and quality of mitochondria.

3. Double the volume of the suspension to 10 times the weight of cells by adding isolation buffer and centrifuge at 3000 g for 5 min, to pellet the cells. Remove supernatant with pipette (do not pour it off, as the slimy digitonin-treated cell pellet will slide out too) and suspend pellet in five times the cell pellet weight in isolation buffer. Transfer to a chilled all-glass Dounce homogenizer and disrupt cells with 10 to 20 passes using a tight-fitting pestle. A correctly paired mortar and pestle is critical. Check microscopically for adequate cell disruption.

4. Triple the homogenate volume to 15 times the cell pellet weight, and centrifuge at 625 g for 5 min. Collect the supernatant with a pipette

[22] R. Moreadith and D. R. Fiskum, *Anal. Biochem.* **137,** 360 (1984).
[23] N. Howell, M. S. Nalty, and J. Appel, *Plasmid* **16,** 77 (1986).

to avoid the loose sediment near the pellet and repeat the centrifugation of the supernatant two more times, or until very little material (unbroken cells, nuclei) is sedimented.

5. Centrifuge the final supernatant at 10,000 g for 20 min at 4°. Collect the mitochondrial pellet by pouring off the supernatant, allowing most of the light material surrounding the buff mitochondrial pellet to pour off. Gently rinse the pellet with a few milliliters cold buffer to remove the remaining light material before suspending the pellet in 25 ml isolation buffer and repeating centrifugation at 10,000 g for 20 min.

6. Pour off the supernatant and suspend the mitochondrial pellet with 0.1 ml isolation buffer per gram of starting cells, giving a net protein concentration of around 10–15 mg/ml. Freeze aliquots of the suspension for later enzyme studies, keeping 100–200 μl on ice for immediate polarographic studies. We have found no deterioration in respiratory capacity of the organelle fraction stored on ice for up to 6 hr. For determination of mitochondrial protein concentration, remove 30 μl of the mitochondrial suspension, centrifuge in an Eppendorf microcentrifuge at 10,000 g for 20 min, suspend the pellet in 300 μl isolation buffer without BSA, and centrifuge again. Suspend the pellet in 30 μl isolation buffer without BSA and determine protein.

2. Assay of Oxidative Phosphorylation Enzyme Complexes

The optimization of the enzymatic assays for skeletal muscle mitochondria has been described elsewhere.[17] We use identical conditions when measuring both skeletal muscle and cultured cell mitochondria respiratory chain enzyme activities. Because of turbidity caused by the addition of decylubiquinone in the complex I, II, and III assays a dual-beam spectrophotometer is essential for those assays. For all assays, an aliquot of mitochondrial suspension is thawed and diluted with isolation buffer to 1 mg/ml mitochondrial protein. For samples from normal subjects, the sample is sonicated with 6 pulses, power level 5 and 20% duty cycle, using a Branson sonicator with microtip. Patient mitochondria may be more fragile than normal subject mitochondria, and the sonication time is reduced proportionately.[17,24] Store the suspension of disrupted mitochondria on ice and use immediately for enzyme assays. Do not measure enzyme activity on refrozen sonicated samples.

For reliable results, it is important not to isolate mitochondria from

[24] D. C. Wallace, X. Zheng, M. T. Lott, J. M. Shoffner, J. A. Hodge, R. I. Kelley, and L. C. Hopkins, *Cell (Cambridge, Mass.)* **55,** 601 (1988).

previously frozen muscle or cell specimens. Such preparations frequently have reduced complex I activity.[17] It is important that samples of isolated mitochondria be stored at $-70°$, and not repeatedly frozen and thawed. For this reason we distribute the freshly isolated mitochondrial suspension into multiple small aliquots and then freeze. Each set of tests then uses a fresh mitochondrial sample.

Table I lists activities of various respiratory chain enzymes in control skeletal muscle mitochondria. This is compared to the control data for the lymphoblast mitochondrial fractions.

Complex I: NADH–Ubiquinone Oxidoreductase, EC 1.6.5.3, NADH Dehydrogenase

Spectrophotometric assays of complex I can follow either the oxidation of NADH or the reduction of ubiquinone analogs or homologs. Specific activity of complex I in either assay should show almost complete ($>90\%$) rotenone sensitivity. The commercial availability of the relatively soluble ubiquinone analog decylubiquinone (DB) improves the reliability of both the complex I and complex III assays. The following assay measures DB reduction at 272 minus 247 nm (extinction coefficient 8 mM^{-1} cm^{-1}).[17] See

TABLE I

ACTIVITIES OF OXIDATIVE PHOSPHORYLATION ENZYME COMPLEXES IN SKELETAL MUSCLE, EBV LYMPHOBLAST, AND WAL-2A-ρ^+ CYBRID MITOCHONDRIAL FRACTIONS FROM CONTROLS[a]

Source		Enzyme complex						
		I	I+III	II	II+III	III	IV	CS
Skeletal muscle	Mean	194	229	—	605	1952	1615	—
mitochondria[b]	SD	60	149	—	164	700	309	—
($n = 11$)								
Lymphoblast	Mean	66	—	153	201	506	513	613
mitochondria	SD	17	—	42	72	84	196	152
($n = 14$)								
WAL-2A-ρ^+	Mean	41	—	—	162	823	396	362
cybrids ($n = 6$)	SD	10	—	—	52	308	72	78

[a] Specific enzyme activities shown as nmol/min/mg protein. I, NADH–DB oxidoreductase; I+III, NADH–cytochrome-c oxidoreductase; II, succinate–DB oxidoreductase; II+III, succinate–cytochrome-c reductase; III, DBH$_2$–cytochrome-c oxidoreductase; IV, cytochrome-c oxidase; CS, citrate synthase.

[b] From D. C. Wallace, J. M. Shoffner, R. L. Watts, J. L. Juncos, and A. Torroni, Ann. Neurol. 32, 113 (1992).

Estornell *et al.*[25] for an NADH–DB assay following NADH oxidation at 340 minus 380 nm and for a useful comparison of various ubiquinone analogs and homologs used to measure complex I activity.

Reagents

Buffer (2×): 500 mM sucrose, 2 mM EDTA, 100 mM Tris-HCl, pH 7.4
Decylubiquinone (DB, Sigma, St. Louis, MO) 1 mM in 100% ethanol
KCN, 0.2 M
NADH, 1 mM (freshly prepared)
Rotenone, 500 μg/ml in 100% ethanol

Procedure. The reduction of DB is monitored at 272 minus 247 nm in a 1-ml quartz cuvette at 30°. The reaction mixture consists of 250 mM sucrose, 1.0 mM EDTA, 50 mM Tris-HCl pH 7.4, 10 μM DB, 2 mM KCN, and 20 μg skeletal muscle mitochondrial protein or 45 μg cell mitochondrial protein. The reaction is initiated by adding 50 μM NADH and monitored through the linear absorbance decrease for 1 min. Rotenone (5 μg) is added, and any rotenone-insensitive activity is measured for 1 min.

Complex I plus III: NADH–Cytochrome-c Oxidoreductase

Owing to high levels of rotenone-insensitive activities in other cell compartments, the complex I + III coupled assay can only be measured reliably on mitochondrial fractions. If respiration studies are not possible because of limited sample, this and the coupled complex II + III assay are sometimes useful. When combined with the specific complex I and III assays, they can provide information on organelle fragility or altered inner membrane function. The NADH-dependent reduction of cytochrome c is followed at 550 minus 540 nm (extinction coefficient 19.0 mM^{-1} cm^{-1}). The rotenone-sensitive activity due to complex I generally varies between 30 and 80% of the total NADH–cytochrome-c reductase activity in mitochondrial fractions from different tissues or cell lines.

Reagents

Buffer (2×): 0.1 M potassium phosphate, pH 7.4
NADH, 1 mM (freshly prepared)
Cytochrome c, 1 mM
Rotenone, 500 μg/ml in 100% ethanol
KCN, 0.2 M

Procedure. The reduction of ferricytochrome c is followed at 550 minus 540 nm in a 1-ml cuvette at 30°. The reaction mixture consists of 50 mM

[25] E. Estornell, R. Fato, F. Pallotti, and G. Lenaz, *FEBS Lett.* **332**, 127 (1993).

potassium phosphate, pH 7.4, 0.1 mM NADH, 80 μM cytochrome c, and 2 mM KCN. Initiate the reaction by adding 30 μg skeletal muscle mitochondrial protein or 50 μg cell mitochondrial protein, and monitor the absorbance increase for 3 min. Repeat with 5 μg rotenone included in the reaction mixture to measure the rotenone-insensitive rate.

Complex II: Succinate–Ubiquinone Oxidoreductase, EC 1.3.5.1, Succinate Dehydrogenase (Ubiquinone)

As complex II contains no mtDNA-encoded subunits, the activity of this complex is not usually decreased in mtDNA diseases. Therefore, complex II can be a useful marker enzyme for comparison with activities of the other OX-PHOS complexes. The following assay measures the reduction of 2,6-dichlorophenolindophenol (DCPIP) when coupled to complex II-catalyzed reduction of DB, monitoring the absorbance at 600 minus 750 nm (extinction coefficient 19.1 mM^{-1} cm^{-1}).[3,26]

Reagents

Buffer (2×): 0.1 M potassium phosphate, pH 7.4
Succinate, 0.5 M, pH 7.4
DCPIP, 5 mM (freshly prepared)
Rotenone, 400 μg/ml in 100% ethanol
Antimycin A, 400 μg/ml in 100% ethanol
KCN, 0.2 M
Decylubiquinone (DB, Sigma), 10 mM in 100% ethanol

Procedure. The reduction of DCPIP is measured at 600 minus 520 nm in a 1-ml cuvette at 30°. Incubate mitochondria (15 μg skeletal muscle mitochondrial protein or 30 μg cell mitochondrial protein) at 30° for 10 min in a mixture of 50 mM potassium phosphate, pH 7.4, and 20 mM succinate. Next add antimycin A (2 μg/ml), rotenone (2 μg/ml), KCN (2 mM), and 50 μM DCPIP and record the blank rate for 1 min. Initiate the reaction by adding 50 μM DB and monitor the change in absorbance for 3 min.

Complex II plus III: Succinate–Cytochrome-c Oxidoreductase

The reduction of cytochrome c by complex III coupled to succinate oxidation through complex II is followed at 550 minus 540 nm (extinction coefficient 19.0 mM^{-1} cm^{-1}). In normal mitochondria, complex II catalyzes

[26] T. P. Singer, *Methods Biochem. Anal.* **22,** 123 (1976).

the rate-limiting step in this coupled assay,[27] but severe inhibition of complex III can reduce the overall rate. Hence, the optimal analysis compares this assay with the specific complex II and III assays.

Reagents

Buffer (2.5×): 0.1 M potassium phosphate, pH 7.4
Succinate, 0.5 M, pH 7.4
EDTA, 15 mM, pH 8.0
KCN, 0.2 M
Cytochrome c, 1 mM

Procedure. The reduction of ferricytochrome c is followed at 550 minus 540 nm in a 1-ml cuvette at 30°. The reaction mixture consisting of 40 mM potassium phosphate, pH 7.4, 20 mM succinate, 0.5 mM EDTA, 2 mM KCN, and 15 μg skeletal muscle mitochondrial protein or 30 μg cell mitochondrial protein is incubated at 30° for 20 min. The reaction is initiated by adding 30 μM cytochrome c; monitor the linear absorbance increase for 3 min.

Complex III: Ubiquinol : Ferricytochrome-c Oxidoreductase, EC 1.10.2.2

The following assay monitors the reduction of cytochrome c at 550 minus 540 nm (extinction coefficient 19.0 mM^{-1} cm^{-1}) catalyzed by complex III in the presence of reduced decylubiquinone (DBH$_2$).[17]

Reagents

Buffer (2×): 500 mM sucrose, 2 mM EDTA, 100 mM Tris-HCl, pH 7.4
Cytochrome c, 1 mM
KCN, 0.2 M
Antimycin A, 1 mg/ml in 100% ethanol
Decylubiquino (DBH$_2$): Add a small crystal of potassium borohydride to 50 μl of 10 mM DB (Sigma) in ethanol. Add 5-μl aliquots of 0.1 M HCl with gentle mixing until the yellow solution becomes colorless. Transfer the DBH$_2$ to a fresh tube, avoiding the borohydride crystal, and add 5 μl of 1 M HCl. Record the final DBH$_2$ concentration. Store on ice while performing assays, and if the color begins to change back to yellow before use repeat the reduction procedure.

Procedure. The reduction of ferricytochrome c is followed at 550 minus 540 nm in a 1-ml cuvette at 30°. Incubate the reaction mixture of 250 mM

[27] R. W. Taylor, M. A. Birch-Machin, K. Bartlett, and D. M. Turnbull, *Biochim. Biophys. Acta* **1181,** 261 (1993).

sucrose, 1 mM EDTA, 50 mM Tris-HCl, pH 7.4, 50 μM cytochrome c, 2 mM KCN, 15 μg skeletal muscle mitochondrial protein or 30 μg cell mitochondrial protein at 30° for 10 min. Initiate the reaction by adding 50 μM DBH$_2$, and monitor the increase in absorbance at 550 minus 540 nm for 1 min. Repeat the experiment, but with 5 μg/ml antimycin A added to the incubation mixture to determine the antimycin-insensitive activity. Results can be calculated using the initial quasi-linear rate or by calculating apparent first-order rate constants.[16]

Complex IV: Ferrocytochrome-c: Oxygen Oxidoreductase, EC 1.9.3.1, Cytochrome-c Oxidase

Complex IV or cytochrome-c oxidase activity is readily measured by following the oxidation of reduced cytochrome c at 550 minus 540 nm (extinction coefficient 19.0 mM^{-1} cm^{-1}).[28] Specific rates can be calculated either from the initial quasi-linear phase of the reaction or by estimation of first-order rate constants.[16] Because the apparent activity varies in proportion to the concentration of cytochrome c, care should be taken to use the same substrate concentration in all assays. Although we do not routinely include laurylmaltoside in the assay, we have found that this detergent at an optimal concentration of 0.025% can approximately double the specific cytochrome oxidase activity of mitochondrial fractions assayed under the following conditions, in agreement with Birch-Machin et al.[3]

Reagents

Buffer: 10 mM potassium phosphate, pH 7.4

Ferrocytochrome c: Dissolve 100 mg cytochrome c (Sigma, from horse heart) in 1 ml of 10 mM potassium phosphate, pH 7.4, and add 1.0 ml of 0.1 M L-ascorbate to reduce the cytochrome. Load mixture onto a 20-ml Sephadex G-25 column prewashed with 100 ml degassed 10 mM potassium phosphate, pH 7.4. Collect the middle three-fourths of the cytochrome band. Check for full reduction of the cytochrome by measuring the absorbance at 550 and 565 nm[29] and calculate the stock concentration from the absorbance at 550 nm using an extinction coefficient of 27.7 mM^{-1} cm^{-1}. Dispense 0.1-ml aliquots, closing the tube caps under a gentle stream of nitrogen gas to displace oxygen from the tube (this helps to minimize oxidation of the cytochrome during storage). The reduced cytochrome stored at $-70°$ is stable for at least 12 months.

[28] S. J. Cooperstein and A. Lazarow, J. Biol. Chem. **189**, 665 (1951).
[29] B. Errede, M. D. Kamen, and Y. Hatefi, this series, Vol. 53, p. 40.

Procedure. The oxidation of ferrocytochrome c is followed at 550 minus 540 nm in a 1-ml cuvette at 30°. The reduced cytochrome is added to 20 μM in 10 mM potassium phosphate, pH 7.4, and the stability of the absorbance is checked for 1 min. The reaction is begun by adding 7.5 μg skeletal muscle mitochondrial protein or 10 μg cell mitochondrial protein. Begin monitoring the reaction as quickly as possible and record the absorbance decrease for 30 sec.

Citrate Synthase EC 4.1.3.7

Citrate synthase is the most commonly used matrix marker enzyme and is a useful inclusion in studies of OX-PHOS defects. The assay follows the reduction of 5,5'-dithiobis(2-nitrobenzoic acid) (DTNB) at 412 minus 360 nm (extinction coefficient 13.6 mM^{-1} cm^{-1}), coupled to the reduction of CoA by the citrate synthase reaction in the presence of oxaloacetate.[30]

Reagents

Buffer: 125 mM Tris-HCl, pH 8.0
Acetyl-CoA, 6 mM
DTNB, 2 mM (in 125 mM Tris-HCl, pH 8.0)
Oxaloacetate, 10 mM (freshly prepared)

Procedure. The reduction of DTNB is followed at 412 minus 360 nm in a 1-ml quartz cuvette at 30°. The reaction mixture of 0.1 M Tris-HCl, pH 8.0, 0.3 mM acetyl-CoA, 0.1 mM DTNB, and 5 μg skeletal muscle mitochondrial protein or 10 μg cell mitochondrial protein is incubated at 30° for 10 min. The reaction is initiated by the addition of 0.5 mM oxaloacetate, and the absorbance change is monitored for 1 min.

3. Polarographic Assay of Mitochondrial Respiration

Polarographic measure of oxygen consumption using freshly isolated mitochondria allows measurement of the impact of OX-PHOS complex defects on maximal rates of respiration and on the coupling of respiration to ATP production. Polarography also permits measurement of the catalytic ability of the H$^+$-translocating ATP synthase (complex V), by comparing the ADP-coupled and protonophore-uncoupled respiration, and measurement of phosphorylation efficiency, ADP/O ratios.[7]

---

[30] P. A. Srere, this series, Vol. 13, p. 3.

Reagents

Except for the respiration buffer, all stocks are made so that a 5-μl addition made with a Hamilton syringe to the 0.65-ml polarograph chamber gives the correct working concentration (shown in parentheses).

Respiration buffer: 225 mM mannitol, 75 mM sucrose, 10 mM KCl, 10 mM Tris-HCl, pH 7.2, 5 mM KH$_2$PO$_4$, pH 7.2; dispense 50-ml aliquots and store at $-20°$

Pyruvate, malate, glutamate, and succinate (5 mM): Make up 10-ml amounts of 0.65 M stocks of each, adjusted to pH 7.2 with KOH; dispense 0.15-ml aliquots and store $-20°$

ADP (125 nmol) and ATP (1 mM): Make 10 ml stock of each, 25 mM for ADP and 130 mM for ATP, pH 6.0–6.8. Adjust pH with dilute KOH to this range, where these nucleotides are most stable on storage. Keep solutions cold. For the ADP solution accuracy is important, because calculation of ADP/O ratios assumes that each microliter of ADP contains 25 nmol. Immediately dispense 0.15-ml aliquots and store at $-80°$

DNP (2,4-dinitrophenol; 50 μM): For 10 ml of 6.5 mM stock, dissolve 12 mg in 4 ml of 1 M NaOH (15-ml polypropylene tube) by heating at 50° for 20 min. Adjust to neutral pH with 1 M HCl by adding 3 ml, and then using 0.2-ml aliquots, checking pH with paper until pH 7.2. Solutions of DNP should be a translucent yellow. Make up to 10 ml final volume, dispense 0.15-ml aliquots, and store at $-20°$

Rotenone, antimycin A, oligomycin (5 μg/ml): Make up 650 μg/ml stocks of each in 10 ml absolute ethanol

KCN (2 mM): Make up 325 mM stock in 10 ml water; dispense 0.15-ml aliquots and store at $-20°$

Equipment

We use an Instech microelectrode (Instech, Plymouth Meeting, PA) and chamber (0.65 ml) fitted with a magnetic stirrer, and a Yellow Springs Instruments (Yellow Springs, OH) YSI 5300 oxygraph. The chamber water jacket is connected to a circulating water bath maintained at 30°. In preparation for assays the system is calibrated as follows. A 50-ml aliquot of respiration buffer is thawed and allowed to equilibrate with O$_2$ by agitation in a glass flask in a shaking water bath at 30° for at least 30 min. Start the circulating water bath (set at 30°), chamber stirrer, YSI polarograph, and chart recorder. Remove the electrode lead from the polarograph jack and set the chart recorder pen to zero. Reconnect the electrode, flush the chamber with a few milliliters water using a 50-ml syringe, then flush with a few milliliters air-equilibrated respiration buffer, and finally fill the chamber with respiration buffer, taking care to exclude any air bubbles. Close the chamber cock. Allow the polarograph reading to stabilize and set desired maximum (oxygen-saturated) point using the polarograph sensitivity control. If the signal is unstable, replace the electrode polyethylene membrane and repeat the calibration.

Assay of Mitochondrial Respiration

1. Thaw aliquots of ADP, ATP, glutamate, pyruvate, malate, succinate, and DNP. Calibrate polarograph as described above.
2. Fill chamber with fresh respiration buffer, then use Hamilton syringes to introduce substrates (see Remarks below). Begin chart and run for several minutes to ensure that the signal is stable. Add mitochondria to between 0.3 and 0.8 mg/ml final protein concentration, keeping the volume of this addition as low as possible by maintaining the mitochondrial stock suspension at greater than 10 mg/ml protein. A low (state IV) oxygen consumption rate should be evident.
3. After 1 min, add 1 mM ATP (see Remarks below). The state IV respiration is briefly stimulated due to small amounts of ADP in the ATP solution, before slowing to a rate that is usually slightly faster than the initial state IV rate (see Fig. 2).
4. Add ADP, usually 100–200 nmol. In well-coupled mitochondrial preparations the resulting state III rate will show an approximately 2.5- to 5-fold stimulation over the state IV rate (or a 5- to 10-fold stimulation over the initial state IV rate) before respiration returns to the state IV rate when the added ADP is phosphorylated (Fig. 2). After allowing state IV respiration to proceed long enough to measure a rate (1–2 min), a second addition of ADP can be made. The resulting state III rate should appear similar to the first. Again, allow the rate to return to state IV for 1–2 min.
5. Add 50 μM DNP and allow the uncoupled respiration to consume all the chamber oxygen in at least one run. This ensures that the rate remains linear and also confirms the zero chart pen position. Otherwise the run can be stopped after 2 to 3 min of uncoupled respiration.

Calculating Respiratory Rates and P:O Ratios

One milliliter buffer is taken to contain 480 ng atom O at 30° when equilibrated against air at 1 atmosphere pressure.[31] In our system, in which the Instech microchamber volume is 0.65 ml, the saturated oxygen content

---

[31] B. Chance and G. R. Williams, *J. Biol. Chem.* **217**, 383 (1955).

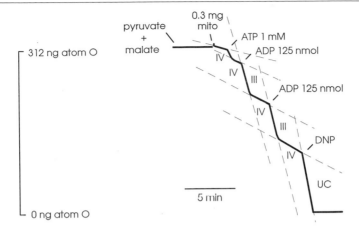

FIG. 2. Typical polarograph trace with well-coupled lymphoblast cell mitochondria, showing ADP-limited (state IV), ADP-stimulated (state III), and DNP-uncoupled (UC) respiration.

is $0.65 \times 480 = 312$ ng atom O. Extrapolate each state III, state IV, and uncoupled rate with pencil and ruler, and calculate the slopes (ng atom O/min; see Fig. 2). Calculate the rates of oxygen consumption for each run, averaging any duplicate runs done with the same substrates. Express specific rates as ng atom O/min/mg mitochondrial protein. The respiratory control ratios (RCRs) are calculated as the ratio of state III respiration to subsequent state IV respiration.[32] Calculate ADP/O ratios by dividing the amount of ADP added (e.g., 100 or 125 nmol) by the measured oxygen consumption during state III (in ng atom O).[31,32]

Remarks

Adjust the amount of mitochondrial suspension added so that a typical run takes about 10 min to complete (see Fig. 2 for typical trace using lymphoblast mitochondria). Avoid runs of less than 5 min, which are too fast to make additions comfortably; also note that experiments extended beyond 25 min usually show some deterioration of mitochondrial function. If material permits, perform repeat runs with all substrates of interest. Try to make ADP and DNP additions to the chamber at around the same oxygen/time coordinates in different runs. Ideally the DNP should be added with around 30% of the chamber oxygen remaining to obtain a reliable uncoupled rate.

Various substrates can be used to test different routes of reducing equivalent supply. Pyruvate plus malate, glutamate plus malate, and succi-

[32] R. W. Estabrook, this series, Vol. 10, p. 41.

nate (all at 5 mM) are oxidized at high rates by both skeletal muscle and transformed cell mitochondria (Table II). Palmitoyl carnitine (0.4 mM) plus malate (1 mM) is a useful substrate pair for testing fatty acid oxidation. To test respiration through complex IV only, ascorbate can be used with a redox mediator, usually tetramethylphenylenediamine (TMPD), to reduce intramitochondrial cytochrome c. Care should be taken to account for autoxidation if these reagents are used.[13]

When using succinate as substrate, rotenone can be included to prevent oxaloacetate production by the Krebs cycle, since oxaloacetate is a strong competitive inhibitor of the succinate dehydrogenase. We find that with transformed cell mitochondria addition of rotenone is not necessary with succinate to obtain well-coupled respiration with linear state III rates, although it is often helpful with skeletal muscle mitochondria. Using the latter fractions, even with rotenone present, we find that succinate oxidation is sometimes problematic to measure due to nonlinear rates.

For efficiently oxidized substrates, the uncoupled respiration is typically somewhat faster than the state III rate (Table II), indicating that the ATP synthase imposes some rate limitation on state III respiratory flux. The optimal DNP concentration for measuring uncoupled respiration is 50 μM. Less than this will not fully uncouple respiration, whereas more may inhibit respiration.[33] This should be confirmed empirically for each batch of DNP.

Specific inhibitors of complex I (rotenone, 5 μg/ml), complex III (antimycin A, 5 μg/ml), complex IV (KCN, 2.5 mM), and complex V (oligomycin, 5 μg/ml) can be used. When rotenone, antimycin A, or oligomycin is used, extensive washing of the chamber afterward is necessary, first with ethanol and then with water and buffer. Residual inhibitors can inhibit subsequent experiments, and we recommend the use of a separate chamber and separate Hamilton syringes for inhibitor experiments.

The respiratory control ratio remains the best indicator of the membrane integrity of the organelle fraction.[16,32] Both skeletal muscle and transformed cell mitochondria isolated from normal subjects by the present methods consistently show RCRs of between 2 and 6. High specific rates of oxidation of both NAD$^+$-linked substrates (pyruvate, glutamate, and malate) and the FAD$^+$-linked succinate are also routinely obtained in mitochondrial fractions of skeletal muscle and all cell lines tested (Table II). Some reports have indicated low rates of NAD$^+$-linked substrate oxidation compared with succinate oxidation in lymphoblast mitochondrial preparations,[34,35]

[33] J. B. Chappell, *Biochem. J.* **90,** 237 (1964).
[34] U. Carpentieri and A. Sordahl, *Cancer Res.* **40,** 221 (1980).
[35] T. Bourgeron, D. Chretien, A. Rotig, A. Munnich, and P. Rustin, *Biochem. Biophys. Res. Commun.* **186,** 16 (1992).

TABLE II

STATE III, STATE IV, AND UNCOUPLED RESPIRATION RATES AND P:O RATIOS OF INTACT MITOCHONDRIA ISOLATED FROM CONTROL CELL LINES[a]

Cell lines[b]	Pyruvate + malate						Glutamate + malate						Succinate					
	III	IV	RCR	UC	III/UC	P/O	III	IV	RCR	UC	III/UC	P/O	III	IV	RCR	UC	III/UC	P/O
Lymphoblasts																		
Mean	152	49	3.2	168	0.86	2.49	156	46	3.5	167	0.89	2.48	202	69	3.0	214	0.88	1.72
SD	32	13	0.6	37	0.05	0.15	32	10	0.7	34	0.04	0.21	40	15	0.3	42	0.08	0.15
WAL-2A-ρ^+																		
Mean	93	34	2.9	101	0.91	2.29	94	33	3.0	102	0.93	2.23	115	45	2.7	137	0.84	1.58
SD	20	9	0.6	20	0.07	0.25	20	10	0.7	19	0.10	0.23	26	12	0.4	29	0.06	0.18
143BTK$^-$-ρ^+																		
Mean	75	20	3.8	86	0.85	2.35	77	21	3.7	82	0.95	2.41	115	33	3.5	127	0.88	1.52
SD	5.6	2.9	0.3	7.8	0.01	0.09	5.7	1.2	0.3	13	0.09	0.04	9.2	0	0.2	32	0.14	0.02

[a] State III, IV, and DNP-uncoupled (UC) rates are shown as ng atom O/min/mg mitochondrial protein. The respiratory control ratio (RCR) is the ratio of state III to subsequent state IV rates, and P/O is the ratio of ADP molecules phosphorylated to oxygen atoms reduced.

[b] Lymphoblasts are control EBV-transformed lymphoblast cell lines, $n = 23$; WAL-2A-ρ^+ control cybrids, $n = 6$; 143B-TK$^-$-ρ^+ control cybrids, $n = 3$.

and we have found this to occur in some suboptimal preparations. Poor coupling (failure of state III respiration to return to state IV) also indicates a suboptimal organelle preparation, usually resulting from problems in mitochondrial isolation or poor quality of the tissue or cell sample. Poor coupling can also result when the mitochondrial protein concentration is suboptimal in the assay.

We have observed during polarographic assay of mitochondria isolated from cell cultures that preincubation of mitochondria with 1 mM ATP can improve coupling, and does not alter state III respiration. Measured ADP/O ratios, especially from the first addition of ADP, are often higher when mitochondria are "loaded" with ATP, and state IV respiration can be decreased, giving higher RCRs. This effect may be due to the restoration of matrix adenine nucleotide pools from the action of the MgATP/P$_i$ transporter of the mitochondrial inner membrane.[36] Many preparations show no improvement in coupling with added ATP, however. In these cases and if using skeletal muscle mitochondria, the ATP pretreatment can be omitted.

4. Production of Transmitochondrial Cybrids

Percoll and Ficoll Gradient Enucleation of Suspension-Growing Cells

Two suspension enucleation approaches have been perfected for enucleation of both suspension and attached cell types, including EBV-transformed lymphoblasts, HeLa S3 cells, and mouse L cells. The isopycnic Percoll gradient, modified from a minicell preparation procedure,[37] is simplest and useful when no separation of cytoplast and karyoplast fractions is needed. The Ficoll step-gradient method of Wigler and Weinstein,[38] while requiring more preparation, an ultracentrifuge, and a swinging-bucket rotor, allows good separation of these fractions.

Procedure for Enucleation Using Percoll Isopycnic Gradient

1. If using lymphoblasts, perform a viability count on a sample of culture in mid-log growth phase. Harvest 2×10^7 viable cytoplast–donor cells by centrifuging at 300 g for 3 min. Thoroughly suspend pellet in 20 ml of a 1:1 mixture of complete medium and Percoll that has been preequilibrated overnight in a 37°, 5% (v/v) CO$_2$ incubator. This helps reduce the alkalinity caused by the Percoll that may otherwise lower the viability of the lymphoblasts. Transfer mixture to

[36] J. R. Aprille, *J. Bioenerg. Biomembr.* **25,** 473 (1993).
[37] J. A. Sanford and E. Stubblefield, *Somatic Cell Mol. Genet.* **13,** 279 (1987).
[38] M. H. Wigler and I. B. Weinstein, *Biochem. Biophys. Res. Commun.* **63,** 669 (1975).

sterile Nalge (Rochester, NY) 30-ml polysulfonate centrifuge tube and add 200 μl of cytochalasin B (from 2 mg/ml stock in DMSO), final concentration 20 μg/ml.

2. Centrifuge for 70 min at 44,000 g_{av}, with the centrifuge temperature set to maintain a rotor temperature of 37°. A fixed-angle rotor is adequate, and we routinely use a Sorvall RC5B with SS-34 rotor at 19,000 rpm set at 25°.

3. At the end of the run collect the cellular band one-third from the bottom of the tube along with the hazy band rich in cytoplasts immediately above (see Remarks below). Dilute 10-fold with complete medium, centrifuge at 650 g for 10 min, and suspend in 10 ml medium. The mixture of washed cytoplasts, karyoplasts, and whole cells is now ready for electrofusion.

Procedure for Enucleation Using Ficoll Step Gradient

1. Prepare a 50% (w/w) stock solution of Ficoll 400 by adding 125 g Ficoll to 125 ml distilled water and stirring overnight with low heat. Filter sterilize the solution (which may take over 24 hr).

2. To prepare two gradients, make up 20 ml of 25% Ficoll by adding 10 ml appropriate culture medium with PSA [penicillin (100 U/ml), streptomycin (0.1 mg/ml), and amphotericin (0.25 μg/ml)] but without serum to 10 ml of 50% Ficoll. Use this solution and culture medium plus PSA to make 5 ml of 17%, 5 ml of 16%, 5 ml of 15%, and 10 ml of 12.5% Ficoll as shown in the tabulation below.

% Ficoll	25% Ficoll (ml)	Medium + PSA (ml)
17%	3.4	1.6
16%	3.2	1.8
15%	3.0	2.0
12.5%	5.0	5.0

When pipetting Ficoll solutions, allow enough time to fully evacuate pipettes, and mix completely. Add cytochalasin B (2 mg/ml in DMSO) to a final concentration of 10 μg/ml to all 5 Ficoll solutions (including the remaining 25% Ficoll).

3. Prepare the step gradients on the afternoon before enucleation (see Fig. 3). Cover tubes with UV-sterilized Parafilm and move to 37° overnight. This will allow some diffusion of the gradient interfaces, to reduce the trauma to cells during centrifugation. Also leave the rotor (SW41TI) at 37° overnight.

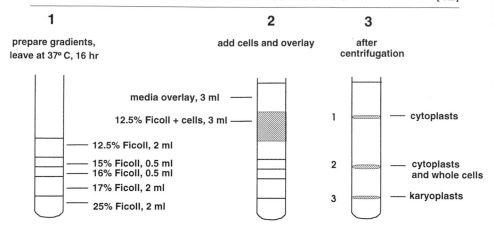

FIG. 3. Ficoll step-gradient enucleation. The middle band after centrifugation (2) is variable with different cell lines, and is sometimes blurred or split into two bands.

4. Suspend 4×10^7 freshly harvested cytoplast donor cells per gradient in 3 ml of 12.5% Ficoll containing 10 μg/ml cytochalasin B, then overlay onto the gradient. Add medium to completely fill the tube. Centrifuge at 85,000 g_{av} (25,000 rpm with Beckman L-80 ultracentrifuge and SW41TI rotor, Beckman, Palo Alto, CA) for 60 min, 31°.

5. Gently remove tubes from the rotor, aspirate the upper cytoplast band (see Fig. 3 and Remarks below), and suspend in 20 volumes complete culture medium. Pellet the cytoplasts by centrifugation at 650 g for 10 min and suspend in 10 ml medium.

Remarks. To assess the efficiency of enucleation and the distribution of cytoplasts and karyoplasts in the gradients, we stain fractions with Hoechst 33258 to identify nuclei and rhodamine 123 to identify mitochondria (both dyes are used at 1 μg/ml and the cells stained for 15 min in normal culture medium at 37°; the cells are washed twice with fresh culture medium before examination in a fluorescent microscope). The mixed fractions from the Percoll gradients should contain few intact cells, indicating efficient enucleation. The number and position of the intermediate bands in the Ficoll step-gradient varies with different cell lines. We find that fraction 1 (Fig. 3) is rich in intact rhodamine-staining cytoplasts and contains very few contaminating nuclei, and we have obtained cybrids using this fraction in fusions with ρ^0 cells. The band(s) of intermediate buoyant density comprises a heterogeneous mixture of larger cytoplasts and many intact cells, and we have also used this fraction in cybrid fusions. The lower fraction at the 17/25% Ficoll interface is comprised mostly of karyoplasts, approximately one-third of which appear to contain no mitochondria while the remaining two-thirds retain some mitochondria as judged by rhodamine staining.

Electrofusion of Cytoplasts and ρ^0 Cells

Preparation of ρ^0 Cells. Cells without mtDNA (ρ^0 cells) are produced according to the method of King and Attardi[9] from the osteosarcoma cell line 143BTK$^-$ (ATCC, Rockville, MD, CRL 8303)[7] and the lymphoblastoid cell line WAL-2A, a HPRT$^-$ clone[8] of the diploid lymphoblast WI-L2 cell line.[39] The 143BTK$^-$ cells are grown in Dulbecco's modified Eagle's medium (DMEM) with 10% fetal calf serum (FCS), 4.5 mg/ml (high) glucose, 1 mM pyruvate, and 50 μg/ml 2,5-bromodeoxyuridine (BrdU). The 143BTK$^-$-ρ^0 clones are grown in DMEM with 10% FCS, 4.5 mg/ml glucose, 1 mM pyruvate, 50 μg/ml uridine, and 50 μg/ml BrdU. The WAL-2A culture is grown in RPMI 1640 with 15% hi-FCS, 2 mg/ml glucose, and 1 μg/ml 6-thioguanine (6-TG). The WAL-2A-ρ^0 cells are grown in RPMI with 15% hi-FCS, 4 mg/ml glucose, 50 μg/ml uridine, 1 mM pyruvate, and 1 μg/ml 6-TG.

Both 143BTK$^-$-ρ^0 cells and WAL-2A-ρ^0 cells grow more slowly than the parental cell lines (mid-log doubling time of about 35 hr for 143BTK$^-$-ρ^0 compared to 17 hr for 143BTK$^-$, and 30 hr for WAL-2A-ρ^0 compared with 18 to 22 hr for WAL-2A). The WAL-2A-ρ^0 cells also exhibit a much lower plateau density than WAL-2A (2 $\times$ 10^6 cells/ml compared to 6 $\times$ 10^5 cells/ml). Both 143BTK$^-$-ρ^0 cells and WAL-2A-ρ^0 cells acidified media more quickly than the parental cell lines, as expected from their complete reliance on glycolytic ATP production and commensurate lactic acid production to reoxidize the glycolytically generated NADH. As previously shown for 143BTK$^-$-ρ^0 cells,[9] both 143BTK$^-$-ρ^0 and WAL-2A-ρ^0 cell lines were dependent on added uridine and pyruvate for growth (Fig. 4). This allows selection against ρ^0 cells in cybrid fusions.

Procedure for Electrofusion of Cytoplasts and 143BTK$^-$ ρ^0 Cells. The electrofusion conditions described below should be used as a guide for empirical determination of optimal conditions for any particular pair of fusion partners. A comprehensive discussion of electrofusion variables is given elsewhere in this series.[40] We use a Biotechnologies and Experimental Research (BTX, San Diego, CA) Electro Cell Manipulator 200 machine equipped with the Optimizor for monitoring chamber electrical pulse characteristics. The machine delivers AC fields of up to 75 V at 1 MHz, and square-wave DC pulses of up to 960 V. The following protocols are designed

[39] J. A. Levi, M. Virolainen, and V. Defendi, *Cancer* **22**, 517 (1968).
[40] G. A. Neil and U. Zimmermann, this series, Vol. 220, p. 174.

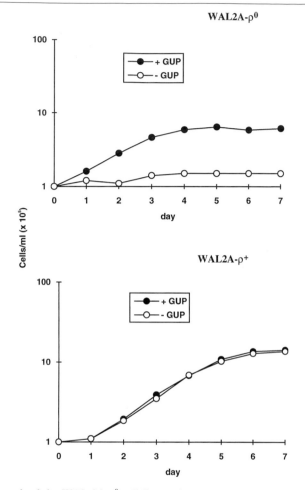

Fig. 4. Growth of the WAL-2A-ρ^0 cell line and a control WAL-2A-ρ^+ cybrid with and without high glucose (4 mg/ml), uridine (50 μg/ml), and pyruvate (1 mM) (GUP). The culture medium was otherwise the same: RPMI with 10% dialyzed FCS and 1 μg/ml 6-TG. The ρ^0 culture undergoes less than one population doubling without the supplements, while growth of the derived cybrid culture is not affected.

for use with the BTX PN453 slide chamber with 3.2 mm electrode gap width and 0.6 ml capacity.

1. Assuming complete enucleation and recovery of cytoplasts from the 2×10^7 cells used in the Percoll gradient or 50% recovery from the upper or middle cytoplast bands of the Ficoll gradient where 4×10^7 cells were used, the number of cytoplasts used for fusion is

approximately 2×10^7. Put aside 1/20 ($\sim 10^6$) of the cytoplasts for a nonfusion control flask. The remaining cytoplasts ($\sim 2 \times 10^7$) are used for the fusion.

2. Harvest 4×10^6 ρ^0 cells and separate into 3×10^6 (fusion) and 10^6 (control) aliquots. These cells are fed fresh medium 3 hr before harvest.

3. Centrifuge all four suspensions at 650 g for 10 min and resuspend each in 3 ml fusion medium (0.3 M mannitol, pH adjusted with a small amount of dilute phosphate buffer to pH 7.4, as the ionic strength of the nonconductive fusion medium should be kept to a minimum). Centrifuge again at 650 g for 10 min. Suspend particles for fusion (3×10^6 ρ^0 cells and 2×10^7 cytoplasts) in 0.6 ml fusion medium, suspending one pellet before combining with second pellet and thoroughly resuspending the mixture. Suspend control cells and cytoplasts in 3 ml fresh fusion medium.

4. Fill the microslide fusion chamber (sterilized prior to use with ethanol) with 0.6 ml fusion medium without cells. Attach electrode leads and set the power source to deliver a 20 sec AC field of 35 V (0.11 kV/cm), followed by two 20-μsec square-wave DC pulses of 800 V (2.5 kV/cm). If available, a pulse recording device is useful for prerunning the machine to confirm actual chamber pulse characteristics. Remove the liquid from the chamber and add the 0.6 ml fusion mixture. Apply the above conditions, and immediately following the fusion pulses reinstate the AC field for 3 min, progressively lowering the voltage during this time from 35 to 15 V. Allow mixture to sit undisturbed for 3 min.

5. Transfer fusion mixture to a 50-ml tube and slowly add 10 ml DMEM with 10% FCS, 4.5 mg/ml glucose, 1 mM pyruvate, and 50 μg/ml uridine, gently suspending with a large orifice pipette. Allow tube to sit at room temperature for 20 min. At this time the control ρ^0 cells and cytoplasts are centrifuged (650 g for 10 min) and suspended in 10 ml each with the same medium.

Procedure for Electrofusion of Cytoplasts and WAL-2A-ρ^0 Cells

1. Prepare cytoplast aliquots as in Step 1 above.

2. Perform viable cell count on WAL-2A-ρ^0 cell culture and harvest 1.1×10^7 viable cells by centrifuging at 300 g for 5 min.

3. Wash fusion (10^7 cells) and control (10^6 cells) aliquots of ρ^0 cells and cytoplasts in mannitol fusion medium as described in Step 3 above.

4. Suspend fusion partners (10^7 ρ^0 cells and $\sim 2 \times 10^7$ cytoplasts) in 0.6 ml fusion medium and add to the chamber. Apply a 50 V AC field (0.16 kV/cm) for 10 sec followed by two 20-μsec DC pulses of 800

V (2.5 kV/cm). No postfusion AC field is used. Allow the mixture to stand undisturbed for 3 min.

5. Transfer fusion mixture to a 75-cm^2 tissue culture flask and slowly add 50 ml RPMI with 15% hi-FCS, 4 mg/ml glucose, 50 μg/ml uridine, and 1 mM pyruvate (RPMI/GUP). Gently suspend the cells by rocking the flask, and place in a 37°/5% CO$_2$ incubator.

Remarks. The efficiency of pearl-chain formation during application of varying AC fields can be assessed visually with the use of simple microslide chambers, as can the degree of cell lysis from both AC and DC fields of increasing strength. These present protocols differ from some conditions considered important by Neil and Zimmermann,[40] namely, mannitol is used instead of other neutral sugars for the fusion medium; the divalent cations Mg^{2+} and Ca^{2+} are not added to the fusion medium; a single wash step to replace the growth medium is used; a 100-fold higher cell particle concentration is used in the fusion mixture; the mixture is not left to stand for 30 min in the fusion chamber after fusion, instead being gently diluted with culture medium after 3 min; and phenol red-containing medium was used throughout. If fusion efficiency is critical, these and other variables should be considered in developing optimal protocols for particular applications.

Selection of Cybrids

Fusion of 143BTK$^-$ ρ^0 Cells and Cytoplasts. Plate out the fusion mixture into 100-mm dishes in DMEM with 10% FCS, 4.5 mg/ml glucose, 1 mM pyruvate, and 50 μg/ml uridine. Use replicate dishes with 2.5 × 10^5, 10^5, and 5 × 10^4 ρ^0 cells per dish. After 24 hr replace the culture medium with select medium: DMEM with 5% dialyzed FCS, 4.5 mg/ml glucose, and 50 μg/ml BrdU. Replace the select medium every other day. After 6 to 8 days cybrid colonies appear at a frequency of around 10^{-4} per ρ^0 cell plated. The desired number of independent clones are ring isolated, expanded, and frozen using standard techniques.

Fusion of WAL-2A-ρ^0 Cells and Cytoplasts. The mixture is allowed to recover for 2 days in 50 ml RPMI with 15% hi-FCS, 4 mg/ml glucose, 50 μg/ml uridine, and 1 mM pyruvate, then is centrifuged at 300 g for 3 min, suspended in 50 ml fresh medium, and returned to the incubator for 2 days. On day 4 postfusion, if the culture appears to be growing, the cells are centrifuged as above and suspended in 50 ml select medium: RPMI with 10% dialyzed FCS, 2 mg/ml glucose, and 1 μg/ml 6-thioguanine. If growth looks poor at this time the culture can be left another 2 to 4 days before changing to select medium. The select medium is replaced every 2 days up to day 12 postfusion, then every 4 days. After 20 to 28 days postfusion a rapid growth of the culture of mixed cybrid clones becomes evident, and the culture is passaged in select medium and aliquots frozen. If independent clones are desired these can be obtained with conventional semisolid media techniques.

Remarks. When studied using the present and other fusion methods with cytoplast donor cells that are heteroplasmic for mtDNA mutants, cybrid clones have been observed to segregate to homoplasmy of either mutant or wild-type mtDNA.[7,12,41] When passaging clones, keep cell pellets frozen for mtDNA analysis so that such genotype shifts can be investigated. Treatment of heteroplasmic clones with ethidium bromide (M. P. King, this volume [30]) or by supplementing the culture medium with uridine[42] may influence this segregation.

Nuclear OX-PHOS differences between cybrid clones[43] should also be considered in cybrid transfer experiments, so that groups of control cybrid clones may be required to determine the significance of some OX-PHOS variants associated with mtDNA mutations. The 2-fold greater specific respiratory rates seen in EBV-lymphoblast mitochondria compared to 143B (osteosarcoma) cell mitochondria (Table II) shows that nuclear differences between cell lines, presumably resulting from differences in differentiation or transformation status, can modulate OX-PHOS capacity.

[41] M. Yoneda, A. Chomyn, A. Martinuzzi, O. Hurko, and G. Attardi, *Proc. Natl. Acad. Sci. U.S.A.* **89,** 11164 (1992).
[42] T. Bourgeron, D. Chretien, A. Rotig, A. Munnich, and P. Rustin, *J. Biol. Chem.* **268,** 19369 (1993).
[43] A. Chomyn, S. T. Lai, R. Shakeley, N. Bresolin, G. Scarlato, and G. Attardi, *Am. J. Hum. Genet.* **54,** 966 (1994).

[43] Cytochemistry and Immunocytochemistry of Mitochondria in Tissue Sections

By Monica Sciacco and Eduardo Bonilla

Introduction

The mitochondria are the primary ATP-generating organelles in all mammalian cells and they contain their own DNA (mtDNA) which is maternally inherited.[1,2] ATP is produced via oxidative phosphorylation through five respiratory complexes located in the inner mitochondrial membrane. These respiratory complexes are multiple polypeptide enzymes whose subunits are encoded by genes of the nuclear DNA (nDNA) and of the mtDNA. The human mitochondrial genome contains genes encoding for thirteen subunits of different respiratory complexes. These include seven subunits of complex I or NADH dehydrogenase–ubiquinone oxidoreductase, one subunit of complex III or ubiquinone–cytochrome-c oxidoreductase, three subunits of complex IV or cytochrome-c oxidase (COX), and two subunits of complex V or ATP synthase.[3] Although mitochondria have their transcriptional and translational machinery, most of the proteins located within mitochondria are encoded by the nDNA. These nuclear gene products are synthesized on cytoplasmic ribosomes and are subsequently imported into the mitochondria.[4,5]

Pathogenic mtDNA mutations have been identified in three of the most prominent mitochondrial encephalomyopathies. First, deletions of mtDNA (Δ-mtDNA) have been associated with sporadic Kearns–Sayre syndrome (KSS) and are often seen in patients with isolated ocular myopathy. Second, myoclonus epilepsy with ragged-red fibers (MERRF) has been associated with two different point mutations, both in the tRNA^{Lys} gene. Third, mitochondrial encephalopathy, lactic acidosis, and stroke-like episodes (MELAS) has been associated with two different point mutations, both in the tRNA$^{Leu(UUR)}$ gene. Other point mutations in tRNAs and structural genes of the mitochondrial genome as well as new disorders apparently due to depletion of mtDNA have now been described.[6,7] Brain and skeletal muscle, whose function is highly dependent on oxidative metabolism, are the most severely affected tissues in these disorders. Consequently, genetic as well as morphologic studies of muscle from patients with mitochondrial encephalomyopathies have proven fundamental to further understand the

[1] S. Anderson, A. T. Bankier, B. G. Barrell, M. H. L. de Bruijn, A. R. Coulson, J. Drouin, I. C. Eperon, D. P. Nierlich, B. A. Roe, F. Sanger, P. H. Schreier, A. J. H. Smith, R. Staden, and I. G. Young, *Nature (London)* **290,** 457 (1981).

[2] R. E. Giles, H. Blanc, H. M. Cann, and D. C. Wallace, *Proc. Natl. Acad. Sci. U.S.A.* **77,** 6715 (1980).

[3] G. Attardi and G. Schatz, *Annu. Rev. Cell Biol.* **4,** 289 (1988).

[4] F. U. Hartl and W. Neupert, *Science* **247,** 930 (1990).

[5] W. Neupert, F. U. Hartl, E. Craig, and N. Pfanner, *Cell (Cambridge, Mass.)* **63,** 447 (1990).

[6] E. A. Schon, M. Hirano, and S. DiMauro, *J. Bioenerg. Biomembr.* **26,** 291 (1994).

[7] D. C. Wallace, *J. Bioenerg. Biomembr.* **26,** 241 (1994).

pathogenesis of respiratory chain enzyme deficiency at the individual muscle fiber level.[8,9]

The purpose of this chapter is to present the cytochemical and immuno-histochemical methods that, in our experience, appear to be the most reliable for the correct identification of mitochondria on frozen tissue sections, to illustrate their potential using specific examples, and to provide an updated version of the methods. Although the described protocols refer to muscle mitochondria, the methods described can be applied to any cell type.[10–12] It is not our intention to cover every study or method related to morphological aspects of mitochondria, but rather to provide enough information to allow investigators to apply these selected tools to a particular scientific or diagnostic question.

Cytochemistry

The visualization of normal and pathological mitochondria on frozen tissue sections can be carried out using a number of cytochemical techniques. These include the modified Gomori trichrome and hematoxylin–eosin stains, and cytochemical methods for the demonstration of oxidative enzyme activity.

The most informative cytochemical alteration in skeletal muscle is the ragged-red fiber (RRF), observed on frozen sections stained with the trichrome method of Engel and Cunningham.[13] The name derives from the reddish appearance of the trichrome-stained muscle fiber as a result of subsarcolemmal and/or intermyofibrillar proliferation of the mitochondria. The fibers harboring abnormal deposits of mitochondria are most often type I myofibers, and they may also contain increased numbers of lipid droplets. As accumulations of materials other than mitochondria may simulate RRF formation, the identification of deposits suspected of being mitochondrial proliferation should be confirmed cytochemically by the application of oxidative enzyme stains.

In our experience, enzyme cytochemistry for the activity of succinate

[8] S. Mita, B. Schmidt, E. A. Schon, S. DiMauro, and E. Bonilla, *Proc. Natl. Acad. Sci. U.S.A.* **86,** 9509 (1989).

[9] M. Sciacco, E. Bonilla, E. A. Schon, S. DiMauro, and C. T. Moraes, *Hum. Mol. Genet.* **3,** 13 (1994).

[10] K. Seki, T. Sato, Y. Ishigaki, and T. Ozawa, *Acta Neuropathol.* **77,** 465 (1989).

[11] M. J. Szabolic, R. Seigle, S. Shanske, E. Bonilla, S. DiMauro, and V. D'Agati, *Kidney Int.* **45,** 1388 (1994).

[12] M. Kumode, T. Yamano, and M. Shimada, *Acta Neuropathol.* **87,** 313 (1994).

[13] W. K. Engel and G. G. Cunningham, *Neurology* **13,** 919 (1963).

dehydrogenase (SDH) and COX have proved to be the most reliable methods for the correct visualization of normal mitochondria, and for the interpretation and ultimately the diagnosis of some of the mitochondrial disorders affecting skeletal muscle.[14]

Succinate Dehydrogenase

Succinate dehydogenase (SDH) is the enzyme that catalyzes the conversion of succinate to fumarate in the tricarboxylic acid cycle. It consists of two large subunits (a 70-kDa flavoprotein and a 30-kDa iron-sulfur containing protein) which form complex II of the mitochondrial respiratory chain along with two smaller subunits, responsible for attaching SDH to the inner mitochondrial membrane.[15,16] Complex II is the only component of the respiratory chain whose subunits are all encoded by the nuclear genome. For this reason, SDH histochemistry is extremely useful for detecting any variation in the fiber distribution of mitochondria, independently of any alteration affecting the mtDNA.

The cytochemical method for the microscopic demonstration of SDH activity on frozen tissue sections is based on the use of a tetrazolium salt (nitro blue tetrazolium, NBT) as electron acceptor with phenazine methosulfate (PMS) serving as intermediate electron donor to NBT.[17,18] The specificity of the method may be tested by performing control experiments in which an SDH inhibitor, sodium malonate (10 mM), is added to the incubation medium.

Using this method for detecting SDH activity in normal muscle sections, two populations of fibers are seen resulting in a checkerboard pattern. Type II fibers, which rely on glycolytic metabolism, show a light blue network-like stain. Type I fibers, whose metabolism is highly oxidative and therefore contain more mitochondria, show a more elaborate and darker mitochondrial network (Fig. 1A). In samples with pathological proliferation of mitochondria (RRF), the RRF show an intense blue SDH reaction corresponding to the distribution of the mitochondria within the fiber (Fig. 1C). This proliferation of mitochondria is associated with most mtDNA defects (deletions, tRNA point mutations, and depletion), but RRF can also be observed in other disorders that are thought to be due to defects of nDNA, such as

[14] S. DiMauro, M. Hirano, E. Bonilla, C. T. Moraes, and E. A. Schon, in "Cytochrome Oxidase Deficiency: Progress and Problems" (A. H. V. Shapira and S. DiMauro, eds.), p. 91. Butterworth/Heinemann, Oxford, 1994.
[15] Y. Hatefi, Annu. Rev. Biochem. 54, 1015 (1985).
[16] H. Beinert, FASEB J. 4, 2483 (1990).
[17] M. M. Nachlas, S. I. Margulies, and A. M. Seligman, J. Biol. Chem. 235, 2739 (1960).
[18] D. Pette, Histochem. J. 13, 319 (1981).

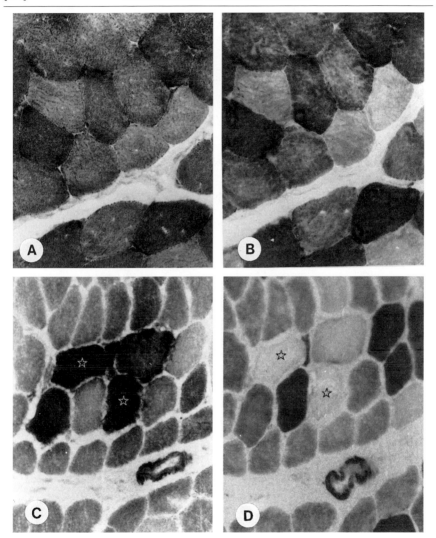

Fig. 1. Histochemical stains for SDH and COX activities in normal and KSS muscle. The normal shows a checkerboard pattern with both enzymes (A and B). The KSS samples demonstrate RRF (white stars) by SDH staining (C), and the same fibers on a serial section (black stars) show lack of COX activity (D). Magnification: ×210.

the depletion of muscle mtDNA, and the fatal and benign COX-deficient myopathies of infancy.[14]

Analysis by SDH cytochemistry is also useful for the diagnosis of complex II deficiency. Two patients with myopathy and complex II deficiency have been reported. In agreement with the biochemical observations, SDH histochemistry showed complete lack of reaction in muscle.[19,20]

Cytochrome-c Oxidase

Cytochrome-c oxidase (COX), or complex IV of the respiratory chain, is a multiple polypeptide enzyme composed of thirteen subunits. The three largest subunits (CO I, CO II, and CO III) are encoded by mtDNA and confer the catalytic and proton pumping activities to the enzyme. The ten smaller subunits are encoded by nDNA and are thought to provide tissue specificity by adjusting the enzymatic activity to the metabolic demands of the different tissues.[21] The dual genetic makeup of COX and the availability of a reliable cytochemical method to visualize its activity have made COX one of the ideal tools for basic investigations of mitochondrial biogenesis, nDNA–mtDNA interactions, and the study of mitochondrial encephalomyopathies at both the light and electron microscopic levels.[22,23]

The cytochemical method to visualize COX activity is based on the use of 3,3'-diaminobenzidine (DAB) as electron donor for cytochrome c.[24] The reaction product on oxidation of DAB occurs in the form of a brown pigmentation corresponding to the distribution of mitochondria in the tissue. The specificity of the method may also be tested by performing control experiments in which the COX inhibitor, potassium cyanide (10 mM), is added to the incubation medium.

As in the case of SDH, staining of normal muscle for COX activity also shows a checkerboard pattern. Type I fibers stain darker due to their mainly oxidative metabolism and more abundant mitochondrial content, and type II fibers show a finer and less intensely stained mitochondrial network (Fig. 1B).

The application of COX histochemistry to the investigation of KSS, MERRF, and MELAS has revealed one of the most important clues for the study of pathogenesis in these disorders. Muscle from these patients shows a mosaic expression of COX consisting of a variable number of

[19] B. Garavaglia, C. Antozzi, F. Girotti, D. Peluchetti, M. Rimoldi, M. Zeviani, and S. Di Donato, *Neurology* **40**(Suppl. 1), 294 (1990).
[20] R. G. Haller, K. G. Henriksson, L. Jorfeldt, E. Hultman, R. Wibom, K. Sahlin, N.-H. Areskog, M. Gunder, K. Ayyad, C. G. Blomquist, R. E. Hall, P. Thuillier, N. G. Kennaway, and S. F. Lewis, *J. Clin. Invest.* **88**, 1197 (1991).
[21] B. Kadenbach, L. Kuhn-Nentwig, and U. Buge, *Curr. Top. Bioenerg.* **15**, 113 (1987).
[22] E. Bonilla, D. L. Schotland, S. DiMauro, and B. Aldover, *J. Ultrastruct. Res.* **51**, 404 (1975).
[23] M. A. Johnson, D. M. Turnbull, D. J. Dick, and H. S. A. Sheratt, *J. Neurol. Sci.* **60**, 31 (1983).
[24] A. M. Seligman, M. J. Karnovsky, H. L. Wasserkrug, and J. S. Hanker, *J. Cell Biol.* **38**, 1 (1968).

COX-deficient and COX-positive fibers.[23,25] In KSS and MERRF, the RRF are invariably COX-deficient, but not all COX-deficient fibers are RRF (Fig. 1D). Before the advent of molecular genetics, it was difficult to understand the reason for the appearance of this mosaic, but when it was discovered that these patients harbored mutations of mtDNA in their muscles, it became evident that the mosaic was an indicator of the heteroplasmic nature of the genetic defects. The mosaic pattern of COX expression in mitochondrial disorders is now considered the "histochemical signature" of a heteroplasmic mtDNA mutation affecting the expression of mtDNA-encoded genes in skeletal muscle.[26,27] It should also be noted that COX deficiency, but showing a more generalized pattern, as well as COX-negative RRF are also observed in infants with depletion of muscle mtDNA or with either the fatal or the benign COX-deficient myopathies of infancy.[14] These observations indicate that cytochemical studies in mitochondrial disorders provide significant information about both the nature and the pathogenesis of mitochondrial disorders. Moreover, they provide useful clues as to which molecular testing is needed to provide a specific diagnosis.

Immunocytochemistry

The unique ability of immunocytochemistry to allow for the detection of specific proteins in single cells makes it a method of choice to study the expression of both mtDNA and nDNA genes in mitochondria of small and heterogeneous tissue samples. Technical advances have greatly increased the scope of immunocytochemistry and made it accessible to a variety of investigators with minimal expertise in immunology.

Several immunological probes are presently available to perform immunocytochemical studies of mitochondria on frozen tissue sections. These include antibodies directed against mtDNA- and nDNA-encoded subunits of the respiratory chain complexes, and antibodies against DNA that allow the detection of mtDNA.[28–30] Because the entire mitochondrial genome

[25] J. Müller-Hocker, D. Pongratz, and G. Hubner, *Virchows Arch. A Pathol. Anat.* **402,** 61 (1983).
[26] E. Bonilla, M. Sciacco, K. Tanji, M. Sparaco, V. Petruzzella, and C. T. Moraes, *Brain Pathol.* **2,** 113 (1992).
[27] E. A. Shoubridge, *in* "Mitochondrial DNA in Human Pathology" (S. DiMauro and D. C. Wallace, eds.), p. 109. Raven, New York, 1993.
[28] C. T. Moraes, S. Shanske, H.-J. Tritschler, J. R. Aprille, F. Andreetta, E. Bonilla, E. A. Schon, and S. DiMauro, *Am. J. Hum. Genet.* **48,** 492 (1991).
[29] A. Oldfors, N.-G. Larsson, E. Holme, M. Tulinius, B. Kadenbach, and M. Droste, *J. Neurol. Sci.* **110,** 169 (1992).
[30] F. Andreetta, H. Tritschler, E. A. Schon, S. DiMauro, and E. Bonilla, *J. Neurol. Sci.* **105,** 88 (1991).

has been sequenced, any mtDNA-encoded respiratory chain subunit is potentially available for immunocytochemical studies, and it is anticipated that the same will soon be true for all the nDNA-encoded subunits of the respiratory chain.[31]

There are several immunocytochemical methods for the study of mitochondria on tissue sections. These include enzyme-linked methods (peroxidase, alkaline phosphatase, and glucose oxidase) and methods based on the application of fluorochromes.[32] We favor the use of fluorochromes because they allow for the direct visualization of the antigen–antibody binding sites, and because they are more flexible for double-labeling experiments on frozen tissue sections.

Immunolocalization of Nuclear DNA- and Mitochondrial DNA-Encoded Subunits of Respiratory Chain

As mentioned earlier, we prefer immunolocalization via immunofluorescence, in particular methods using different fluorochromes for double-labeling studies. The main advantage of this approach is that it allows for the visualization of two different probes in the same mitochondrion and in the same plane of section. These methods also eliminate the inferences that must be made with studies on serial sections, and they are particularly indicated in immunocytochemical investigations of mitochondria in nonsyncytial tissues such as heart, kidney, and brain.

In our laboratory, we routinely use a monoclonal antibody against COX IV as probe for a nDNA-encoded mitochondrial protein and a polyclonal antibody against COX II as probe for a mtDNA-encoded protein. For these studies, the sections are first incubated with both the polyclonal and the monoclonal antibodies at optimal dilution, that is, the lowest concentration of the antibody giving a clear particulate immunostain corresponding to the localization of the mitochondria in normal muscle fibers. Subsequently, the sections are incubated with goat anti-rabbit immunoglobulin G (IgG)–fluorescein (to visualize the mtDNA probe in "green") and goat anti-mouse IgG–Texas Red (to visualize the nDNA probe in "red"). We carry out these studies with unfixed frozen sections, but with some antibodies it may be required to permeabilize the mitochondrial membranes to uncover the antigenic sequences or to facilitate the penetration of the probes into the inner mitochondrial compartment. In agreement with Johnson *et*

[31] B. Kadenbach and P. Merle, *FEBS Lett.* **135,** 1 (1981).

[32] H. M. Reisner and M. R. Wick, *in* "Monoclonal Antibodies in Diagnostic Immunohistochemistry" (M. R. Wick and G. P. Siegal, eds.), p. 1. Dekker, New York and Basel, 1988.

al.[33] we have also found that fixation of fresh frozen sections with 4% formaldehyde in 0.1 M $CaCl_2$, pH 7, followed by dehydration in serial alcohols (outlined in the final section of this chapter) provides the most reproducible and successful results.

Using unfixed muscle sections from normal samples, a checkerboard pattern resembling the one described for histochemistry is usually observed, type I fibers appearing brighter due to their higher mitochondria content (Fig. 2A,B). In muscle sections from patients with KSS harboring a documented Δ-mtDNA, both COX II and COX IV are normally present in nonaffected fibers. However, in COX-negative RRF we have observed a lack or marked reduction of COX II, associated with normal or enhanced COX IV staining (Fig. 2C,D). Because COX II was decreased or absent in COX-negative fibers when the Δ-mtDNA did not remove the COX II gene, we concluded that a deletion anywhere in the mtDNA can affect translation, even of genes not encompassed by the deletion. This is presumably because the deletion eliminates essential tRNA genes that are required for mitochondrial translation of all thirteen mtDNA-encoded subunits of the respiratory chain.[34,35]

Immunolocalization of Mitochondrial DNA

Immunocytochemistry using anti-DNA antibodies has been applied as an alternative method to *in situ* hybridization for the studies of localization and distribution of mtDNA in normal and pathological conditions.[30,36] The advantages of this method are that both mitochondrial and nuclear DNA are detected simultaneously, at the single cell level, and that the nuclear signal can be used as an internal control.

For detection of mtDNA using immunological probes, we also carry out double-labeling experiments with different fluorochromes. We utilize polyclonal antibodies against COX IV for immunolabeling of mitochondria in one color (green), and a monoclonal antibody against DNA for immunostaining of mtDNA and nDNA in another color (red). In frozen muscle sections from normal controls and from patients without depletion of

[33] M. A. Johnson, B. Kadenbach, M. Droste, S. L. Old, and D. M. Turnbull, *J. Neurol. Sci.* **87**, 75 (1988).

[34] H. Nakase, C. T. Moraes, R. Rizzuto, A. Lombes, S. DiMauro, and E. A. Schon, *Am. J. Hum. Genet.* **46**, 418 (1990).

[35] C. T. Moraes, E. Ricci, V. Petruzzella, S. Shanske, S. DiMauro, E. A. Schon, and E. Bonilla, *Nat. Genet.* **1**, 359 (1992).

[36] H.-J. Tritschler, F. Andreetta, C. T. Moraes, E. Bonilla, E. Arnaudo, M. J. Danon, S. Glass, B. M. Zelaya, E. Vamos, N. Telerman-Toppet, B. Kadenbach, S. DiMauro, and E. A. Schon, *Neurology* **42**, 209 (1992).

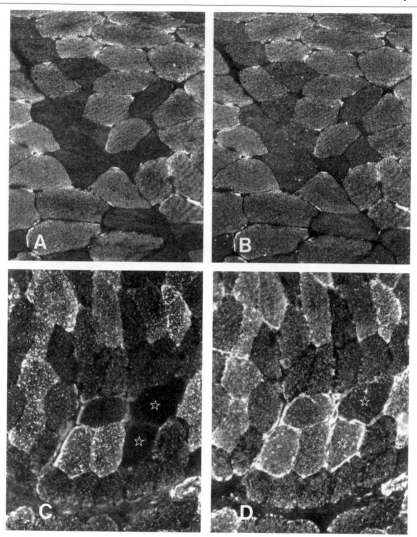

Fig. 2. Immunolocalization of COX II and COX IV in normal (A and B) and KSS (C and D) muscle. The normal muscle shows an identical checkerboard pattern for both subunits of COX. The KSS sections show reduced COX II immunostaining in some fibers (white stars) with normal or enhanced staining for COX IV. Magnification: ×110.

mtDNA, these antibodies show an intense staining of both the nuclei and a cytoplasmic network correlating with mitochondrial localization (Fig. 3A,B). Conversely, when muscle biopsies from patients with mtDNA depletion are analyzed, the particular immunostaining of mtDNA is not detect-

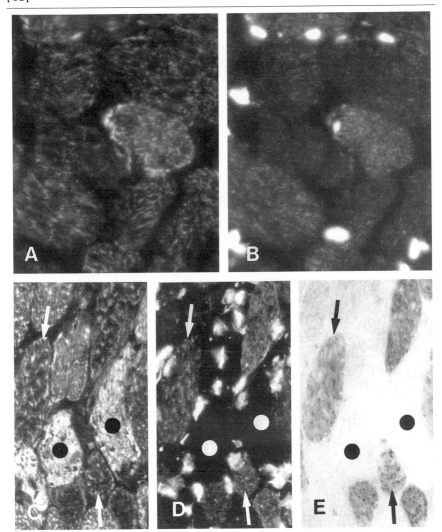

FIG. 3. Immunolocalization of COX IV and mtDNA in normal (A and B) and in muscle from a patient with mtDNA depletion (C, D, and E). The normal muscle shows the mitochondrial network with both antibodies, but the nuclei are stained (B) with the anti-DNA antibody. Magnification: ×320. The muscle sections from the depleted patient show RRF (C) stained with COX IV antibodies, but the same fibers (white and black circles) lack the mitochondrial stain (D) with anti-DNA antibody and COX enzymatic activity (E). Also notice that fibers with COX activity (black arrows) show immunostaining of the mitochondrial network with both antibodies (white arrows). Magnification: ×210.

able, or it is present in only a small number of COX-positive fibers (Fig. 3C–E). The intensity of nDNA immunostaining shows no alteration compared with nondepleted controls.

Immunocytochemistry utilizing antibodies against DNA is useful for the rapid evaluation of the distribution of mtDNA in normal cells and for the detection of depletion of muscle mtDNA. This method is particularly precise for the diagnosis of mtDNA depletion when it is confined to only a subpopulation of fibers.

Cytochemical Methods

Succinate Dehydrogenase

Collect 8-μm-thick cryostat sections on poly(L-lysine)-coated (0.1%) coverslips. Dissolve the following in 10 ml of 5 mM phosphate buffer, pH 7.4:

5 mM Ethylenediaminetetraacetic acid (EDTA)
1 mM Potassium cyanide (KCN)
0.2 mM Phenazine methosulfate (PMS)
50 mM Succinic acid
1.5 mM Nitro blue tetrazolium (NBT)

Adjust to pH 7.6 and filter solution with Whatman No. 1 filter paper. Incubate sections for 20 min at 37°. For control sections, sodium malonate (10 mM) is added to the incubation medium. Rinse sections three times, 5 min each, in distilled water, at room temperature. Mount on glass slides with warm glycerin gel.

Cytochrome-c Oxidase

Collect 8-μm-thick cryostat sections on poly(L-lysine)-coated (0.1%) coverslips. Dissolve the following in 10 ml of 5 mM phosphate buffer, pH 7.4:

0.1% (10 mg) 3,3'-Diaminobenzidine (DAB)
0.1% (10 mg) Cytochrome c (from horse heart)
0.02% (2 mg) Catalase

Adjust to pH 7.4. Do not expose solution to light. Filter solution with Whatman No. 1 filter paper. Incubate sections 1 hr at 37°. For control sections, potassium cyanide (10 mM) is added to the incubation medium. Rinse sections three times with distilled water, 5 min each time, at room temperature. Mount on glass slides with warm glycerin gel.

Immunocytochemical Methods

Double-Labeling for Simultaneous Visualization of Mitochondrial DNA- and Nuclear DNA-Encoded Subunits of Respiratory Chain Using Different Fluorochromes

Collect 4-μm-thick cryostat sections on poly(L-lysine)-coated (0.1%) coverslips. Incubate the sections 2 hr at room temperature (in a wet chamber) with anti-COX II polyclonal antibody and with anti-COX IV monoclonal antibody at optimal dilutions (1:100 to 1:500) in phosphate-buffered saline containing 1% bovine serum albumin (PBS/BSA). Control sections are incubated without the primary antibodies. Rinse the samples three times with PBS, 5 min each time, at room temperature. Incubate the sections 1 hr at room temperature (in a wet chamber) with anti-rabbit IgG–fluorescein and with anti-mouse IgG–Texas Red diluted 1:100 in 1% BSA/PBS. Rinse the samples three times with PBS, 5 min each time, at room temperature. Mount on slides with 50% glycerol in PBS.

Double-Labeling for Simultaneous Visualization of Mitochondria and Mitochondrial DNA Using Different Fluorochromes

Collect 4-μm-thick cryostat sections on poly(L-lysine)-coated (0.1%) coverslips. Fix the sections in 4% formaldehyde in 0.1 M CaCl$_2$, pH 7, for 1 hr at room temperature. Dehydrate the sections in 70, 80, 90% (v/v) ethanol, 5 min each, and in 100% ethanol, 15 min. Rinse the samples with PBS, three times, 5 min each time, at room temperature. Incubate the sections 2 hr at room temperature (in a wet chamber) with anti-DNA monoclonal antibody (1:100) and with anti-COX IV polyclonal antibody (1:500) in 1% BSA/PBS. Control sections are incubated without the primary antibodies. Rinse the samples with PBS, three times, 5 min each time, at room temperature. Incubate the sections for 30 min at room temperature (wet chamber) with biotinylated anti-mouse IgG (1:100) in 1% BSA/PBS. Rinse the samples with PBS, three times, 5 min each time, at room temperature. Incubate the sections for 30 min at room temperature (wet chamber) with streptavidin–Texas Red (1:250), and with anti-rabbit IgG–fluorescein isothiocyanate (FITC) (1:100) in 1% BSA/PBS. Rinse the samples with PBS, three times, 5 min each time, at room temperature. Mount on slides with glycerol–PBS (1:1, v/v).

Acknowledgments

Research was supported by Grant NS-11766 from the National Institutes of Health, and by grants from the Muscular Dystrophy Association (E. B.) and the Italian Telethon (M. S.)

[44] Detection and Analysis of Mitochondrial DNA and RNA in Muscle by *in Situ* Hybridization and Single-Fiber PCR

By Carlos T. Moraes *and* Eric A. Schon

Detection of Mitochondrial DNA and RNA by *in Situ* Hybridization

Overview

The power of *in situ* hybridization (ISH) in providing spatial information of gene location and expression is unique. In the last few years, ISH has provided a wealth of information on the intercellular distribution of heteroplasmic mitochondrial DNA (mtDNA) populations, with its usefulness particularly evident in the study of human pathologies. The technique has been used extensively to correlate mitochondrial abnormalities with the presence of mutated mtDNAs, an analysis that provides strong support for a pathogenetic role of a specific mitochondrial genotype. Because the method relies on sequence homology, it has been used mainly to differentiate mtDNA targets with extensive sequence variations (e.g., distinguishing mtDNAs with large-scale deletions from wild-type sequences[1-5]). Easier and more powerful techniques, such as single-cell polymerase chain reaction (PCR) (see second part of this chapter) have been devised for the study of the intercellular distribution of mtDNA populations differing by single nucleotide substitutions.[6]

Despite its qualitative nature, ISH can be used for semiquantitative determinations, provided that certain precautions are taken. This approach has been used to determine the ratio of rRNA and heavy-strand-encoded mRNAs in muscle of patients with mitochondrial encephalomyopathy, lactic acidosis, and stroke-like episodes (MELAS). *In vitro* studies suggested

---

[1] C. T. Moraes, E. Ricci, V. Petruzzella, S. Shanske, S. DiMauro, E. A. Schon, and E. Bonilla, *Nat. Genet.* **1,** 359 (1992).

[2] S. R. Hammans, M. G. Sweeney, D. A. G. Wicks, J. A. Morgan-Hughes, and A. E. Harding, *Brain* **115,** 343 (1992).

[3] A. Oldfors, N.-G. Larsson, E. Holme, M. Tulinius, B. Kadenbach, and M. Droste, *J. Neurol. Sci.* **110,** 169 (1992).

[4] E. A. Shoubridge, G. Karpati, and K. E. M. Hastings, *Cell (Cambridge, Mass.)* **62,** 43 (1990).

[5] S. Mita, B. Schmidt, E. A. Schon, S. DiMauro, and E. Bonilla, *Proc. Natl. Acad. Sci. U.S.A.* **86,** 9509 (1989).

[6] C. T. Moraes, E. Ricci, E. Bonilla, S. DiMauro, and E. A. Schon, *Am. J. Hum. Genet.* **50,** 934 (1992).

that an A → G transition at mtDNA position 3243 could alter rRNA/mRNA ratios because of a change in the binding site of a transcription termination factor.[7] Semiquantitative measurements of mitochondrial transcripts by ISH showed that transcription termination was not significantly impaired in muscle of MELAS patients.[6] These results were confirmed in studies of rRNA and mRNA isolated from transmitochondrial cybrids.[8,9] Because the transmitochondrial cybrid system, while powerful, is a somewhat limited model system for human pathologies, it is important to perform experiments with patients' tissues, even when the amount available is too small for standard purification of RNA or proteins (e.g., small muscle biopsies). Moreover, mtDNA heteroplasmy and the focal nature of the defect make the study of mtDNA pathogenesis difficult if not impossible in tissue homogenates. Therefore, ISH may provide the only means to perform an informative analysis.

Although ISH utilizing RNA probes is more widely used in typical cell and molecular biology applications, most of our experience derives from the use of DNA probes.[1,5,6] We found that the high copy number and the consequent high levels of transcription products of mtDNA provide an easy target for ISH experiments, and both DNA and RNA probes would give satisfactory results. In this chapter we describe both protocols, but the reader should keep in mind that the use of RNA probes, although requiring special care, may yield better signal-to-background results when the levels of the target sequence are low.

Preparation of Tissue Sections

Although tissue samples can be frozen or fixed in paraffin, we describe a procedure optimized for frozen specimens, and thoroughly tested on muscle sections. Several ISH protocols for paraffin-embedded tissues can be found in technical articles. Glass slides should be treated with polylysine to ensure adherence of the sample to the glass. Clean slides are dipped in a 0.2 mg/ml polylysine solution and allowed to dry in a vertical position. Because of the focal pattern of distribution of mutant and wild-type mtDNA in muscle fibers, be sure that serial sections above and below the ones used for ISH are phenotyped [e.g., by staining for cytochrome-c oxidase (COX) and succinate dehydrogenase (SDH) activities]. Although tissue preparation is described first in this chapter, radiolabeled probes should be prepared in advance.

[7] J. F. Hess, M. A. Parisi, J. L. Bennett, and D. A. Clayton, *Nature* (*London*) **351,** 236 (1991).

[8] M. P. King, Y. Koga, M. Davidson, and E. A. Schon, *Mol. Cell. Biol.* **12,** 480 (1992).

[9] A. Chomyn, A. Martinuzzi, M. Yoneda, A. Daga, O. Hurko, D. Johns, S. T. Lai, I. Nonaka, C. Angelini, and G. Attardi, *Proc. Natl. Acad. Sci. U.S.A.* **89,** 4221 (1992).

Preparation of Muscle Sections

1. Eight-μm-thick cryostat-cut muscle sections from patients and controls should be mounted on the same slide.
2. Sections are fixed in 4% (w/v) paraformaldehyde in 1× phosphate-buffered saline (PBS, pH 7.4) for 10 min. This and the next steps should be performed in Coplin jars.
3. Sections are treated with proteinase K (5 μg/ml in 1× PBS) to facilitate tissue penetration. The treatment is performed at room temperature for 5 min. Although this time could be modified to adjust to the sample, longer treatments may lead to undesirable morphological changes.
4. Wash with 1× PBS for 5 min.
5. Treat slides with a freshly prepared 0.25% v/v solution of acetic anhydride in 0.1 M triethanolamine at room temperature for 10 min. Acetylation helps reduce nonspecific background.
6. Wash with 1× PBS for 5 min.
7. If detection of mtRNA is desired, proceed to step 10 without further treatment.
8. For detection of mtDNA, samples should be treated with DNase-free RNase (3.2 U/ml of 10 mM Tris-HCl, pH 8.0, 50 mM NaCl) at 37° for 30 min. To conserve enzyme, this step can be performed in damp boxes, with the slides lying horizontally and the sections facing up; 500 μl solution is enough to cover several sections. The damp box (a covered plastic box with damp paper inside) can be placed in a dry incubator.
9. Wash the RNase out with 1× PBS for 5 min.
10. Prehybridize sections (see composition of prehybridization solution in Appendix) at 42° for 1–3 hr in a box damped with "box buffer" (4× SSC, 50% v/v formamide); 100 μl prehybridization solution is usually enough to cover all sections on the slide. The prehybridization solution will stay on top of the sections if the unused region of the slide surrounding the sections is dried with Kimwipes tissue. Drip the prehybridization solution on top of the sections, and carefully spread the solution with the side of the pipette tip to cover all the desired area. The viscous prehybridization and hybridization solutions will not spread to the dried region of the slide.
11. The next step (denaturation) is specific for DNA detection. Prepare a solution of 70% v/v deionized formamide, 2× SSC and place it in a glass Coplin jar kept in a 75° water bath. After the liquid has equilibrated at 75°, the slides are placed in the hot solution for 4 min, and transferred quickly to a second jar containing 70% (v/v)

ethanol precooled at $-70°$ (5 min). Sometimes 70% ethanol will freeze at temperatures below $-70°$. In these cases remove the jar from the freezer 1–2 min earlier to allow thawing.

12. Dry sections by transferring sequentially to a 90% (5 min) and 100% (5 min) ethanol (v/v). When performing RNA detection (denaturation step omitted) start drying the sample with 70% ethanol.

In Situ Hybridization with DNA Probes

Preparation of DNA Probes. The size of the labeled probe is very important and should balance specificity and tissue penetration. Sizes between 100 and 500 nucleotides are optimum, even though we have obtained excellent results with 600-nucleotide probes.[6] DNA probes can be prepared in different ways. Most of our experience derives from the use of probes prepared by primer extension of single-strand M13 constructs harboring specific human mtDNA regions as inserts (a kind gift of M. P. King and G. Attardi). After primer extension in the presence of [^{35}S]dATP, the double-stranded segment is cleaved at flanking sites of the insert with restriction endonucleases, and purified by agarose gel electrophoresis. The detailed protocol is given below. We have also obtained good results labeling PCR fragments by the random primer method. In these cases, PCR fragments are purified by electrophoresis in an agarose gel, followed by isolation by silica matrix binding. The purified template ($>$200 bp) is labeled with a random primer labeling kit following the manufacturer's recommendations. The labeled fragments can be purified from the reaction mix by ethanol precipitation in the presence of 0.3 M ammonium acetate, or by using a silica matrix purification kit.

Primer Extension of Mitochondrial DNA Inserts in M13 Clones

1. Mix 2 μg of single-strand M13 (containing the desired insert), 10 ng of M13 universal primer, and 2 μl of 10$\times$ buffer M [from Boehringer Mannheim (Indianapolis, IN) restriction endonucleases; final concentrations (diluted 1:10) of 10 mM Tris-HCl, 10 mM MgCl$_2$, 50 mM NaCl, 1 mM dithioerythritol; pH 7.5 at 37°]. Complete the volume with distilled water to 20 μl total.
2. Heat the sample at 55° for 5 min, followed by 37° for 10 min.
3. Add 1 μl of each dCTP, dGTP, and dTTP (0.5 mM stocks), 5 μl [^{35}S]dATP, and 1 μl sequencing grade Klenow enzyme (5 U/μl). Incubate at 37° for 20 min.
4. Add 1 μl dATP (from 0.5 mM stock). Incubate at 37° for 5 min.
5. Incubate mixture at 68° for 10 min to inactivate the Klenow enzyme.

6. The labeled insert is excised from the vector by digestion with restriction endonucleases that recognize sites flanking the insert. We commonly use *Eco*RI and *Hin*dIII double digestion. Digestion can be performed in the same tube after adding 1 μl of a 10× *Eco*RI buffer.
7. The digestion mixture is electrophoresed through a 1.5% agarose gel. The labeled fragment is visualized with ethidium bromide and excised from the gel with a scalpel.
8. The fragment can be purified from the agarose plug using a silica matrix kit [e.g., GeneClean from BIO 101 (La Jolla, CA) or Quiex from Qiagen (Chatsworth, CA)] or by phenol/ethanol purification (if the latter method is used, the desired band should be purified from low-melting agarose gel).
9. Dithiothreitol (DTT, 10 mM final concentration) is immediately added to the purified probe to avoid oxidation. Store probe at $-70°$. ^{35}S-Labeled probes are stable and sufficiently active for up to 2 months. The minimum specific activity should be 10^7 counts/min (cpm)/μg of plasmid. For random primer probes, specific activity should be between 10^8 and 10^9 cpm/μg.

Hybridization and Washes

1. To the dried prehybridized slides add a solution containing 1 μl of 1 M DTT and 1–2 × 10^6 cpm of boiled-denatured probe per 100 μl of hybridization solution (see Appendix). The volume of hybridization mix necessary to cover the specimens will depend on the area taken by the sections (usually 50 μl would be enough for each slide).
2. Hybridization is performed at 42° overnight in a closed box containing paper damped with box buffer (see Appendix).
3. Next day (after 12–16 hr), place the slides in a plastic rack that can be hung at the edge of a 2-liter beaker. Wash with 2× SSC (with gentle stirring on a magnetic stirrer) at room temperature for 1 hr.
4. Change the washing solution to 0.1× SSC, 14.4 mM 2-mercaptoethanol, and 0.05% w/v sodium pyrophosphate. This solution should be kept at 50° with gentle stirring. Because the temperature should be carefully controlled, we do not recommend the use of a hot plate to heat the solution. A dipped heating element with a thermostat control will maintain a constant temperature more effectively. Wash for 3 hr.
5. Dry slides through 70, 80, and 90% (v/v) ethanol containing 0.3 M ammonium acetate for 2 min each, and finally through 100% ethanol for 2 min.
6. Dry completely in air.

In Situ Hybridization with RNA Probes

Preparation of RNA Probes. Kits and polymerases for RNA labeling are available from most molecular biology companies. The protocol described below has also been used successfully on brain tissues.[10]

1. Cleave 1 μg plasmid at the 3' end of the insert. Purify plasmid by phenol extraction/ethanol precipitation (with carrier) or silica matrix kits. Add 30 μl [^{35}S]UTP (800 Ci/mmol).
2. Evaporate to dryness in a Speed-Vac. Immediately proceed with labeling.
3. At room temperature add to the tube containing the restricted plasmid and [^{35}S]UTP the following: 20 μl of 5× transcription buffer, 10 μl of 1 M DTT, 5 μl of 10 mM ATP, 5 μl of 10 mM CTP, 5 μl of 10 mM GTP, 2.4 μl of 1 mM UTP (sum of hot and cold UTP, 25 μM), 1 μl RNasin (Promega, Madison, WI) and 1 μl promoter-compatible RNA polymerase (e.g., T3, SP6). Add water to 100 μl final volume.
4. Incubate 30 min at 37°. Add 0.5 μl of RNasin and 0.5 μl of DNase I (0.5 U, RNase-free). Incubate 10 min at 37°.
5. Add 422 μl TE (10 mM Tris-HCl, 1 mM EDTA; pH 8.5) and 22.5 μl of 4 M NaCl. Extract the aqueous phase with phenol/chloroform/isoamyl alcohol (50:49:1, v/v). Reextract the aqueous phase with chloroform/isoamyl alcohol (49:1, v/v). To the aqueous phase add 2.5 volumes of ethanol. Place on ice for 10 min. Spin in microcentrifuge for 10 min. Remove supernatant and rinse pellet with 70% (v/v) ethanol.
6. Remove ethanol, dry and add 500 μl TE, 0.2% SDS, and 5 μl of 5 M DTT. Count 1 μl and freeze the probe until ready for use.
7. If the transcripts are larger than 400 nucleotides, the RNA probe should be sheared to allow better penetration (optimum around 150 nucleotides). Shearing can be made by the following protocol. Mix 50 μl of RNA suspended in water, 30 μl of 0.2 M Na$_2$CO$_3$, and 20 μl of 0.2 M NaHCO$_3$. Incubate at 60° for a period of time calculated as follows.

$$t(\text{minutes}) = (L_o - L_f)/(KL_oL_f)$$

where L_o is the starting length (kb); L_f, final length (kb); and K, 0.11. Stop the reaction by adding 3 μl of 3 M sodium acetate, pH

---

[10] W. S. I. Young, *in* "Handbook of Chemical Neuroanatomy" (A. Björklund, T. Hökfelt, F. G. Wouterlood, and A. N. van den Pol, eds.), Vol. 8, p. 481. Elsevier, Amsterdam, 1990.

6.0, and 5 μl of 10% v/v glacial acetic acid. Precipitate with ethanol in the presence of 0.3 M sodium acetate. The average size of labeled RNA can be determined in glyoxyl gels.[11]

Hybridization and Washes. Using the RNA probe hybridization solution described in the Appendix, prehybridize for 1 hr at 55° without the probe and hybridize at 55° overnight in humid chambers with 1–2 × 10⁶ cpm of riboprobe as described earlier. Some investigators have found that the prehybridization step could be omitted without undesirable increase in background signal.

After hybridization, wash the slides in 4× SSC at room temperature for 15 min. Wash four more times in 4× SSC, 1 mM DTT. Incubate 30 min with 20 μg/ml RNase A at 37°, in RNase buffer (for 500 ml: 62.5 ml of 4 M NaCl, 2.5 ml of 2 M Tris-HCl, pH 8.0, 0.5 ml of 0.25 M EDTA). Wash twice, 5 min each, with 1× SSC, 1 mM DTT, once in 0.5× SSC, 1 mM DTT, and once in 0.1× SSC, 1 mM DTT. Wash twice (30 min each) in 0.1× SSC, 1 mM DTT at 65°. Cool the slides by washing them in 0.1× SSC, 1 mM DTT at room temperature for 5 min. Dry slides through 70, 80, and 90% ethanol containing 0.3 M ammonium acetate for 2 min each, and in 100% ethanol. Air dry.

Analysis of Results. Tape the border of the slides to a hard support (e.g., old X-ray film) and expose to an X-ray film for 16–24 hr. Radioactive signals in test samples should be stronger than the signal in negative controls. The X-ray images can be used for semiquantitative analysis. For this purpose, slides should be exposed with a set of radioactive standards (e.g., serial 1× dilutions of a known amount of ³⁵S blotted onto a piece of paper) to be used in image analysis. A number of image analysis software packages are available for such analysis, and the final results are usually expressed as cpm/mm². Figure 1 exemplifies a semiquantitative analysis using the X-ray image of ISH experiments. Using an RAS-DG 1000 autoradiography analysis system (Amersham, Arlington Heights, IL), standard curves were constructed with coexposed ³⁵S-labeled markers. Second-order correlations of 0.97–0.99 were obtained for ³⁵S markers spanning 20-fold increments (with sample signals within this range). Muscle section images were digitalized (Fig. 1A) and quantified according to the manufacturer's guidelines (Fig. 1B). With the help of a computer mouse the boundaries of the quantitation were restricted to large but intact areas of the specimen image.

After obtaining film images, slides can be dipped in the dark in Kodak (Rochester, NY) NTB-2 nuclear emulsion (diluted 1 : 1 with water). The emulsion (extremely thick when cold) should be warmed at 42° before the desired amount is pipetted out. The dilution should be made with pre-

---

[11] J. Sambrook, E. F. Fritsch, and T. Maniatis, "Molecular Cloning: A Laboratory Manual." Cold Spring Harbor Laboratory, Cold Spring Harbor, New York, 1992.

warmed distilled water directly into a Coplin jar in the dark. Because dipping is also done in the dark, blank slides should be dipped first to remove bubbles. Place ISH slides in red plastic slide grips and dip into emulsion. Blot the dipped end quickly in paper towels, invert, and let them dry for several hours in the vertical position. Transfer slides to black slide boxes with desiccant capsules or Drierite. Tape edges of the boxes with a dark tape and store them at 4° for 1–2 weeks.

Develop the slides in D-19 developer (agitating every 30 sec) for 5 min; water for 15 sec; and Kodak rapid fix (without hardener) for 5 min. Rinse in running tap water for 10 min. Counterstain, if desired, for 30 sec in 0.4% toluidine blue, and rinse with water to remove excess stain. Different counterstains can be used, but some of them may interfere with the radioactive signal, particularly if the staining is performed before the ISH procedure. Cover the sections with coverslips using glycerin gel. The silver grains can be visualized as black dots on top of the counterstained tissue. Dark-field condensors allow better visualization of the signal but not of the counterstained background (Fig. 2). Semiquantitative analysis can be performed by visual or automatic grain counting.[1]

Genotyping Mitochondrial DNA from Single Muscle Fibers

Overview

Although ISH can provide powerful information on the cellular distribution of deletion mutant and wild-type mtDNAs, it cannot discriminate mitochondrial genomes differing by a single nucleotide. Moreover, only semiquantitative measurements can be achieved by ISH. Because of these limitations, we and others have developed a PCR-based system to study heteroplasmic mtDNA mutations in single muscle fibers. In brief, phenotypically characterized muscle fibers are microdissected under the microscope and subjected to PCR amplification. Molecules originating from mutant and wild-type mtDNAs are distinguished either by size or by restriction fragment length polymorphism (RFLP). The presence of higher levels of mutant mtDNAs in cells with an abnormal phenotype is one of the most compelling pieces of evidence of pathogenicity for a mtDNA polymorphism.[11]

Equipment

Microdissection of muscle fibers does not require unusual equipment such as micromanipulators or especially designed microscopes, just a steady hand and a few hours of practice. Other investigators have used thin tungsten needles to "hook" out the fibers.[4] We have used a simple mouth suction

A

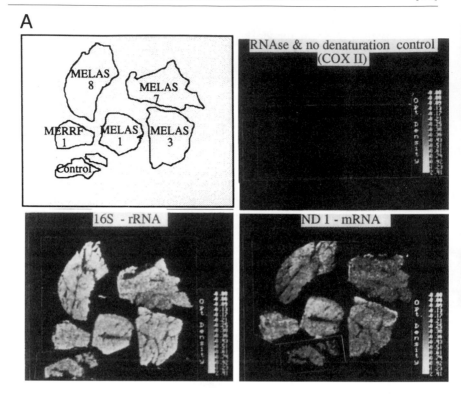

B

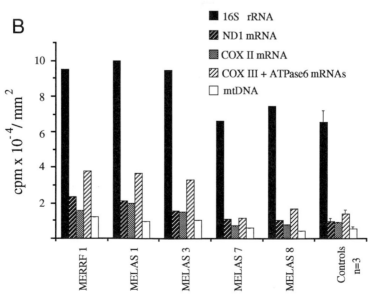

SDH   COX

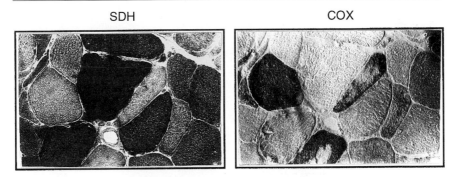

mtDNA

FIG. 2. Cellular localization of mtDNA by *in situ* hybridization. Serial muscle sections from a patient with MERRF were stained for succinate dehydrogenase (SDH) and cytochrome-*c* oxidase (COX) activity and for mtDNA localized by ISH. The ISH signal is seen as white spots under dark-field illumination. Note the strong mtDNA signal in fibers containing increased numbers of mitochondria (fibers staining dark for SDH activity).

FIG. 1. Semiquantitative analysis of *in situ* hybridization images. Slides of muscle sections subjected to *in situ* hybridization were exposed to X-ray films for appropriate times. Radioactive markers were exposed together with the sections for quantitation. Each X-ray image was digitized and quantitated using an RAS-DG 1000 autoradiography analysis system (Amersham). (A, *top left*) Muscle sections shown in the remaining panels. (A, *top right*) Background levels obtained after RNase treatment but no denaturation. (A, *bottom*) Digitized image obtained for the denoted probes used for detection of RNA. DNA detection was also performed but is not shown in (A). The image of 16 S rRNA was obtained from an X-ray film exposed for 6 hr. The images of the different mRNAs were obtained from an X-ray film exposed for 16 hr. DNA images as well as the RNase and nondenatured control images were obtained from a film exposed for 20 hr. (B) Semiquantitative image analysis results expressed as cpm/mm². Modified from Ref. 6 with permission from The University of Chicago Press.

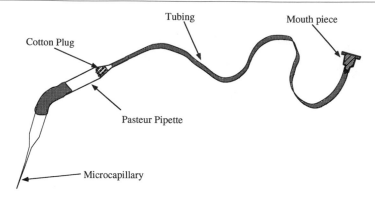

FIG. 3. Microdissection device. This simple device can be constructed with materials commonly available in a research laboratory. The mouthpiece is connected to a cotton-plugged Pasteur pipette that connects to the dissecting microcapillary. The microcapillaries are changed constantly, while the rest of the device can be used for undetermined periods of time.

device depicted in Fig. 3. The tubing and mouthpieces can be found inside boxes of microcapillaries. It is important to add a Pasteur pipette with a cotton plug into the system to avoid contamination by the operator (Fig. 3). Capillaries are prepared by flaming Pasteur pipettes at the narrow end and pulling the ends apart when the glass starts to melt. The pulled glass is broken by rubbing another small piece of glass against it. After preparing several microcapillaries, siliconize the tips by dipping them into a silicone solution (the same used for siliconizing glass plates for electrophoresis), blowing it out (with the cotton plug-protected tubing device), and letting them dry. Dissection is performed under an inverted microscope commonly found in cell culture laboratories.[6]

Preparation of Muscle Sections and Isolation of Single Fibers

Although the thickness of the section may vary, we found 30 μm to be a convenient size. Thicker sections stain too strongly with the histochemical methods, and it becomes difficult to distinguish cellular phenotypes. Slides should be pretreated with polylysine to avoid movement during the dissection. Sections from test and controls are placed side by side on the same slide and stained for enzyme activity of succinate dehydrogenase, cytochrome-c oxidase, both, or any other staining appropriate for the experiment.[12,13] Stained sections provide a visual phenotype that will direct the

[12] A. M. Seligman, M. J. Karnovsky, H. L. Wasserkrug, and J. S. Hanker, *J. Cell Biol.* **38**, 1 (1968).
[13] V. K. Dubowitz and M. H. Brooke, "Muscle Biopsy: A Modern Approach." Saunders, Philadelphia, Pennsylvania, 1973.

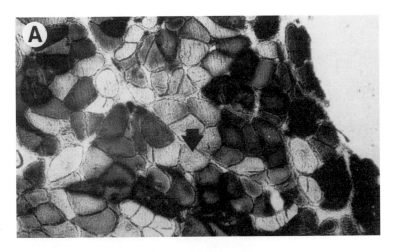

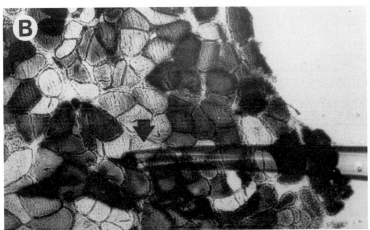

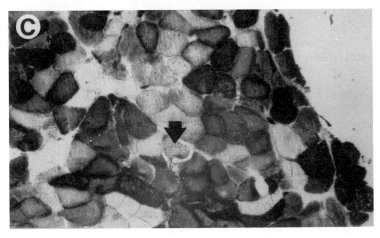

Fig. 4.

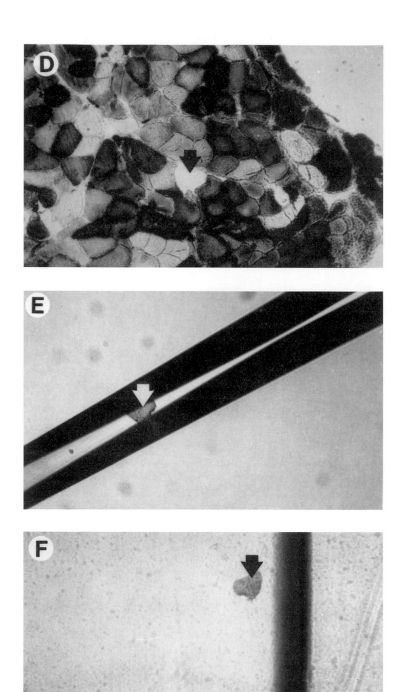

Fig. 4. (*continued*)

dissection. Slides are then immersed in a 100-mm^2 cell culture dish containing 50 ml of 50% ethanol. Single fibers are isolated by aerosol-protected mouth suction with siliconized microcapillaries under an inverted microscope (Fig. 4). Sometimes, muscle fibers are too small for easy isolation from neighboring fibers. In such cases, isolate the desired fiber by pulling away adjacent fibers surrounding the targeted one, and then pick up the targeted fiber. Most often the operator has to hold the slide in place to get a better grip. A clean powder-free glove-protected finger from the nondissecting hand can be used for this purpose. Fibers are isolated with as little liquid (ethanol, 50% v/v) as possible and placed directly into 50 μl of water in a 250-μl Eppendorf tube. With the help of the microscope, try to locate the fiber inside the tube to ensure that it was expelled from the pipette and not lost. Also pick a few fibers from the control section. As a negative control for the PCR reaction, place an equivalent volume of the overlaying liquid (ethanol, 50%) in an Eppendorf tube containing 50 μl of distilled water.

DNA Amplification from Single Fibers

After all fibers are picked, spin tubes in a microcentrifuge at maximum speed for 5 min. Remove the supernatant and treat the fiber to release the DNA. We have used a modification of the procedure by Li et al. developed for single-spermatozoa PCR.[14] It involves adding 5 μl of an alkaline mix (see Appendix) to the fiber and incubating it at 65° for 30 min. The solution is then neutralized by adding 5 μl of a buffered neutralizing solution (see Appendix). Aliquots of the resulting 10 μl solution are ready for PCR. For the detection of single nucleotide changes, a standard PCR reaction is performed, and successful amplification checked in an agarose gel. If the correct product is obtained, PCR mixes are subjected to an additional cycle after the addition of 10 μCi of [α-^{32}P]dATP, 100 pmol of each primer, and 2 U of Taq polymerase in 1× PCR buffer. The addition of ^{32}P in the last cycle avoids the detection of heteroduplexes that can skew the RFLP result.[6] Restriction endonuclease-digested fragments are separated by native

[14] H. Li, X. Cui, and N. Arnheim, *Methods (San Diego)* **2,** 49 (1991).

FIG. 4. Microdissection procedure. The microdissection procedure is described in detail in the text. In brief, phenotypically characterized muscle fibers are selected (A), loosened from the surrounding fibers (B), and pulled out with the help of the device depicted in Fig. 3 (C–E). The fiber is placed into an Eppendorf tube (F), spun down, treated to expose the DNA, and subjected to PCR. Reproduced from Ref. 6 with permission from The University of Chicago Press.

12% polyacrylamide gel electrophoresis and exposed to an X-ray film. Figure 5 illustrates the results obtained with three different pathogenic mtDNA mutations.[15,16]

Single-fiber PCR can also be used for the quantification of mtDNA deletions.[4,17] In this case, three or more primers have to be used during the amplification. We have developed a three-primer PCR procedure for the quantification of the "common deletion."[17] After careful standardization, coamplification products can be directly distinguished by polyacrylamide gel electrophoresis (Fig. 6). Single-fiber PCR can also be used to estimate the absolute amounts of mtDNA,[17] an application that can be particularly useful in the study of mtDNA depletions.[18]

Appendix

Solutions for in Situ Hybridization

All solutions must be sterile, preferably prepared with 0.1% DEPC (diethyl pyrocarbonate)-treated water. Use DEPC-treated and autoclaved Coplin jars in all pretreatment steps.

$2\times$ Prehybridization solution for DNA probes (25 ml)[a]

Reagent	Volume (ml)	Final concentration (after 1:1 dilution)
DEPC-treated water	13.35	
5 M NaCl	6	0.6 M
1 M Tris-HCl, pH 7.5	0.5	10 mM
6% w/v Ficoll	0.165	0.02%
6% w/v Polyvinylpyrrolidone (PVP)	0.165	0.02%
6% w/v BSA	0.835	0.1%
250 mM EDTA	0.1	0.5 mM
Sonicated salmon sperm DNA (10 mg/ml)	2.5	0.5 mg/ml
Total yeast RNA (20 mg/ml)	1.25	0.5 mg/ml
Yeast tRNA (50 mg/ml)	0.05	0.05 mg/ml
Total	25	

[a] Dilute 1:1 with deionized formamide before use.

[15] C. T. Moraes, F. Ciacci, E. Bonilla, C. Janse, M. Hirano, N. Rao, R. E. Lovelace, L. P. Rowland, E. A. Schon, and S. DiMauro, *J. Clin. Invest.* **92,** 2906 (1993).

[16] C. T. Moraes, F. Ciacci, E. Bonilla, V. Ionasescu, E. A. Schon, and S. DiMauro, *Nat. Genet.* **4,** 284 (1993).

[17] M. Sciacco, E. Bonilla, E. A. Schon, S. DiMauro, and C. T. Moraes, *Hum. Mol. Genet.* **3,** 13 (1994).

[18] C. T. Moraes, S. Shanske, H.-J. Tritschler, J. R. Aprille, F. Andreetta, E. Bonilla, E. Schon, and S. DiMauro, *Am. J. Hum. Genet.* **48,** 492 (1991).

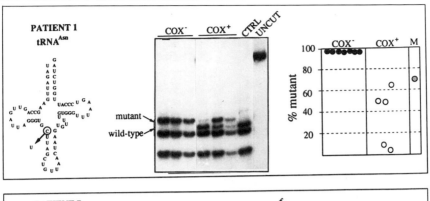

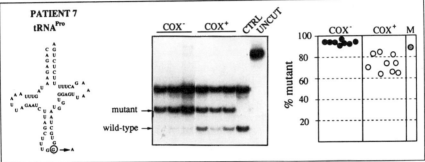

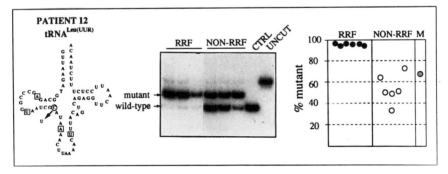

FIG. 5. Analysis by RFLP of microdissected muscle fibers. Muscle fibers from three different patients harboring potentially pathogenic mtDNA polymorphisms were microdissected and selected regions of their mtDNA directly amplified by PCR. The PCR fragments were analyzed by RFLP after digestion with the following enzymes: *Dde*I (patient 1), *Msp*I (patient 7); *Hin*PI (patient 12). Representative autoradiograms of polyacrylamide gels are shown in the central portion of each panel. COX⁻, COX-negative (affected) fibers; COX⁺, COX-positive (normal) fibers; RRF, ragged-red (affected) fibers; NON-RRF, weak SDH-staining (normal) fibers; CTRL, control fiber; UNCUT, PCR fragment not digested with restriction enzyme. The total determinations of affected (black circles) and normal (white circles) fibers are plotted on the right-hand side of each panel. The levels of mutant mtDNA observed in total muscle DNA is also shown (M, shaded circle). Modified from *J. Clin. Invest.* (1993) **92,** 2906 by copyright permission of The Society for Clinical Investigation.

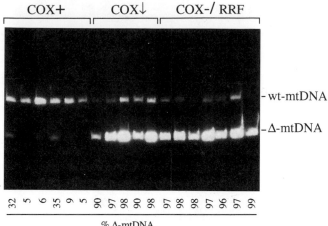

FIG. 6. Quantification of partially deleted (Δ) and wild-type (wt) mtDNA in single muscle fibers. Autoradiogram shows three-primer PCR amplification of single muscle fibers from a patient harboring the "common deletion" of mtDNA. The percentages obtained by this procedure are normalized to known standards,[17] and the corrected percentages are shown below each lane. COX$^+$, COX-positive fibers; COX$\downarrow$, COX deficient fibers; COX$^-$, COX-negative fibers; RRF, ragged red fibers. Reproduced from Ref. 17 by permission from Oxford Press.

$2\times$ Hybridization solution for DNA probes (25 ml)^a

Reagent	Volume (ml)	Final concentration (after 1:1 dilution)
DEPC-treated water	16.56	
5 M NaCl	6	0.6 M
1 M Tris-HCl, pH 7.4	0.5	10 mM
6% Ficoll w/v	0.165	0.02%
6% Polyvinylpyrrolidone (PVP) w/v	0.165	0.02%
6% BSA w/v	0.835	0.1%
250 mM EDTA	0.1	0.5 mM
Dextran sulfate	5 g	0.1 g/ml
Sonicated salmon sperm DNA (10 mg/ml)	0.5	0.1 mg/ml
Total yeast RNA (20 mg/ml)	0.125	0.5 mg/ml
Yeast tRNA (50 mg/ml)	0.05	0.05 mg/ml
Total	25	

^{a} Dilute 1:1 with deionized formamide before use.

Hybridization solution for RNA probes

To 4 μl nucleic acids mix add 8 μl probe (1–2 $\times$ 10^6 cpm) plus water. Heat at 65° for 5 min, place on ice for 1 min. Add 84 μl hybridization solution, 2 μl of 5 M DTT, 1 μl of 10% sodium thiosulfate w/v (freshly prepared), and 1 μl of 10% sodium dodecyl sulfate w/v (SDS). The latter three stock solutions should be filter sterilized.

Nucleic acids mix (1 ml)

Reagent	Volume (μl)	Final concentration (after final dilutions)
10 mg/ml Salmon sperm DNA	250	100 μg/ml
20 mg/ml Yeast total RNA	313	250 μg/ml
25 mg/ml Yeast tRNA	250	250 μg/ml
Water	187	
Total	1000	

Hybridization solution (40 ml)

Reagent	Volume (ml)	Final concentration (after final dilutions)
1 M Tris-HCl, pH 7.4	0.95	20 mM
250 mM EDTA	0.19	1 mM
4 M NaCl	3.57	0.3 M
100% Formamide	23.8	50%
50% Dextran sulfate w/v	9.52	10%
50$\times$ Denhardt's solution	0.95	1$\times$
Water	1.00	
Total	40	

Box buffer: 4 $\times$ SSC, 50% formamide v/v

Solutions for Single-Fiber PCR

Alkaline lysis solution: 200 mM KOH, 50 mM DTT. Make stocks of 222 mM KOH (made fresh weekly) and 500 mM DTT. Mix 9 volumes of KOH stock and 1 volume of DTT stock for working solution.

Neutralizing solution: 900 mM Tris-HCl, pH 8.3, 200 mM HCl. Make stock of 1 M Tris-HCl, pH 8.3. Make a working mix containing 450 μl of 1 M Tris-HCl, 8.6 μl concentrated HCl (11.6 N), and 40 μl water.

Acknowledgments

We thank Jessie Singer for protocols and suggestions on *in situ* hybridization with RNA probes. This work was supported by grants from the Muscular Dystrophy Association, the

National Institutes of Health, the Myoclonus Foundation, and the Procter & Gamble Company. Figures 1 and 4 were adapted from Ref. 6 with permission from University of Chicago Press (© The American Society of Human Genetics. All rights reserved. 0002-9297/92/5005-0007$02.00). Figure 5 was adapted from Ref. 15 with permission from The Rockefeller University Press (reproduced from the *Journal of Clinical Investigation,* 1993, 92, 2906–2915, by copyright permission of The Society for Clinical Investigation). Figure 6 was reproduced from Ref. 17 by permission from Oxford Press.

[45] Cytochemistry and Immunocytochemistry of Cytochrome-c Oxidase at Electron Microscope Level

By J. MÜLLER-HÖCKER and S. SCHÄFER

Introduction

Ultracytochemical staining reactions exist for various mitochondrial enzymes including the respiratory chain complexes I (NADH dehydrogenase), II (succinate dehydrogenase), IV (cytochrome-c oxidase), and V (ATP synthase).[1] The cytochemical reaction for complex I is not specific because it cannot discriminate between rotenone-sensitive (complex I) and rotenone-insensitive activity. Disorders of complex II and V are rare, and experience at the electron microscopical level is limited to a few reports dealing with alteration of complex V.[2-5] Electron microscopical investigation of complex IV activity, however, has been performed more extensively for the detection of tissue-specific or cell-specific defect manifestation and especially for the study of intracellular heterogeneity in mitochondrial diseases and imaging.[6-14] In this chapter methods for the cytochemical and immunocytochemical visualization of complex IV are described.

[1] P. R. Lewis and D. P. Knight, "Practical Methods in Electron Microscopy" (A. M. Glauert, ed.), Vol. 5, Part 1. North-Holland Publ., Amsterdam, New York, and Oxford, 1977.
[2] J. Müller-Höcker, I. Paetzke, D. Pongratz, and G. Hübner, *Virchows Arch. A Pathol. Anat.* **45,** 125 (1984).
[3] J. Müller-Höcker, I. Paetzke, D. Pongratz, and G. Hübner, *Virchows Arch. B* **48,** 185 (1985).
[4] J. Müller-Höcker, S. Stünkel, D. Pongratz, and G. Hübner, *J. Neurol. Sci.* **69,** 27 (1985).
[5] J. Müller-Höcker, D. Pongratz, and G. Hübner, *J. Neurol. Sci.* **74,** 199 (1986).
[6] J. Müller-Höcker, D. Pongratz, G. Hübner, *Virchows Arch. A Pathol. Anat.* **402,** 61 (1983).
[7] J. Müller-Höcker, A. Johannes, M. Droste, B. Kadenbach, D. Pongratz, and G. Hübner, *Virchows Arch. B* **52,** 353 (1986).
[8] J. Müller-Höcker, *Am. J. Pathol.* **134,** 1167 (1989).
[9] J. Müller-Höcker, *J. Neurol. Sci.* **100,** 14 (1990).
[10] J. Müller-Höcker, G. Hübner, K. Bise, C. Förster, S. Hauck, I. Paetzke, D. Pongratz, and B. Kadenbach, *Arch. Pathol. Lab. Med.* **117,** 202 (1993).

Cytochemistry of Cytochrome-c Oxidase

Cytochrome-c oxidase is demonstrated best using 3,3'-diaminobenzidine tetrahydrochloride.[15–19] Diaminobenzidine tetrahydrochloride (DAB) is oxidized by the enzyme and undergoes oxidative polymerization to produce water-insoluble material which is strongly osmiophilic. By postfixation with aqueous osmium tetroxide, the characteristic high contrast at the electron microscopical level is achieved. Solutions of DAB are readily darkened by light. Incubation media should therefore be freshly made up and protected from light during incubation. The reaction works best with unfixed fresh or frozen tissue, even after storage at $-70°$ for long periods. The tissue is chopped into slices (<1 mm thick) either by a vibratome or with a razor blade.

The incubation medium is made up as follows: 35 ml distilled water is added to 150 mg DAB. While stirring, 10 ml of 0.1 M Tris base is added, and the pH is adjusted to pH 8.5 with 1 N NaOH. The volume is adjusted to 50 ml, and the solution is filtered to remove undissolved DAB. KCN added to 1 mM (6.5 mg/10 ml medium) makes a negative control solution.

The tissue slices are incubated in the dark (e.g., in petri dishes) for 3–4 hr at room temperature. The postincubation procedure is as follows:

Wash with 0.15 M Tris-HCl, pH 7.4.

Fix with glutaraldehyde (3% in 0.1 M sodium cacodylate buffer, pH 7.4) for 2 hr at $4°–10°$.

Wash with 0.15 M Tris-HCl, pH 7.4.

Postfix with osmium tetroxide (1% in distilled water) for 1 hr.

Dehydrate in ethanol (50%, 70%, 80%, 90%, 95%, three times 100%, each for 30 min).

Treat with propylene oxide, 20 min.

Treat with propylene–Epon, 1 : 1, 2 hr.

[11] I. Nonaka, Y. Koga, E. Ohtaki, and M. Yamomoto, *J. Neurol. Sci.* **92,** 193 (1989).

[12] K. Haginoya, S. Miyabayashi, K. Iinuma, and K. Tada, *Acta Neuropathol.* **80,** 642 (1990).

[13] I. Nonaka, Y. Koga, A. Kikuchi, and Y. Goto, *Acta Neuropathol.* **82,** 286 (1991).

[14] K. Haginoya, S. Miyabayashi, K. Iinuma, E. Okino, H. Maesaka, and K. Tada, *Pediatr. Neurol.* **8,** 13 (1992).

[15] A. M. Seligman, M. J. Karnovsky, H. L. Wasserkrug, and J. S. Hanker, *J. Cell Biol.* **38,** 1 (1968).

[16] F. Roels, *J. Histochem. Cytochem.* **22,** 442 (1974).

[17] S. Hanker, "Osmiophilic Reagents in Electron Microscopic Histochemistry." Gustav Fischer Verlag, Stuttgart and New York, 1979.

[18] J. A. Litwin, *Folia Histochem.* **17,** 3 (1979).

[19] S. Angermüller and H. D. Fahimi, *Histochemistry* **71,** 33 (1981).

Embed in Epon 812.
Polymerize overnight in oven at 60°.

Prior to the study of ultrathin sections it is desirable to study semithin sections (1–2 μm) for better orientation. The semithin sections may be counterstained by toluidine blue–azure II stain.

Defective cells lacking cytochrome-c oxidase reaction product are easily localized (Fig. 1). Ultrathin sections are best studied after counterstaining with uranyl acetate. Counterstaining with lead citrate is not necessary. Ultracytochemistry clearly shows that, for example, in the heart, defects of cardiomyocytes are strictly restricted to single cells, ending exactly at the intercalated disk of the neighboring heart muscle cell (Fig. 2). Owing to the higher resolution capacity, ultracytochemistry may detect defects of small single cells such as smooth muscle cells in the vessel wall[10] which easily would be missed in light microscopy. It is the only method to detect intracellular heterogeneity of the enzyme activity with certainty (Fig. 3). Because of the high resolution capacity, a better correlation between morphology and function is possible. It may be shown, for example, that in

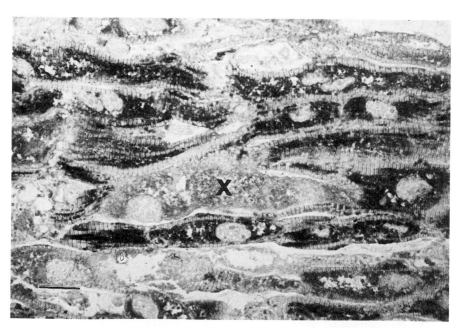

Fig. 1. Cytochemical detection of cytochrome-c oxidase activity with DAB. In the heart muscle of a patient with a mitochondrial cardiomyopathy, randomly distributed defective heart muscle cells are seen (X). Bar: 13 μm.

Fig. 2. Ultracytochemistry of cytochrome-*c* oxidase showing a defective heart muscle cell with well-preserved ultrastructure but lacking cytochrome-*c* oxidase activity in all the mitochondria of the cell. The defect ends exactly at the intercalated disc (↑). Bar: 3 μm.

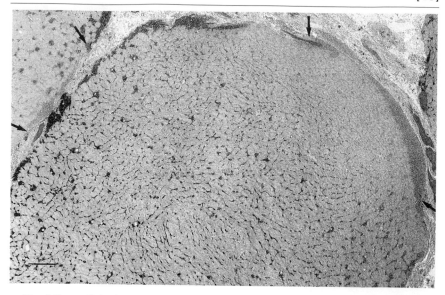

FIG. 3. Intracellular heterogeneity of cytochrome-*c* oxidase in a muscle fiber with clusters of mitochondria lacking activity (↑) neighboring normally active mitochondria. In the interstitial tissue (vessels, fibrocytes) the mitochondria are intact. Also the myofibrillary apparatus of the partially defective muscle fiber is well preserved. Bar: 7 μm.

mitochondrial diseases enzymatic defects are expressed in muscle fibers with a normal or abnormally high content of mitochondria showing a well-preserved myofibrillary apparatus (Figs. 3 and 4).

The enzyme reaction provides a high signal/background ratio. There is, however, minimal diffusion of the reaction product so that discrimination of the inner and outer mitochondrial membrane usually is obscured.

In our experience tissue penetration of the incubation medium may differ from organ to organ. Penetration usually represents no problem in skeletal muscle, but may be of more concern in heart muscle. The reaction in intervening tissues is usually positive and can serve to monitor the extent of tissue penetration (Figs. 3 and 4).

In rare cases where only cryostat sections are available, ultrastructural studies may also be performed. After the cytochemical reaction and fixation, Epon is polymerized on the slide and lifted from it using boiling water. With this procedure, however, the fine structural details generally are less well preserved.

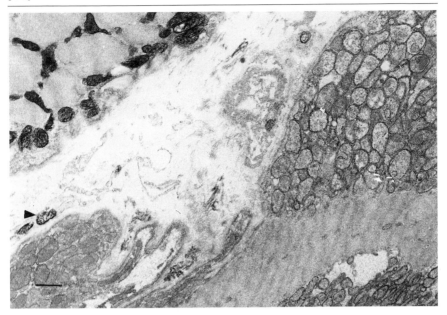

FIG. 4. High magnification of an abnormal muscle fiber lacking enzyme activity in subsarco-lemmal aggregates of mitochondria. In the intervening tissue (▲) and the adjacent muscle fiber the reaction intensity is well preserved. Bar: 1 μm.

Ultraimmunocytochemistry

Various techniques are available for electron immunolabeling.[20-25] Basic regimens are well established. However, because tissue processing affects epitopes in different ways, the methodology has to be varied to avoid loss of antigenicity. For the immunolabeling of cytochrome-c oxidase, postembedding techniques are well suited.[7,26]

[20] J. M. Polak and I. M. Varndell (eds.), "Immunolabelling for Electron Microscopy." Elsevier, Amsterdam, New York, and Oxford, 1984.
[21] J. M. Polak and S. van Noorden, "An Introduction to Immunocytochemistry: Current Techniques and Problems." Oxford Univ. Press, Royal Microscopical Society, Oxford, 1992.
[22] J. M. Polak and J. V. Priestly (eds.), "Electron Microscope Immunocytochemistry." Oxford Univ. Press, London, 1992.
[23] G. Griffiths, in "Fine structure immunocytochemistry." Springer, Berlin, 1993.
[24] J. Roth, J. Microsc. 143, 125 (1986).
[25] J. W. Stirling, J. Histochem. Cytochem. 38, 145 (1990).
[26] T. Sato, S. Nakamura, and H. Hirawake, in "Mitochondrial Encephalomyopathies" (T. Sato and S. DiMauro, eds.), Vol. 7, p. 195. Raven, New York, 1991.

The standard system consists of the following reaction steps: (1) fixation, (2) dehydration, (3) resin embedding, (4) resin polymerization, (5) sectioning, (6) antigen localization, and (7) antibody visualization. Incubations are carried out in moist chambers either at room temperature for 2–4 hr or at 4°–10° overnight. The ultrathin sections are placed upside down on drops of incubation medium on Parafilm.

Fixation

Various fixation protocols for postembedding immunolabeling employ acids such as picric acid, periodate–lysine in combination with paraformaldehyde (PLP), and/or glutaraldehyde.[25] The PLP fixative of McLean and Nakane,[27] to which 0.5% glutaraldehyde may be added, has proved especially suitable for the demonstration of cytochrome-*c* oxidase. The fixative stabilizes carbohydrates, lipids, and proteins.

To prepare PLP fixative, to 9 ml of 0.2 M lysine hydrochloride (dissolved in distilled water, pH correction with 0.1 M Na$_2$HPO$_4$) add the following: (a) 9 ml sodium phosphate buffer (0.1 M, pH 7.4); (b) 6 ml formaldehyde (16%, freshly prepared from paraformaldehyde in boiling water, with clearance of the medium by the addition of drops of 1 M NaOH); and (c) 256 mg sodium metaperiodate. Equivalent retention of antigenicity is achieved by a fixative consisting of 4% paraformaldehyde in 0.1 M phosphate buffer, pH 7.4, with 0.5% glutaraldehyde. Fixation must be performed for at least 2–4 hr, but may also be done overnight. In contrast to routine procedures, postfixation with osmium tetroxide should be avoided, although certain antigens may be rescued by a saturated aqueous solution of sodium metaperiodate (8%).[28,29]

Resin Embedding and Polymerization

A wide range of immunocompatible resins are now available.[30] The main factors determining the performance of these resins are polarity, degree of cross-linking, and hydrophilicity. Aliphatic epoxy resins such as Epon 812, the normal embedding medium, are generally unsuitable for ultraimmunolabeling because the immunoreacting substances cannot penetrate the medium. Before immunostaining Epon-embedded tissues, the

[27] I. W. McLean and P. K. Nakane, *J. Histochem. Cytochem.* **22,** 1077 (1974).
[28] M. Bendayan and M. Zollinger, *J. Histochem. Cytochem.* **31,** 101 (1983).
[29] J. A. Litwin, A. Völkl, J. Müller-Höcker, T. Hashimoto, and H. D. Fahimi, *Am. J. Pathol.* **128,** 141 (1987).
[30] B. C. Causton, *in* "Immunolabelling for Electronmicroscopy" (J. M. Polak and I. M. Varndell, eds.), p. 29. Elsevier, Amsterdam, New York, and Oxford, 1984.

resin has to be removed, and sodium alkoxide solutions are best suited to this purpose.[31,32] Whereas ultrathin sections are difficult to stabilize without the resin, semithin sections treated with sodium ethoxide provide excellent results (Fig. 5).

Semithin plastic sections (1–2 μm) are submerged in a solution of saturated NaOH in absolute ethanol (sodium ethoxide) diluted 1 : 1 with fresh absolute ethanol for 15 min at room temperature. Saturated NaOH solution is made by adding 15 g NaOH pellets to 100 ml absolute ethanol, stirring for 1 hr, and allowing the solution to age at room temperature for 5 days before using the supernatant. After performing the immunolabeling, the semithin section may be reembedded in Epon, removed from the slide with boiling water, and used to cut ultrasections.[32] In our hands this technique also proved feasible (Fig. 6). However, it is difficult to obtain significant immunolabeling.

LR white, a hydrophilic acrylic resin which can be cured at 50° or at room temperature with amine accelerator, in our hands, did not allow immunolabeling of cytochrome-c oxidase, either at the semithin or at the ultrathin section level. In contrast, using Lowicryl K4M (Chemische Werke LOWI, 84478 Waldkraiburg, Germany; Polysciences, Inc., Warrington, PA), a low viscosity polar resin with 50% hydrophilic hydroxypropylmethacrylate, good immunolabeling results are obtained. The weak adherence between Lowicryl and biological material caused by the lack of covalent linkages opens up cleavages between the resin and the tissue, thus facilitating immunostaining.[33–35] Lowicryls are particularly suitable for the ultrastructural demonstration of heat-sensitive epitopes.

Preparation of the embedding medium has to occur under a stream of nitrogen gas to prevent the incorporation of oxygen. Polymerization may be carried out below 0° with an initiator which starts a free radical reaction or above 0° with benzoin ethyl ether replacing the initiator. For detection of cytochrome-c oxidase, polymerization above 0° is sufficient. Polymerization is achieved by exposure to long-wavelength ultraviolet light overnight. The ultraviolet lamp-to-tissue distance is about 10 cm. The blocks are hardened further by exposure to daylight before sectioning. Lowicryl can easily be sectioned with a water content of 4%. However, because it is

[31] H. Mar and T. N. Wight, *J. Histochem. Cytochem.* **36,** 1387 (1988).
[32] H. Mar, T. Tsukada, A. M. Gown, T. N. Wight, and D. G. Baskin, *J. Histochem. Cytochem.* **35,** 419 (1987).
[33] E. Carlemalm, R. M. Garavito, and W. Williger, *J. Microsc.* **126,** 123 (1981).
[34] E. Kellenberger, M. Dürrenberger, W. Williger, E. Carlemalm, and U. Wurz, *J. Histochem. Cytochem.* **35,** 959 (1987).
[35] L. G. Altman, B. G. Schneider and D. S. Papermaster, *J. Histochem. Cytochem.* **32,** 1217 (1984).

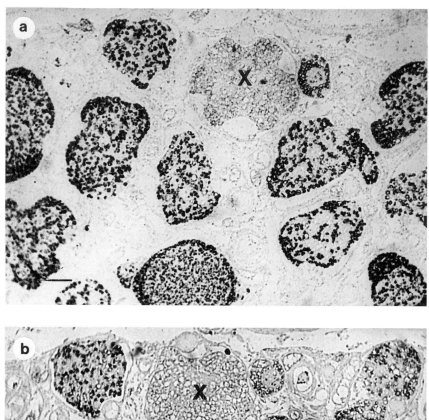

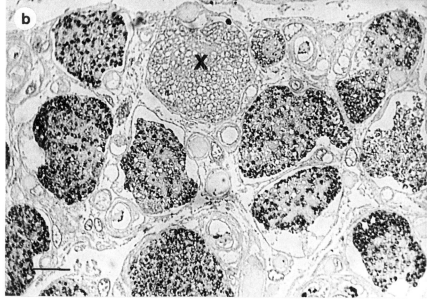

FIG. 5. Immunochemical detection of cytochrome-*c* oxidase subunit Vab in Epon-embedded external eye muscle. After removal of Epon with sodium ethoxide, the enzyme protein is easily detected in the mitochondria. In one fiber (X) the protein is lacking. (a) Immunogold detection and silver enhancement. (b) Avidin–biotin detection (peroxidase labeled) with 3-amino-9-ethylcarbazole as chromogen showing a somewhat lower signal compared to the immunogold–silver detection method. Bars: 15.6 μm.

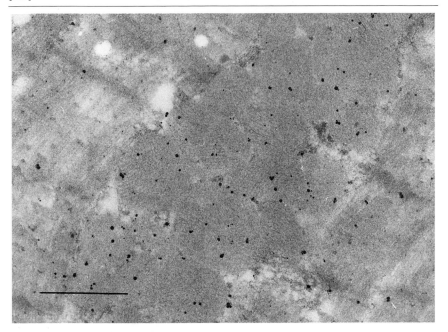

FIG. 6. Heart muscle. Immunostaining of semithin sections (cytochrome-*c* oxidase, subunit II/III) after removal of Epon with sodium ethoxide and reembedding after immunostaining (ultrasmall immunogold and silver enhancement). Silver grains of different sizes are concentrated over the mitochondria. Bar: 0.9 μm.

hydrophilic, for ultrasections the level of the water bath should be lowered to a minimum to discourage wetting of the block face.

Lowicryl K4M Embedding Method. After fixation the specimens are rinsed in two changes of 0.1 *M* phosphate buffer, pH 7.4, and sequentially dehydrated in 50, 70, 80, 90, 95, and 100% ethanol (each step 30 min). The polymerization medium is prepared by mixing 2.4 g cross-linker A, 12.6 g monomer B, and 0.075 g benzoin ethyl ether. The polymer should be stored protected from light (e.g., in brown glass). For infiltration of the tissue with the Lowicryl mixture use Lowicryl : absolute ethanol, 1 : 1 (v/v) for 4 hr with intermittent stirring under nitrogen gas (gently to avoid bubbling), then 2 : 1 overnight at 4°; 100% Lowicryl for 4 hr with intermittent inflow of nitrogen gas; 100% Lowicryl with inflow of nitrogen at the beginning and UV-polymerization at 4° overnight. Methylmethacrylate polymerized by UV light has also been used with success.[26]

Antigen Localization

The ultrathin sections are mounted on uncovered nickel grids. For better stabilization of the ultrathin sections, Formvar-coated grids may also be used. The incubation is performed in a humidity chamber. The grids are floated upside down on drops of immune reagents on a sheet of paraffin and are washed on larger drops of buffer or jet washed. The transfer of the grids can be performed with a pair of fine forceps.

Incubation with the primary antibody should last at least 2 hr at room temperature but may also last overnight at 4°. Generally the primary antibodies are more concentrated for immunoelectron microscopy compared to light microscopy. Whenever a permeabilization step is necessary, incubation with the primary antibody may be preceded by a pretreatment with Triton X-100, 0.5% in PBS (10 mM phosphate buffer, 150 mM NaCl, pH 7.4, for 30–60 min). In our experience labeling is enhanced and the ultrastructure sufficiently well preserved.

To reduce nonspecific background staining bovine serum albumin (BSA) may be added (up to 5%). The nonionic surfactant polyoxyethylene sorbitan monolaurate (Tween 20) is especially effective in preventing nonspecific gold probe reactions and can be used in conjunction with BSA in similar concentrations.

Antibody Visualization

Colloidal gold is the probe of choice for localizing specific sides of antigen–antibody reactions in ultrathin sections.[36-42] Furthermore, silver enhancement procedures may be used for intensification[43-46] and for light microscopic visualization of gold particles. As for the cytochemical demonstration of cytochrome-c oxidase, immunolabeling of semithin Lowicryl sections is recommended prior to ultraimmunocytochemical studies (Fig. 7); thus, the desired area may easily be chosen.

The most commonly used antibody visualization technique is the indirect method in which gold-labelled markers [immunoglobulin (Ig), protein A

[36] J. Roth, M. Bendayan, and L. Orci, *J. Histochem. Cytochem.* **26,** 1074 (1978).
[37] J. Roth and P. U. Heitz, *Ultrastruct. Pathol.* **13,** 467 (1989).
[38] M. M. Silver and S. A. Hearn, *Ultrastruct. Pathol.* **11,** 693 (1987).
[39] M. J. Warhol, *Am. J. Anat.* **185,** 301 (1989).
[40] G. A. Herrera, *Ultrastruct. Pathol.* **13,** 485 (1989).
[41] A. J. Verkley and J. L. M. Leunissen (eds.), "Immunogold Labelling." CRC Press, Boca Raton, Florida, 1989.
[42] M. A. Hayat (ed.), "Colloidal Gold." Academic Press, San Diego, California (1989); J. E. Beesley, "Colloidal Gold." *Welcome Res. Lab. Kent.* (1989).
[43] P. M. Lackie, R. H. Hennessy, G. W. Hacker, and J. M. Polak, *Histochemistry* **83,** 545 (1985).
[44] G. Danscher, *Histochemistry* **71,** 81 (1981).
[45] L. Scopsi, *in* "Colloidal Gold: Principles, Methods, and Applications" (M. A. Hayat, ed.), Vol. 1, p. 251. Academic Press, San Diego, 1989.
[46] J. D. Stierhof, B. M. Humbel, R. Hermann, M. T. Otten, and H. Schwarz, *Scanning Microsc.* **6,** 1009 (1992).

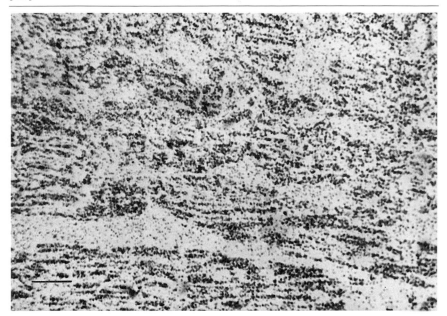

FIG. 7. Heart muscle. Immunolabeling of cytochrome-c oxidase in Lowicryl-embedded tissue on semithin sections. After ultrasmall immunogold (0.8–1 nm) detection and silver amplification of the immunoreaction, the mitochondria are visualized as small dots. Bar: 13 μm.

and G, avidin–gold, streptavidin–gold, and biotin-gold conjugates] are used for detection of the antigen–antibody reaction. Protein A and G bind avidly to the FC portion of the immunoglobulins and have a high affinity for rabbit, pig, and human IgG, but they do not bind to mouse and rat immunoglobulins and therefore are unsuitable for monoclonal antibodies. For ultrastructural visualization without amplification the gold particle size should be in the range of 10–15 nm (Fig. 8). Owing to steric hindrance the density of gold markers usually is inversely proportional to the size of the gold particles. Ultrasmall gold particles of about 0.8–1 nm (Aurion, Costerweg 5, 6702 AA Wageningen, The Netherlands, Fax: +31-8370-15955; Electron Microscopy Sciences, 321 Morris Road, PO Box 251, Fort Washington, PA 19034, Fax: +215-646893) therefore provide even better results.[46] For visualization, however, silver identification is necessary (Fig. 9).

Immunolabeling: Actual Procedure for Ultrathin Sections

1. Wash in PBS (10 mM phosphate buffer, 150 mM NaCl, pH 7.4), 5 min.

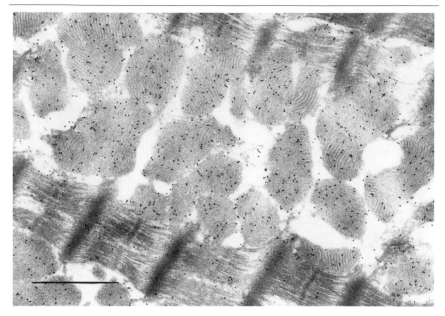

Fig. 8. Heart muscle. Ultracytochemical detection of cytochrome-*c* oxidase (subunit II/III) using protein A–gold. The gold grains (15 nm) accumulated over the mitochondria of heart muscle cells. Bar: 0.9 μm.

2. Place in 50 mM lysine in PBS, 15 min, to inactive aldehyde groups after aldehyde fixation.
3. Rinse two times in PBS.
4. Treat with normal goat serum, 5% in PBS with BSA (5%), 30 min.
5. Treat with primary antibody, diluted in PBS plus 0.1% BSA, overnight at 4°.
6. Wash in PBS.
7. Stain with immunogold (1:50) in protein A–gold (1:20) in PBS with 0.1% BSA, 4 hr.
8. Wash three times in PBS, 5 min each.
9. Fix with glutaraldehyde, 2% in PBS, 10 min.
10. Wash with PBS 5 min.
11. Wash in distilled water 4 times.
12. Conduct silver intensification (for ultrasmall gold) with Aurion RGent (Aurion, Netherlands) for 20 min in daylight. The intensification mixture consists of equal parts of developer, enhancer, and gum arabic solution. Gum arabic (one part) is dissolved in two parts (weight) of sodium citrate buffer (100 mM) adjusted to pH 5.5 with citric acid.

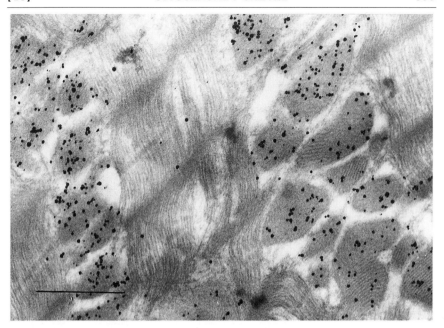

FIG. 9. Heart muscle. Ultraimmunolabeling of cytochrome-c oxidase (subunit Vab) with ultrasmall gold (0.8–1 nm) and silver intensification (10 min) resulting in a high labeling index of mitochondria. Counterstaining was with uranyl acetate (20 min) and lead citrate (1 min). Bar: 0.9 μm.

13. Counterstain ultrathin sections with uranyl acetate for 20 min and (facultatively) with lead citrate for 1 min.

The same protocol (steps 1–12) may be used for semithin sections embedded in Lowicryl or Epon (after removal of the resin) and when gold is used for antigen visualization.

Mammalian cytochrome-c oxidase consists of 13 subunits.[47] Depending on the quality of the antisera and the presentation of the epitopes after fixation and embedding, there may exist a great variety in the intensity of the immunoreaction. In our hands antisera against mitochondrial subunits II/III and the nuclear coded subunits IV and Vab provided the best results. Ultraimmunolabeling of cytochrome-c oxidase can be performed on long-term frozen muscle tissue, resulting in an acceptable ultrastructure (Fig.

[47] B. Kadenbach, L. Kuhn-Nentwig, and U. Büge, *Curr. Top. Bioenerg.* **15,** 113 (1987).

10). Furthermore, ultraimmunolabeling is also possible for other complexes of the respiratory chain, for example, the complexes III and V (J. Müller-Höcker, unpublished results).

Conclusion

The cytochemical and immunochemical detection of cytochrome-c oxidase represents a valuable method for the study of mitochondrial disorders. Both techniques are suited best for the combined study of morphological and functional aspects in mitochondrial diseases. Generally, before performing ultracytochemical investigations, cytochemical and immunocytochemical studies on semithin sections should be done to make certain that tissue preservation is appropriate and areas of interest are chosen for ultrastructural investigation.

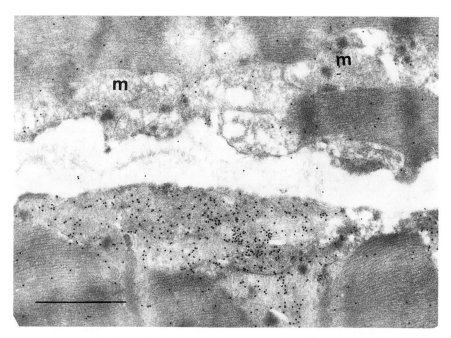

FIG. 10. Postmortem, frozen heart muscle of a patient with a mitochondrial cardiomyopathy. A defective heart muscle cell (*top*) is seen without staining of mitochondria (m) adjacent to a normally reacting cell with a high density of gold particles over the mitochondria (holoenzyme antibody). Bar: 0.9 μm.

Acknowledgments

The author is indebted to S. Dolling and M. Wittmaier for preparing the manuscript and to B. Stadler for language corrections. The present study was supported by grants from the Deutsche Forschungsgemeinschaft (MU 755/1-3).

[46] Electrophoretic Techniques for Isolation and Quantification of Oxidative Phosphorylation Complexes from Human Tissues

By Hermann Schägger

Introduction

Blue native electrophoresis (BN-PAGE) is a method for isolation of enzymatically active membrane proteins from biological membranes, which was successfully applied to bovine mitochondria,[1] for the preparation of highly pure oxidative phosphorylation (OX-PHOS) proteins for production of antibodies, N-terminal protein sequencing,[2] and for the analysis of molecular masses and oligomeric states of native complexes.[3] Practical details regarding BN-PAGE and related native and denaturing electrophoresis techniques are given in Ref. 4. The focus of this chapter is on the use of BN-PAGE and second-dimension Tricine-sodium dodecyl sulfate (SDS)–PAGE for the microscale isolation and quantification of the protein subunits of OX-PHOS complexes from milligram amounts of human tissues, for the determination of OX-PHOS defects in mitochondrial encephalomyopathies, and for the analysis of possible OX-PHOS defects in Alzheimer's and Parkinson's diseases. In contrast to other methods used for studies of human OX-PHOS enzymes, the electrophoretic techniques provide information about the quantity of correctly assembled multiprotein complexes.

Starting with 10–20 mg skeletal muscle, two-dimensional electrophoresis (BN-PAGE/Tricine-SDS–PAGE) allows separation of OX-PHOS complexes and their protein subunits. A quantity of tissue that can be obtained by needle biopsy is sufficient for a sensitive and relatively simple identifica-

---

[1] H. Schägger and G. von Jagow, *Anal. Biochem.* **199,** 223 (1991).
[2] H. Schägger, this series, Vol. 260 [12].
[3] H. Schägger, W. A. Cramer, and G. von Jagow, *Anal. Biochem.* **217,** 220 (1994).
[4] H. Schägger, *in* "A Practical Approach to Membrane Protein Purification" (G. von Jagow and H. Schägger, eds.), p. 59. Academic Press, San Diego, 1994.

tion of OX-PHOS defects in mitochondrial encephalomyopathies.[5] Because the method is also applicable to tissue from other organs (e.g., brain tissue), diseases suspected to involve proteins of the OX-PHOS system, such as Parkinson's disease, Alzheimer's disease, late-onset diabetes, and also aging, can be studied directly by protein analysis.

The preparation of mitochondria, especially from liver and brain, is problematic. To avoid varying protein losses with different samples, a method was developed for direct use of homogenized tissue. The experiments testing reproducibility and near-quantitative recovery of proteins after two-dimensional separation (BN-PAGE/Tricine-SDS–PAGE) as a prerequisite for the quantification of OX-PHOS complexes, and the characteristics and limitations of the electrophoretic techniques, are described in detail in Ref. 6. Protocols for the analysis of platelets and cell lines are given below.

Materials and Methods

Chemicals

Dodecyl-β-D-maltoside is purchased from Boehringer Mannheim (Indianapolis, IN), 6-aminocaproic acid from Fluka (Ronkonkoma, NY), acrylamide and bisacrylamide (the commercial, twice-crystallized products) and Serva blue G (Coomassie blue G-250) from Serva Biochemicals (Hauppauge, NY). All other chemicals are from Sigma (St. Louis, MO).

First-Dimension: Blue Native Electrophoresis

Sample Preparation Starting from Homogenized Tissue from Heart, Skeletal Muscle, and Brain. The procedures for preparing samples for application to BN-PAGE are simliar for tissue from heart, skeletal muscle, and brain. A protocol is summarized in Table I. The indicated milligram quantities of tissue (wet weight) are those for application to 10 × 1.6 mm gel wells in BN-PAGE. If larger quantities of tissue are available, it is advisable to homogenize on a larger scale, to divide the homogenized sample into aliquots after step 2, and store the sediments after step 3 at −80° for repeated analysis.

Sample Preparation from Liver Homogenates. The processing of liver tissue, also performed at 4°, differs somewhat from the protocol for heart,

[5] H. A. C. M. Bentlage, R. de Coo, H. ter Laak, R. C. A. Sengers, F. Trijbels, W. Ruitenbeek, W. Schlote, K. Pfeiffer, S. Gencic, G. von Jagow, and H. Schägger, *Eur. J. Biochem.* **227,** 909 (1995).

[6] H. Schägger, *Electrophoresis* **16,** 763 (1995).

TABLE I
PROTOCOL FOR SAMPLE PREPARATION FROM HUMAN TISSUES AT 4°

Step	Heart (5 mg)	Muscle (20 mg)	Brain (10 mg)
1. Add buffer (440 mM sucrose, 20 mM MOPS, 1 mM EDTA, pH 7.2) and add 0.2 mM phenylmethylsulfonyl fluoride (PMSF) from a 0.5 M stock in dimethyl sulfoxide (DMSO) shortly before use	250 μl	250 μl	250 μl
2. Homogenize using a tightly fitting glass–Teflon homogenizer			
3. Centrifuge 20 min at 20,000 g; discard supernatant			
4. Add 1 M aminocaproic acid, 50 mM Bis–Tris-HCl, pH 7.0, and homogenize by twirling with a tiny spatula	20 μl	40 μl	40 μl
5. Add dodecylmaltoside (10%)	3 μl	15 μl	20 μl
6. Centrifuge 15 min at 100,000 g; collect supernatant			
7. Add Serva blue G (5% in 1 M aminocaproic acid)	1.5 μl	7.5 μl	10 μl
8. Apply total volumes to 10 × 1.6-mm gel wells			

muscle, and brain. Liver samples (50 mg; wet weight) are homogenized in 500 μl sucrose buffer (see steps 1 and 2 in Table I). Then an additional step is introduced: The homogenate is centrifuged 15 min at 10,000 g, and the sediment is homogenized in 500 μl of 500 mM NaCl, 10 mM Na$^+$/MOPS, pH 7.2. The procedure then continues with step 3 and 4 of Table I, but 150 μl of the 1 M aminocaproic acid, 50 mM Bis–Tris-HCl buffer, pH 7.0, and 20 μl from a 10% Brij 35 detergent solution is added in step 4. After an additional centrifugation step (15 min at 100,000 g), 40 μl of the 1 M aminocaproic acid, 50 mM Bis–Tris-HCl buffer, pH 7.0, is added to the sediment, and the sample is homogenized. Steps 5 to 8 of Table I then follow, with 20 μl of 10% dodecylmaltoside added in step 5, and 10 μl of Serva blue G (5% in 1 M aminocaproic acid) added in step 7.

Sample Preparation from Cell Lines and from Blood. Primary fibroblasts, amnion cells, lymphoblasts, and HeLa cervix carcinoma epithelial cells are collected by centrifugation starting with 10 mg sedimented cells (1–5 × 10^6 cells). A platelet/mononuclear cell fraction is prepared from 4 ml of EDTA- or citrate-treated blood by centrifugation on Histopaque 1.077 (Sigma). The total mass of sedimented cells again is around 10 mg. The sediments are stored at −80° because direct use of nonfrozen cells leads to a considerably reduced yield of mitochondrial membrane proteins. The yield from frozen cells is in the 60–70% range.

Ten milligrams of cells are suspended in 500 μl of a Hypotone buffer

(83 mM sucrose, 10 mM MOPS, pH 7.2), homogenized in a tightly fitting glass–Teflon homogenizer, and mixed with 500 μl of 250 mM sucrose, 30 mM MOPS, pH 7.2. After a 15-min centrifugation at 600 g (4°) to remove sedimented broken cells, a mitochondrial fraction is collected by a 15-min centrifugation at 15,000 g (4°). Starting with 10 mg cells, sediments are suspended in 15 μl of 1 M 6-aminohexanoic acid, 50 mM Bis–Tris-HCl, pH 7.0, and the membrane proteins are solubilized by the addition of 5 μl of 10% (w/v) dodecylmaltoside. After a 15-min centrifugation at 100,000 g (4°), 5 μl of 5% (w/v) Serva Blue G dye in 1 M 6-aminohexanoic acid is added to the supernatant, and the total volume is applied to 5 × 1.6 mm gel wells in BN-PAGE.

Casting of Acrylamide Gradient Gels. Gel dimensions are 14 × 14 × 0.16 cm. Combs with 5 or 10 mm teeth are usually used. Gradient separation gels are cast at 4° and maintained at room temperature for polymerization. Essentially linear 5–13% polyacrylamide gradient gels, overlaid by a 4% sample gel, are cast as summarized in Table II. Buffers for gel casting and BN-PAGE are listed in Table III. After casting of the 4% gel at room temperature, and removal of the combs, gels are overlaid with gel buffer (1×) and stored at 4°.

Electrophoresis. Blue native PAGE using cathode buffer A (Table III) is performed at 4° in a vertical apparatus. After sample application the electrophoresis is started at 100 V until the sample has completely entered the sample gel, and continued for several hours with voltage and current limited to 500 V and 15 mA. For a better detection of faint bands, cathode buffer A can be removed after one-third of the run and electrophoresis

TABLE II
CASTING OF GRADIENT GELS FOR BLUE NATIVE
ELECTROPHORESIS[a]

Component	Stacking gel (4%)	Gradient gel	
		5%	13%
AB mix	0.5 ml	1.8 ml	4 ml
Gel buffer (3×)	2 ml	6 ml	5 ml
Glycerol	—	—	3 g
APS (10%)	50 μl	100 μl	70 μl
TEMED	5 μl	10 μl	7 μl
Total volume	6 ml	18 ml	15 ml

[a] For AB mix and gel buffer (3×) see Table III; APS (10%) is freshly prepared ammonium persulfate solution; TEMED, tetramethylethylenediamine.

TABLE III
STOCK SOLUTIONS FOR BLUE NATIVE ELECTROPHORESIS

Component	Composition
AB mix[a]: 49.5% T, 3% C, stored at 7°	48 g acrylamide and 1.5 g bisacrylamide/100 ml
Cathode buffers	50 mM Tricine, 15 mM Bis–Tris-HCl, pH 7.0 (4°),
A	+ 0.02% Coomassie Blue G-250[b]
B	+ 0.002% Coomassie Blue G-250
Gel buffer (3×)	150 mM Bis–tris, 1.5 M aminocaproic acid, adjusted to pH 7.0 with HCl (4°)
Anode buffer	50 mM Bis–Tris-HCl, pH 7.0 (4°)

[a] %T represents the total concentration of both monomers; %C, percentage of cross-linker to total monomer concentration.

[b] Serva blue G (highly pure Coomassie Blue G-250) was used exclusively. The dye was dissolved by stirring for several hours at room temperature, and cathode buffer A was stored at room temperature to prevent formation of large dye micelles not penetrating the sample gel. Cathode buffer B with 10-fold lower dye concentration can be stored at 4°.

continued with cathode buffer B, containing a 10-fold lower dye concentration.

Electroelution of Native Proteins. The method for the electroelution of native proteins after BN-PAGE, and enzymatic tests, has been described previously.[1] For quantification of protein recovery see Refs. 2 and 6.

Second-Dimension: Tricine-SDS–PAGE

After BN-PAGE individual lanes are excised, laid on a glass plate at the sample gel position, and soaked with 1% SDS, 1% mercaptoethanol for 2 hr. Spacers, slightly thinner than those used for the first dimension, are then positioned, the second glass plate is put on top, and mercaptoethanol is removed thoroughly. The second-dimension SDS gel is then cast in three steps, by passing acrylamide mixtures sequentially through the gaps between the native gel and spacers. First, a 16.5% acrylamide Tricine-SDS gel[7] is cast, leaving a 2 cm gap to the native gel, and overlaid with water. After polymerization a 10% acrylamide Tricine-SDS gel mixture is added that should not reach the native gel and water is overlaid, which may contact the native gel. After polymerization, another 10% acrylamide gel is used to surround the first-dimension gel without covering it. However, for this 10% acrylamide gel the buffer composition of native gels is used (see Table II).

[7] H. Schägger and G. von Jagow, *Anal. Biochem.* **166,** 368 (1987).

Electrophoresis Conditions. Electrophoresis of $14 \times 14 \times 0.16$ cm gels in the second dimension is performed at 120 V overnight at room temperature.

Staining and Quantification. All gels are stained by Coomassie Blue G-250 as described previously.[7] Densitometric quantification using (i) the "Personal Densitometer" (from Molecular Dynamics), which allows quantification of the staining intensity of total bands, and (ii) a "Quick Scan Jr." filter densitometer (Desaga, Heidelberg, Germany), for one-dimensional scanning through the most intensely stained part of the polypeptide patterns, gives similar results (see Refs. 5 and 6).

Application to Studies of Oxidative Phosphorylation Proteins in Human Diseases

Complex V Deficiency in Alzheimer's Disease

Autopsy samples (10 mg) from cortical areas from a control and from a patient with Alzheimer's disease (histological classification by Prof. Dr. Wolfgang Schlote, Frankfurt, Germany) were analyzed by BN-PAGE. Faint blue bands, representing complexes V, III, and IV, could be detected, even during BN-PAGE (not shown) when cathode buffer A with 0.02% Coomassie Blue G was exchanged after one-third of the run for cathode buffer B with 10-fold lower dye concentration. The detection of bands was better after fixation and Coomassie staining (Fig. 1). Only one band intensity, namely, that of complex V, was much higher in the control than in the patient sample, indicating a complex V deficiency in Alzheimer's disease. Expanded studies as described in Ref. 8 followed, which confirmed a complex V deficiency in most of the cases of Alzheimer's disease studied.

Other nonfixed lanes from the same BN-PAGE separation were used for identification and quantification of the complexes. Complete lanes were cut out after BN-PAGE and resolved by second-dimension Tricine-SDS–PAGE (Fig. 2). All complexes, except complex II, and most protein subunits could be detected after Coomassie staining. The staining intensities of complex I, III, and IV from the patient sample were somewhat lower than the control; however, the intensity of complex V was reduced dramatically to about 30%. Defining the ratio of staining intensities of complexes I/V/III/IV in the control sample as 1/1/1/1, the ratio in the patient sample was $1/{<}0.5/1/1$. The selective reduction of complex V can easily explain the "partially uncoupled state" of mitochondria in Alzheimer's disease. Possible primary or secondary defects of complex V cannot be discerned at this point.[8]

---

[8] H. Schägger and T. G. Ohm, *Eur. J. Biochem.* **227,** 916 (1995).

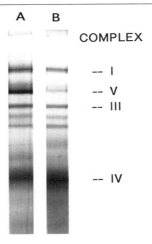

FIG. 1. Complex V deficiency in Alzheimer's disease revealed by blue native electrophoresis. Lane A shows analysis of a 10-mg autopsy specimen from a cortical area from a control person. Lane B shows analogous analysis from an Alzheimer's disease patient, demonstrating a considerable reduction in quantity of complex V. The separation of OX-PHOS complexes was performed as described in the text. The complexes separated in the native state were then fixed and stained with Coomassie dye. The complexes were identified by second-dimension Tricine-SDS–PAGE from further unfixed lanes as shown in Fig. 2.

To identify possible differences in subunit composition of complex V from the control and patient, BN-PAGE was also performed on a preparative scale, using 100-mg samples of brain tissue. Bands of complex V from patient and control samples were electroeluted and the subunits resolved by SDS–PAGE. No differences in polypeptide patterns were observed (gels not shown).

Studies of Oxidative Phosphorylation Complexes in Parkinson's Disease

Autopsy samples from substantia nigra, tegmentum, and cerebellum of Parkinson's patients and controls were analyzed by BN-PAGE/Tricine-SDS–PAGE. The samples, matched for age and for time of death to refrigeration and autopsy, were kindly provided by Prof. A. H. V. Schapira, London, UK.

All quantities of complexes I, III, and IV from a total of five Parkinson's patients were in the normal range of the controls, but a general slight reduction of complex IV (80% of that of the control group) was observed.[6] The normal quantities of complex I are in agreement with normal polypep-

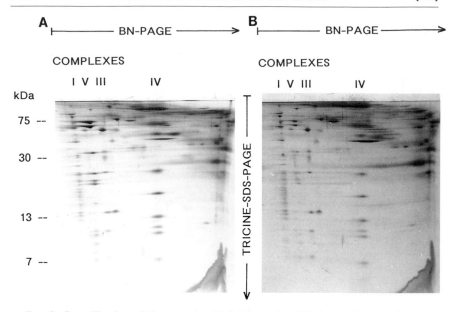

FIG. 2. Quantification of the complex V deficiency in Alzheimer's disease after two-dimensional electrophoretic analysis. (A) A complete lane from BN-PAGE of a control sample (analogous to lane A in Fig. 1, but unfixed) was processed in the second dimension by Tricine-SDS–PAGE, using a 16.5% acrylamide gel, as described in the text. (B) Analogous two-dimensional analysis of a 10 mg cortical sample from an Alzheimer's disease patient (cf. lane B in Fig. 1). The complexes were identified by their characteristic polypeptide patterns after Coomassie staining. The densitometric analysis of the most prominent protein subunits confirmed the presence of a complex V deficiency in the Alzheimer's disease sample, tentatively assigned after first-dimension BN-PAGE (Fig. 1), and allowed quantification of the defect (see text).

tide concentrations in immunological studies by Schapira et al.[9] However, they conflict with significant reductions of complex I activities reported by several groups.[9–12] One of the five samples from Parkinson's patients showed a 50% reduction of complex V in substantia nigra, in tegmentum, and in cerebellum[6] similar to that observed in Alzheimer's disease.[8]

Oxidative Phosphorylation Defects in Mitochondrial Encephalomyopathies

Studies of mitochondrial encephalomyopathies usually were performed on an analytical scale using 10–20 mg (wet weight) skeletal muscle samples.

[9] A. H. V. Schapira, J. M. Cooper, D. Dexter, J. B. Clark, P. Jenner, and C. D. Marsden, *J. Neurochem.* **54,** 823 (1990).

[10] W. D. Parker, S. J. Boysson, and J. K. Parks, *Ann. Neurol.* **26,** 719 (1989).

[11] D. Krige, M. T. Carroll, M. T. Cooper, C. D. Marsden, and A. H. V. Schapira, *Ann. Neurol.* **32,** 782 (1992).

[12] R. Benecke, P. Strümper, and H. Weiss, *Brain* **116,** 1451 (1993).

One example comparing the resolution of OX-PHOS complexes of a control and of an isolated complex I defect is shown in Fig. 3.

The two-dimensional resolution of complete lanes from BN-PAGE by Tricine-SDS–PAGE allowed detection and quantification of most of the more than 80 OX-PHOS proteins. Protein subunits comprised within native complexes were identified as bands arranged in a row.

The results of the analysis of different forms of mitochondrial myopathies, described in more detail in Ref. 5, can be summarized as follows: (i) BN-PAGE detects OX-PHOS defects by reduced assemblage of the complexes and by altered ratios of the staining intensities of the complexes within one gel. (ii) Defects of single OX-PHOS complexes in almost all cases of mitochondrial encephalomyopathies were characterized by simultaneously reduced catalytic activities and reduced concentrations of correctly assembled complexes. (iii) The electrophoretic technique was especially useful for discrimination between a defect of a specific complex and a generally low content of mitochondrial proteins. (iv) The MELAS disease (mitochondrial myopathy, encephalopathy, lactic acidosis, and stroke-like episodes), caused by a point mutation in a mitochondrial $tRNA^{Leu(UUR)}$ gene, was characterized by reduced ratios of complexes I/V, IV/V, and III/V, with the extent of reduction decreasing in the order complex I > complex IV > complex III. Complex V was used as internal reference, because complex V was always present at the highest absolute concentration and seemed to be least affected. The effect of the point mutation could be demonstrated directly on the protein level. (v) The electrophoretic technique allows one to identify defects of complex V that can hardly be detected by enzymatic measurements.

For a better comparison of the polypeptide patterns of complexes from patients and controls, studies were also performed on the "preparative scale" using samples of 50–200 mg, depending on the tissue used. After native electroelution of blue bands separated by preparative BN-PAGE (i.e., at a 10- to 20-fold higher protein load, using 1.6 and 2.8 mm gels, respectively, and 10-cm sample wells), the resolution of subunits of complexes could be performed on regular SDS gels in the second dimension. The subunit composition of complexes from different patients could thus be analyzed at high protein load, and at almost identical quantities of electroeluted complexes. In Fig. 4 the polypeptide patterns of complexes V from five human hearts without OX-PHOS defects were compared, showing the reproducibility of the technique.[13,14]

[13] J. E. Walker, R. Lutter, A. Dupuis, and M. J. Runswick, *Biochemistry* **30,** 5369 (1991).
[14] T. R. Collinson, M. J. Runswick, S. K. Buchanan, T. M. Fearnley, T. M. Skehel, M. T. von Raaij, D. E. Griffith, and J. E. Walker, *Biochemistry* **33,** 7971 (1994).

COMPLEXES   I   V III   IV

A

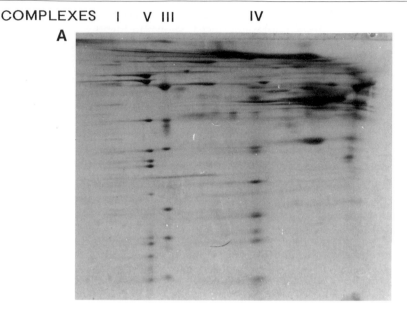

B

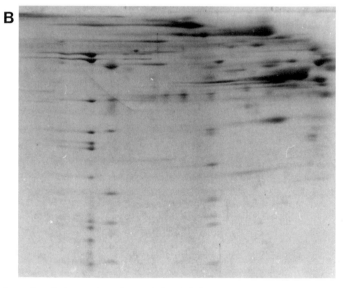

Fig. 3. Two-dimensional electrophoretic analysis of defects of OX-PHOS complexes in the diagnosis of mitochondrial encephalomyopathies. (A) Typical protein patterns starting from 20 mg undiseased samples from frozen autopsy or biopsy skeletal muscle. At least the most prominent subunits of complexes I, III, IV, and V can be detected and quantified after Coomassie staining. Complex II usually can be detected only after silver staining (not shown). (B) Polypeptide patterns showing an isolated defect of complex I in one case of a mitochondrial encephalomyopathy.

SUBUNIT

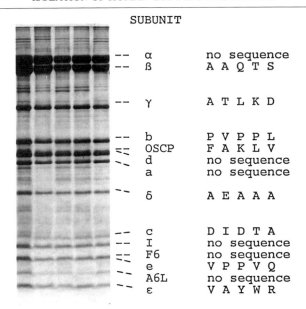

α	no sequence
ß	A A Q T S
γ	A T L K D
b	P V P P L
OSCP	F A K L V
d	no sequence
a	no sequence
δ	A E A A A
c	D I D T A
I	no sequence
F6	no sequence
e	V P P V Q
A6L	no sequence
ε	V A Y W R

FIG. 4. Comparison of polypeptide patterns of complexes V from human hearts isolated by preparative BN-PAGE. The OX-PHOS complexes from 50–100 mg specimens of human heart were resolved by preparative BN-PAGE. The visible bands of native complexes were excised and electroeluted. Complexes V were resolved by Tricine-SDS–PAGE using 16.5% T, 3% C gels, and stained by Coomassie blue. Subunits were identified by N-terminal protein sequencing. Subunit c seems to be present in dimeric form. N-Terminally blocked subunits α, d, a, I, F6, and A6L were tentatively assigned according to Walker et al.[13] by comparison with the migration distances of the bovine protein subunits. Bands of subunits f and g, located between A6L and F6 subunits,[14] could not be detected in this gel type.

Outlook

In future analyses the electrophoretic techniques might help to identify OX-PHOS defects in human diseases other than mitochondrial encephalo-myopathies or Alzheimer's disease. Especially for the analysis of complex V defects, which can hardly be identified with the required accuracy by measuring oligomycin-sensitive ATPase activity, the electrophoretic analysis seems to be advantageous. Combined with the use of antibodies against individual subunits prepared according to Schägger,[2] the electrophoretic techniques might also help to identify fetal or organ-specific isoforms of protein subunits.

Using silver staining allows analysis of OX-PHOS complexes II to V from human cell lines and from the platelet/mononuclear cell fraction from blood (gels not shown). Complex I usually is not detected in silver-stained two-dimensional gels, but it can be detected by complex I-specific antibodies.

Hints for the localization of a genetic defect of a single protein subunit, leading to reduced assembly of the holo complex, cannot be expected to be obtained from the portion of correctly assembled complex, but such information potentially will be accessible by immunological screening for free protein subunits on electroblotted two-dimensional (free subunits present in the mitochondrial matrix) or by immunological screening of the previously neglected supernatant of step 3 of the sample preparation protocol (Table I; free subunits present in the cytosolic fraction).

Genetically modified protein subunits, not impairing assembly but reducing catalytic activity, cannot be expected to be identified by altered migration behavior in the second-dimension SDS–PAGE if, for example, only one amino acid is exchanged for another one. However, in the rare cases that can be identified by reduced catalytic activity at normal concentration of a correctly assembled complex in two-dimensional gels, mass spectrometric techniques might help to identify the altered subunit and to find a direct access to the altered gene in the mitochondrial as well as in the nuclear genome.

Acknowledgments

These investigations were supported by a grant of the Deutsche Forschungsgemeinschaft to H. S., S. Gencic, and G. von Jagow. The excellent assistance of Kathy Pfeiffer and Anna Königer is gratefully acknowledged. Studies on Alzheimer's disease were performed in collaboration with Prof. W. Schlote (Frankfurt, Germany), and with Prof. Dr. T. G. Ohm (Berlin, Germany), studies on Parkinson's disease in collaboration with Prof. A. H. V. Schapira, Dr. M. Cooper, and Dr. V. Mann (London, UK), and studies on mitochondrial myopathies in collaboration with H. Bentlage, R. De Coo, H. ter Laak, R. Sengers, F. Trijbels, and W. Ruitenbeck (Nijmegen, NL).

[47] Assays for Characterizing URF13, the Pathotoxin and Methomyl Receptor of cms-T Maize

By David M. Rhoads, H. Carol Griffin, Barbara Brunner Neuenschwander, Charles S. Levings III, and James N. Siedow

Introduction

Maize (*Zea mays* L.) carrying the Texas cytoplasm (*cms-T*) is male sterile. Cytoplasmic male sterility (CMS) in *cms-T* maize is characterized by pollen abortion and failure to exert anthers, but female fertility is unaffected. Two nuclear genes, *Rf1* and *Rf2*, act together to restore fertility to *cms-T* maize.[1,2] Maize containing the *cms-T* cytoplasm was used extensively in hybrid seed production in the 1950s and 1960s because it eliminated the need for manual emasculation. By the late 1960s *cms-T* maize constituted the vast majority of the total U.S. corn acreage.[1,2] In 1969 and 1970, Southern corn leaf blight, which is caused by the fungus *Bipolaris* (formerly *Helminthosporium*) *maydis* race T, occurred in epidemic proportions in the Southern and Corn Belt regions of the United States. Because *B. maydis* race T has only mild effects on maize plants containing normal (N) cytoplasm, it was readily apparent that the unique fungal susceptibility was related to the Texas cytoplasm, and large-scale use of *cms-T* maize for hybrid seed production was discontinued. Maize carrying *cms-T* is also susceptible to *Phyllosticta maydis,* the fungus that causes yellow corn leaf blight. Because susceptibility to fungal pathogens is a cytoplasmically inherited trait, it was suspected that susceptibility might be associated with mitochondrial dysfunction. This was confirmed when mitochondria isolated from *cms-T* maize plants and exposed to the host-specific toxins (T-toxins) produced by *B. maydis* race T or *P. maydis* (BmT toxin and Pm toxin, respectively) were shown to exhibit swelling, uncoupling of oxidative phosphorylation, inhibition of malate-driven electron transfer, and leakage of NAD^+ and other ions.[1-5] These results indicated that the inner mitochondrial membrane becomes permeable to molecules of M_r 660 or less in the presence of added T-toxin.

A 13-kDa protein, URF13, is the product of a mitochondrial gene called T-*urf13*, which is specific to *cms-T* maize and arose from a novel series of rearrangements within the mitochondrial genome.[6] URF13 is an integral membrane protein that is located in the inner mitochondrial membrane.[7,8] *Escherichia coli* cells expressing URF13 are also sensitive to T-toxin. The effects of T-toxin on *E. coli* cells are analogous to those observed when T-toxin interacts with isolated *cms-T* mitochondria: cell respiration is inhib-

[1] C. S. Levings III, *Science* **250,** 942 (1990).
[2] C. S. Levings III and J. N. Siedow, *Plant Mol. Biol.* **19,** 135 (1992).
[3] M. J. Holden and H. Sze, *Plant Physiol.* **75,** 235 (1984).
[4] D. E. Matthews, P. Gregory, and V. E. Gracen, *Plant Physiol.* **63,** 1149 (1979).
[5] R. J. Miller and D. E. Koeppe, *Science* **173,** 67 (1971).
[6] R. E. Dewey, C. S. Levings III, and D. H. Timothy, *Cell (Cambridge, Mass.)* **44,** 439 (1986).
[7] R. E. Dewey, D. H. Timothy, and C. S. Levings III, *Proc. Natl. Acad. Sci. U.S.A.* **84,** 5374 (1987).
[8] E. Hack, C. Lin, H. Yang, and H. T. Horner, *Plant Physiol.* **95,** 861 (1991).

ited, spheroplasts swell, and ion leakage is induced.[9-11] Methomyl {S-methyl-N-[(methylcarbamoyl)oxy]thioacetimidate}, which is the active ingredient in the DuPont insecticide Lannate, mimics the effects of T-toxin (at 10^4- to 10^5-fold higher concentrations) when added to *cms-T* mitochondria or *E. coli* cells expressing URF13, even though it has no obvious structural similarity to T-toxins.[11-13] Methomyl is often used in assays and in selective screens for the presence of functional URF13 because it is readily available (as Lannate), whereas T-toxin is more difficult to obtain. The effects of T-toxins or methomyl on *cms-T* mitochondria have led to the hypothesis that the interaction between these compounds and URF13 results in the formation of hydrophilic pores that permeabilize the inner mitochondrial membrane.[2,13]

Detection of Methomyl Sensitivity

Mutants of *cms-T* maize plants that are male sterile and resistant to *B. maydis* race T and *P. maydis* have never been found. In contrast, mutants have arisen spontaneously from *cms-T* maize tissue culture[14-16] that are male fertile and resistant to *B. maydis* race T and *P. maydis*. We have developed a technique for screening maize plants regenerated from tissue culture for sensitivity to methomyl using an oxygen electrode to determine the respiratory capacity in whole leaf tissue after methomyl treatment. This method provides a quick and accurate means of detecting T-*urf13* mutations that alter methomyl or T-toxin sensitivity. This technique avoids more time-consuming methods in which oxygen uptake is assayed using isolated mitochondria or methods that involve possible destruction of the plant such as inoculation with fungal pathogens or direct application of methomyl.

Assay for Methomyl Sensitivity in Intact Plant Tissue

A single leaf from a 3- to 4-week-old maize plant regenerated from tissue culture is washed briefly on the top and bottom with sterile, distilled

[9] C. J. Braun, J. N. Siedow, M. E. Williams, and C. S. Levings III, *Proc. Natl. Acad Sci. U.S.A.* **86,** 4435 (1989).
[10] C. J. Braun, J. N. Siedow, and C. S. Levings III, *Plant Cell* **2,** 153 (1990).
[11] R. E. Dewey, J. N. Siedow, D. H. Timothy, and C. S. Levings III, *Science* **239,** 293 (1988).
[12] A. Bervillé, A. Ghazi, M. Charbonnier, and J.-F. Bonavent, *Plant Physiol.* **76,** 508 (1984).
[13] R. R. Klein and D. E. Koeppe, *Plant Physiol.* **77,** 912 (1985).
[14] R. I. S. Brettell, E. Thomas, and D. S. Ingram, *Theor. Appl. Genet.* **58,** 55 (1980).
[15] I. K. Dixon, C. J. Leaver, R. I. S. Brettell, and B. G. Gengenbach, *Theor. Appl. Genet.* **63,** 75 (1982).
[16] R. P. Wise, D. P. Pring, and B. G. Gengenbach, *Proc. Natl. Acad. Sci. U.S.A.* **84,** 2858 (1987).

water and removed from the plant. The leaf is then submerged in sterile 20 mM HEPES/0.2 mM CaCl$_2$, pH 6.5, at room temperature.[17] Using a standard, hand-held hole punch (~6 mm diameter) that has been sterilized, eight leaf disks are excised from the leaf while it is submerged. Alternatively, the disks can be punched from a submerged leaf that is still attached to the plant. The leaf disks are then transferred to a 100-ml beaker that contains 20 ml of sterile 20 mM HEPES/0.2 mM CaCl$_2$, pH 6.5, and 1% (v/v) dimethyl sulfoxide (DMSO). Lannate (available at stores selling agricultural products), which contains 1.3 M methomyl, is added to the solution such that the final concentration of methomyl is 20 mM. Alternatively, pure methomyl is available in limited quantities by request from DuPont (Wilmington, DE). The disks are incubated at room temperature with vigorous stirring for 16 to 20 hr in the dark. The lengthy incubation is necessary in order for methomyl to penetrate the maize leaf disks. The leaf disks are then rinsed with excess 20 mM HEPES/0.2 mM CaCl$_2$, pH 6.5, to remove any residual Lannate or methomyl, transferred to the sample chamber of a Clark-type oxygen electrode (YSI, Yellow Springs, OH, or Rank Brothers, Cambridge, UK) containing air-saturated 20 mM HEPES/0.2 mM CaCl$_2$, pH 6.5, at 25°, where the rate of oxygen consumption is measured.[18] After attaining the steady-state respiration rate, the leaf disks are removed from the chamber, blotted dry, and weighed. The rate of oxygen consumption is calculated[18] and expressed as nanomoles O$_2$ per minute per gram fresh weight.

Representative results from one experiment are shown in Fig. 1. In this case, we used leaf disks from three separate plants: Va35N, which does not contain the T-*urf13* gene; Va35*cms-T*, which expresses URF13; and a plant regenerated from tissue culture derived from a cross between lines A188 *cms-T* and T204. This cross was selected because plants readily regenerate from its cell culture and because it carries the T-*urf13* gene. During regeneration of plants from tissue culture, spontaneous mutations of T-*urf13* often arise.[14-16] Although the regenerated plants arose from a methomyl-resistant callus culture, it is necessary to confirm that the regenerated plant originated from a methomyl-resistant cell. Methomyl-resistant calli frequently contain a mixture of resistant and susceptible cells.[19] Incubation with methomyl did not affect respiration of the normal maize plant (Fig. 1A); however, respiration of the *cms-T* plant was greatly reduced after incubation with methomyl (Fig. 1B). The A188*cms-T* × T204 mutant plant regenerated

[17] D. A. Day, O. C. De Vos, D. Wilson, and H. Lambers, *Plant Physiol.* **78,** 678 (1985).
[18] R. W. Estabrook, this series, Vol. 10, p. 41.
[19] A. R. Kuehnle and E. D. Earle, *Plant Cell, Tissue Organ Cult.* **28,** 129 (1992).

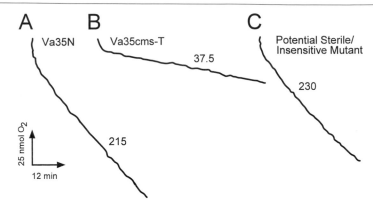

Fig. 1. Respiration assays using whole leaf disks. The rate of oxygen consumption was measured in a Clark-type oxygen electrode as described in the text. Samples: (A) eight leaf disks from Va35N, a normal line of maize that does not have the T-*urf13* gene; (B) eight leaf disks from VA35*cms-T*, a male-sterile line of maize that expresses URF13; (C) eight leaf disks from a plant that is the result of the cross A188T*cms-T* × T204 and was determined to be male sterile. The number associated with each trace represents the rate of oxygen consumption in units of nanomoles oxygen reduced to water per minute per milligram protein.

from tissue culture was also insensitive to methomyl (Fig. 1C), confirming that this plant carried a mutation in T-*urf13*. This technique has also been applied successfully to anther tissue from normal and *cms-T* maize plants.

Assay of Methomyl Sensitivity of Escherichia coli Cells

Stationary phase cultures of *E. coli* cells are diluted 1:50 into 100 ml Luria–Bertani broth[20] containing the appropriate antibiotics in a 250-ml flask and grown at 37° for 3 hr. Expression of URF13 (mutant or unaltered) is induced from either pKK13T, which is the pKK223-3 vector (Pharmacia, Piscataway, NJ) containing the T-*urf13* gene, or pET5b13T, which is the pET5b vector (Novagen, Madison, WI) containing the T-*urf13* gene, after addition of isopropyl-β-D-thiogalactoside (IPTG) to 0.2 mM and growth for 1.5 hr at 37°. Cells are harvested by centrifugation at 5000 g for 5 min at 4° in 250-ml centrifuge bottles and suspended in 100 ml cold M9 salts (42 mM Na$_2$HPO$_4$/22 mM KH$_2$PO$_4$/8.6 mM NaCl/19 mM NH$_4$Cl). The cells are pelleted again using the same centrifugation as above, and suspended in 0.5 ml cold M9 salts. Cells (between 100 and 200 μg) are added to the chamber of a Clark-type oxygen electrode containing M9 salts at 25°, 0.2

[20] J. Sambrook, E. F. Fritsch, and T. Maniatis, "Molecular Cloning: A Laboratory Manual," 2nd Ed. Cold Spring Harbor Laboratory, Cold Spring Harbor, New York, 1989.

M glucose is added to a final concentration of 10 mM, and the steady-state rate of oxygen consumption is measured.[18] After the rate of oxygen uptake has been established, Lannate is added to achieve a final methomyl concentration of 8.0 mM or higher. If the cells are sensitive, the rate of oxygen consumption decreased dramatically after the addition of Lannate.

Chemical Cross-Linking of Membrane Proteins

Expression of URF13 in *E. coli* cells has been used to (1) show that URF13 interacts with T-toxin or methomyl to produce pores in the bacterial plasma membrane[11]; (2) study the binding of T-toxin to URF13[10]; (3) study the topography of URF13 in the *E. coli* plasma membrane[21]; (4) perform site-directed mutagenesis to determine amino acid changes that result in insensitivity to T-toxin or methomyl[9]; and (5) carry out chemical cross-linking to show that URF13 forms an oligomeric protein complex in the membrane.[21,22] We have also used a combination of site-directed mutagenesis and chemical cross-linking to study the structure of the URF13 oligomers.[23] The use of site-directed mutagenesis and chemical cross-linking to study the structure of membrane proteins expressed in *E. coli* cells may be a generally useful combination of technologies for studying the structure of other oligomeric proteins or protein complexes. If the primary amino acid sequence of the protein(s) of interest has been used to develop a model for the tertiary and/or quarternary structure of the protein(s) within the membrane, it is possible to make predictions about amino acids on one molecule that are juxtaposed in such a manner to be cross-linked either to amino acids in other domains of the same protein (intramolecular cross-linking) or to amino acids on adjacent molecules within an oligomeric complex (intermolecular cross-linking). These predictions can be tested by expressing the protein(s) of interest in *E. coli* cells, followed by attempts to cross-link the protein(s) using the appropriate chemical cross-linker(s).

For example, it was known that N,N'-dicyclohexylcarbodiimide (DCCD) treatment of *cms-T* mitochondria or *E. coli* cells expressing URF13 resulted in insensitivity to T-toxin or methomyl.[9,24,25] This was thought initially to result from covalent modification of aspartate-39.[9] Close inspection of the three-helix model for the disposition of URF13 in the membrane

[21] K. L. Korth, C. I. Kaspi, J. N. Siedow, and C. S. Levings III, *Proc. Natl. Acad. Sci. U.S.A.* **88,** 10865 (1991).

[22] C. I. Kaspi and J. N. Siedow, *J. Biol. Chem.* **268,** 5828 (1993).

[23] D. M. Rhoads, C. I. Kaspi, C. S. Levings III, and J. N. Siedow, *Proc. Natl. Acad. Sci. U.S.A.* **91,** 8253 (1994).

[24] P.-Y. Bouthyette, V. Spitsberg, and P. Gregory, *J. Exp. Bot.* **36,** 511 (1985).

[25] M. J. Holden and H. Sze, *Plant Physiol.* **91,** 1296 (1989).

(Fig. 2A) and helical wheel representations of helix II (Fig. 2B), however, revealed that aspartate-39 on one URF13 monomer and lysine-32 and lysine-37 of a second URF13 monomer can be juxtaposed in such a way as to allow direct chemical cross-linking by DCCD of one of these two lysines to aspartate-39. We combined site-directed mutagenesis and DCCD cross-linking to show that lysine-37 is specifically cross-linked to aspartate-39 by DCCD.[23] From these results, we concluded that there is a central core of helices II in URF13 oligomers (Fig. 2B).

Alternatively, different hydrophobic chemical cross-linkers (in the case of membrane proteins) can be tested, one at a time, for their ability to cross-link the protein(s) of interest expressed in *E. coli* cells. The observation of cross-linking with specific reagents can be used to develop a model of the tertiary/quarternary structure of the protein–protein complex that can be tested using site-directed mutagenesis and the procedures outlined here.

In this section, we provide (1) the basic methodology for the isolation

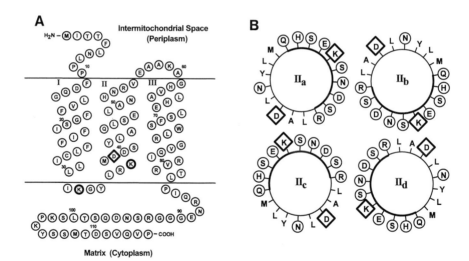

FIG. 2. Proposed topology of an URF13 monomer in the membrane (A) and representation of the putative central core of helices II within an URF13 tetramer (B). (A) Amino acids are designated by the standard single-letter code, and Arabic numerals indicate the number of each amino acid residue. The three transmembrane α helices are designated by Roman numerals I, II, and III starting with the helix closest to the amino terminus. Horizontal lines indicate boundaries of the membrane. Aspartate-39, lysine-32, and lysine-37 are each outlined with bold lines. (B) Helical wheel representation in which the amino acids of each helix II are plotted as a two-dimensional projection of the helix parallel to the plane of the membrane. The hydrophilic face of each helix is represented by a bold arc and the hydrophobic face by a thin arc. Circled amino acids are hydrophilic. Lysine-37 and aspartate-39 are boxed with bold lines.

of *E. coli* spheroplasts (because cross-linking is generally more efficient when the bacterial cell wall has been removed), (2) examples and suggestions for cross-linking plasmid-expressed proteins in the plasma membrane of the spheroplasts, and (3) suggestions for cross-linking mitochondrial proteins. Methods for site-directed mutagenesis are not discussed because these well-established techniques have been described elsewhere.[20,26]

Escherichia coli Spheroplast Preparation

Production of the recombinant protein of interest is induced by the method appropriate to the vector used. For our studies, expression of URF13 from pKK13T in 400-ml bacterial cultures is induced by adding 0.2 mM IPTG for 1.5 hr at 37°. Spheroplasts are prepared from the cells by a modification of the procedure of Witholt *et al.*[27] Cells are harvested by centrifugation at 5000 g for 5 min at 4° in two 250-ml centrifuge bottles, suspended in 100 ml of 0.2 M Tris-HCl, pH 8.0, for each tube, pelleted by centrifugation, and suspended in 50 ml of 0.5 M sucrose/0.2 M Tris-HCl, pH 8.0, at room temperature for each tube. The cells are kept in 250-ml centrifuge bottles, and 0.5 ml of 50 mM ethylenediaminetetraacetic acid (EDTA) is added while swirling the sample. Next, 0.5 ml lysozyme at 4.0 mg/ml is quickly added, and the sample is swirled vigorously. An equal volume (51 ml) of sterile, deionized water at room temperature is added while the sample is swirled vigorously, and the sample is incubated at room temperature for 3 to 5 min. The incubation time for optimum digestion varies depending on the strain of *E. coli* and the number of cells in the sample and should be determined experimentally. The spheroplasts are pelleted by centrifugation at 10,500 g for 5 min at 4°. If the spheroplasts do not pellet well in this step, it is likely that the sample was overdigested in the previous step. The resulting spheroplasts are suspended in 5 ml phosphate-buffered saline (140 mM NaCl/2 mM KCl/10 mM Na$_2$HPO$_4$/2 mM KH$_2$PO$_4$, pH 7.2). This buffer contains enough osmoticum that the spheroplasts do not burst and avoids the use of buffers, such as Tris, that contain primary amines which may interfere with the reactions of cross-linking agents that react with the primary amine groups of lysine residues. It is often necessary to use a ground-glass tissue grinder in order to thoroughly homogenize the sample. The protein concentration of the sample is then determined by a standard protocol, and a portion of the sample is added to phosphate-buffered saline to give 1.0 mg total protein/ml in a volume of 10 ml.

[26] T. Kunkel, *Proc. Natl. Acad. Sci. U.S.A.* **82,** 488 (1985).
[27] B. Witholt, M. Boekhout, M. Brock, J. Kingma, M. van Heerickhuizen, and L. De Leij, *Anal. Biochem.* **74,** 160 (1976).

Chemical Cross-Linking of Proteins Expressed in Escherichia coli Cells

From the 10-ml sample, aliquots of 0.5 ml are transferred to 1.5-ml Eppendorf microcentrifuge tubes. One 0.5-ml aliquot in an Eppendorf microcentrifuge tube remains untreated and serves as an un-cross-linked control. Chemical cross-linking is accomplished by adding a concentrated stock of the chemical cross-linker of choice to another 0.5-ml aliquot of spheroplasts in an Eppendorf microcentrifuge tube and incubating on a nutator mixer (VWR Scientific, West Chester, PA) for 20 min at room temperature. This is followed by the addition of an appropriate quencher in some cases. The cross-linked products are analyzed by sodium dodecyl sulfate–polyacrylamide gel electrophoresis (SDS–PAGE) and protein immunoblotting according to standard procedures.[23]

We have cross-linked the oligomeric URF13 using a variety of hydrophobic chemical cross-linkers including bismaleimidohexane (BMH), DCCD, ethylene glycol bis(succinimidyl succinate) (EGS), and succinimidyl-4-(N-maleimidomethyl)cyclohexane 1-carboxylate (SMCC) (all from Pierce Chemical Co., Rockford, IL). When using BMH, a homobifunctional, cysteine-specific cross-linker, a concentrated stock of 25 mM in dimethyl sulfoxide (DMSO) is added to the sample to give a final concentration of 250 μM. The reaction is quenched by adding 2-mercaptoethanol to a final concentration of 10 mM. When using DCCD, a heterobifunctional, carboxyllysine-specific cross-linker, a concentrated, fresh stock of 50 mM DCCD in anhydrous, absolute ethanol is added to the sample to a final concentration of 500 μM. The DCCD solution must be anhydrous because the reaction between DCCD and water rapidly inactivates DCCD.[28] Anhydrous ethanol is obtained by using a freshly opened bottle of absolute ethanol. When using EGS, a homobifunctional and lysine-specific cross-linker, a concentrated stock of 500 mM in DMSO is added to the sample to a final concentration of 5.0 mM, and the reaction is quenched by adding glycine to a final concentration of 40 mM. When using SMCC, a heterobifunctional, lysine–cysteine-specific cross-linker, a concentrated stock of 100 mM in DMSO is added to a final concentration of 1.0 mM, and the reaction is quenched by the addition of glycine to a final concentration of 10 mM. After treatment with any given cross-linker, an equal volume of 2× protein gel sample buffer (1× is 0.0625 M Tris-Cl, 2% SDS, 10% glycerol, 5% 2-mercaptoethanol, pH 6.8) is added to the sample, which is immediately boiled for 5 min and either put on ice for 5 min and analyzed by SDS–PAGE, or stored at −20° and later thawed, reboiled for 5 min, and analyzed by SDS–PAGE.

---

[28] M. J. Nałęcz, R. P. Casey, and A. Azzi, this series, Vol. 125, p. 86.

Notes. A potential result of treatment with a chemical cross-linker is the chemical modification of the epitope recognized by antibodies to the protein of interest. One solution to this problem is to use an expression vector, such as the pET series (Novagen), that adds an epitope to the amino terminus of the protein. The added epitope must, however, lack any amino acids that can react with the specific cross-linker being used, or it may be altered in such a way that the antibody will not recognize it. This approach may require alteration of a cDNA clone encoding the protein of interest to insert the gene in-frame into one of the pET vectors. Alternatively, polymerase chain reaction (PCR) technology can be used to add a specific epitope to either the amino or carboxyl terminus of the protein of interest.

Cross-Linking of Mitochondrial Membrane Proteins

Mitochondria are isolated by an established protocol.[29] The concentration of total mitochondrial protein is determined by standard procedures.[30,31] A portion of the sample is added to phosphate-buffered saline such that the final concentration is 1.0 mg total protein/ml and the volume is 0.5 ml. The cross-linking reagent of choice is added from a concentrated stock solution to the final concentration described above or determined from published procedures. The reactions are quenched as described above, and the results are analyzed by SDS–PAGE and immunoblotting.

URF13:T-Toxin Binding Studies

Quantitative studies of the interaction between membrane-bound receptors and their associated ligands have provided a useful approach to understanding the details of ligand–receptor interactions in a wide range of biological systems. Measurement of the binding of radiolabeled T-toxin to its receptor protein, URF13, has provided several insights into the interaction between these two species.[10,22] The binding of reduced, tritium-labeled Pm toxin to URF13 expressed in *E. coli* cells is specific and saturable.[10] The apparent K_d for the interaction is in the range of 50 to 70 nM, whereas the maximum amount of Pm toxin bound varies somewhat but consistently falls between 200 and 450 pmol toxin bound/mg *E. coli* protein. The K_d for the interaction of Pm toxin with URF13 is in the concentration range required to observe inhibition of respiration by added BmT toxin in intact *E. coli* cells.[11] The binding of Pm toxin to URF13 expressed in *E. coli* cells displays a distinct positive cooperativity.[10] Hill coefficients ranging between

[29] D. A. Day, M. Neuburger, and R. Douce, *Aust. J. Plant Physiol.* **12**, 219 (1985).
[30] M. M. Bradford, *Anal. Biochem.* **72**, 248 (1976).
[31] O. H. Lowry, N. J. Rosebrough, A. L. Farr, and R. J. Randall, *J. Biol. Chem.* **193**, 265 (1951).

1.4 and 2.0 have been observed. Being able to generate mutations of the URF13 primary structure in *E. coli* cells permits the characterization of changes in Pm toxin binding resulting from different amino acid modifications. Pm toxin binding experiments to URF13 with primary amino acid mutations has shown that the mutants each fall into one of several specific categories. These categories include mutants showing (1) reduced levels of Pm toxin binding,[10,22,31a] (2) a loss of binding cooperativity,[10,22] and (3) an apparent increase in the K_d for toxin binding.[31a] The first and third categories include some URF13 mutants that remain sensitive to Pm toxin. Although fewer in number, mutants in the second category are insensitive to Pm toxin.

Binding of labeled Pm toxin to isolated *cms-T* mitochondria has not been as informative as the binding studies using URF13-expressing *E. coli* cells. Even though the observed K_d for Pm toxin binding is similar to that of bacterial cells (70 nM), the maximum level of Pm toxin binding is only around 20 pmol Pm toxin bound/mg mitochondrial protein, less than 10% of the value observed using *E. coli* cells.[10] In addition, the binding of Pm toxin to *cms-T* mitochondria did not display binding cooperativity, but this could be due to a very low level of Pm toxin binding. The ability to distinguish cooperativity to Pm toxin binding is particularly dependent on the points in the initial part of the binding curve, where the amount of bound Pm toxin is very low. With *cms-T* mitochondria, the maximum Pm toxin binding levels are too low to discriminate accurately between a sigmoidal and a hyperbolic binding pattern in this initial region of the binding curve.

Pm Toxin Labeling

Tritium-labeled, reduced Pm toxin is obtained by treating 10 mg purified Pm toxin with 1.0 Curie of tritiated sodium borohydride (Amersham, Arlington Heights, IL) according to Frantzen *et al.*[32] The tritiated Pm toxin is separated from the remainder of the reactants and unreacted material by gel filtration through a Sephadex LH-20 (Pharmacia) column (20 × 1 cm) using anhydrous, reagent-grade methanol as the solvent. The tritiated toxin elutes from the column as a sharp peak of radioactivity just after the void volume. The fractions in the radioactive band are pooled, divided into several aliquots, and stored at −80°. For reasons that are not understood, a large percentage of the total radioactivity loaded onto the column adsorbs to the Sephadex beads and does not elute from the column.

The specific activity of the reduced, tritium-labeled Pm toxin is calcu-

[31a] G. C. Ward, M. E. Williams, and C. S. Levings III, in preparation.
[32] K. A. Frantzen, J. M. Daly, and H. W. Knoche, *Plant Physiol.* **83,** 863 (1987).

lated by extensively drying a carefully measured liquid aliquot *in vacuo* in a standard glass desiccator, over silica gel desiccant until a constant weight is achieved, as measured on a Cahn 25 Automatic Electrobalance (Cahn Instruments, Inc., Cerritos, CA). The radioactivity of a second liquid aliquot is determined by liquid scintillation counting. Counting efficiency is determined using a tritium standard (LKB Biochrom Ltd., Cambridge, UK). Each procedure is repeated at least three times to ensure a constant measurement, and the specific activity of the resulting Pm toxin is calculated using the pooled measurements of radioactivity, Pm toxin weight, and a molecular weight of the tritiated, reduced component C of Pm toxin of 639.[32] Because the half-life of tritium is 12.3 years, it is important to recalculate the specific activity at regular intervals.

Pm Toxin Binding to Escherichia coli

Prior to carrying out the binding assay two cultures of *E. coli* cells, one harboring a plasmid vector that contains the T-*urf13* gene and one harboring the vector alone, are diluted 1 : 50 from stationary phase cultures into Luria–Bertani broth supplemented with 100 mg/ml ampicillin and subjected to an appropriate treatment to induce the expression of URF13.[9,21] After induction, cells are harvested by centrifugation (5000 *g* for 10 min), washed once with M9 salts, and finally suspended in M9 salts to 1% of the initial volume. The protein concentration of the *E. coli* cells is determined using a standard protocol.[30,31] *Escherichia coli* cells equivalent to 0.5 mg protein are added to M9 salts to give a final volume of 1.0 ml in a 1.5-ml microcentrifuge tube. Two stock concentrations of tritiated Pm toxin are prepared containing approximately 5 and 50 μM Pm toxin, respectively. Pm toxin is added to each of the microcentrifuge tubes, with no addition exceeding a final ethanol concentration of 1.25%. To obtain a reasonable number of points for the subsequent regression analysis, seven to nine tubes containing Pm toxin concentrations distributed between 5 and 500 nM are prepared. To assess the presence of binding cooperativity, a disproportionate number of points (i.e., six of the nine) should be in the range 5 to 100 nM.

Immediately after adding Pm toxin, the tubes are closed and the contents mixed by vortexing briefly. The samples are incubated with continuous agitation on a nutator mixer (VWR Scientific) at room temperature for 20 to 30 min. After the incubation period, 0.7 ml of the mixture is removed and added to a second 1.5-ml microcentrifuge tube containing 0.5 ml of Wacker AR200 silicone oil ($\rho = 1.04$; Wacker Silicone Corp., Adrian, MI). The sample is centrifuged 60 sec in a microcentrifuge to pellet the bacterial cells through the silicone oil layer. Control experiments mixing aqueous solutions of labeled Pm toxin and silicone oil followed by centrifugation to separate the two liquids indicated that the Pm toxin has no tendency to partition into the nonaqueous layer.

After centrifugation, 50 μl is transferred from the upper aqueous phase into a 10-ml scintillation vial containing a standard water-compatible scintillation cocktail such as DuPont Biofluor. This fraction contains the Pm toxin that did not associate with the bacterial cells and is used subsequently to calculate the final concentration of unbound (free) Pm toxin. After removal of the 50-μl aliquot from the binding mixture, the solution remaining in the microcentrifuge tube is rapidly poured off, and the residual silicone oil in the tube is aspirated, taking care not to disturb the bacterial cell pellet. The microcentrifuge tube containing the bacterial cells is then subjected to a second centrifugation for 120 sec to ensure that the cells are firmly pelleted, and all remaining silicone oil is removed by a second round of careful aspiration. The bacterial pellet is completely suspended with 100 μl distilled water, and the entire mixture is transferred to a scintillation vial. This sample represents the Pm toxin associated with the bacterial cells and is used subsequently to calculate the amount of bound Pm toxin.

Analysis of Binding to Escherichia coli Cells

For each assay tube, a pair of values, corresponding to the unbound (free) concentration and total amount of Pm toxin bound, are obtained. In addition, a complete set of binding data corresponding to the binding of Pm toxin to *E. coli* cells not expressing URF13 are obtained. A low level of nonspecific Pm toxin binding to *E. coli* cells not expressing URF13 was observed in control experiments, and this nonspecific binding must be subtracted from the total Pm toxin binding to cells expressing URF13 to calculate the amount of Pm toxin bound specifically to URF13.

Using the values obtained from scintillation counting, the nonspecific binding of Pm toxin to *E. coli* cells that are not expressing URF13 is determined by graphing bound Pm toxin (pmol/mg *E. coli* protein) versus the unbound Pm toxin concentration (nM). The resulting curve is inevitably biphasic, and is best approximated as two straight lines, one of which applies for unbound Pm toxin concentrations below 50 μM and a second that applies to unbound toxin concentrations above this value. The nonspecific Pm toxin binding is subtracted from the value obtained for the total amount of Pm toxin bound to the cells that express URF13. This is accomplished by using the two regression lines cited above and determining the value of nonspecific binding expected for each concentration of unbound Pm toxin obtained with the *E. coli* cells expressing URF13. The determined value of nonspecific binding is then subtracted from the observed value for the total amount of Pm toxin bound to the URF13-expressing cells, and the difference is taken as the amount of Pm toxin bound specifically to URF13.

Pm Toxin Binding to cms-T Mitochondria

The binding of Pm toxin to mitochondria isolated from *cms-T* maize is carried out using the same general procedure outlined above for the binding of Pm toxin to *E. coli* cells. Mitochondria are isolated from 5- to 7-day-old etiolated maize mesocotyls and purified by sucrose density gradient centrifugation as described by Stegink and Siedow.[33] Purification of plant mitochondria using Percoll density gradient centrifugation[29] is a reasonable alternative to the more time-consuming and laborious sucrose gradient procedure. Pm toxin binding to isolated *cms-T* mitochondria involves essentially the same procedure outlined above for Pm toxin binding to *E. coli* cells except (1) the protein concentration in the binding assay is increased to 1.0 mg/ml to compensate for the lower overall level of specific binding of Pm toxin to *cms-T* mitochondria, (2) a 9:1 mixture of Wacker AR200 and AR20 ($\rho = 1.01$) silicone oils is used to separate the free and bound Pm toxin, and (3) a standard mitochondrial reaction buffer (0.4 M mannitol, 1.0 mM K_2HPO_4, 10 mM KCl, 5.0 mM $MgCl_2$, 10 mM HEPES buffer, pH 7.2) is used for the binding reaction instead of M9 salts. In these studies, mitochondria isolated from normal (N) maize plants were used as the control lacking URF13 expression to determine the level of nonspecific binding of the Pm toxin to maize mitochondria.

Data Analysis of Binding to Escherichia coli Cells

Analysis of both specific and nonspecific Pm toxin binding is carried out using nonlinear regression analysis.[34] Although any of the numerous commercially available nonlinear regression analysis software packages can be used for this purpose,[35] we have used the program ENZFITTER.[36] Initially, attempts were made to fit the relationship between the unbound and specifically bound to Pm toxin to a rectangular hyperbola:

$$[URF13\text{-}PmT] = [URF13\text{-}PmT]_{max}[PmT]/(K_d + [PmT]) \qquad (1)$$

where [URF13-PmT] and $[URF13\text{-}PmT]_{max}$ represent the measured amount of specific Pm toxin–URF13 complex formation and maximum

[33] S. J. Stegink and J. N. Siedow, *Plant Physiol.* **80,** 196 (1986).
[34] P. J. Munson and D. Rodbard, *Anal. Biochem.* **107,** 220 (1980).
[35] R. J. Leatherbarrow, *Trends Biochem. Sci.* **15,** 455 (1990).
[36] R. J. Leatherbarrow, "Enzfitter." Elsevier Biosoft, Amsterdam, 1987.

possible Pm toxin–URF13 complex formation, respectively (in units of pmol/mg *E. coli* protein), [PmT] is the concentration of unbound Pm toxin, and K_d is value for the dissociation constant of the Pm toxin–URF13 complex. Nonlinear regression analysis using the data obtained for the binding of Pm toxin to URF13 was unable to give a reasonable fit to the observed data for specific Pm toxin binding to URF13 (Fig. 3A, dotted line). The inability to fit the observed data to a hyperbolic function was particularly evident at the lower concentrations of unbound Pm toxin, where there appeared to be a distinct lag associated with Pm toxin binding.

A better fit was obtained when the data were analyzed using an equation that assumed a cooperativity to the specific interaction between the labeled Pm toxin and URF13:

$$[\text{URF13-PmT}] = [\text{URF13-PmT}]_{\max}[\text{PmT}]^n/(K_d^n + [\text{PmT}]^n) \qquad (2)$$

where n is the value of the Hill coefficient, which reflects the degree of cooperativity observed with the binding. When the data obtained with the binding of Pm toxin to URF13 in *E. coli* cells were analyzed using Eq. (2),

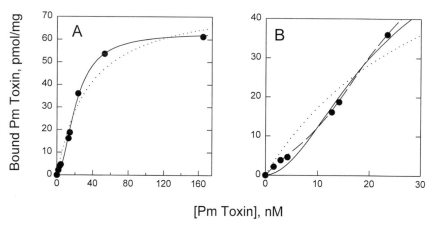

FIG. 3. Binding of reduced, tritiated Pm toxin to URF13 expressed in *E. coli* cells. The filled circles in each graph represent the data obtained in a representative binding experiment as described in the text. The lines represent curves obtained using Eq. (1), dotted line; Eq. (2), solid line; and Eq. (3), dashed line. (A) The dotted line represents the fit involving hyperbolic binding of Pm toxin to URF13 [Eq. (1)] when K_d = 35 nM and [URF13-PmT]$_{\max}$ = 78 pmol/mg *E. coli* protein. The solid line represents the fit involving cooperative binding of Pm toxin to URF13 [Eq. (2)] when K_d = 21 nM, [URF13-PmT]$_{\max}$ = 63 pmol/mg *E. coli* protein, and the Hill coefficient (n) = 1.9. (B) The solid and dotted lines are as in (A). The dashed line represents a fit involving a combination of hyperbolic and cooperative binding [Eq. (3)] where K_d = 21 nM, [URF13-PmT]$_{\max}^H$ = 30 pmol/mg *E. coli* protein, [URF13-PmT]$_{\max}^C$ = 35 pmol/mg *E. coli* protein, and n = 3.4.

a consistently better fit was obtained (Fig. 3A, solid line) relative to that observed with the hyperbolic function (Fig. 3A, dotted line).

The values obtained for the observed Hill coefficient have been somewhat variable over the several years that these binding studies have been carried out and have ranged from as low as 1.3 to as high as 1.9. We subsequently carried out a detailed analysis of the binding of Pm toxin to URF13 during the time course after induction of the URF13 protein and observed that the calculated Hill coefficient showed an extremely high value, above 3.0, during the earliest time after URF13 induction but dropped to the more commonly observed value (1.3 to 1.7) as the induction time increased and additional URF13 protein was expressed (C. I. Kaspi, 1991, unpublished observation). Using this observation as a starting point, data for a number of previous binding studies were reanalyzed using the assumption that two species of URF13 existed, one of which bound Pm toxin cooperatively and had a Hill coefficient greater than 3.0 and a second species that bound Pm toxin with the same dissociation constant but did not display any binding cooperativity (i.e., $n = 1.0$). This relationship is represented by Eq. (3):

$$[\text{URF13-PmT}] - [\text{URF13-PmT}]_{max}{}^{H}[\text{PmT}]/(K_d + [\text{PmT}]) + [\text{URF13-PmT}]_{max}{}^{C}[\text{PmT}]^n/(K_d{}^n + [\text{PmT}]^n)$$

(3)

where the superscripts H and C associated with the two $[\text{URF13-PmT}]_{max}$ species refer to the maximum amount of binding by the hyperbolic and cooperatively binding URF13 species, respectively. As shown for a representative set of data (Fig. 3B, using expanded axes for the data shown in Fig. 3A), the regression line obtained using Eq. (3) (Fig. 3B, dashed line) provides a better simulation to the observed binding data than that seen with purely cooperative (Fig. 3B, solid line) or hyperbolic (Fig. 3B, dotted line) binding. At present, the structural nature of the two URF13 species suggested by this analysis is not known, but this observation does point out the potential complexity associated with successfully analyzing binding data for a system as complex as that involving the binding of a ligand to an oligomeric protein complex expressed in intact bacterial cells.

Acknowledgments

This research was supported by grants from the U.S. Department of Energy (J. N. S.) and the National Science Foundation (C. S. L.) and by a National Science Foundation Postdoctoral Fellowship in Plant Biology (D. M. R.).

Author Index

Numbers in parentheses are footnote reference numbers and indicate that an author's work is referred to although the name is not cited in the text.

A

Abelson, J., 162
Aboujaoude, E. N., 442, 450(11), 452(11)
Abraham, J., 108
Ackerman, S. J., 230
Adachi, S., 442, 448(8), 449(8), 452(8)
Adler, B. K., 100
Adlover, B., 514
Aebersold, R., 4, 130–131, 159, 165, 166(12), 169–170, 173(12)
Agarwal, S., 446
Agrawal, H. P., 174
Akashi, K., 280, 285(11)
Albert, V. R., 167
Alberts, B., 160
Albring, M., 47
Alexander, N. J., 286, 290(66)
Allen, L. A., 41
Allen, N., 265
Allmaier, M., 81
Altman, L. G., 547
Altura, R., 84
Alziari, S., 358
Amaldi, F., 184
Amalric, F., 50, 56(4)
Ames, B. N., 442, 446, 450(11), 452(11)
Ammini, C. V., 12, 23, 33(3)
Anderson, 183
Anderson, R. S., 279
Anderson, S., 14, 129, 141–142, 162, 173, 197, 218, 221(7), 423, 424(5), 434, 510
Andreetta, F., 515, 517(30)
Andreetta, F., 218, 227(13), 515, 517, 536
Angelini, C., 122, 159, 197, 280, 433, 436, 523
Angermüller, S., 541
Antozzi, C., 514
Anziano, P. Q., 265, 269(5), 271, 274(17), 281, 292(30), 357

Appel, J., 489
Applegate, E. F., 158, 159(8)
Aprille, J. R., 502, 515, 536
Aquadro, C. F., 43
Areskog, N.-H., 514
Armaleo, D., 268
Arnaudo, E., 517
Arnberg, A. C., 67, 72(9)
Arnberg, A. C., 273
Arndt-Jovin, D. J., 160, 163(16)
Arnheim, N., 421, 424(2–5), 425(2, 4), 431(4)
Aits, G. J., 121
Atlung, T., 89
Attardi, B., 122, 175, 185, 202, 203(22)
Attardi, G., 3–4, 4(7), 22, 35–36, 43–44, 44(1), 45(21), 46(19–21), 47, 47(21), 49(1), 50, 52, 54(20), 55(11), 56(3, 4, 9, 20), 122, 126, 129–130, 130(8), 131, 132(8, 21), 133(24), 135(21), 137(20, 24), 158–159, 159(2, 6), 162(11), 163, 164–165, 166(11, 12), 169–170, 170(23), 171(23), 173(12), 175, 178–179, 181, 183–184, 184(5), 185, 185(7), 186, 187(2, 7, 8), 188(7), 189(7), 191(8), 192(8), 195(8), 197–198, 198(8, 9), 200, 200(7), 201–202, 202(17), 203(22), 204(3, 4, 8), 207(7), 209(17), 211, 214, 216–218, 219(3), 221(3–6, 8, 11), 222(3, 11), 223(3), 226(3, 10), 227, 227(3, 4, 10, 11), 247, 265, 280, 294(23), 298, 304–305, 306(2, 11), 307, 307(2), 308, 309(2), 310, 310(2), 311(26), 312–313, 313(26), 315, 316(11, 12), 322(12, 15), 327(11, 12), 328, 328(12), 334, 335(1, 2), 339–340, 341(1, 10), 343, 357, 364(3), 369, 403, 432–433, 434(6), 435–436, 455, 473, 474(10), 475–478, 480, 483(8), 485, 505(9), 509–510, 523
Augenlicht, L. H., 43, 44(10)
Augustin, S., 79
Ausenda, C., 122, 202

Subject Index

A

Agarose gel electrophoresis
  polymerase chain reaction products, 423,
    434–435, 439
  RNA transcripts, 55–56
Allotopic expression, *see* ATPase; bI4
    RNA maturase
Alzheimer's disease, complex V deficiency,
    560–562, 565
Antibody, *see* Immunoprecipitation
Antimycin, inhibition of ubiquinol–
    cytochrome-*c* oxidoreductase, 477, 500
ATPase, *Saccharomyces cerevisiae* mito-
    chondria
  allotopic expression of subunit 8
    assembly assay, 383
    cell growth, 374
    cell maintenance, 375–376
    codon considerations, 370–371
    evaluation, 380–383
    genome stability testing, 376–377
    growth rate assessment, 382–383
    leader sequence
      duplication, 383–384, 386
      selection, 371–372
    regulated expression with *GAL1* pro-
      moter, 386–388
    site-directed mutagenesis, 372, 374
    structure–function applications,
      388–389
    subunit depletion, 388
    transformation, 379
    vectors, 377–378
    yeast strains, 374–375
  leader sequence and subunit import in
    other species, 369
  subunit 9 gene, 369
ATP synthase, *see also* ATPase
  blue native gel electrophoresis, 560–562
  deficiency in Alzheimer's disease,
    560–562
  inhibition by oligomycin, 477, 500

ATP synthesis assay
  cultured skin fibroblasts, 461–462
  lymphoblast mitochondria, 462, 464
Azidothymidine, effect on mitochondrial 8-
  oxo-2′-deoxyguanosine levels, 451
AZT, *see* Azidothymidine

B

bI4 RNA maturase
  allotopic expression
    construction of universal code equiva-
      lent for mitochondrial gene
      multi-site mutagenesis, 394–396
      polymerase chain reaction, 394,
        396–397
      product analysis, 397
    expression vector selection, 399–400
    mitochondrial targeting sequence selec-
      tion, 397–400
    protein import studies, 401–403
    biological importance in yeast, 390–391
    complementation assay, 391, 393
    precursor, 391
Biolistic gun
  helium gun design, 266–267
  transformation of *Chlamydomonas* chloro-
    plasts
    bombardment, 293
    efficiency, 293
    markers, 295–296
    targets, 295
    transformant selection, 292–294
  transformation of yeast mitochondria
    cell bombardment, 267
    COX1 deletion repair, 269, 273–274
    COX2 deletion repair, 271–273
    DNA attachment to microprojectiles,
      268–269
    efficiency, 265–266, 269
    functional analysis of transformed se-
      quences

high-performance liquid chromatography, 250, 257–259
  mitochondria isolation, 253–254
  yield, 261
Embryo, heteroplasmic mouse production
  cytoplast transfer, 350–351, 353–354
  efficiency, 355–356
  electrofusion, 347, 353, 356
  embryo isolation, 350
  equipment, 346–347
  implantation of pseudopregnant foster females, 354–355
  micropipette manufacture, 348–349
  mouse strains, 347
  reagents, 347–348
  segregation versus mutation rate studies, 345–346, 357
  Sendai virus as fusing agent, 353–354, 356
Emetine, cytosolic protein synthesis inhibition, 223
Enzyme-linked immunosorbent assay, mitochondrial translation product antibodies, 223
Epstein-Barr virus, lymphoblast transformation in cell culture, 488
Ethidium bromide
  accumulation in mitochondria, 296–297
  effects on mitochondrial DNA
    mutation in yeast, 340–341
    synthesis inhibition, mechanism, 340
  mutagenic effects, 310
  mutated/wild type DNA ratio manipulation
    cell growth, 342–343
    human cell lines, 342
    rationale, 341
    solution preparation, 342
    transfer hybridization analysis, 343–344
    treatment duration, 341–342
  rho^0 mutant generation
    avian cells, 296–297, 299–301
    human cell lines, 305–306, 308–310, 312

F

FACS, *see* Fluorescence-activated cell sorting

Fluorescence-activated cell sorting, human satellite cells, 472–473
Footprinting, *see also* Genomic footprinting
  dimethyl sulfate protection, 4, 7–9, 13, 15–16
  mitochondrial transcription termination factor, 4, 11, 22, 166–167, 173
  polymerase chain reaction
    in organello footprinting reactions, 14–15, 19, 21
    in vivo footprinting reactions, 3, 10–11, 24–25, 29–31, 33
  Southern analysis, 3, 13–14, 16, 21–24

G

Galactose, metabolism in respiratory chain-defective cell culture, 455
Gel electrophoresis, *see* Agarose gel electrophoresis; Blue native electrophoresis; Polyacrylamide gel electrophoresis
Genomic footprinting
  ligation-mediated polymerase chain reaction, 3, 24
  mitochondrial DNA *in organello*
    control sample methylation, 17, 19
    dimethyl sulfate methylation, 13, 15–16
    mitochondria lysis, 16
    piperidine cleavage of DNA, 13
    primer extension with polymerase chain reaction, 14–15, 19, 21
    restriction site selection, 14
    Southern hybridization analysis, 13–14, 16, 21–22
  mitochondrial DNA *in vivo*
    buffers, 7
    cell lysis, 8
    dimethyl sulfate protection, 4
    methylation *in vivo*, 7–8
    methylation of naked DNA, 8–9
    piperidine cleavage of DNA, 9
    primer
      end labeling, 9–10, 29
      extension with polymerase chain reaction, 10–11, 24–25, 29–31
    principle, 5
    visualization, 5, 11
    Xenopus laevis egg, 25–29, 31, 33, 35–36

Glucose, metabolism in respiratory chain-defective cell culture, 454–455

Group II intron
biolistic transformation of yeast mitochondria, 273–274, 276–277
conserved secondary structure, 66
self-splicing reaction
assay, 82–85
broken lariat identification, 76
cis reactions, 74–75
exon length effects, 73–74
high salt reaction conditions, 74–77
intron types, 69
linear intron identification, 76
low salt reaction conditions, 72–74
mechanism, 68, 75–76
mutation effects, 78–79
pH optimum, 77
reverse splicing, 79–81
spliced exons reopening reaction, 75–76, 85
stereochemistry, 83–84
stopping, 77–78
temperature optimum, 72, 76–77
trans reactions, 73, 81–86
subgroups, 66
transcript preparation
DNA-mediated ligation, 72
polymerase chain reaction, 70
purification, 71–72
quantitation, 71
radiolabeling, 71

H

Heteroplasmy
biological importance, 345
Drosophila, production
cytoplasmic transplantation, 364
efficiency of donor mitochondria incorporation, 362, 364
establishment and maintenance of heteroplasmic lines, 364–365
germ plasm transplantation, 360–362
mitochondria microinjection, 364
nuclear genome effects, 367–368
rationale, 359
strains, 360–361, 366–368

temperature dependence, mode of DNA transmission, 365–366, 368
mouse embryo, production
cytoplast transfer, 350–351, 353–354
efficiency, 355–356
electrofusion, 347, 353, 356
embryo isolation, 350
equipment, 346–347
implantation of pseudopregnant foster females, 354–355
micropipette manufacture, 348–349
mouse strains, 347
reagents, 347–348
segregation versus mutation rate studies, 345–346, 357
Sendai virus as fusing agent, 353–354, 356
mutated/wild type DNA ratio manipulation with ethidium bromide
cell growth, 342–343
human cell lines, 342
rationale, 341
solution preparation, 342
transfer hybridization analysis, 343–344
treatment duration, 341–342
myotubules, 465–466
natural causes, 359–360
sequence differences in, types, 345

High-performance liquid chromatography
initiation factor 2, 250, 255–256
mitochondrial elongation factors, 250, 257–261
8-oxo-2′-deoxyguanosine, 443, 446, 449–452

HPLC, *see* High-performance liquid chromatography

I

Immunoprecipitation, mitochondrial translation products
antibody
enzyme-linked immunosorbent assay, 223
immunization of rabbits, 222
preparation, 222–223
specificity, 227
control reactions, 225
incubation, 225